EARTHQUAKE ENGINEERING FRONTIERS IN THE NEW MILLENNIUM

PROCEEDINGS OF THE CHINA–U.S. MILLENNIUM SYMPOSIUM ON EARTHQUAKE
ENGINEERING, BEIJING, 8–11 NOVEMBER 2000

Earthquake Engineering Frontiers in the New Millennium

Edited by

B.F. Spencer, Jr
University of Notre Dame, Notre Dame, Indiana, USA

Y.X. Hu
China Seismological Bureau, Beijing, China

A.A.BALKEMA PUBLISHERS   LISSE / ABINGDON / EXTON (PA) / TOKYO

Published by: A.A.Balkema, a member of Swets & Zeitlinger Publishers
        www.balkema.nl and www.szp.swets.nl

ISBN 90 2651 852 8

Printed in the Netherlands

Table of contents

Foreword

The year 2000 marked the 20th anniversary of Cooperative Studies in Earthquake Engineering and Hazard Mitigation between the United States of America and the People's Republic of China under Annex III to the US/PRC Protocol for Scientific and Technical Cooperative Research. This Cooperation, initiated in 1980, has been active and fruitful, and has promoted mutual understanding and collaboration between earthquake scientists and engineers in the two countries.

To commemorate this 20th anniversary, the China–U.S. Millennium Symposium on Earthquake Hazard Mitigation was held in Beijing, P.R. China on November 8–11, 2000. The Symposium was planned as a forward-looking intellectual endeavor involving multiple agencies and major universities in both countries. It brought together leading international researchers in earthquake hazard mitigation to deliver topical presentations and to develop a clear picture of the current worldwide status of the field.

The Symposium program consisted of plenary sessions, working group sessions, and presentations of technical papers in parallel sessions. This proceedings of the Symposium describes the motivation for the organization of the symposium and the working groups and their tasks, and documents the working group reports and recommendations, as well as the technical papers presented at the symposium.

Based on the recommendations by the working group, a formal Symposium resolution has been prepared to promote and enhance the cooperative research and exchange activities. This resolution, representing the collective and unanimous opinion of the participants, is provided for the consideration of the Chinese and U.S. agencies involved in supporting and conducting research in earthquake hazard mitigation. It is presented on the page following this foreword.

The financial support of the Symposium by the US National Science Foundation, the China Seismological Bureau, the Chinese Ministry of Construction, and the National Natural Science Foundation of China is gratefully acknowledged. The expert work of Xuefen Han, Guangqiang Yang, Aaron Vorwerk, Yong Gao, Hyuang Jo Jung, Saang Bum Kim, Manuel Sandoval and Nelson Duran in editing the manuscripts for publication is also appreciated. However, the major credit for the success of the Symposium belongs to the authors, whose excellent presentations and papers make this volume possible.

B.F. Spencer, Jr.
Yuxian Hu

November 2000

Symposium resolution

Whereas the People's Republic of China (P.R.C.) and the United States of America (U.S.A.) face similar risk of catastrophic earthquakes, and whereas the aggregation of science and experience within the two countries has great potential for combining symbiotically to solve major problems leading to a safer environment for society, it is resolved that:

A joint technical coordination committee comprising two co-chairmen and five members from each side (P.R.C. and U.S.A.) be formed with the mandate of encouraging and coordinating joint research towards common objectives, and with the requirement that this committee meet at least three times within the first year and at least once each following year. Potential tasks for the joint technical coordination committee include:

- Development of research directions and goals
- Coordination of research conducted
- Assessment of research products
- Development and maintenance of a website

The P.R.C. side, mainly represented by the Ministry of Construction (MOC), the China Seismological Bureau (CSB), and the National Natural Science Foundation of China (NSFC), is requested to jointly create a coordinated funding mechanism for the P.R.C.– U.S.A. collaborative research projects.

The U.S.A. side is requested to create a similarly coordinated program involving the participation of additional agencies to develop new joint research initiatives toward achieving the cooperative research objectives of the protocol.

It was the consensus of the participants of the Millennium Symposium that these actions be taken without delay.

Earthquake Engineering Frontiers in the New Millenium, Spencer & Hu (eds),
© *2001Swets & Zeitlinger, ISBN 90 2651 852 8*

Cooperative earthquake engineering and hazard mitigation research: Challenges and opportunities in the new millennium

A report of the Working Groups at the China–U.S. Millennium Symposium on Earthquake Engineering November 9–11, 2000, Beijing, China

1 INTRODUCTION

The purpose of the Millennium Symposium was to bring together leading international researchers in earthquake hazard mitigation to present their work and to assess the current status of the field. The main objective of the Symposium was to chart a clear course and develop a concrete plan for future U.S.–China cooperative earthquake engineering and hazard mitigation research. This plan is expected to become a model for the U.S.–China Earthquake Protocol Annex III Program, in anticipation of future implementation by the government agencies involved as well as other entities and participating researchers.

It is intended that the plan be inclusive in terms of anticipated continuing addition of new agency partners to the Annex III Protocol, as well as increased participation by university researchers from U.S., China, and other countries, as individuals, or as small teams or research centers. The following three working groups were formed to deliberate on these issues at the Symposium and suggest ways to formulate the future research plan:

I. New Areas for Cooperative Research Activities

II. Bilateral Program/Research Coordination

III. Multinational Cooperative Research Activities

After the deliberations were completed, the working group Chairs prepared a report about the discussions and opinions expressed, and recommendations made in the sessions. The working group reports were presented before all Symposium participants. The final report, including the deliberations of the Symposium participants, prepared by each working group are given in the sequel.

2 REPORT OF WORKING GROUP I: NEW AREAS FOR COOPERATIVE RESEARCH

2.1 *Introduction*

Current needs for the mitigation of earthquake hazards and continuing research activities in the U.S., China and other countries have resulted in new research frontiers that must be explored for efficient and economical hazard mitigation and for quick response and recovery after a damaging earthquake event. Significant research strides have been made in the areas of passive and active structural control, and exciting applications are being continuously discovered for advanced sensors, condition monitoring and assessment technologies, smart materials and structures, information processing and technology, robotics, and remote sensing and wireless communications. Advances have also been made in the use of GIS and other research tools in hazard mitigation, emergency recovery, and risk assessment.

Working Group I was responsible for developing and proposing areas for cooperative research in earthquake engineering. The charge for the working group included identifying new technological developments and implementations that are needed to design, construct, maintain, manage and renew the built environment for reduced earthquake risk. The working group was to outline thrust areas for future research for the bilateral (China-U.S.) program, as well as areas that could take advantage of multi-lateral cooperation.

As a first step, Working Group I discussed trends in research for earthquake engineering, with an emphasis on identifying bilateral and multi-lateral components of the research and application of research results. Through this discussion, the working group identified four important trends.

The first trend is the worldwide appreciation for the unique characteristics of strong ground motion close (within 15 km) to a fault rupture. Since the 1994 Northridge earthquake, near-fault ground motion has come under increased scrutiny. The 1995 Kobe earthquake and the 1999 earthquakes in Turkey and Taiwan have further highlighted the damaging effects of near-fault ground motion and associated permanent fault displacements on urban infrastructure. The vulnerability of densely populated urban regions is an international concern, not just in terms of human safety but also considering regional, national, and international economies. There are numerous research areas related to improving understanding of the physics of fault rupture and near-fault ground motion, to finding new ways to characterize the effects of motion on infrastructure systems, and to developing new ways to mitigate the special characteristics of near-fault ground motion. Research on these topics is underway in China, the United States, and many other countries.

A second trend is the development of technologies to improve structural resistance to earthquakes. Advances in passive and hybrid control and new materials (often characterized as "smart" materials) have shown much promise in preliminary research. These technologies include magneto-rheological fluid dampers, shape memory alloys, and composite materials. As earthquake engineering applications are developed, there are many opportunities for multi-disciplinary research with material scientists, control experts, and mechanical engineers to create technologies that are effective, reliable, and economical. An important multi-lateral effort would be to examine applications of new technologies in different countries to understand how successes in research laboratories can be translated to successful implementations.

A third trend in earthquake engineering research is the integration of experimental and computational methods of research. Experiments provide fundamental information about materials, components, and how systems behave. Computation with mathematical models, when properly validated, provide the opportunity of "virtual experiments" over a wider range of parameters and cases. When experimental and computational research methods are combined, improved understanding and new discoveries can result. The rapid advancement in information technology will allow researchers to improve measurements, visualization, communication, data mining and archiving, and interpretation of experimental and computational simulations. Related technologies have application to real-time sensing and health monitoring of individual structures and also entire urban regions.

A fourth trend, which is motivating a new era in earthquake engineering research, is the growth of mega-cities, huge urban regions in highly seismic zones that are exposed to increasing risk from earthquakes. Enormous future losses are expected in mega-cities, and reduction of losses requires a new type of research that has direct application to this problem. In addition to technology, research is needed on how policies are developed to reward risk reduction, how emergency response systems can save lives, and how post-earthquake recovery plans can reduce impacts on the economic health of a mega-city.

2.2 Framework for Cooperative Research

Working Group I identified criteria for selecting areas for bilateral and multi-lateral research in earthquake engineering. The following criteria were proposed:

- *Need* -- The research must address bilateral or multi-lateral needs for increasing earthquake safety and reducing earthquake losses in the countries participating in the research program.

- *Interest* -- Researchers in China and the United States, and those in other participating countries, must have interest in the area and have the required experience to pursue the research.

- *Opportunity* -- There must be unique opportunities, not available in uni-lateral research programs, created by combining international expertise, research facilities, and opportunities for field and case studies.

- *Time Frame* -- Research should have the potential to reduce earthquake risk in a ten-year time frame. Shorter-term research goals are most likely appropriate for individual countries to pursue, and longer-range goals may be too speculative for cooperative research.

In addition to these criteria, the working group identified three components of cooperative research programs in earthquake engineering. The first component is fundamental research encompassing basic scientific and engineering research. Fundamental research in earthquake engineering provides the foundations for increasing the knowledge base and emphasizes new discoveries. The second component is new technologies that provide possibilities for increasing seismic safety in an effective and economical ways. New technologies are often developed in a multi-disciplinary environment of science and engineering research. The third component is earthquake engineering application of fundamental research and new technologies. Implementation of research findings, particularly in different countries, is essential for testing new concepts and transferring knowledge to industry, the private sector, governments, and other stakeholders in seismic safety.

2.3 Areas for Cooperative Research

With the framework for cooperative research discussed in the preceding section, the working group identified three major categories for cooperative research, with potential thrust areas in each category.

2.3.1 Technological "push" research

Technological advances often push a new technology into the earthquake engineering field, with researchers identifying applications appropriate for the technology. Although it is sometimes viewed as solutions looking for problems, technological research can leap beyond constraints of conventional applications by transferring knowledge from other disciplines to earthquake engineering. The working group identified two major topics on technology for earthquake engineering:

- *Auto-adaptive materials for civil engineering applications* — auto-adaptive materials are new classes of materials that sense and adapt themselves in an optimal way to loads and other influences.

- *Self-sensing structural and geotechnical systems* – rapid advances in sensing technology provide new opportunities for monitoring and controlling large structural systems and the behavior of soil deposits, foundations, and other geotechnical systems.

2.3.2 Societal "pull" areas

Examining societal needs for earthquake risk reduction reveals problems that can be addressed by pulling specific knowledge, methods, and technology to address such needs. The working group discussed five potential research areas for cooperative research, namely:

- *Advanced response management for natural disasters in urban areas* — New methods and technologies are required for rapid response to disasters in urban regions.

- *Design for near-fault ground motions* -- As discussed previously, recent large earthquakes have shown the need for new design methods for structures to resist the special characteristics of near-fault ground motions.

- *Control of large civil systems* -- Changing structural systems from giving passive resistance to loads to providing active resistance has proven to be challenging. For certain critical structures with high performance requirements, control of behavior can be effective. New research is needed to control large civil systems in a reliable manner.

- *Disaster resistance in large-scale built infrastructure development* -- Mega-cities under rapid development in the Pacific Rim urgently need new methods for constructing communities that are disaster-resistant. The cost of retroactively

increasing disaster resistance has proven to be expensive and less effective.

- *Disaster resistant residential construction* -- The largest vulnerability worldwide is to the residential housing stock. Residential structures are typically not engineered and research is needed to improve seismic safety of homes. Disaster resistance of residential structures depends on a country's typical construction and contracting methods and materials, and upon building code enforcement regulations and practices.

2.3.3 Research methodology areas

The third category focuses on methodologies for earthquake engineering research. The working group identified two areas for cooperative research in this area:

- *Advanced earthquake engineering experimentation* – Improved earthquake experimental facilities, including earthquake simulators, and new testing methodologies, such as real-time pseudo-dynamic testing and hybrid methods, and new instrumentation can improve simulation of earthquake effects in the laboratory. Advanced experimental facilities provide opportunities for discoveries on how structures and geotechnical systems behave during earthquakes.

- *Model based simulation* — Rapid advances in computational methods, high-performance computing, networking, databases, and visualization can be combined for improved computational simulation of systems during earthquakes. While simulation models must be validated by experiment, model based simulation can provide a virtual laboratory for examining behavior and testing new ideas.

2.4 High Priority Areas for the U.S.–China Program

After much discussion of this framework, Working Group I identified three high priority areas for cooperative research under the U.S.–China Program. The following sections provide more information about each area.

2.4.1 Auto-adaptive and Sensing Materials for Disaster Resistant Construction

The vision of research in auto-adaptive and sensing materials is to create a new generation of sensors, actuators, and functional materials that will enhance the seismic performance of the built environment. There are advances in China, the United States, and other countries on smart materials, in which properties can be altered with small amounts of energy. Examples include magneto-rheological fluids and shape memory allows. Preliminary studies have shown important earthquake engineering applications for such materials. New sensing technologies, notably micro-electromechanical systems (MEMS), allow sensors to be distributed and embedded in materials for improved control of behavior.

There are three objectives for cooperative research in auto-adaptive and sensing materials. The first is to specify the needs for new materials for civil applications. The second objective is to identify the breakthrough materials technologies from chemistry, physics, and material science that can be applied to civil infrastructure and earthquake resistant construction. The third objective is to foster multi-disciplinary and multi-lateral research between scientists and engineers to speed applications of new materials.

The justification for cooperative research is that new materials will improve the performance and safety of the built environment in a cost effective and environmentally friendly manner. Furthermore, successful applications of research will promote the growth of new materials industries, with large-scale production and new global markets, for civil infrastructure applications in China, the United States, and other countries.

The cooperative research should include several aspects. The first is material design for seismic resistance, including modification of traditional materials and development of new materials with sensing and self-actuating properties. More future oriented is research into materials that exhibit self-healing or repair when damaged either by degradation or by energy dissipation during seismic loading. The second aspect of research is the development of structural components and systems that are effective with auto-adaptive and sensing materials. This may require new ways to connect and configure structural systems, such as new bracing and connection strategies, that optimally use the properties of new materials. Research should encompass both new construction and upgrading seismic performance of existing structures. Finally, verification, prototyping, and testbed studies should be conducted in China, the United States, and other countries.

2.4.2 Infrastructure Design for Ground Motions in Urban Areas: What are the Enemies?

The damaging earthquakes attacking Los Angeles, Kobe, Taichung, and Kocaeli have dramatically shown the need to understand near-fault earthquake motions in a fundamental way. This need is accentuated by the rapid urbanization and growth of mega-cities in highly seismic Pacific Rim countries. Earthquake risk is increasing because a larger population and greater variety of buildings and lifelines are being exposed to seismic hazard.

Multi-lateral research is urgently needed in all affected countries to focus on the specific characteristics of earthquakes and how they damage the built environment. In reality, there are few records available in the near field of large earthquakes, and engineers and scientists are just beginning to learn about near-fault motion and to understand directivity, fling, and permanent ground displacements. As more records are becoming available, particularly since 1999, more questions are being raised about the characteristics of near-field ground motion. In addition, effects of large sedimentary basins and weak surface soils found in many urban coastal regions requires research because of the large effects they may have on ground motion.

A cooperative research agenda on infrastructure design for ground motions in urban regions should include the following components. The first is seismological and geotechnical characterization of near-fault ground motions. The research includes ground motion simulation methods capable of representing long and short period components and permanent displacements. The second component is identifying the parameters of near-fault ground motions that cause damage in different types of infrastructure components and systems. Such knowledge will allow improved attenuation and hazard modeling of ground motions. The third component is creating an archival database of recorded and simulated ground motions and geotechnical site data for (1) verifying ground motion simulations, and (2) cataloguing damaging characteristics. The database should be accessible over the World Wide Web using GIS technology. The final research component is the development of new structural components and systems resistant to specific characteristics of near-field ground motions, including adaptive and control strategies that are optimized for near-fault ground motions.

2.4.3 Advanced Disaster Response Management

The increasing urbanization in mega-cities requires new technologies for rapid and effective response to human and economic needs after a natural disaster. Rapid urbanization, with large-scale infrastructure development and economic exposure, is overcoming the ability to respond after a disaster, yet society expects rapid recovery and reconstruction.

The objectives of cooperative research are to develop state-of-art technologies, analytical tools, and policy instruments to characterize urban systems, to assess vulnerability, to estimate the spectrum of losses, and to respond rapidly to save lives and enhance recovery. There are many opportunities for cooperative research because of the common aspects of vulnerability in many countries and the increasingly interdependent economic effects in global markets.

Components of the research program should define regional risk levels and performance standards, and a framework for establishing loss reduction goals in urban regions. Economics research can be applied to hazard analysis, such as adaptation of general equilibrium models and evaluating recovery strategies. Real-time information systems, such as GIS and remote sensing, have many applications for emergency response and post-disaster recovery by providing real-time impact information to emergency officials in a deployable command and control network. There is an urgent need for research into new technologies for rapid location of casualties and rescue after a disaster. Another important research area is the development of policy instruments to promote disaster resistant communities, such as incentives, insurance, and financial mechanisms.

3 REPORT OF WORKING GROUP II: BILATERAL RESEARCH AND PROGRAM COORDINATION

3.1 *Introduction*

The program of Cooperative Studies in Earthquake Engineering and Hazard Mitigation sponsored under Annex III of the US/PRC Scientific and Technical Protocol on Cooperative Earthquake studies has been very effective in advancing cooperative research and mutual understanding between the two countries for the last two decades. This working group was charged to re-examine the existing U.S.–China Bilateral Cooperative Program and to discuss the involvement and activities of the research institutions in China and the U.S. for the purpose of identifying issues that affect the U.S.–China Program. The main objective was to draft a resolution with recommendations to enhance the Program and to promote cooperative research activities recommended by Working Group I. It was also desired to identify the organizations in each country that might be involved in future research projects under the U.S.–China Bilateral Cooperative Program.

The Working Group started the discussions by identifying government agencies that do research and/or fund research in the area of earthquake hazard mitigation. Some examples of cooperative and exchange activities were cited. Barriers and issues affecting the cooperative research activities were discussed. Finally, the recommendations were made to promote the cooperative and exchange activities. These discussions and recommendations are reported in the following.

3.2 *Research in Earthquake Hazard Mitigation*

The Working Group first tried to identify the organizations conducting and supporting research in earthquake hazard mitigation in both China and the U.S.

3.2.1 *Organizations in China*
- Ministry of Construction (MOC)
- Office of Earthquake Resistance (OER)
- China Seismological Bureau (SSB)
- National Natural Science Foundation of China (NNSFC)
- Department of Architectural Environment and Structural Engineering
- Ministry of Education
- Ministry of Energy
- Science and Technique Development Foundation of Power Industry (STDFP)

3.2.2 *Organizations in the United States*
Organizations primarily responsible for the earthquake hazard mitigation are:
- Federal Emergency Management Agency (FEMA)
- National Science Foundation (NSF)
- U.S. Geological Survey (USGS)
- National Institute of Standards and Technology (NIST)

Other U.S. Government agencies that have responsibilities and interests in earthquake engineering relevant to their specific missions:
- U.S. Bureau of Reclamation in the Department of Interior
- Nuclear Regulatory Commission
- Department of Energy
- Department of Housing and Urban Development
- Army Corps of Engineers
- Department of Transportation

3.3 *Issues Raised by Working Group II*

The overall issue is to find the processes and the finances for enhancing the cooperation between China and the U.S. in earthquake engineering and earthquake hazard reduction. That is, we must find the best, most efficient way to effect cooperation between China and the U.S. and to secure the finances required to achieve this goal. Within this context, the discussion pursued the following questions and answers:

- *What kind of mechanisms and organizations can we develop to promote collaboration between China and the U.S.?* Collaboration cannot be forced. There must be identification of mutual, common interests.

- *How to exchange researchers effectively between the two sides?* As an example, bilateral collaboration has existed between universities in Hong Kong and Mainland China universities and research institutes since the mid 80's. For example, Hong Kong does not have a large shaking table, and therefore collaborates with Mainland China on research projects involving shaking table tests. In addition, there is joint supervision of students; students go to Hong Kong for a year and then return to China. Financial support is available. In these cooperative projects joint funding applies between Hong Kong and Mainland China.

- *Collaboration with U.S. is difficult because it is so far away.* The collaboration is largely limited to joint workshops or conferences (e.g., Earthquake Engineering Related to Moderate Seismicity Regions, which is a collaboration with the MCEER Center in the U.S.)

- *There is a need for faster and more regular exchange of information and a need for regular organization of joint workshops.* China and the U.S. share common mutual interests, but collaboration has to start naturally. Communication via computers (e-mails, internet) helps, but person-to-person contacts are invaluable. Interactive meetings need more participants and deeper discussions. Fifteen-minute technical paper presentations are not enough for mutual understanding. There is a need for both more exchanges and more participants. There is presently a very limited number of Chinese and U.S. professors collaborating on joint research projects (less than ten in the last two years).

- *To improve collaboration, we need to find more effective ways to exchange researchers.* A stay of at least one year and preferably more in the U.S. is very helpful in getting to know the frontiers of knowledge in a research topic. Visiting scholars sent to the U.S. have been very fruitful in developing successful research programs when they return to China, carrying out high level research on specific topics. The intent of such cooperative research exchanges are sometimes hindered by the fact that Chinese researchers sent to the U.S. do not return to their home laboratory.

- *Features of the cooperation between Hong Kong and France may be helpful.* An exchange program exists between France and Hong Kong that allows exchange of student researchers between the respective countries (France pays half and Hong Kong pays half from research projects). About 15 Hong Kong universities send senior undergraduate students to France, and France sends undergraduate students to Hong Kong. This collaboration has a history of substantial success. One possible expansion of this program would be to extend it at the graduate level between Hong Kong and the U.S. However, the problem of credit transfer would need to be resolved; that is, how do we recognize the degree status of such students?

- *Financial barriers.* Some Chinese participants indicated that in China, there is presently no mechanism to support individual researchers. Obtaining financial support from NNSFC, the Chinese counterpart of NSF, is very difficult, because NNSFC typically supports larger research projects with a group of researchers. Even for larger projects, obtaining financial support is challenging. Thus, in collaborating with the U.S., a stumbling block is securing financial support on the Chinese side.

The current state of funding in Hong Kong is not encouraging. Many researchers are fighting for a small piece of the pie. Only about 2,000 USD is available per researcher per year to cover traveling expenses.

3.3.1 *Examples of U.S.–China and other successful collaborations*

Among the several successful collaborative and exchange activities in earthquake engineering between the engineers and scientists from China and the U.S., and also other countries, the following examples were cited by the working group participants:

- One Chinese researcher had a very fruitful experience of research collaboration with the U.S. He spent two years (1978-80) with Prof. Ray Clough at U.C. Berkeley working on dam engineering with support by the Chinese government. In this project, they completed six successful field measurements on arch dams with joint financial support from China (Dept. of Water and Power) and U.S. (NSF). In addition, they organized a successful workshop on arch dams in China in 1986, with participation of 50 international experts from all over the world.

- A visiting scholar supported by Hong Kong University has spent an extensive period of time working with Prof. B.F. Spencer, Jr. at the University of Notre Dame to work on benchmark problems in structural vibration control. Collaboration between the two institutions has increased substantially.

- A visiting scholar from Dalian University/Shenyang Institute of Architectural Engineering supported by Chinese Government and U.S. National Science Foundation spent more than two years at Virginia Tech working with Prof. M.P. Singh. Since then, the cooperative and exchange activities have continued between the two institutions.

- There have also been several other joint research projects with the U.S., including one with Professor Lagorio at U.C. Berkeley, Professor Shah at Stanford University, and Professor George Lee at SUNY Buffalo. The U.S. side was supported by the National Science Foundation. The Chinese side obtained its financial support from the Ministry of Construction.

- Collaborations have also existed between Chinese researchers and Professor T.T. Soong at SUNY Buffalo and Professor J.N. Yang at U.C. Irvine.

- There has also been successful cooperation between Tsinghua University and Cincinnati University in the area of fiber reinforced polymers, including exchange of Ph.D. students.

- A joint research project has been carried out between China and Japan, with financial support from China's Ministry of Construction. Japan contributed heavily to the funding of the research and the international travel. In addition, collaboration has existed between China and the University of Tokyo.

3.4 *Working Group II Recommendations*

Working Group II made the following recommendations for improvement of the bilateral program in the three areas of exchange of researchers, their research activities, and organization of workshops and meetings.

3.4.1 *Exchanges of students and researchers*

The members of Working Group II felt strongly that the Bilateral Program should emphasize exchange of students and young researchers in the future.

- The Bilateral Program needs to develop a financial support mechanism for sending bright young professors, lecturers and researchers to U.S. universities. Several months may not be sufficient to develop meaningful collaboration. One or several years are needed to improve significantly the research level of the visiting scholars. When these Chinese visiting scholars return to China, their research level will be significantly higher, and this will help to keep close collaborative relationships with U.S. researchers.

- There is a strong need to secure Chinese government funds for the exchange of students.

- There should be a way under the bilateral program to send Chinese researchers to the U.S. for a few months. For example, in joint China-U.S. projects, part of the U.S. NSF budget should be used to support the Chinese researchers while in the U.S. Funding sources also need to be identified in China.

- In addition to government support, it was recommended that a special foundation be established to support exchange of visiting scholars between China and the U.S. Opportunities within the Fullbright program in the U.S. should also be investigated.

3.4.2 *Topics for collaborative research*

There is a need to identify specific targets for collaboration under the Bilateral Program. Working Group III believes that some areas of research collaboration have great potential in the next 5–10 years. Although Working Group I was primarily responsible for identifying the research needs, the participants offered the following for collaborative activities.

- Joint research projects should be developed around specific real needs in certain regions of China, e.g., novel design and construction of base-isolated structures and energy dissipation protective systems in the very high seismicity ($M = 9$) region of Jansu Province.

- Although collaboration has to start naturally from common, mutual interests, it can be fostered and instigated through the set-up of benchmark problems (e.g., benchmark problems in structural control and health monitoring). Benchmark problems bring various researchers to work on the same set of problems and challenges.

- The participants should identify facilities or other opportunities not available at one side but available at the other. It would be useful to identify and catalog all resources available from both sides.

- Research cooperation between the U. S. and China should focus on experimental studies and methods, to take advantage of the strengths of the respective countries.

3.4.3 *Activities and organization of the Program*

The working group also made several recommendations aimed at improving the organization of the Bilateral Program:

- U.S. and Chinese researchers should organize joint conferences and workshops. There is a need to jointly organize workshops at regular intervals on selected specific topics.

- Joint technical committees should be set up within the Program for different sub-areas of research.

- U.S.-Japan cooperative research in earthquake engineering is very well structured and has well-defined steps for calling for proposals, for collecting and evaluating them, and for their funding. It would be useful to examine the formal cooperative activities between the U.S. and Japan and to consider a similar model for cooperative research between the U.S. and China.

- To better promote cooperation between the China and the U.S., research directors (e.g., Dr. Shih-Chi Liu in U.S. and a similar one in China) are needed. It is also necessary for the U.S.–China Bilateral Program to have a standing Joint Coordinating Committee.

4 REPORT OF WORKING GROUP III: MULTINATIONAL COOPERATIVE RESEARCH ACTIVITIES

4.1 *Introduction*

There already exist formal bilateral cooperative programs between the U.S. and China and between the U.S. and Japan. Scientists and engineers from Taiwan, Korea and European countries also participate in cooperative activities with the U.S., although in a less formal manner. Extension of bilateral cooperative activities to trilateral or multinational cooperative activities is thus a natural next step to integrate and utilize the multinational resources in earthquake engineering and hazard mitigation more effectively.

Working Group III was charged to explore the feasibility of developing a trilateral, regional, or multinational cooperative research program in earthquake engineering and hazard mitigation and to articulate the motivation for forming such a program, *i.e.*, identify the research, educational, and societal challenges and opportunities that such a program would offer.

The working group was supportive of the opportunities presented by trilateral or multinational research programs in earthquake engineering. Recent earthquakes in densely-populated urban areas emphasize the opportunities and needs for multinational efforts to reduce earthquake hazards worldwide. The working group cited successful examples of multilateral research programs where the results influenced practice and were adopted by each of the countries involved.

For multilateral programs to be successful, a number of factors are required to be considered:

- The research program needs a clear mission.

- Realistic goals must be established for any multilateral program, and these goals must be long-term because of the nature of the cooperation and coordination involved.

- The program should include balanced participation from the countries involved.

- The research community should initiate multilateral research programs. Bottom-up initiatives are typically more successful than top-down research directives.

- Balanced funding is required from each country at a level appropriate to the country's activities.

- Governmental cooperative agreements supporting the cooperative research can be very useful, but are not essential, particularly for smaller projects.

- Participants from each country must perceive something of value coming from the cooperation.

- Opportunities offered by the cooperative research must be sufficiently attractive that top people will want to initiate, guide, and apply the results of the research.

- Research should provide important opportunities for young researchers. International research programs would benefit from enhanced use of existing programs for exchanges of young scholars.

- If governmental support is required, the right agencies must be involved from each country.

- Research funds should have flexibility so the program leaders can address the problems that arise without undo organizational and political constraints.

4.2 *Features of Multinational Research Programs*

The working group identified several features of multinational research programs that distinguish them from bilateral programs; these features may affect the organization and eventual success of multinational programs in earthquake engineering and earthquake risk reduction.

- The potential benefits of multilateral programs can be larger than bilateral programs because the results are likely to impact more countries, and the researchers may benefit from a broader research experience.

- Multilateral cooperative research is inherently more difficult than bilateral research, because of the interplay of more social cultures, languages, and technical environments.

• It is easier to collaborate internationally on topics related to fundamental research and advanced technology because of their independence of individual cultures. On the other hand, applied programs and application of research results can be more difficult because of their dependence on local practices.

• Many countries possess key national resources that are appropriate for cooperative research programs. In addition to facilities, technical expertise and communication facilities are important resources for international research programs.

• A framework for multilateral research can provide a mechanism for individual bilateral or trilateral research programs, as appropriate. Conversely, trilateral or multinational cooperative research programs could evolve from new or existing bilateral research efforts.

• Center-to-Center cooperation may provide a mechanism for some multilateral cooperative programs. Those that are multi-disciplinary would benefit the most from this level of interaction.

• Attempts should be made to develop international funding sources for multinational earthquake engineering research programs. Such funds would greatly enhance the possibilities of implementing cooperative research programs and adopting the research results in developing countries.

• Multinational cooperation among individuals at the level of the exchange of information does not require further funding or cooperative agreements. Enhanced use of the Internet provides a link for cooperating at this level.

5 CLOSURE

The working sessions at the Millennium Symposium have developed a vibrant and workable research and implementation plan for a future cooperative protocol on earthquake engineering and hazard mitigation between the U.S. and China. This future plan has considered (1) program expansion in terms of technical scope, new technologies and participation of additional government agencies and entities, and (2) scalability of the cooperative protocol in the sense that it could easily and conveniently accommodate activities that are deemed timely and important for East Asia regional cooperation in the new millennium. The participants of the Millennium Symposium are to be congratulated for their outstanding efforts.

ACKNOWLEDGEMENTS

The efforts of P.C. Jennings, M.P. Singh and B.F. Spencer, Jr. in editing the contents of this manuscript are greatly appreciated.

APPENDIX

The co-chairs and co-recorders of the Working Groups were as follows:

Working Group I
• Co-Chairs:
  Prof. Xiyuan Zhou (China)
  Prof. Gregory L. Fenves (U.S.)

• Co-Recorders:
  Prof. Yifan Yuan (China)
  Prof. Lawrence A. Bergman (U.S.)

Working Group II
• Co-Chairs:
  Prof. Li-Li Xie (China)
  Prof. Mete A. Sozen (U.S.)

• Co-Recorders:
  Prof. Yayong Wang (China)
  Prof. Joel P. Conte (U.S.)

Working Group III
• Co-Chairs:
  Prof. Houqun Chen (China)
  Prof. Paul C. Jennings (U.S.)

• Co-Recorders:
  Prof. Xilin Lu (China)
  Prof. Sharon L. Wood (U.S.)

China-US Millennium Symposium on Earthquake Engineering
Beijing, China, November 8-11, 2000

Plenary lectures

Earthquake Engineering Frontiers in the New Millennium, Spencer & Hu (eds),
© 2001 Swets & Zeitlinger, ISBN 90 2651 852 8

Twenty years of field work on dam–foundation–reservoir interaction

Ray W. Clough
Professor Emeritus, University of California, Berkeley

1 INTRODUCTORY REMARKS

I am happy to take part in this Symposium, celebrating twenty years of the U.S. – China Protocol for Cooperation in Earthquake Studies, because a major part of the research support for the latter stage of my academic career at Berkeley was provided from this source. Although I am well satisfied with the title of this talk shown above as provided by the Symposium organizers, the title that I originally had thought of using was "Seismic Safety of Concrete Dams in China". The plan was for the talk then to give answers to the following three questions:

(I) Why Seismic Effects?
(ii) Why Concrete Dams?
(iii) Why China?

In putting the talk together, however, I decided to present the questions in reverse order, as shown in the following paragraphs.

1.1 *Why China?*

To provide an answer to this question, I must first mention my brother, Ralph, who is four years older than I am. His interest in China was initiated by his taking part in a Boy Scout Jamboree that was held in Australia in 1933. In order to visit Australia in those days, he had to travel by ship, of course, and his ship was scheduled to stop at several South China ports, including Canton and Hong Kong, on the way to Sydney, so he had an interesting early view of China. Following that experience, he decided to join the U.S. Foreign Service aiming toward service in China, and accordingly he began to study the Chinese language both spoken and written. He also applied for and received an Exchange Fellowship with which he went to Lingnan University in Canton for a year's study — at the same time a student from Lingnan University was sent to the University of Washington.

After this experience of language study, Ralph graduated from the University of Washington and then spent a year at the Fletcher School of Law and Diplomacy in Boston. Subsequently he was appointed to serve as a Foreign Service Officer with the U.S. State Department in China. After the end of the war with Japan, he was serving with General Marshall in Chungking when I next had contact with him. At that time I had been on active duty in an Engineer Aviation Battalion on Okinawa, but when our Battalion was de-activated, I was left with nothing to do. So I was appointed to assist the Provost Marshall in managing a large group of Japanese prisoners of war; and this was intended to be my job until I had acquired enough "points" to be sent home and discharged from the Air Force.

While I was in this rather gloomy situation, I became somewhat of a celebrity in my Regiment on Okinawa when I received orders from General Marshall in Chungking by way of General MacArthur in Tokyo, directing me to report to Foreign Service Officer Ralph N. Clough at the U.S. Consular Office in Chungking. This was quite a surprise to me, but I soon learned the reason for the move. General Marshall was aware, of course, of the vast mountains of supplies that had been accumulated in China in preparation for the invasion of Japan, and he thought some of this material should be used to re-supply the Consular offices in China, which had been stripped bare during the Japanese occupation. So he asked my brother to take charge of this job, and the result was my unexpected order to head for Chungking as soon as possible.

Ralph told General Marshall that he did not have the technical expertise to do this job properly, but that he could recommend someone to help with the work. Within hours I was packed and flying to Shanghai, with no regrets for my sudden departure from the Aviation Engineer Regiment in Okinawa. However, the final stage of my transfer to Chungking was delayed by bad weather, so I was stuck in the Broadway Mansion Hotel in Shanghai until the weather improved. I must admit that this was a great improvement over the tent in which I had been living in Okinawa, so I had no complaint. Each morning I would contact the air transport office to see if a plane would be flying to Chungking that day, but the bad weather continued. Then one morning while I was having breakfast, I was surprised to see my brother come in and sit at the table next to mine — his travel also had been interrupted by the weather. So the end result was that I didn't have to go to Chungking to meet Ralph, and we were able to begin at once the allocation of supplies to the Consular Offices.

We began with Shanghai, of course, and then went on to Canton and Hong Kong. By the time we had finished with Hong Kong, however, my discharge number had come up, so I said goodbye to my brother with only minor regrets and was transferred back to Okinawa to board the troopship for the trip back to the States. This interlude was of trivial importance, but it provided my introduction to China — the beginning of an interaction which continues now, 55 years later.

The other person who has made a major impact on my interest in China is Professor K.T. Zhang of Tsinghua University. I first met Professor Zhang at a Conference on Arch dams held in Southhampton, England in 1964. At that time, I was on a sabbatical leave from Berkeley, at Churchill College in Cambridge, so the cost of travel to the conference was trivial, and I was glad to join a meeting on the field which had become my major research interest. The most important event for me at the Conference was when I was approached by a man of Chinese origin who shook my hand and told me he was an alumnus of U.C. Berkeley. He introduced himself as Professor K.T. Zhang of Tsinghua University, and said he had been a graduate student in Structural Engineering at Berkeley in the 1930's. We quickly learned that we had many mutual interests in the field of concrete dam design.

The next time I saw Professor Zhang was in 1978, when I was sent to China as a member of the U.S. Delegation on Earthquake Engineering which was under the leadership of Professor George W. Housner. When this 1978 delegation arrived at Tsinghua University, I was very pleased to see Professor Zhang waiting for our group on the front steps of the building. I was especially happy to see him on this occasion because when I visited Tsinghua University on a previous trip to China in 1974, as a member of the U.S. Earthquake Prediction Delegation headed by Professor Frank Press, I was disappointed that I did not see Professor Zhang. The problem at that time was that the Cultural Revolution was in progress, and the leading Tsinghua University Professors were not allowed to meet with foreign visitors.

The most important development of my meeting with Professor Zhang in 1978 was his asking me to come back to Tsinghua University at some future date to give a series of lectures on Earthquake Engineering and on the Finite Element Method. I told him I would be glad to come in 1980 to give the lecture series to an advanced group of Civil Engineering students, but that it would be necessary to obtain funding for my trip. Fortunately, the U.S. – China Protocol had just been put into effect, so that made it possible for me to have an extended visit to China in 1980, working with Professor Zhang in presenting lectures on Finite Elements and Earthquake Engineering.

1.2 Why concrete dams?

My interest in concrete dams first was initiated by my professors of undergraduate Civil Engineering at the University of Washington, several of whom had worked for the U.S. Corps of Engineers or the Bureau of Reclamation. In fact, the most intensive research project in Civil Engineering at Washington at that time was on the properties of the concrete designed for use in construction of Hoover (Boulder) Dam. As students, we went on several field trips to observe the construction of concrete dams or to see them in operation; and actually my first professional engineering job was as an Engineering Aide working for Bonneville Power on the Columbia River Project. Possibly even more important, was the fact that one of my colleagues at Berkeley, Professor J.M. Raphael, had worked for both the Corps of Engineers and the Bureau of Reclamation before he joined the Berkeley faculty. He was continually advising me of projects he thought would be worthy of my attention.

An example of a project he suggested to me was the Norfork Dam located near Little Rock, Arkansas. This dam is a simple gravity structure, but as a result of uncontrolled cooling during construction a tension crack developed near its mid-length, and this crack was of such concern to the design engineers that they did not allow the reservoir to be filled to more than half the design depth. This limitation greatly impacted the usefulness of the dam, of course, and the Corps of Engineers allocated research funds to me to study the state of stress that would be developed in the cracked dam with full reservoir load. Based on the results of that study, I was able to convince the dam owners that it would be safe even with full reservoir. I was very pleased with this solution of a practical problem, because it provided an excellent demonstration of the refined stress analysis work that could be done by the finite element method, and it encouraged me to continue pushing the use of the method to study the behavior of concrete dams at any state of their history.

1.3 Why seismic effects?

This question goes back to a conversation I had with my doctoral thesis advisor at MIT — Professor C.H. Norris — as I was saying goodbye to him at the end of my graduate studies in 1949. I had completed all requirements for my degree and I was about to embark on my teaching career in Berkeley, so I asked Professor Norris what aspects of structural engineering I should emphasize in my research at Berkeley. Professor Norris looked at me in surprise and said "Earthquake Engineering, of course! What else for an institution located adjacent to the San Andreas Fault?"

In describing this situation, I should mention that Earthquake Engineering was not yet a recognized part of any Civil Engineering curriculum in the United States in 1949. The limited recognition of the field at that time is evidenced by the fact that the first World Conference on Earthquake Engineering did not take place until 1956 when it was held in Berkeley. But Professor Norris certainly was wise in making that suggestion.

2 TWENTY YEARS FIELD WORK ON DAM–RESERVOIR SYSTEMS

Throughout my career I have always been concerned with the dynamic behavior of structures, and in Civil Engineering by far the most important dynamic loading results from earthquake excitation. So for many years I have concentrated on the earthquake response behavior of concrete dams. The primary complicating factor in this dynamic response is that it is influenced strongly by dynamic interaction of the dam with its environment — specifically by dam– reservoir interaction and by dam–foundation interaction. The most important indicator of these interaction effects is in their influence on the vibration mode shapes and frequencies; so my research mainly involved measurements of the dam vibration properties and comparison of these with corresponding values obtained by analysis. As a matter of interest, it may be noted that it was in such analyses of the complicated geometric forms selected by dam designers as well as the dramatic variations of material properties that are inherent in dam systems that the finite element method first demonstrated its value.

In our earliest work done on measuring the vibration properties of concrete dams, the motions were excited by bolting a rotating mass shaker to the crest of the dam, and then operating it over the entire range of frequencies expected to be present. In such a test, the resonant frequencies were easily identified as those which induced maxima in the curve expressing the relationship between response amplitude and frequency. Of course, in performing such a test (both experimentally and analytically) the interaction of the dam with the reservoir and/or foundation had to be taken into consideration.

Running a complete vibration survey in this way was a rather tedious process. First, it was necessary to transport the very heavy shaking machine to the dam site, and in China this often involved use of a quite primitive transport system to get to a remote location in the mountains. Usually, the freight was transported by train to a location near the site, and then the journey was completed by truck. When the shaking machine arrived at the dam, it had to be bolted to the crest at a prescribed position, usually at about one-third of the crest length from one end. Then the shaker had to be operated through the entire frequency range expected to be excited by an earthquake in that geographic area. The complete response recording system was employed during this operation, and the vibration shapes were noted for each response peak in the response curve.

At Berkeley we used this shaking machine test procedure for all of the dams we studied before we entered into the U.S. – China Protocol for Cooperative Earthquake Studies, and it was also used for the first two dams that we tested in China with support from the Protocol — namely, Xiang Hong Dian and Quan Shui dams. In addition, we tested Monticello Dam in California with financial support from the Protocol and with manpower assistance from the Protocol partners.

Finally, we got the idea that an easier way to obtain the dam dynamic response information was by an explosion generated ground shock that would excite all the important vibration modes of the dam and also would induce the desired dam–reservoir– foundation interaction. The owners of the dams we tested in the United States were not willing to allow the use of explosives to test their structures, but one of the major advantages we gained from the U.S. – China Protocol was that our Chinese partners did not object to explosion testing. So we began evaluating the blast test procedure with the next dam we tested in China — which was Dongjiang Dam.

It must be mentioned now that I had supervised all the dam tests done in this program up to and including the study of Quan Shui Dam. But at that time I retired from active teaching at Berkeley, and beginning with Dongjiang Dam, the U.S. side of the cooperative studies was supervised by my former doctoral student — Dr. Yusof Ghanaat of Quest Structures in Orinda, California, while I was merely an observer. Dr. Ghanaat was thoroughly familiar with the cooperative dam test program in China because he had served as my assistant during all of the previous tests done with support from the Protocol.

For the test of Dongjiang Dam, bore holes were drilled into the foundation rock to a depth of forty meters at a location 800 meters downstream of the dam and centered in the canyon. The blast procedure proved to be very successful in that it provided information on the mode shapes and frequencies of the first several symmetric vibration modes and the first several antisymmetric modes in a single operation. In addition, it was possible to get a good measure of the percentage of critical damping developed in these modes, taking account of the energy losses due to interaction as well as internal material damping. Because little was known about the effectiveness of blast generated ground shock in exciting the vibration response of a large concrete dam, several sizes of blast charges were tried — a single 100 kg charge, a single 300 kg charge, and a double 300 kg charge.

In a secondary test series on Dongjiang Dam, the system was excited by small explosive charges suspended in the reservoir. The results of this test program demonstrated that dam system vibrations can be excited most effectively by explosives detonated in the reservoir and only small charges are needed to obtain responses of the dam system that are significantly greater than those resulting from resonance testing using a rotating mass vibration generator.

The final dam tested in this U.S.–China program of cooperative research was Longyangxia, located near the headwaters of the Yellow River. The first attempt by our test team to reach this dam site was not successful because the construction road that had been used in building the dam was washed out by spring floods just weeks before the test of this dam was to be done. Clearly, it was necessary to abandon the test at that time, and as a consolation prize, the U.S. team decided to visit the site where the Three Gorges Dam is to be built. To make this trip we boarded the Yangtze River boat at Chungking and cruised past the gorges to the dam site at Sandouping. And one of the most impressive features of the trip was a drive on the new super-highway that is being built along the side of the canyon at about 550 feet above the present Yangtze River channel — at the level of the future reservoir surface. This visit clearly demonstrated the size of this tremendous project.

When we returned to the Longyangxia dam site in the following year, it was quite evident why we could not get through on our preceding visit — many rock slides had blocked the road. But we had no trouble getting through on this occasion, and were able to carry out the test of the dam using exci-

tation caused by small charges detonated in the reservoir. We were very happy that we did not have to use the rotating mass shaker because it would have been a major task to transport it to the dam crest.

3 CLOSURE

I close these informal comments now by expressing my gratitude to the U.S. – China Protocol for Cooperation in Earthquake Studies. During this history of the cooperative research effort, I was involved in the study of four dams at widely different locations in China, as well as one in California that was included to make this a truly cooperative effort. This work was done by a team of researchers from China working in close collaboration with our team from the USA. When we first began the studies, the U.S. team provided the computer system, the shaking machine, and most of the instrumentation and recording equipment, while the principal contribution from the Chinese side was providing the test structure and the manpower to install the instrumentation. However, as we progressed through the years, our Chinese colleagues took over more and more of the work. So by the time the last test was done it was essentially a completely Chinese job. As was mentioned earlier, the other major contribution from the Chinese side was being able to do the tests by means of blast excitation which we were not able to do in the U.S.

Another major benefit derived from the Protocol was our having continued support for a long period of time. Clearly, the work we had in mind could not be carried out by a short term effort, and it was essential to carry on step by step as we learned how best to study the dynamic interaction mechanisms that determine how a major concrete dam responds to severe earthquake excitation. So I say thanks again to the U.S. – China Protocol for Cooperation in Earthquake Studies for making possible the extensive research effort that Professor Zhang and I had in mind when we initiated our research program twenty years ago.

Earthquake Engineering Frontiers in the New Millennium, Spencer & Hu (eds),
© 2001 Swets & Zeitlinger, ISBN 90 2651 852 8

Some challenges to earthquake engineering in a new century

Li-Li Xie
Institute of Engineering Mechanics, China Seismological Bureau, Harbin, 150080, China
Harbin Institute of Technology, Harbin, 150001, China

ABSTRACT: The recent destructive earthquakes occurred around the world revealed that the existing knowledge and techniques are still not sufficient to achieve safety against earthquakes at an effective cost. It is believed that in-depth research to earthquake engineering is urgently needed. It provides a profitable field for China-U.S. bilateral cooperation and international cooperation as well, particularly in the new era of the coming Century. In this paper an extensive research program to earthquake engineering both in traditional and non-traditional approaches for future cooperation are recommended. The application of modern advanced technologies and its potential roles in seismic disaster reduction are emphasized.

1 INTRODUCTION

1.1 *A short review of past twenty years cooperative studies in earthquake engineering and hazard mitigation between the People's Republic of China and the United States of America.*

It has been twenty years since the joint protocol for Scientific and Technical Cooperative Research in Earthquake Studies between the Peoples' Republic of China and the United States of America was established in 1980. Of this Protocol, the Annex III covered the research area on Earthquake Engineering and Hazards Mitigation. The primary objective of Annex III is to develop safe and cost-effective engineering methods and construction practices and other countermeasures to improve seismic safety. Initially the emphasis of this Annex was on the application of engineering knowledge of seismic hazards and strong seismic ground motions, including its measurement and effects on structures, and the dynamic behavior of soils and sites. Since the late 80s', as a response of the International Decade for National Disaster Reduction (IDNDR), both China and United States agreed to include in the Annex III other aspects of earthquake disaster reduction, such as disaster reduction measures, emergency management, social and economic effects, earthquake insurance and education. It is unanimously recognized from both sides that the twenty years cooperation between the PRC and USA has been active and fruitful, produced excellent results in expanding knowledge and technologies in earthquake engineering and greatly promoted mutual understanding and collaboration between earthquake scientists and engineers in the two countries and benefited to both sides.

1.2 *New situations need upgrading of Annex III*

During the past twenty years some great changes happened to the two countries. It resulted in an urgent need for seismic safety and updating the content of the annex III to the China-US Protocol of joint studies in earthquake. Among them the most exciting ones are as follows.

(1) As a result of the IDNDR campaign, the awareness of publics and governments to disaster mitigation and demand for sustainable development are greatly raised in the both countries. For examples, in China the Central Government set immediately after the Hanshin, Japan Earthquake of 1995 a "Ten Years Goal of Earthquake Preparedness and Reduction for Moderate and Major City". The Ten Years Goal states that during the coming ten years, all efforts should be taken towards a safer city prone to earthquake with the capabilities against earthquake of magnitude of six. Furthermore, "The Law of the Peoples' Republic of China on Protecting Against and Mitigating Earthquake Disaster" was approved in the December of 1997 and has been enforced since March 1st, 1998. It regulated that the governments at each level are liable for protecting people for seismic safety.

(2) Both China and United States has ambitious program for large scale construction of their infrastructures in the coming Century. In the China, a grand plan of developing its West-Northern Area is initiated. This area consists of 6,500,000 km^2 i.e. 68% of nation's territory and 308 million popula-

tions. Specially, this area is located in the high seismic regions. A great number of major projects such as high dams, long distance oil and gas pipelines, electricity systems, modern transportation systems, high dams, new urban area etc. will be constructed that new technologies and methods for seismic safety at an effective cost are particularly in urgent need.

(3) Recent devastating earthquakes occurred in the past twenty years, such as Loma Prieta U.S. earthquake of 1989, Northridge U.S. earthquake of 1994, Hanshin Japan earthquake of 1995, Jiji Taiwan, China earthquake of 1999 and Izmid Turkey earthquake of 1999 hit the urban area that raised a series of problems: intolerable economic losses and tremendous life losses and revealed that the existing knowledge is still not sufficient for seismic safety at an effective cost. It is recognized that to achieve the goal of controlling the seismic risks in our urban areas, reducing them to socially acceptable levels there is still a long way to go. In particular, more efforts are needed to improve our earthquake resistance design and earthquake resistance construction.

(4) The rapid development of new technologies such as smart materials and intelligent structures, advanced sensors, super computing power, information technologies, wireless communication, geographical information system, remote sensing technology, structural control, etc. will provide not only an unprecedented opportunity for improving seismic risk control but also some new tools for solutions of better understanding of seismic damages that cannot be solved by traditional approaches.

All the above-mentioned new situations will greatly increase opportunities and broaden the scopes for future cooperation in earthquake engineering both in traditional and non-traditional approaches between the PRC and USA.

2 TRADITIONAL RESEARCH AREAS RECOMMENDED FOR FUTURE COOPERATION

In spite of the great progress obtained in the field of earthquake engineering during the past fifty years, recent destructive earthquakes occurred around the world revealed that the existing knowledge and techniques are still not sufficient to achieve safety against earthquakes at an effective cost. It is believed that among all natural hazards earthquakes are still number one disaster for which in-depth research, particular in traditional approaches to earthquake engineering is still needed. A better understanding of all aspects of devastating earthquake can expand our knowledge and strengthen our defenses more rapidly than if each country works in isolation. To this regard, following potential areas for joint research might be appropriate.

2.1 *Strong ground motion measurement and analysis*

The measurement and analysis of earthquake ground motion and its effects on structures are one significant area for which there is obvious benefit in bilateral and/or international cooperation. Such cooperation could be of great mutual benefit to all countries over the world. Strong earthquake motion data must be obtained from a variety of earthquake sources, wave propagation paths and site conditions as quick as possible in order to influence the retrofit of existing structures and the design of new structures in earthquake prone regions of the world.

China is one of the most seismically active regions in the world. There have been about 300 earthquakes with magnitudes greater than six in the continent of China since 1900 and seven of these have had magnitudes greater than eight. This level of seismic activity is much higher than that in the United States.

The largest earthquakes in China generally occur in one of five zones: 1) the Himalayan zone, 2) the Central Asia zone, extending northeast from Pamir, through Altai in western Mongolia to Baikal, 3) the North-South zone, extending along the eastern margin of the Qinhai-Tibet Plateau, 4) the North china Plane zone, which includes the Fenwei Zone, the Hebei Plane and the Tanlu Zone, along the Pacific Ocean. (Ding G. 1988).

As the beginning of the new Century, a five-year plan of strong motion instrumentation consisting of 2000 accelerographs is being reviewed and this plan is likely to be approved and will be implemented during the 10th Five-Years (2001-2005) period of Chinese National Plan for Social and Economic Development. This provides an exceptional opportunity for strong-motion studies in China and also for China-U.S. cooperation in the new century as well. The potential topics for such cooperation might be:

- Installation of digital strong motion network and array by using the new technologies,

- Study and analysis of observation data of near fault strong ground motions and structure reaction,

- Construction of different kind of arrays for observing structural responses, ground motion attenuation, site effect, etc.,

- Construction of strong earthquake motion and seismic damage database,

- Establish internet network to make data available to the world,

- Comparative study of strong motion data from different area over the world aim at using the data collected from one region in other regions lacking of strong motion data.

2.2 *Seismic hazard analysis and seismic zoning*

- Study on ground motion attenuation for regions lacking of strong ground motion records,
- Site effect on ground motion and site classification,
- Effect of fault on characteristics of strong ground motion,
- In-situ geo-technical test technology and devices,
- Prediction of spatial distribution of strong ground motion.

2.3 *Seismic safety of critical structures*

In China, a number of large dams (concrete arch dams with the height of 250 to 300 meters and concrete gravity dams with the height of 200 meters) are being planned for construction in known high-seismic areas (earthquake intensity M7.0 or greater) that will offer a unique opportunity for both China and US dam research workers to work together, Topics for cooperation might be:

- Evaluation of seismic performance of high dam,
- Seismic safety of geo-technical systems, such as earth dams, retaining walls, performance of loess structures during earthquakes, etc.,
- Seismic safety of long-span bridge and underground tunnel,
- Seismic analysis of extensively buried pipe-line and its seismic performance.

2.4 *Study on earthquake disaster mitigation for cities*

- How to define a cities' capability in earthquake disaster resistance,
- How to develop a methodology to assess a cities' capability of earthquake resistance,
- What measure should be taken for reducing earthquake risks,
- Earthquake damage assessment for urban areas,
- Vulnerability analysis and strengthening methodology for existing masonry structures, frame structures and high rise structures etc.
- Vulnerability analysis for highway, bridge and pipeline network (including buried pipeline, erected pipeline, etc),
- Quick evaluation of post earthquake damage to a city,
- Development of high efficient emergency response technologies.

2.5 *Basic research in earthquake engineering and hazard mitigation*

In spite of the great progress achieved in the field of earthquake engineering during the past fifty years, the basic research on earthquake engineering, particularly for those devastating earthquakes, is still in need to expand our knowledge and strengthen our defenses. The research areas are as follows:

- Better prediction of future ground motion,
- Foundations of performance based design and performance based engineering,
- Developing optimum seismic design criteria to control structural performance and even economic and life losses at an effective costs,
- Better understanding, quantifying and minimizing uncertainties in all aspects involving in seismic design procedures,
- Reliability based seismic design theory and practice,
- Pile-soil-structure interaction,
- Similarity rule of dynamic structural testing,
- Study on seismic performance of geo-technical structures. Study on seismic performance of geo-technical structures.

2.6 *Research on seismic design codes*

It is noteworthy that all major earthquake disasters during the past twenty years have occurred in countries where the seismic design code were available, so it is clear that having a seismic code no sufficient to prevent earthquake disasters. Examples of such disastrous earthquakes are the 1999 Izmid, Turkey Earthquake; the 1999 Jiji, Taiwan China Earthquake; the 1995 Henshin, Japan Earthquake; the 1994 Northridge, California U.S. Earthquake; the 1976 Tangshan, China Earthquake; and many others. In the past, the usual procedure for upgrading seismic code has been to wait until a destructive earthquake occurred and then to change the building code to strengthen the demonstrated weakness, and then wait for the next earthquake to demonstrate other weakness. This is not an efficient way of reducing earthquake disasters. It would be more advantageous to improve the seismic code as new knowledge is developed by research and experience, rather than to build under the existing code while waiting for the next earthquake. In drafting a new seismic code, or revising an existing code, it is necessary to balance the cost of seismic design against the reduction of future losses from earthquakes. The estimation of future losses must recognize not only structural damage but also the economic and social impacts that can be very severe. (Housner 1996).

In China, there has been forty years since the first draft seismic code was prepared in 1959. The latest code was revised in 1989. It is not sufficient to have only one seismic code in so vast territory of China where seismicity, construction technologies and materials are very different from regions to regions. To change this situation, as the first step it is necessary to develop a "model code" as United States and other countries do. The model code would serve as an educational document for design engineers and seismic code compliers for preparing the local code. It is important to learn experiences and lessons from United States and other countries. The possible areas for cooperation are:

- Research and development of the seismic model code,
- Comparative study on different seismic codes over the world,
- The cost-benefit and cost effective analysis of seismic codes.

2.7 *Research related social-economic aspect*

Socio-economic policy research should address the evaluation of risk associates with various socio-economic consequences of seismic hazard applied to large cities. The interaction between subsystems within cities and the effects of both direct and indirect losses should be assessed. The comparison of methods and techniques for assessing the socio—economic risks associated with multiple hazards to large cities with different backgrounds and systems.. Possible cities for comparison include Beijing, San Francisco, Seoul, Shanghai and Tokyo.

3 NON-TRADITIONAL RESEARCH AREAS RECOMMENDED FOR FUTURE COOPERATION

In recognition of the recent rapid advancement of technologies related to earthquake engineering, new approaches to hazards mitigation based on innovative technologies need to be developed and validated. Emphasis shall be placed on broad-based, multidisciplinary activities that would accelerate application and implementation of research results. Topics suggested for cooperative research are as follows.

3.1 *Innovations for high rise structures*

The rapid increase in the construction of tall structures in seismic zones requires special attention to their safety under natural hazards, such as earthquakes, wind and soil failures. In particular attention should be directed at:

- Development of new materials and methods of construction, fabrication and manufacture, for example, composite materials adapted from aero-

space applications may prove advantageous for all structures. Similarly, new applications of reinforced concrete may be still suitable for continuing cooperation and structural steel applications may be of particular interests for future joint research between China and the USA,

- Development of improved methods for controlling structural response.

New methods for active, passive and hybrid structural control should be developed and the applicability of existing methods should be extended. In both cases, the performance, practicability and reliability of the methods should be thoroughly investigated through analysis as well as laboratory and field tests. Earthquake, wind and other dynamic excitations should be considered.

- Development of improved damping characteristics.

Viscous damping is an important characteristic of tall structures affecting dynamic response. Improved methods are needed to estimate the values of viscous damping inherent in various types of tall structures. Similarly, devices or construction techniques should be developed to increase damping values to artificially high levels and new design methods need to be formulated and verified accounting for these new characteristics.

3.2 *Facilitating the application of advanced technologies*

The rapid development of new technologies such as smart materials and intelligent structures, advanced sensors, super computing power, information technologies, wireless communication, geographical information system, remote sensing technology, structural control, etc. will provide not only an unprecedented opportunity for improving seismic risk control but also some new tools for solutions of better understanding the seismic damages that cannot be solved by traditional approaches, such as, to detect and diagnose the hidden and/or localized damage by health monitoring system, to improve structural performance by functional materials, to measure seismic displacement-time history curve of structures during earthquake with Global Positioning Systems, to prepare an efficient post quake emergency plan by using the Geographical Information System technology and so on. These technologies will improve the science and practice of earthquake engineering, and will allow better communication, nationally and internationally, with the public and decision-makers responsible for earthquake risk reduction.

Furthermore, the author wishes hereon to emphasize the prospects of the application of satellite remote sensing technology and the so-called Digital Disaster Reduction System (DDRS) in earthquake disaster reduction.

4 APPLICATION OF RECENT SATELLITE REMOTE SENSING TECHNOLOGY IN EARTHQUAKE DISASTER REDUCTION

With the high-speed development of satellite remote sensing technology, it has played significant roles in reducing various kinds of natural disasters, for examples, in forecasting and controlling of flood, forecasting hurricane, monitoring landslides and forest fire and so on (see Table 1). Regretfully, as we understand that the satellite remote sensing technology is rarely applied both in domestic and international for earthquake disaster reduction. It is because that on the one side, earthquake is a very complicated natural phenomenon with its indistinct genesis mechanism and occurrence of very low probability and on the other side, the resolution of satellite remote sensing image is too low and satellite re-visit period is too long that constrain this technique to be used in earthquake disaster reduction.

Table 1. Application of satellite remote sensing in disaster reduction

Disaster	Monitor	Prediction	Prewarning	Emergency
Drought	OK	OK	OK	OK
Flood	OK	OK	OK	OK
Bush fire	OK	OK	OK	OK
Hurricane	OK	OK	OK	OK
Landslide	OK	Limited	Limited	Limited
Volcano	OK	OK	OK	OK
Earthquake	NO	NO	NO	NO

Fortunately, the new development of satellite remote sensing technology, such as, successfully launching a series of radar satellites, emergence of international open market for high revolution remote sensing satellite and rapid development of micro-satellite and constellation technique, can greatly shorten the satellite re-visit period and reduce cost by a large amount, and then can make it possible to play an important role in earthquake disaster reduction.

The satellite remote sensing technologies can be used in rapid assessing the seismic damage for effective post quake emergency action and in monitoring crustal movement for better understanding of seismic risk. (Xie and Zhang 2000).

4.1 *Rapid evaluation of seismic damage by satellite remote sensing*

An efficient emergency plan should be based upon a rapid and accurate estimation of seismic data, such as the extent and distribution of building structures damaged and destructed, the damage to urban lifeline systems, large reservoirs and highways, passable capacity of urban traffic paths, casualties and injured and even the outline of overall damage to the city and its vicinities. As we estimate for an emergency activity, such information should be provided no later than 8-10 hours after the occurrence of a devastating earthquake. It is convinced that no other technologies like satellite remote sensing technologies can provide rapidly such data without any limitation to time (day or night), weather and location of the city. However, to meet such demand we need very high space resolution and short re-visit cycle period of satellite remote sensing technologies. At present time the United States has launched several commercial satellites, such as Orbview-3, Quickbird and Ikonos-3 with high resolution of 1-3 meter that are quite accurate for post-quake damage assessment. But their revisit cycle periods are about 10-15days still too long to be sufficient for earthquake emergency plan and action. The way to shorten the revisit cycle period is to develop the micro-satellite constellation technologies. It could expect that such constellation will be launched very soon from the United States, Canada, Europe and China. As earthquake engineers, we should prepare the appropriate techniques and methods to analyze the data collected from remote sensing imagines. It is needless to say that is one of the exciting areas for bilateral collaboration.

4.2 *Improvement of hazard analysis by monitoring crustal movement with satellite remote sensing technologies*

It is known that the available seismic hazard analysis methods could be used in seismic mapping or assessing the possible peak ground motions for seismic design. It is based on the existing knowledge of tectonic geology. However such knowledge is quite not sufficient for accurate assessment of future earthquake. For example, many earthquakes like Hanshin Japan earthquake were caused by hidden strike faults that were unknown by both geologists and seismologists. To avoid repeat of such tragedy we should monitor and understand the changes of crustal movement. The radar satellite provides a powerful tool for monitoring such kind of crustal movement. The radar satellite (SAR or InSAR) is an active microwave remote sensing tool. It can penetrate through cloud, fog, rain, and snow, and work in all weather and full day. It can receive data from very rough configuration of geometric structure and echo nature of ground substance. The crustal deformation can be easily detected by such technology.

4.3 *Potential area of satellite remote sensing technologies for cooperation*

- Exchange of remote sensing data,
- Development of imagine processing and analysis methods,

- Identification of seismic damage to the city from remote sensing data, and

- Developing rapid analytical method for damage identification.

5 THE CONCEPT OF THE DIGITAL DISASTER REDUCTION SYSTEM (DDRS)

In the history of earthquake engineering development, the past fifty years can be characterized as a growing period during which knowledge was built and the earthquake engineering manpower base developed. Now the earthquake engineering has developed into an integral part of the engineering profession that is built on a solid foundation of theoretical, analytical, field and laboratory experimental knowledge through well-balanced research. However, the existing knowledge is still quite not sufficient for the requirement of seismic safety. The science and technique of earthquake engineering cannot provide efficient technologies to reduce the seismic losses at an effective cost. Now earthquake scientists and engineers have little knowledge about the time and place of occurrence of earthquakes. The rapid elapse of earthquakes that occurred suddenly, provide little opportunities for researchers to study response and damage of structures in detail and in depth. The devastating earthquakes had damaged and collapsed countless buildings and various structures, even very modern infrastructures, However we cannot reproduce such damage completely because we do not fully understand the mechanism and whole process of various types of damages. It is believed that in case seismic damage could be well duplicated on the screens of computers, earthquake engineers will have a powerful tool with which performance of structures during earthquakes could be identified and appropriate criteria for earthquake resistance design and strengthening of existing hazardous structures could be well developed. However, it is rarely possible to solve this problem by traditional approaches. It needs not only the integrated knowledge involved in earthquake engineering but also the new technology such as Virtual Reality incorporated (Xie and Wen 2000). Development of the Digital Disaster Reduction System might be one of the best choices for this purpose.

5.1 *What is the digital disaster reduction system (DDRS)*

The Digital Disaster Reduction System (DDRS) would be a specially designed system to study the virtual seismic damages that may happen to real structures during real earthquakes. The DDRS is constituted by integrating of computer hardware and software, supported by the large-scale database,

Remote Sensing, Global Positioning System, Geographic Information System and Virtual Reality technology, with rational mathematical and physical models of disasters and high-fidelity simulation as the core of the system to simulate the whole process of seismic disasters. The proposed DDRS could be applied as a powerful tool not only for seismic disasters study but also for other natural disaster research.

5.2 *The use of the DDRS*

DDRS is a virtual reality computer system designed to simulate the occurrence and propagation of disaster and whole process of damages caused by natural disasters. In the frame of the DDRS, digital earthquake, digital flood, and other digital natural hazards are the research objects of the DDRS. Taking earthquake as an example, with the real accelerograms and the real seismic damages to the ground surface and structures as the final goals of simulation, we can understand the damage process step by step through adjusting the constitution parameters and the physical and mathematical models of virtual environments (such as faults, sites, structures and so on) within the DDRS. Similarly, This system can be used to simulate independently different processes of disaster, such as source mechanism and source rupture, strong ground motions and response of and damage processes of the structures. DDRS is ultimately expressed by using Virtual Reality (VR). VR is specific expression of visualization in scientific computing, a process in which large numbers of data obtained from scientific experiments or numerical computation are converted to something perceptible through computer system, which create a better virtual environment for the researchers (Xie and Wen 2000).

6 CONCLUSIONS

The bilateral cooperation under the Annex III of Earthquake Engineering and disaster mitigation to the China-US Protocol for joint studies in Earthquake was initiated in 1980. There has been a worthwhile exchange of information and excellent results worked out over these twenty years and we can look forward to more active and fruitful collaboration in the New Century. For this purpose, some research areas to earthquake engineering both in traditional and non-traditional approaches for future cooperation are recommended. As one of the exciting research topics in earthquake engineering, "Reproduction of Seismic Damage" is proposed as an initiative project that needs international concerted efforts from various disciplinary areas. It is anticipated that this will stimulate bilateral and international cooperation in engineering and science related earthquake as well as other natural disasters.

REFERENCES

Ding, G. 1988. A brief introduction to recent strong earthquake activity in the continent of China. *Proceedings of the Sino-American Workshop on Strong–Motion Measurement.* December 13-15, 1988. California Institute of Technology, Pasadena, California, U.S.A. 15–19.

Housner, G.W. 1996. Preface. *Proceedings of the PRC-U.S.A. Bilateral Workshop on Seismic codes.* December 3-7, 1996. Guangzhou, China.

U.S. Panel on the Evaluation of the U.S.-P.R.C. Earthquake Engineering Program and Commission on Engineering and Technical Systems of U.S. National Research Council 1993, Printed in the United States of America.

Xie, L. & Wen, R. 2000a. Digital disaster reduction system (text in Chinese with English abstract). *Journal of Natural Disasters.* (ISSN 1004-4574), 9(2):1–9.

Xie, L. & Wen, R. 2000b. Application of satellite remote sensing technology in earthquake disaster reduction (text in Chinese with English abstract). *Journal of Natural Disasters,* (ISSN 1004-4574), 9(4):1–8.

Earthquake Engineering Frontiers in the New Millennium, Spencer & Hu (eds),
© 2001 Swets & Zeitlinger, ISBN 90 2651 852 8

Seismic performance of transportation structures

J. Penzien
University of California, Berkeley, California, USA
International Civil Engineering Consultants, Inc., Berkeley, California, USA

ABSTRACT: Revolutionary changes which have taken place in the USA over the past 50 years in earthquake engineering as applied to transportation structures (highway bridges, and transit and high-speed rail aerial structures) are reviewed, giving emphasis to seismic design criteria, characterization of seismic ground motions, dual strategy of seismic design, modelling and dynamic analysis, assessment of seismic performance, design detailing, and retrofitting of existing structures. To reduce the large uncertainties which still remain, it is pointed out that research is needed for improving the predictions of future ground-motion characteristics, developing better seismic design criteria in support of the dual strategy of design, establishing more realistic analytical models of components and systems based on laboratory and field-test results, advancing nonlinear dynamic analysis procedures and associated computer programs, developing more effective measures of active and passive control, and implementing concepts of performance-base design.

1 SEISMIC DESIGN CRITERIA

Revolutionary changes have taken place over the past 50 years in earthquake engineering as applied to transportation structures. This becomes apparent when one reviews the changes in seismic design criteria specified by the American Association of State Highway Officials (AASHO) in its Standard Specifications for Highway Bridges, First (1931) through Eleventh (1973) Editions, and by the American Association of State Highway and Transportation Officials (AASHTO) in its 1973 Interim Specifications for Highway Bridges and the subsequent Standard Specifications for Highway Bridges, Twelfth (1975) through Sixteenth (1996) Editions, and in AASHTO's LRFD Bridge Design Specifications, First (1994) and Second (1999) Editions. All of the above-mentioned specifications apply to Ordinary Bridges having span lengths under 500 feet.

1.1 *Standard specifications, 1949-1961*

The first reference to considering earthquake effects on bridges came in the Fifth (1949) Edition of the Standard Specifications which stated that earthquake stresses should be considered; however, no guidelines for doing so were given. This same reference was stated again in the Sixth (1953) and Seventh (1957) Editions.

1.2 *Standard specifications, 1961-1975*

The Eighth (1961) Edition of Standard Specifications was the first to specify an earthquake loading for design (*EQ*), namely

$$EQ = CD \tag{1}$$

which was to be applied statically in any horizontal direction as part of a Group VII load combination given by

$$Group\ VII = D + E + B + SF + EQ \tag{2}$$

in which D, E, B, and SF denote dead load, earth pressure, buoyancy, and stream flow, respectively. The numerical values of C were specified to be 0.02 for structures supported on spread footings where the soil bearing capacity was rated to be greater than 4t/ft^2, 0.04 for structures supported on spread footings where the soil bearing capacity was rated to be less than 4 t/ft^2, and 0.06 for structures founded on piles. The Group VII load combination was to be used in the working-stress design (WSD) with a 33-1/3 percentage increase in allowable stress because of the presence of the earthquake loading *EQ*. No seismic zone factors were provided in the specifications.

The above seismic loading provisions of the Eighth (1961) Edition of Standard Specifications were repeated, without modification, in the Ninth

(1965), Tenth (1969), and Eleventh (1973) Editions. It should be noted that these seismic loading provisions were based mainly on the lateral force requirements for buildings developed prior to 1961 by the Structural Engineers Association of California (SEAOC).

1.3 Standard specifications, 1975-1992

As a result of the 1971 San Fernando, California earthquake during which many highway bridges were severely damaged, some of which even collapsed, the California Department of Transportation (Caltrans) issued new seismic design criteria for bridges in 1973, which formed the basis of the 1975 AASHTO Interim Specifications for Highway Bridges. The equivalent static lateral force loading specified in this document for bridges having supporting members of approximately equal stiffness was of the form

$$EQ = CFW \qquad (3)$$

which was to be applied in any horizontal direction as part of the same Group VII load combination given by Eq. (2) in a working stress design with a 33 percent increase in allowable stress. In this equation, W represents dead load, F is a framing factor assigned the values 1.0 for single columns and 0.8 for continuous frames, and C is a combined response coefficient as expressed by

$$C = ARS/Z \qquad (4)$$

in which A denotes maximum expected peak ground acceleration (PGA) as shown in a seismic risk map of the United States, R is a normalized (PGA = 1g) acceleration response spectral value for a rock site, S is a soil amplification factor, and Z is a force reduction factor depending upon structural-component type which accounts for the allowance of inelastic deformations. The numerical values specified for A were 0.09g, 0.22g, and 0.50g in seismic zones numbered I, II, and III, respectively. The cities of San Francisco, CA, St. Louis, MO, and Charleston, SC, all having experienced large earthquakes in the past, were shown to be located in separate zones numbered III. Numerical values for R, S, and Z were not provided in the 1975 Interim Specifications; rather, four plots of C as functions of period T were given for discrete values of A. Each of these plots represents a different depth range of alluvium to rock-like material, namely 0-10', 11-80', 81-150', or >150'.

Figure 1 shows the plot for depth range 11-80'. Period T was to be evaluated using the single-degree-of-freedom (SDOF) relation

$$T = 0.32\sqrt{\frac{W}{P}} \qquad (5)$$

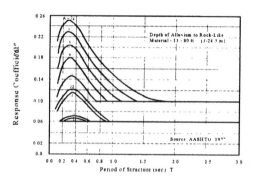

Figure 1. Response coefficient "C" for values of peak rock acceleration "A"

in which P equals the total uniform static loading required to cause a 1-inch horizontal deflection of the whole structure.

For complex or irregular structures, the Interim Specifications required use of the modal response-spectrum analysis method to generate design loads; and, in special cases of such structures having fundamental periods greater than 3 seconds, it required that they be designed using "current seismicity, soil response, and dynamic analysis techniques."

The same seismic design criteria in the 1975 Interim Specifications were repeated in the Twelfth (1977), Thirteenth (1983), and Fourteenth (1989) Editions of AASHTO's Standard Specifications; however in these editions, the designer was given, for the first time, the choice of working-stress design (WSD) or load-factor design (LFD). When using the WSD, the same Group VII load combination given by Eq. (2) was specified to be used along with a 33 percent increase in allowable stress; however, when using the LFD, the Group VII load combination was changed to the form

$$\text{Group VII} = \gamma[\beta_D D + \beta_E E + B + SF + EQ] \qquad (6)$$

in which load factor γ was assigned the value 1.3, β_D was assigned the values 0.75, 1.0, and 1.0 when checking columns for minimum axial load and maximum moment or eccentricity, when checking columns for maximum axial load and minimum moment, and for flexure and tension members, respectively, and β_E was assigned the value 1.3 for lateral earthpressure and 0.5 for checking positive moments in rigid frames.

1.4 Standard specifications, 1992-1999

In 1981, the Applied Technology Council (ATC) issued its ATC-6 Seismic Design Guidelines for Bridges under the sponsorship of the Federal Highway Administration, Department of Transportation. These guidelines were reviewed and revised slightly by the National Center for Earthquake Engineering Research (NCEER) under sponsorship of the National Cooperative Highway Research Program (NCHRP) Project 20-7/45 to form the basis of AASHTO's Fifteenth (1992) and Sixteenth (1996) Editions of the Standard Specifications. In these editions, each bridge structure must first be classified as either "Essential" or "Other" in accordance with given definitions and then be assigned to one of four Seismic Performance Categories (SPC) A, B, C, or D as defined in Table 1 below

Table 1. Seismic performance categories in which the acceleration coefficient, A, for a given bridge site is taken from contour maps provided.

Acceleration Coefficient	Bridge Classification	
	Essential	Other
$A \leq 0.09$	A	A
$0.09 \leq A \leq 0.19$	B	B
$0.19 \leq A \leq 0.29$	C	C
$0.29 \leq A$	D	C

No dynamic analysis is required in these editions for bridges having single spans, regardless of the value of the site acceleration coefficient A, and for all bridges in SPC A. All other bridges, regular or irregular, having two or more spans must be analyzed by at least one of two dynamic analysis procedures, namely, the single-mode spectral method (SMSM) or the multi-mode spectral method (MMSM). The SMSM is specified as minimum for regular bridges in SPC B, C, and D; while the MMSM is specified as minimum for irregular bridges in these same categories. An "irregular" bridge is defined as one having abrupt or unusual changes in mass, stiffness, and/or geometry from abutment to abutment; a "regular" bridge is one not meeting the definition of an "irregular" bridge.

The seismic input in any horizontal direction to be used in each of these minimum dynamic analysis procedures is specified in terms of an elastic seismic response coefficient, C_{sm}, as expressed by

$$C_{sm} = \frac{1.2AS}{T_m^{2/3}} \qquad (7)$$

in which T_m is the period of vibration of the mth mode, S is a site coefficient having the values 1.0, 1.2, 1.5, and 2.0, respectively, for soil profile Types S_1, S_2, S_3, and S_4 ranging from hard (S_1) to very soft (S_4), and A is an acceleration coefficient taken from the contour map prepared by the U.S. Geological Survey for the 1988 Edition of NEHRP Recommended Provisions for the Development of Seismic Regulations for New Buildings. The values of A in this map represent peak ground accelerations having a mean return period of 475 years.

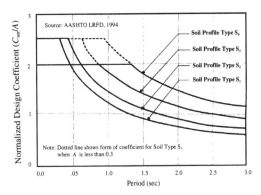

Figure 2. Seismic response coefficient for various soil profiles, normalized with respect to acceleration coefficient "A"

A plot of C_{sm} / A for each of the four soil profile types is shown in Figure 2 as a function of period T_m.

The constant cut-off levels in the lower period range of this figure represent exceptions to the use of Eq. (7). Note that the ratio C_{sm} / A corresponds to the acceleration response spectrum ratio $S_a(T)/S_a(0)$ which is commonly referred to as the normalized acceleration response spectrum; thus, it can be used as such in carrying out the minimum dynamic analysis procedures mentioned above.

Since each of these analysis procedures generates internal force components in members caused by only a single component (x or y) of horizontal seismic input, the procedure selected must be repeated using the same response-spectrum seismic input applied horizontally in the orthogonal direction to the first. The corresponding pairs of internal force components (Q_x and Q_y) produced by both inputs must then be combined using the "30 percent" rule into the two combined forms $Q_x + 0.3Q_y$ and $Q_y + 0.3Q_x$ with the larger of those two used for design. It is more rational however to use the "40 percent" rule when the two orthogonal inputs are of the same intensity as specified in the AASHTO Standard Specifications. The square-root-of-the-sum of squares (SRSS) method, which is the basis for both the "30 percent" and "40 percent" rules, can be used directly to combine pairs of force components regardless of whether or not the orthogonal inputs are of the same intensity.

Since inelastic deformations are allowed in ductile bridge elements, the combined elastic force components are then divided by appropriate response modification factors, R, as specified in Table 2 below to obtain modified values, EQM.

These modified values, EQM, replace the values EQ in Eq. (2) for use in WSD of structures in Categories, B, C, and D allowing a 50 percent increase in allowable stresses for structural steel and a 33-1/3 percent increase for reinforced concrete.

Table 2. Response modification factors (R)

Substructure	R	Connections	R
Wall-Type Pier	2	Superstructure to Abutment	0.8
Reinforced Concrete Pile Bents a. Vertical piles only b. One or more battered piles	 3 2	Expansion joints within a span of the superstructure	0.8
Single Columns	3	Columns, piers, or pile bents to cap beam or superstructure	1.0
Steel or Composite Steel & Concrete Pile Bents a. Vertical piles only b. One or more battered piles	 5 3	Columns or piers to foundations	1.0
Multiple-Column Bent	5		

1.5 LRFD specifications, first (1994) and second (1999) editions

The working-stress design (WSD) philosophy, which requires that calculated design stresses not exceed specified levels, underwent adjustment in the 1970's through the introduction of load factors reflecting the variable predictabilities of different load types, a philosophy referred to as load factor design (LFD). During the period 1988 to 1993, the AASHTO LRFD Bridge Design Specifications was developed using statistically based probability methods. The load and resistance factor design (LRFD) philosophy makes use of load and resistance factors developed through statistical analyses (Kulicki, 1999).

The AASHTO LRFD Bridge Design Specifications, First (1994) and Second (1999) Editions, requires that each bridge component and connection satisfy all limit states in accordance with the relation

$$\eta \sum \gamma_i \, Q_i \leq \phi \, R_n \qquad (8)$$

in which η is a factor related to a ductility factor η_D, a redundancy factor η_R, and an operational importance factor η_i in accordance with $\eta = \eta_D \eta_R \eta_i$, γ_i is a statistically-based load factor applied to force effect Q_i, and ϕ is a statistically-based resistance factor applied to the nominal resistance R_n. The numerical values to be used for these factors can be found in the LRFD Specifications (AASHTO LRFD, 1994 and 1999).

The value of Q_i for that value of i representing an extreme seismic event, designated EQ, is found using the same procedure described above for Stan-

dard Specifications, Fifteenth (1992) and Sixteenth (1996) Editions.

An additional bridge classification, "Critical," has been added to the LRFD Specifications; and the number of substructure response modification factors R, have been increased to cover all three classifications, "Critical," "Essential," and "Other" as indicated in Table 3 below.

Table 3. Response modification factors (R)

Substructure	Importance Category		
	Critical	Essential	Other
Wall-type piers-larger dimension	1.5	1.5	2.0
Reinforced concrete pile bents ♦ vertical piles only ♦ with batter piles	 1.5 1.5	 2.0 1.5	 3.0 2.0
Single columns	1.5	2.0	3.0
Steel or composite steel and concrete pile bents ♦ vertical piles only ♦ with batter piles	 1.5 1.5	 3.5 2.0	 5.0 3.0
Multiple column bents	1.5	3.5	5.0

1.6 Advances over the past 50 years

The revolutionary changes in seismic design criteria, as noted above, are the result of technological advances made over the past 50 years, namely (1) developing digital computers, (2) advancing numerical methods applicable to linear and nonlinear modelling and dynamic analysis of structures, (3) improving the quality, quantity, and processing of strong motion recordings, (4) understanding and applying the concept of allowing controlled inelastic deformations to occur in structural components during seismic events, (5) changing design detailing to satisfy strength/ductility requirements and to avoid brittle failures, (6) applying statistical and probabilistic methods to characterizing expected ground motions and structural behavior, and (7) recognizing and quantifying uncertainties in all aspects of bridge engineering.

The greatest stimulant to implementing these technological advances in bridge engineering was the 1971 San Fernando, California earthquake, which, for the first time, demonstrated the vulnerability of bridges to seismically-induced vibratory motions. The more recent seismic events, such as the 1989 Loma Prieta and 1994 Northridge, California earthquakes, the 1995 Kobe, Japan earthquake, and the 1999 Taiwan Chi-Chi earthquake, have further added to this change toward improving the seismic performance of transportation structures.

In the subsequent sections of this paper, I will

discuss in more detail some of the issues previously touched upon, with emphasis given to current procedures and related problems needing further development and research.

2 DUAL STRATEGY OF SEISMIC DESIGN

The design of transportation structures to perform satisfactorily under expected seismic conditions requires that realistic earthquake loadings during their life times be specified and that the structural components be proportioned to resist these and other combined loadings within the limits of certain expected performance requirements. In regions of high seismicity, earthquake loading is often critical among the types of loading that must be considered because a great earthquake will usually cause greater stresses and deformations in the various critical components of a structure than will all other loadings combined; yet, the probability of such an earthquake occurring within the life of the structure is very low. On the other hand, a moderate earthquake is very likely to occur during the same period of time having the potential to produce damage unless controlled. Considering both types of earthquakes, a dual-criteria strategy of two-level design is usually adopted for Ordinary Bridges as follows:
• Functional Evaluation Earthquake (FEE) - A functional evaluation earthquake is defined as one, which has a relatively high probability of occurrence during the lifetime of a structure. The structure should be proportioned to resist the intensity of ground motion produced by this event without significant damage to the basic system, thus allowing it to remain functional immediately following the FEE event.
• Safety Evaluation Earthquake (SEE) - A safety evaluation earthquake is defined as the most severe event which can reasonably be expected to ever occur at the site. Because this earthquake has a very low probability of occurrence during the life of a structure, significant structural damage is permitted; however, collapse and serious personal injury or loss of life should be avoided.

The challenge is to set seismic design criteria which will satisfy this dual-criteria strategy in a cost-effective manner.

Important bridges located on major heavily-traveled routes, where no convenient alternative routes exist, are now being designated as LIFELINE BRIDGES. These bridges are expected to remain functional immediately following an SEE event; therefore, they must be proportioned to resist the intensity of this event without experiencing significant damage. Because of this specified high-level of performance during an SEE event, response under the FEE condition is of minor concern.

3 CHARACTERIZATION OF SEISMIC GROUND MOTIONS

It is my contention that at least one-half of the bridge engineer's overall problem in designing either a new bridge or retrofit measures for an existing bridge is to establish appropriate design ground motions which, along with other specified design criteria, will satisfy the dual-criteria strategy described above. In the past, it has been common practice to represent the design ground motions using acceleration response spectra developed through statistical averaging of such spectra generated for families of recorded accelerograms representing different site conditions. A deficiency of these spectra has been that they do not represent the same probability of exceedance, for a specified period of time, over the full spectral period (or frequency) range of interest. Further, the probability of exceedance of the spectral value at any specified period is not well known.

Because of these deficiencies, Probability Risk Assessment (PRA) methodologies have emerged having the objective of providing uniform hazard response spectra for a given site with each spectrum curve representing the same numerical probability of exceedance for a specified duration of time over the entire spectral period range of interest. Usually, these spectra are generated for the "rock-outcrop" condition at the site and then modified either through standard site response analyses using a computer program such as SHAKE (Schnabel et al., 1972) or by applying published site amplification factors (NEHRP, 1997, Dobry et al., 2000).

The procedure for generating "rock-outcrop" uniform hazard response spectra requires establishing (1) the contributing seismic source zones in the near region of the site, based on available geological and seismological information, (2) an upper-bound earthquake magnitude, M_u, for each source zone, (3) median attenuation relations for acceleration response spectral values, $S_a(T_i)$, at discrete values of period, T_i, and their associated standard deviations to account for the dispersions (usually assumed to be lognormal) about the median values, (4) a magnitude-recurrence relation for each source zone, usually based on geologic slip rates or dating of prehistoric earthquakes, and (5) a fault-rupture-length vs. magnitude relation. Having established these requirements, uniform hazard response spectra can be generated by a consistent probabilistic approach (Cornell, 1968, Coppersmith and Youngs, 1986, Der Kiureghian and Ang, 1977, Frankel et al., 2000, and McGuire, 1993). To illustrate the form of such results, Figure 3 shows a set of uniform hazard curves generated by Geomatrix Consultants which represents a single horizontal component of "rock-outcrop" motion at the base of a marl layer resting

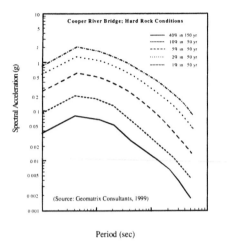

Figure 3. Equal-hazard response spectra for hard rock (5% damped)

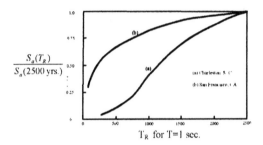

Figure 4. Acceleration response spectral ratio $S_a(T_R)/S_a(2500$ yrs.) vs. mean return period T_R for T=1 sec.

on hard rock at the site of the planned Cooper River Bridges in Charleston, South Carolina (Geomatrix Consultants, 2000).

Once a set of uniform hazard response spectra has been generated for a specific site, a decision must be made as to which two probabilities of exceedance in a specified period of time (or mean return periods, T_R, in years) are the proper choices to represent the FEE and SEE events in satisfying the dual-criteria strategy of design previously discussed. The proper choices will, of course, depend upon the nature of the uniform hazard spectra generated. When considering a site in the eastern part of the U.S., one finds the uniform hazard spectral values for a mean return period equal to the life of a structure, say 150 years, will be very low in comparison with the corresponding values having a mean return period of say 2,500 years. On the other hand, when considering a site in the western part of the U.S., the 150-year mean-return-period spectral values will usually be much higher in comparison with and much closer to the corresponding 2,500-year values (Leyendecker et al., 2000). To illustrate these differences, consider the acceleration response spectrum ratio $S_a(T_R)/S_a$ (2,500 years) for a fixed spectral period, say T =1.0 sec., as shown in Figure 4 where the response spectral ratios are plotted as a function mean return period T_R in years for two specific sites: namely, the Charleston, S.C. Cooper River Bridge site and the Richmond-San Rafael Bridge site in the San Francisco Bay Area (Geomatrix Consultants, 1993 and 2000).

If a mean return period T_R equal to 2,500 years should be specified to represent the SEE event in designing a bridge for the Cooper River Bridge site using the AASHTO response modification factors shown in Table 3, the structural response due to this event will be extremely high in comparison with the

response of an FEE event having an assigned mean return period T_R equal to say 300 years. Therefore, in this case, the SEE event will totally control the design. On the other hand, in designing a bridge for the Richmond-San Rafael site using the same mean return periods for the two events and the same response modification factors applied to the SEE event, the FEE event for which limited damage is specified would most likely control the design since the ratio $S_a(300\text{years})/S_a(2,500\text{years})$ is approximately 0.6. The above comparisons show the importance of assessing the performance of a final bridge design under both the FEE and SEE conditions to insure that the specified dual performance criteria have been met.

In selecting uniform hazard response spectra to represent the FEE and SEE events, one should take into consideration the epistemic uncertainties involved. The uncertainty bands for such spectra representing sites on the east coast of the U.S. are considerably wider than those for corresponding spectra representing sites on the west coast, due to the fact that the lower seismic activity in the East limits the definition of seismic source zones and the establishment of corresponding recurrence and attenuation relations.

Figure 5 shows different uniform hazard hard-rock response spectra, representing a common mean return period of 500 years, for sites in the New York City region. Curves Nos. 2 and 3 were generated for the Bronx-Whitestone Bridge site (Geomatrix Consultants, 1995) and the Queensboro Bridge site (Jacob et al., 1994), respectively, Curve No. 4 was generated for the John F. Kennedy Airport site (Idriss, 1996), and Curve No. 1 was developed by a panel of experts retained by Weidlinger Associates (Weidlinger Associates, 1998) to recommend an appropriate spectrum to be adopted uniformly for the New York City region. Curve No. 5 is the design AASHTO/NYSDOT rock spectrum intended to represent a 475-year mean return period event, which was developed primarily using California earthquake data. It is difficult to rationalize the large differences in spectral ordinates shown by Curves 2, 3, and 4;

and, it is somewhat surprising that the panel's spectrum (No. 1) is so low relative to the other three uniform hazard spectra. Curve No. 5, the design AASHTO/NYSDOT spectrum, is very high relative to the other spectra (Nos. 1, 2, 3, and 4). It is clear that epistemic uncertainties have played a major role in the development of these spectra resulting in the differences shown in Figure 5.

In addition to the spectra shown in Figure 5, the 1998 Weidlinger Associates report presents uniform hazard hard-rock response spectra for the Bronx-Whitestone and Queensboro bridge sites and a corresponding panel spectrum, all representing a mean return period of 2,500 years; see Figure 6. In addition, the National Earthquake Hazard Reduction Program 1997 MCE (Maximum Credible Earthquake) Hard Rock spectrum (NEHRP, 1977) is presented in Figure 6 for comparison purposes. All 2,500-year spectra are, of course, much higher than the corresponding 500-year spectra bringing them much closer to

the design AASHTO/NYSDOT rock spectrum, however large differences still exist among them.

It is clear that uncertainties have played a major role in the above-discussed differences among the various response spectra generated for sites in the New York City region; thus, making it most difficult to set spectra for the FEE and SEE events which will be appropriate for satisfying the dual performance criteria of design. It is my personal view that the 2,500-year mean return period is too long for representing the SEE event considering the inherent margin of safety against collapse which is usually provided by good design; a 1,500-year return period seems more appropriate.

In addition to setting "rock-outcrop" response spectra representing the FEE and SEE events at a specific site, corresponding response-spectrum-compatible time histories of motion must be generated for use in the design of bridges expected to experience inelastic deformations. These motions should be obtained by modifying "rock-outcrop" recorded time histories using the time-domain procedure of adjustment, rather than the frequency-domain procedure, since it results in modified motions closely resembling the initial recorded motions selected for modification should initially possess durations and peak ground accelerations, velocities, and displacements similar to those of the target design seismic event. For a near-field seismic event, e.g., within about 10 km of the site, the recorded motions selected should contain a definite velocity pulse or so-called "fling."

When generating components of motion to be used as seismic inputs at multiple-pier locations, they should reflect realistic spatial variations produced by wave-passage and wave-scattering effects and be response-spectrum-compatible. Thus, an appropriate wave-passage velocity (usually shown to be in the range 2,000 to 3,000 m/sec by instrument-array recording data), an established set of coherency functions (Abrahamson, 1991) characterizing wave-scattering, and the initial response-spectrum-compatible time histories of motion should be used to generate the desired "coherency-compatible and response-spectrum-compatible" time histories of acceleration to be used as inputs at multiple-pier locations (Tseng and Penzien, 1999). Corresponding time histories of velocity and displacement should also be generated. The frequency-domain procedure shown in Figure 7 should be used in interpolating motions at intermediate pier locations.

Standard site-response analyses using the above-described "rock-outcrop" motions as inputs to soil columns representing conditions at multiple-pier locations should be carried out to obtain corresponding soil motions at discrete elevations over the foundation depths. These motions are needed to assess soil-foundation-structure interaction (SFSI) effects (Tseng and Penzien, 1999).

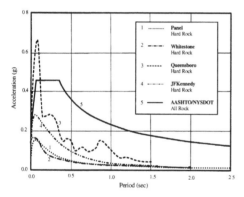

Figure 5. NYC - Rock acceleration response spectra - 5% damping 500-year return period = 10% in 50 years probability of exceedance

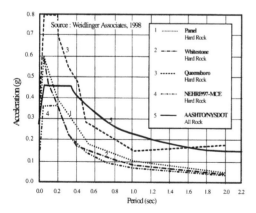

Figure 6. NYC - Rock acceleration response spectra - 5% damping 2500-year return period = 2% in 50 years probability of exceedance

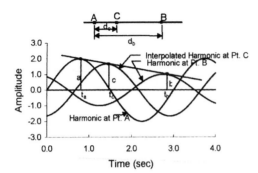

Figure 7. Interpolation in the frequency domain

4 MODELLING AND DYNAMIC ANALYSIS

The development of digital computers and the finite element method of analysis in the 1950s and 60s made possible the shift from static to dynamic methods of evaluating seismically induced internal forces in bridge structures. Specifications indicating this shift started with the 1973 Caltrans design criteria, which formed the basis for the 1975 AASHTO Interim Specifications. While regular bridge structures could still be treated statically at that time, the specified equivalent static loading depended upon the structure's fundamental period, T. The modal response-spectrum analysis method was required however for complex or irregular structures. The internal elastic forces so obtained could then be used in either the working-stress design procedure or the load-factor design procedure.

Over the years, finite element modelling and analysis, along with associated computer programs, have advanced greatly. Modelling can now include nonlinear elements such as force/velocity and force/displacement dependent hysteretic elements, and can account for nonlinear geometric effects. Much remains to be done however in defining elements which can realistically represent in three-dimensional forms the variety and interaction of nonlinearities present in bridge structures, including those involved in soil-foundation interaction.

For determining low-level seismic response of complex or irregular bridge structures, the governing equations of motion used in evaluating seismically-induced internal forces can be expressed in the linear time-domain matrix form

$$\begin{bmatrix} M_{ss} & M_{sf} \\ M^T_{sf} & [M\mbox{'}_{ff}+\overline{M}_{ff}] \end{bmatrix}\begin{Bmatrix} \ddot{u}_s(t) \\ \ddot{u}_f(t) \end{Bmatrix} + \begin{bmatrix} C_{ss} & C_{sf} \\ C^T_{sf} & [C\mbox{'}_{ff}+\overline{C}_{ff}] \end{bmatrix}\begin{Bmatrix} \dot{u}_s(t) \\ \dot{u}_f(t) \end{Bmatrix} + \quad (9)$$

$$\begin{bmatrix} K_{ss} & K_{sf} \\ K^T_{sf} & [K\mbox{'}_{ff}+\overline{K}_{ff}] \end{bmatrix}\begin{Bmatrix} u_s(t) \\ u_f(t) \end{Bmatrix} = \begin{bmatrix} 0 \\ [\overline{K}_{ff}\overline{u}_f(t)+\overline{C}_{ff}\dot{\overline{u}}_f(t)+\overline{M}_{ff}\ddot{\overline{u}}_f(t)] \end{bmatrix}$$

in which all of the M, K, and C letters denote mass, stiffness, and damping matrices, respectively, all u quantities denote time-dependent total-displacement

vectors, subscript s denotes the number of DOF in the structure, excluding its f DOF located at the structure/foundation interface, and a bar placed above a letter indicates that the quantity applies to the f DOF of the foundations when isolated from the structure and subjected to the seismic free-field soil environment. For a bridge having multiple pile foundations, the structure/foundation interface is normally specified to be at the lower surface of each footing having six degrees of freedom (three translations and three rotations). Thus, the footing masses are included in the structural system; however, to satisfy pile-head boundary conditions, rigid massless footings are included in the isolated foundation system. The total displacement vector $u_f(t)$ in Eq. (9), and corresponding velocity and acceleration vectors, $\dot{u}_f(t)$ and $\ddot{u}_f(t)$, respectively, represent motions in the f DOF of the isolated foundations. These motions have been referred to in the literature as the "scattered" foundation motions (Tseng and Penzien, 1999).

The foundation stiffness, $\overline{K}_{ff}$, damping, $\overline{C}_{ff}$, and mass, $\overline{M}_{ff}$, matrices in Eq. (9) which have constant coefficients represent, collectively, approximations of the complex frequency-dependent foundation impedance matrices. These approximations have been made to remove frequency-dependent parameters in the equations of motion, thus allowing a time-domain solution of the equations of motion. If the same equations of motion were expressed in the frequency domain, then such approximations would not be necessary as the complex frequency-dependent impedance functions are fully compatible with a frequency-domain solution. The damping and mass terms on the right-hand side of Eq. (9) usually have small effects on the solution; however, their importance should be checked.

The viscous damping matrix on the left hand side of Eq. (9), excluding matrix $\overline{C}_{ff}$, is usually expressed in Rayleigh form (Clough and Penzien, 1993) with the two constants involved being assigned numerical values to limit the modal damping ratios to levels within acceptable bounds over the range of frequencies dominating seismic response. It is my position that changes are needed to the Rayleigh form of damping used, and that structure analysis computer programs should be changed accordingly.

The above equations of motion have been expressed in the time domain, as it is necessary to do so when nonlinearities in the structural system are represented. These nonlinearities usually occur in the form of hysteretic force-displacement relations of individual components, thus requiring that the linear forms represented in the third term on the left-hand side of Eq. (9) be changed to the appropriate nonlinear hysteretic forms. Special damping devices having nonlinear viscous properties will require modifications to the second term in this equation.

Having established all nonlinear forms, the corresponding coupled equations of motion can be solved for total displacements $u_s(t)$ and $u_f(t)$ using step-by-step numerical integration procedures. The use of total displacements, rather than relative displacements, is required to avoid superposition of solutions, which is invalid when treating nonlinear systems. To complete the dynamic analysis of the overall bridge system, the time histories in vector $u_f(t)$ must be applied as inputs to the f DOF of the isolated soil/foundation model in a separate "feedback" analysis (Tseng and Penzien, 1999).

Since the low-level seismic response produced by the FEE event remains essentially elastic, the linear equations of motion, Eq. (9), can be used directly yielding reliable results, even for structures having very large numbers of DOF; however, when these equations are modified to represent the variety of nonlinear forms of component behavior occurring under SEE conditions, the predicted response results are much less reliable. These less reliable results are due primarily to the lack of realistic modelling of the nonlinear components under their three-dimensional time-dependent deformation conditions.

It should be realized that increasing the number of degrees-of-freedom in modelling a particular structure does not necessarily improve the global response results obtained therefrom, especially when nonlinearities develop in the system. Often better predictions of global response can be obtained using wisely-chosen super elements resulting in fewer degrees-of-freedom.

5 ASSESSMENT OF SEISMIC PERFORMANCE

The procedure one should use in assessing seismic performance of an ordinary transportation structure depends upon (1) type of structure, regular or irregular, (2) level of seismic excitation, FEE or SEE, and (3) stage of the design process, preliminary or final.

5.1 Regular structure

In treating a regular structure such as a single-deck bridge supported on equally-spaced piers having nearly constant heights, a generalized single-degree-of-freedom (SDOF) model can be used to represent an isolated foundation/pier/tributary-deck system subjected to either the FEE or SEE level of excitation, even when assessing the performance of a final design. Starting with an initial design, generalized forces EQ_{xu} and EQ_{yu} in the x and y (usually, longitudinal and transverse) directions are evaluated using the relations

$$EQ_{xu} = W S_{ax}(T_x)/R_x \; ; \quad EQ_{yu} = W S_{ay}(T_y)/R_y \qquad (10)$$

in which W is structural weight of the tributary system, $S_{ax}(T_x)$ and $S_{ay}(T_y)$ are the x and y SEE acceleration response spectral values for structural periods T_x and T_y, respectively, and R_x and R_y are the corresponding response modification factors as specified for the substructure types shown in Table 3. These forces are distributed in the horizontal plane located at the center of mass elevation in proportion to mass per unit length times the corresponding generalized SDOF displacements. Internal force components in the substructure are then evaluated for each of the two generalized force distributions, corresponding pairs of components are combined using the "30 percent" or whatever other rule is specified, and the resulting combined components of force are used to modify the initial design based on ultimate-strength considerations to arrive at a preliminary design.

To assess the performance of the preliminary design under the SEE condition, the focus should shift from evaluating internal forces to evaluating deformations of those components allowed to yield inelastically as a result of using response modification factors, R, greater than unity. The maximum absolute values of the relative global displacements of the generalized SDOF systems in the x and y directions, namely, $|Y_x(t)|_{max}$ and $|Y_y(t)|_{max}$ as denoted by $|Y_{x,y}(t)|_{max}$, should be evaluated using

$$|Y_{x,y(t)}|_{max} = \frac{4\pi^2}{T_{x,y}^2} S_{ax,y}(T_{x,y}) \begin{cases} \left(1 - \frac{1}{R_{x,y}}\right)\dfrac{T^*}{T_{x,y}} + \dfrac{1}{R_{x,y}} & T_{x,y} \le T^* \\ 1 & T_{x,y} \ge T^* \end{cases} \qquad (11)$$

in which each subscript x,y denotes use of either x or y, $T_{x,y}$ represents structural periods T_x and T_y of the modified (preliminary) design, and T^* represents the predominant period in the seismic ground-motion inputs (ATC-32, 1996). If it is desired to control the maximum global displacement ductility demands, $\mu_{dx,y} \equiv |Y_{x,y}(t)|_{max} / Y_{x,y}$ (yield), to be constant at specified values, $\mu_{x,y}$, over the entire period range as recommended in ATC-32, 1996, then the response modification factors R_x and R_y, denoted $R_{x,y}$, should be period dependent in accordance with

$$R_{x,y} = \begin{cases} 1.0 + (\mu_{x,y} - 1)\dfrac{T_{x,y}}{T^*} & 0 \le T_{x,y} \le T^* \\ \mu_{x,y} & T_{x,y} \ge T^* \end{cases} \qquad (12)$$

The specified maximum allowable global ductility demands, $\mu_{x,y}$, should be set at levels below their corresponding ductility capacities by margins of safety which will reasonably insure that the SEE performance criteria, usually based on significant damage but no collapse, will be satisfied.

It is now standard practice to evaluate capacities by conducting an inelastic static (pushover) analysis under controlled monotonic displacement and/or force conditions, noting the formation of plastic hinges, etc. as they take place up to the point of maximum allowable performance. The main focus of this analysis is on deformation capacities. If the structural system can be modelled adequately with only one independent degree of freedom, e.g., a transverse frame supporting a single deck, then the pushover analysis is straightforward. In this case, one controls incrementally the single displacement at deck level and, as it increases monotonically, one follows the local member deformations and corresponding forces. If the frame is supporting decks at two levels, one has essentially two independent degrees of freedom. The proper control of increasing monotonically the displacements and/or forces in these two degrees of freedom cannot be rigorously specified; thus, considerable guidance must be provided to the designer. The more independent degrees-of-freedom contributing to the seismic response of the structure, the more difficult it is to perform a meaningful pushover analysis for the entire system. In such cases, pushover analyses for local elements or subsystems can be more meaningful.

If the ratio of the controlling acceleration response spectral value, $S_{ax}(T)$ or $S_{ay}(T)$ for the SEE event to the corresponding acceleration response spectral value for the FEE event is considerably greater than the design modification factor, R_x or R_y applied to the SEE event, then the SEE event will control the design in which case no detailed assessment of performance to the FEE event is necessary, as it is obvious the system will respond in essentially an elastic manner with no significant damage. If, on the other hand, the controlling spectral ratio is significantly less than the response modification factor applied to the SEE event, the FEE event may well control the design if the response under this seismic condition is to remain essentially elastic.

5.2 Irregular structure

When assessing the SEE performance of an irregular structure, i.e., one having abrupt changes in mass, stiffness, and/or geometry, a separate linear response-spectrum modal analysis of a multi-degree-of-freedom (MDOF) finite element model is usually carried out first for each of three (x, y, and z) rigid-boundary inputs as defined by their corresponding acceleration response spectra. Maximum values of internal force components are evaluated and divided by their corresponding strength capacity values to establish force demand/capacity ratios for members of the initial design. Demand/capacity ratios greater than unity are, of course, fictitious since they cannot occur; however, they do provide an indication of where inelastic deformations may occur first and, to a limited extent, some measure of the magnitudes of these inelastic deformations. The accuracy of this information depends very much on the amount of redundancy in the structural system. If the system is highly redundant, the distribution of internal forces

will change each time an individual component undergoes yielding, which will continue until a collapse mechanism is reached. Nevertheless, the results of the linear response spectrum analysis will provide guidance toward making effective modifications to the initial design, leading to an improved (preliminary) design in terms of meeting the SEE performance criteria.

Again, as in the case of a regular structure, assessing the performance of the preliminary design of an irregular structure under the SEE condition should focus primarily on evaluating global displacements and deformations in those individual components which experience yielding. A response-spectrum modal analysis, along with response modification factors, should not be used at this stage of the design process. Rather, nonlinear finite-element modelling of the overall system, including foundations, should be established for use in carrying out nonlinear time-history analyses having the objective of determining maximum values of component deformations, which can be compared with their corresponding deformation capacities. Deformation capacity of a member is defined as that deformation level at which the member's resistance starts to decrease with increasing deformation.

In carrying out these nonlinear time-history analyses, simultaneous three-dimensional (x, y and z) response-spectrum-compatible time-histories of seismic input should be used, since superposition of separate solutions is no longer valid due to the nonlinear character of response. Further, for long strongly-coupled structures along its alignment, multiple-span segments of the total structure should be modelled; and, simultaneous three-component time histories of seismic input should be applied at each pier location. From pier-to-pier, these inputs should possess appropriate spatial characteristics reflecting realistic wave-passage, wave-scattering, and local site-response effects; and, as mentioned previously, if located in the near field to a controlling seismic source, each input should possess an appropriate velocity pulse (or fling). The critical nonlinear response of a transportation structure in such a location will most likely be dominated by such velocity pulses.

In assessing the performance of a final design under the SEE condition, it is recommended that a minimum of three independent sets of three-component seismic inputs be applied to the nonlinear model separately and that the largest of the resulting maximum values of any critical response be used in assessing performance. This recommendation is made because of the large variations in critical response which usually occur due to nonlinear effects.

6 DESIGN DETAILING

The previously discussed procedures for assessing the performance of transportation structures under a specified SEE condition are reliable only if (1) the modelling of individual components is realistic, particularly for those components responding nonlinearly under three-dimensional cyclic deformation conditions, and (2) the dynamic time-history analyses are carried out with reasonable accuracy. Because of structural deficiencies which were present in the older structures designed prior to the 1971 San Fernando, California earthquake, major changes in design detailing have been made since then. These changes have focused on providing (1) necessary levels of toughness (strength and ductility) in those components designed intentionally to deform inelastically, (2) adequate strengths in other components to avoid brittle failures, (3) continuity of strength through connections, and (4) sufficient seat widths and/or restrainers at deck supports to prevent unseating of spans.

The most critical deficiencies present in reinforced concrete bridges designed prior to 1971, which will allow failures to occur under moderate to high-level seismic conditions, are (1) insufficient shear strength in columns, (2) lack of confinement steel in columns and joints, allowing local buckling of the main reinforcing bars under yielding condition, (3) inadequate anchorage of column main reinforcing bars into footings and cap beams, permitting pull-out prior to developing full yield strengths, (4) insufficient strengths of deck support bearings and shear keys, permitting overturning and brittle fracture, respectively, (5) lack of seat widths at deck supports, allowing the falling of deck spans, and (6) inadequate clearance at expansion joints to prevent pounding action. Also, the foundations of these bridges lack the necessary strength to develop full plastic hinges in the pier columns.

Figure 8 shows the shear failures which occurred in columns of the Foothill Boulevard Undercrossing at Foothill Freeway during the 1971 San Fernando earthquake; while Figure 9 shows shear failures in the connections between upper and lower decks of the Cypress Street Viaduct during the 1989 Loma Prieta earthquake.

The lack of confinement steel is very apparent in both of these cases. The confinement steel used in pre-1971 column designs was typically #4 spiral hoops at 12-inch spacing with the terminal ends not anchored into the core concrete. Current designs of transverse column reinforcement reflect actual confinement and strength demands resulting in transverse reinforcement 8 to 12 times higher than provided pre-1971, as can be seen in the example of Figure 10 which uses #6 spiral hoops at 3-inch spacing throughout the height of a column.

Figure 8. Column failures of Foothill Boulevard undercrossing during 1971 San Fernando earthquake (Source: Roberts, J. and Maroney, B., 1999)

Figure 9. Cypress Street viaduct collapse during 1989 Loma Prieta earthquake (Source: Governor's Board of Inquiry, 1990)

Figure 10. Column reinforcing steel cage (Source: Roberts, J. and Maroney, B.,1999)

Figure 11. Pull-out of main reinforcing bars from footing (Source: Roberts, J. and Maroney, B., 1999)

Figure. 12 Narrow joint (Source: Roberts, J. and Maroney B., 1999)

Figure 11 shows the pull-out of a column's main reinforcing bars from its footing which occurred at the Golden State Freeway and Foothill Freeway Interchange during the 1971 San Fernando, California earthquake. Lack of anchorage of these bars into the footing is very apparent. Current detailing however provides sufficient anchorage to develop full yield flexure capacity of the column under combined axial load and biaxial bending.

Figure 12 shows a deck expansion joint of pre-1971 design having a seat width of only 5-1/2 inches. Such short seat lengths at joints having no restrainer ties are very likely to allow unseating of deck spans during an SEE event.

Figure 13 shows a deck expansion joint of current design having a seat width of 24 inches. Along with larger seat widths, restrainers are also provided across deck joints to insure that the necessary support bearing capability will remain during an SEE event. Vertical cables are also provided to prevent uplift at the supports.

Figure 14 shows typical fixed hinge and expansion-rocker bearings used to transmit loads from a bridge superstructure to its substructure (FHA, 1995). These bearings have been shown to be very vulnerable during major earthquakes, especially the rocker bearings, which often overturn due to the eccentricity produced by relative horizontal displacement. Because of this unstable feature of such bearings, elastomeric bearing pads are now being used in new and retrofit designs.

With the above-mentioned, and other, improvements to design details since 1971, modelling of complete bridge systems, including foundations, has

Figure 13. Wide joint (Source: Roberts, J. and Maroney, B., 1999)

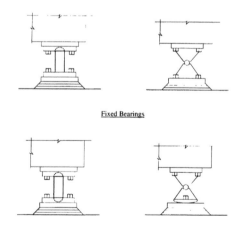

Fixed Bearings

Expansion Rocker Bearings

Figure 14. Seismically vulnerable bearings (Source: FHA, 1995)

become more reliable than previously; however, improvements are still needed in modelling the nonlinear hysteretic behavior of components under the three-dimensional cyclic deformation conditions imposed by the SEE event.

7 RETROFITTING OF EXISTING STRUCTURES

As pointed out previously, bridges designed prior to the 1971 San Fernando, California earthquake possess serious deficiencies making them vulnerable to extreme seismic-loading conditions. Retrofitting to remove such deficiencies has thus become a high priority issue in many states. California, which has approximately 24,000 bridges in its inventory, started its retrofit program shortly after the San Fernando earthquake. Since then other states have initiated their own retrofit programs.

Since bridge retrofit measures have been described in considerable detail in the literature (Federal Highway Administration, 1995, Keady, Alameddine, and Sardo, 1999, and Roberts and Maroney, 1999), I will not discuss them in any detail herein; however, I would like to show a few pictures taken from the reference by Roberts and Maroney, 1999 to illustrate some of the more common retrofit measures used in California.

Figure 15 shows a hinge joint restrainer installed to prevent excessive longitudinal joint separation; while Figure 16 shows a hold-down device intended to prevent uplift of a superstructure from its supporting substructure. Such retrofit measures were shown to be effective during the 1989 Loma Prieta earthquake. The installation of joint restrainers and hold-down devices on approximately 1260 bridges in California during the period 1971-1989 has cost over $55 million.

Retrofitting of columns to provide the needed confinement to insure ductile performance is illustrated in Figures 17 and 18. These figures illustrate a reinforcement-confinement retrofit and a steel-jacket retrofit, respectively. Fiberglass-composite jackets have also been used to provide the needed confinement. The effectiveness of such jacketing of columns in providing the needed confinement has been demonstrated through full-scale column tests at the University of California, San Diego (Priestly, Seible, and Chai, 1991) and through the performance of actual bridge columns, so retrofitted, during the 1994 Northridge, California earthquake (Housner, 1994) in an area that experienced peak ground accelerations at the 0.5g level. A typical footing and pile cap modification now being used in the Caltrans retrofit program is shown in Figure 19.

Figure 20 shows a double-deck structure in the Alemany Interchange on U.S. 101 in South San Francisco after having been retrofitted. The retrofit makes use of independent edge beams alongside the bridge deck as shown in Figure 21. Field installation of the joint reinforcing steel is shown in Figure 22. This arrangement of steel is designed to provide confinement as needed to prevent failure within the joint.

Figure 15. Hinge joint restrainer (Source: Roberts, J. and Maroney, B., 1999)

Figure 16. Hold down devices for vertical acceleration (Source: Roberts, J. and Maroney, B., 1999)

Figure 17. Reinforcement confinement retrofit (Source: Roberts, J. and Maroney, B., 1999)

Figure 18. Steel jacket column retrofit (Source: Roberts, J. and Maroney, B., 1999)

Figure 19. Typical footing and pile cap modification (Source: Roberts, J. and Maroney, B., 1999)

Figure 20. Completed retrofitted structure (Source: Roberts, J. and Maroney, B., 1999)

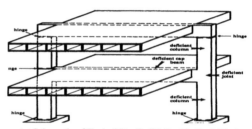

(a) Schematic of Typical "As-Built" Double-Deck Viaduct

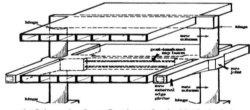

(b) Schematic of retrofit scheme for double-deck viaduct

Figure 21. San Francisco double-deck viaduct proof-test (Source: Seible, F., 1999)

Figure 22. Field installation of joint reinforcing steel (Source: Roberts, J. and Maroney, B., 1999)

28

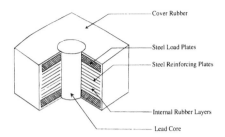

Figure 23. Lead-core rubber bearing isolation device (Source: Zhang, 1999)

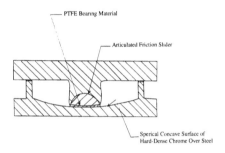

Figure 24. Friction-pendulum isolation device (Source: Zhang, 1999)

Seismic isolation and energy dissipation devices are now being used selectively in retrofitting bridges to enhance their performance characteristics under seismic conditions. The isolation devices are most commonly used to replace vulnerable support bearings. In doing so, the bridge-system's flexibility can be increased considerably, lengthening the fundamental periods resulting in reduced horizontal seismic forces but increasing superstructure displacements. Various types of isolation devices are now in use, including the lead-core rubber bearing device shown in Figure 23 and the friction pendulum bearing device shown in Figure 24.

The energy absorption characteristics of these devices are used to advantage in controlling seismic response. The characteristics of the various types of isolation devices now in use are described in detail in the reference by Zhang, 1999.

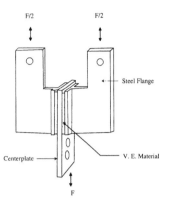

Figure 26. Viscoelastic damper device (Source: Zhang, 1999)

Energy dissipation devices now being used in retrofitting older bridges include dampers of the viscous-fluid, viscoelastic, friction, solid-state, and tuned-mass types. Each of these dampers has its own unique characteristics as described by Zhang. Figs. 25 and 26 show details of a viscous-fluid damper (Taylor device) and a viscoelastic damper, respectively.

Since the use of seismic isolation and energy dissipation devices in retrofitting bridges is relatively new, actual field data regarding their performance under severe earthquake conditions are lacking. For this reason, a major earthquake simulation testing program, sponsored by the Federal Highway Administration and the California Department of Transportation is being conducted to experimentally determine such characteristics. The tests in this program are being carried out primarily at the University of California, Berkeley and the University of California, San Diego.

Following the testing program, guidelines are to be developed regarding the use of isolation and energy dissipation devices. In developing these guidelines, the large relative displacements which can occur across isolation devices should be carefully considered in relation to specific functional requirements. For example, their use in isolating high-speed-rail decks from their supporting piers is gen-

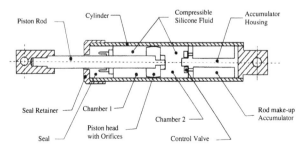

Figure 25. Taylor viscous fluid damper device (Source: Zhang, 1999)

29

erally considered unacceptable because of the excessive transverse track irregularities which develop endangering safe train operation at high speeds.

While I find the use of isolation and energy dissipation devices acceptable and indeed appropriate for retrofitting some older bridges, I feel their use in new bridge designs should be limited to only those cases where they can be fully justified based on performance and cost considerations. The reliability of some devices functioning as intended over the life of a bridge, say 100 years, is not yet known. High maintenance costs can be expected. It is clear that most new bridges can be designed to meet seismic performance requirements in a cost-effective manner without the use of such devices.

8 UNDERGROUND STRUCTURES

Seismic effects on underground structures, such as bored tunnels, cut-and-cover tunnels, and subway stations which are important elements in our transportation systems, need proper consideration. This topic is much too broad to be covered in any detail herein; however, I would like to discuss very briefly the performance of tunnel linings under seismic conditions.

The most critical deformation produced in a tunnel lining during a seismic event is racking of the cross-section caused by vertically propagating shear waves. Because the dimensions of a typical lining cross-section are small compared with the wave lengths in the dominant ground motions producing the racking, the lining cross-section can be assumed to be interacting with soil having a uniform shear-strain field. Further, because inertia effects in both the lining and its surrounding soil as produced by soil-structure interaction are relatively small, the seismically induced racking deformations in a lining take place in essentially a quasi-static fashion. These two conditions form the basis of the procedure summarized herein for evaluating the racking of rectangular and circular tunnel linings. For a detailed treatment of this procedure, see the references Penzien and Wu, 1998 and Penzien, 2000.

A circular lining of diameter D will rack into the oval shape shown in Figure 27(a) producing diameter changes Δ_ℓ in the principal directions ($\pm 45°$) when it interacts, without slippage, with a uniformly-strained (γ_{ff} =constant) free-field soil produced by vertically propagating shear waves. The corresponding diameter changes Δ_{ff} of an imaginary circle of diameter D, as it racks compatibly with the free-field soil strain γ_{ff}, are $\gamma_{ff} D / 2$; and, the corresponding diameter changes Δ_c of a cylindrical cavity of diameter D, as it responds to the free-field environment, can be evaluated using

$$\Delta_c = 4(1 - v_s)\Delta_{ff} \tag{13}$$

A lining-soil racking ratio R_r defined as $\Delta_\ell / \Delta_{ff}$ can now be evaluated using

$$R_r = \frac{4(1 - v_s)}{(1 + \dfrac{1}{F})} \tag{14}$$

where v_s is Poisson's ratio of the soil, and F is a dimensionless parameter defined by

$$F \equiv k_s / k_\ell (3 - 4v_s) \tag{15}$$

in which k_s is the generalized stiffness of the soil displaced by the lining and k_ℓ is the generalized stiffness of the lining. These stiffnesses can be evaluated using the relations

$$k_s = 2G_s / D; \; k_\ell = 48E_\ell I_\ell / D^4 (1 - v^2_\ell) \tag{16}$$

In these equations, G_s is the soil shear modulus, and v_ℓ, E_ℓ, and I_ℓ denote the circular lining's Poisson's ratio, Young's modulus, and circumferential cross-section moment of inertia per unit of longitudinal length, respectively.

A rectangular lining of height H and width W will rack into the shear-type mode shown in Figure 27(b) when it interacts with the uniformly-strained (γ_{ff} =constant) free-field soil. This mode consists of a relative displacement Δ_ℓ between top and bottom of the lining, and a counter-clockwise rigid body rotation. The corresponding racking relative displacement Δ_{ff} of the soil displaced by the lining, as it racks compatibly with the free-field soil strain γ_{ff}, is $\gamma_{ff} H$; and, the corresponding racking relative displacement Δ_c of a rectangular cavity of dimensions W and H as it responds to the free-field environment can be evaluated using Eq. (13). Likewise, Eqs. (14) and (15) can be used for this rectangular-lining case in the same way they were used for the circular-lining case. However, the generalized stiffness of the soil displaced by the lining is now given by

$$k_s = G_s / H \tag{17}$$

and the generalized lining stiffness k_ℓ can be obtained through a simple static analysis of the lining under the plane-strain condition using any standard computer program.

The racking ratio R_r as represented by Eqs. (14) and (15) is plotted as a function of the soil/lining stiffness ratio k_s / k_ℓ in Figure 28 for two discrete values of v_s, namely, 0.4 and 0.5. Note that the racking ratio R_r equals unity, i.e., Δ_ℓ equals Δ_{ff} only when the lining stiffness, k_ℓ, equals the stiffness of the soil, k_s, it displaced. As the lining stiffness k_ℓ approaches zero, the racking ratio approaches $4(1 - v_s)$, i.e., the lining will rack compatibly with the cavity.

The triangular points in Figure 28 correspond to the published results of Wang, each of which was

30

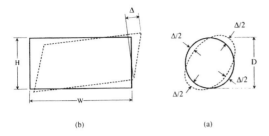

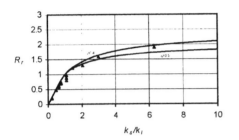

(b)   (a)

Figure 27. Racking of circular and rectangular linings (Source: Penzien, 2000)

Figure 28. Racking ratio Δ_l / Δ_{ff} for circular and rectangular linings

obtained through a separate finite-element dynamic solution (Wang, 1993).

Tunnel linings will, of course, deform globally when subjected to a spatially varying free-field soil deformation environment, i.e., will experience axial and flexural deformations. It has been standard practice in the past to evaluate these deformations assuming the passage of shear waves moving at 45° to the longitudinal axis of the lining without the presence of wave scattering. Using a typical apparent wave-passage velocity, say 2500 m/sec, one finds the maximum values of axial force and bending moment on the lining cross-section to be relatively small. However, if one assumes the spatially varying ground motions are produced predominantly by the passage of slower Rayleigh waves, then the maximum values of axial force and moment will be considerably higher. If in addition, one accounts for local incoherencies in the free-field ground motions, these force components will be further increased. For a general discussion of present procedures being used in evaluating the global response of tunnel linings, a reference by M. Power is recommended (Power, 2000).

9 FUTURE IMPROVEMENTS

The designs of transportation structures have greatly improved over the past 30 years in terms of providing specified levels of seismic safety in a cost-effective manner; however, large uncertainties still remain which need to be reduced. With this objective in mind, research is needed to continue improving the predictions of future ground-motion characteristics, developing better seismic loading criteria for the dual strategy of design, establishing more realistic analytical models of components and systems based on laboratory and field test results, advancing nonlinear dynamic analysis procedures and associated computer programs, and updating seismic design criteria.

9.1 *Ground-motion characteristics*

To improve the predictions of future ground-motion characteristics, improvements are needed in our understanding of seismic sources, energy release mechanisms, rupture-directivity effects, energy transmission source to site, attenuation relations and associated dispersions about median (or mean) values, recurrence rates, spatial variations due to wave passage and wave scattering, and local site effects.

In some regions of the U.S.A., especially in the East, large uncertainties exist in identifying seismic sources for inclusion in hazard analyses. As previously indicated, these uncertainties greatly affect the reliability of design response spectra. In identifying seismic sources, the question arises as to whether or not a particular fault should be treated as being active. This decision can significantly affect the results of a hazard analysis carried out for sites located nearby. As an example, the Chelungpu fault in Taiwan had been treated as being inactive in conducting seismic hazard analyses prior to its rupture on September 21, 1999 producing the devastating Magnitude 7.6 Chi-Chi earthquake (Loh, 2000). Treating this fault as an active seismic source results in a substantial increase in the seismic hazard for sites located nearby, such as sites located along the alignment of the high-speed rail system now being designed. Considering that the mean return period of a Magnitude 7.6 event on this fault is now considered to be approximately 1,000 years and that the life of the high-speed rail system will probably be less than 100 years, it is very unlikely that structures of this system will ever experience the ground motions of another Chi-Chi earthquake caused by rupture along this same fault. However, occurrence of the September 21, 1999 event has undoubtedly increased the potential for large earthquakes on nearby faults.

The above-mentioned Taiwan experience raises the issue as to whether or not particular faults should be treated as being inactive when conducting seismic hazard analyses; and, it suggests that elapsed time since occurrence of the last large seismic event on a fault be taken into consideration when setting seismic design criteria for a particular project. The very intense velocity pulse (fling) produced in the near-

31

field of the Chi-Chi earthquake is another feature of great concern since selecting an appropriate velocity pulse for design purposes is critical to assessing the seismic performance of transportation structures.

Looking to the future, it is my firm belief that the probabilistic approach to setting seismic loading criteria for design purposes should be actively pursued, eventhough presently large uncertainties exist. While this approach is now being used to develop site-specific uniform hazard spectra for special projects, its general use in setting design response spectra is very limited. It is important that the USGS (United States Geological Survey) effort in developing maps of the United States showing rock-outcrop response-spectrum-intensity contours for discrete values of structural period T and return period T_R be continued with the objective of eventually providing sufficient information to allow the direct development of uniform hazard response spectra for any site in the United States (Frankel, A., et al., 1996).

Hopefully, through research, seismologists and geologists will be able to significantly reduce the uncertainties associated with developing uniform hazard response spectra so that civil engineers can set reliable seismic loading criteria. In doing so, the randomness of ground-motion parameters should be taken into consideration. I urge strong interaction among the seismologists, geologists, and engineers involved in the above overall effort.

9.2 Dual strategy of seismic design

The dual strategy of design calls for proportioning Ordinary Bridges to respond to the functional evaluation earthquake (FEE) without experiencing significant damage to the basic system and to respond to the safety evaluation earthquake (SEE) without experiencing collapse eventhough major structural damage is permitted. Specifying mean return periods T_R for the FEE and SEE events is now an issue which needs further study.

In California, a mean return period of 1,500 years is considered appropriate for the SEE event, while in the eastern part of the United States 2,500 years is being specified. Considering that the life of a typical transportation structure is usually in the range 75 to 150 years, it is my judgment that a 2,500-year period is too long and that a 1,500-year period is more appropriate. I express this position knowing that response spectral values representing sites in the East increase much more slowly with increasing values of mean return period T_R than do the corresponding spectral values representing sites in the West., see Figures 3 through 6.

It seems to me that in order to set the SEE mean return period on a rational basis, an appropriate cost-benefit analysis should be carried out with the objective of setting T_R values for different regions of the country which will minimize total direct and indirect

life-time costs for the entire population of new bridges to be built in each region while at the same time meeting specified performance requirements. I believe this analysis would be of greatest benefit to regions in the East where the SEE event totally controls the design of transportation structures. Because of the nature of this analysis, it could be an extension of the type of study now being carried out by the Multidisciplinary Center for Earthquake Engineering Research (MCEER) under sponsorship of the Federal Highway Administration (Freidland, 1999; Werner, S.D., et al., 1999).

Since the SEE event fully controls structural design in the East, there is no need to specify a mean return period for the FEE event, unless it is needed for geotechnical reasons, e.g., in assessing risks associated with liquefaction. However, it needs to be specified in the West since, as pointed out earlier, the FEE event may well control the design of Ordinary Bridges. This return period should also be established using a rational probabilistic approach considering the risks involved. Once established, the FEE design response spectrum (preferably a uniform hazard spectrum) should correspond to this return period. No force reduction factors should be used in assessing structural performance to the FEE event.

Since a Lifeline Bridge is to perform in essentially an elastic manner under the SEE condition, its performance under the FEE condition is of lesser concern; however, it should be checked.

9.3 Modelling and analysis

The uncertainties in modelling structural components are associated mainly with characterizing their nonlinear hysteretic behavior under the three-dimensional cyclic-response condition imposed by the SEE event. Improved modelling is especially needed for columns, which normally are the members in which controlled yielding is allowed in transportation structures. Characterizing their local behavior in the yield zones under the biaxial-bending and axial-load cyclic condition needs further research so as to permit more reliable capacity analyses. Because of the complexities involved, it is clear that laboratory test results must form the basis for improved modelling of structural components.

As pointed out earlier, the capacity (or pushover) analysis of a limited structural system, such as a single bent, is conducted under controlled monotonic displacement and/or force conditions noting the formation of plastic hinges, etc. as they take place up to the point of maximum allowable performance. This capacity analysis is straight forward for a structural system which can be modelled adequately with only one degree of freedom; however, as the number of coupled degrees of freedom increase, monotonic control of the multiple boundary inputs is no longer realistic.

Consider, for example, a strongly coupled three-dimensional subassemblage of members removed from a large structural system. It is suggested that time-dependent displacements, corresponding to those determined from a global time-history demand analysis of the complete structural system, be used as boundary inputs to the subassemblage in verifying performance or in determining capacity. In conducting a capacity analysis of the subassemblage subjected to these inputs, the modelling of individual members must be more rigorous than required in conducting the global time-history demand analysis. Refined modelling of members, as required for a capacity analysis, is not required for the global demand analysis. The number of members in, or size of, the subassemblage should be limited so that inertia and viscous damping forces can be ignored in conducting the capacity analysis, thus allowing a quasi-static solution. This approach should be investigated in a research mode to determine its feasibility.

Previously, little has been said about modelling of bridge foundations (mats, piles, shafts, and caissons), even though such modelling is important in assessing the performance of bridge systems. The so-called "p-y" method is most applicable to the modelling of slender-pile foundations, but it has serious deficiencies in modelling large foundations such as caissons. On the other hand, the so-called elasto-dynamic method is most applicable to the modelling of large foundations, but it has serious deficiencies in modelling slender-pile foundations. Neither of these two methods alone can effectively treat large-diameter shaft foundations. Because of the above-mentioned deficiencies, a so-called "hybrid" method has been formulated which basically connects the p-y and elasto-dynamic models in series; thus, creating a model which is reasonably effective for all sizes of foundations (Tseng and Penzien, 1999). By so combining the p-y and elasto-dynamic methods, the presence of the p-y element treats the nonlinear near-field soil region next to the foundation, while the elasto-dynamic element treats the linear far-field soil outside the near-field region. The p-y element of the hybrid model provides major control over the soil-foundation interaction associated with slender foundations while the elasto-dynamic element provides major control over such interaction associated with large foundations. For intermediate size foundations, such as large-diameter shafts, both elements have significant influence on soil-foundation interaction. Further research is needed to improve all three of the above-mentioned modelling methods, especially under extreme three-dimensional seismic conditions.

The interaction of tunnel linings (rectangular and circular) with the free-field soil environment has been discussed earlier, particularly with regard to racking of the cross-sections. While reasonably reliable analytical methods are available for treating the racking response of linings, significant uncertainties still exist with regard to current methods of treating global response. It is clear that further research is needed to allow realistic assessments of this latter-type of response. In doing so, wave-scattering effects, as characterized by established coherency functions (Abrahamson, 1991) should be investigated.

9.4 Laboratory and field tests

Extensive quasi-static laboratory tests are needed to provide the basis for improving the analytical modelling of structural components under general three-dimensional nonlinear-cyclic deformation conditions, especially for use in conducting the three-dimensional capacity analysis of a subassem-blage as suggested above. Preferably, these tests should be conducted on large-scale specimens using realistic three-dimensional boundary-displacement inputs.

As an example of such laboratory testing, Figure 29(b) shows a half-scale proof-test model in the Charles Lee Powell Structural Research Laboratories at the University of California, San Diego which represents a half-width half-span portion of the lower deck system with upper and lower columns as shown in Figure 21(b). This model was subjected to fully reversed cyclic deformation patterns using the fourteen (14) hydraulic actuators indicated in Figure 29(a). Clearly, such testing was absolutely essential to understanding the performance this system under

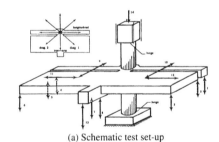

(a) Schematic test set-up

(b) Half-scale proof-test model

Figure 29. 3-D San Francisco double-deck viaduct proof-test (Source: Seible, 1999) extreme seismic conditions. For a complete description of this test, see reference Seible, 1999.

Shaking table tests should continue to be conducted using smaller-scale specimens subjected to three-translational rigid-boundary seismic inputs; and, field tests of real bridges during future earthquakes should be carried out. Extensive instrumentation is required to obtain the necessary field data for conducting thorough correlation studies. The results of such studies can be invaluable in checking the validity of previously established analytical models.

9.5 Seismic design criteria

Over the past 40 years, the seismic design criteria of AASHO/AASHTO have periodically undergone major changes. The first of these changes occurred in 1961 when specific seismic design provisions were initially introduced into the Standard Specifications (Eighth Edition). These provisions, based primarily on the lateral force requirements developed for buildings by the Structural Engineers Association of California, continued in succeeding editions (Ninth - Eleventh) until 1975 when major changes were introduced by AASHTO's Interim Specifications. These changes, based primarily on the new seismic design criteria issued by the California Department of Transportation (Caltrans) in 1973, remained in effect until 1992 when again major changes were introduced into the Standard Specifications (Fifteenth Edition). These new changes, based primarily on the ATC-6 Seismic Design Guidelines for Bridges published by the Applied Technology Council under sponsorship of the Federal Highway Administration (ATC-6, 1981), remained in the 1996 issue of Standard Specifications (Sixteenth Edition). In 1994, AASHTO initiated the load and resistance factor design philosophy in its First Edition of LRFD Specifications for Highway Bridges, a philosophy that has been repeated in the current 1999 Second Edition.

In view of the above-mentioned periodic changes, further changes can be expected soon in future editions of both the Standard and LRFD Specifications. These changes will undoubtedly be based on results of the following recent activities:

(1) Applied Technology Council ATC-32 Project entitled "Improved Seismic Design Criteria for California Bridges," initiated by Caltrans in 1991 to recommend revisions to its seismic design standards, performance criteria, specifications, and practice (ATC-32, 1996).

(2) Multidisciplinary Center for Earthquake Engineering Research (MCEER) Projects 106 and 112 entitled "Seismic Vulnerability of Existing Highway Construction" and "Seismic Vulnerability of New Highway Construction," respectively, which were initiated by the Federal Highway Administration in 1992 to carry out an extensive research program aimed at improving the vulnerability of U.S. highway systems. The final product of this overall effort will be three volumes (I, II, and III) of seismic evaluation and retrofitting guidelines entitled "Seismic Retrofitting manuals for Highway Systems" to be published officially in year 2000 (Friedland, 1999).

(3) Caltrans in-house program to develop "displacement-based" seismic design criteria for standard bridges (Duan and Chen, 1999; Duan and Li, 1999; Caltrans, 1999).

(4) MCEER Project TEA-21 entitled "Seismic Vulnerability of Highway System," initiated by the Federal Highway Administration in 1999 under authorization of the 1998 Transportation Equity Act for the 21st Century.

(5) United States Geological Survey's program on probabilistic seismic hazard assessment for the United States (Frankel et al., 1996).

(6) Numerous research projects in universities having the objective of advancing earthquake engineering for transportation structures.

Hopefully the next major changes in AASHTO's seismic design criteria will be more performance-focused by requiring stronger implementation of validated state-of-the-art modelling and analysis procedures directly into the design process; thus providing better assurance that structures so designed will meet specified performance criteria during future seismic events. This implementation should reflect a shift of the major focus from concentrating on seismically-induced forces to seismically-induced global displacements and corresponding local member deformations. The new Caltrans' "displacement-based" criteria represent an appropriate shift in this direction (Caltrans, 1999).

10 ESTHETICS

While beautiful bridges have been constructed in the past, consideration of esthetics in the conceptual design of major bridges has often been lacking. To bring about increased emphasis on esthetics, architects should be involved in the conceptual design of major bridges and the general public should be given an opportunity to express its views on the selection of final concepts. An example where this dual involvement contributed greatly to a successful outcome is the famous Golden Gate Bridge. When the general public was informed of the initial conceptual design shown in Figure 30, it expressed such strong dissatisfaction that designer Joseph Strauss was forced to develop a new design. In doing so, he brought to his team an architect, Irving F. Morrow, who contributed much to the esthetics of the final design, see Figure 31.

Figure 30. Initial conceptual design for a Golden Gate Bridge

Figure 31. Golden Gate Bridge as-built

11 CONCLUDING STATEMENT

This paper reflects my own experience in earthquake engineering as applied to transportation structures; therefore, many important aspects related to seismic-resistant design have not been treated, e.g., detail-design procedures and geotechnical considerations. Further, the listed references related to the topics discussed are very limited. Hopefully those individuals who may have expected a more complete coverage of the issues and those whose important contributions to the topics discussed have not been referenced will understand such lack of completeness.

12 ACKNOWLEDGEMENT

This paper is based on the author's 2000 Earthquake Engineering Research Institute (EERI) Distinguished Lecture presented at the Annual Meeting on June 3, 2000, in St. Louis, MO, and on lectures presented to EERI regional and student chapters during spring 2000. Essentially, the same paper is to be submitted to EERI for publication in EARTHQUAKE SPECTRA.

REFERENCES

AASHTO, 1996, *Standard Specifications for Highway Bridges*, Sixteenth Edition, American Association of State Highway and Transportation Officials.

AASHTO, 1994, *LRFD Bridge Design Specifications*, First Edition, American Association of State Highway and Transportation Officials.

Abrahamson, N.A., Schneider, J.E., and Stepp, J.C., 1991, Empirical spatial coherency functions for application to soil-structure analyses, *Earthquake Spectra*, **7**, 1.

ATC-6, 1981, Seismic Design Guidelines for Highway Bridges, *Applied Technology Council*, funded by Federal Highway Administration.

ATC-32, 1996, Improved Seismic Design Criteria for California Bridges: Provisional Recommendations, *Applied Technology Council*, funded by California Department of Transportation.

Caltrans, 1999, Seismic Design Criteria, Version 1.1, California Department of Transportation, Sacramento.

Clough, R.W. and Penzien, J., 1993, Dynamics of Structures, Second Edition, McGraw-Hill, Inc.

Coppersmith, K.J. and Youngs, R.R., 1986, Capturing uncertainty in probabilistic seismic hazard assessment in intraplate tectonic environments, *Proceedings, Third, U.S. National Conference on Earthquake Engineering*, Earthquake Engineering Research Institute, Vol. 1.

Cornell, C.A., 1968, Engineering Seismic Risk Analysis, *Bull. of the Seismological Society of America*, Vol. 58, No. 5.

Der Kiureghian, A. and Ang, A.H-S., 1977, A Fault-Rupture Model for Seismic Risk Analysis, *Bull. of the Seismological Society of America*, Vol. 67, No. 4.

Dobry, R. et al., "New Site Coefficients and Site Classification System used in Recent Building Seismic Code Provisions," Earthquake Spectra, Vol. 16, No. 1.

Duan, L. and Chen, Wai-Fah, 1999, Seismic Design Methodologies and Performance-Based Criteria, *Proceedings of International Workshop on Mitigation of Seismic Effects on Transportation Structures*, National Center for Research on Earthquake Engineering, R.O.C.

Duan, L. and Li, F., 1999, Seismic Design Philosophies and Performance - Based Design Criteria, Chapt. 37, *Bridge Engineering Handbook*, Edited by Chen, W-F. and Duan, L., CRC Press.

FHA, 1995, *Seismic Retrofitting Manual for Highway Bridges*, Federal Highway Administration, U.S. Department of Transportation, Edited by Buckle, I.G. and Freidland, I.M.

Frankel, A.D., et al., 2000, USGS National Seismic Hazard Maps, Earthquake Spectra, Vol. 16, No. 1.

Freidland, I.M., 1999, Seismic Evaluation and Retrofitting of Highway Systems and Structures, *Proceedings*, International Workshop on Mitigation of Seismic Effects on Transportation Structures, National Center for Research on Earthquake Engineering, R.O.C.

Geomatrix Consultants, Inc., 1993, Seismic Ground Motion Study for Richmond-San Rafael Bridge, Contra Costa and Marin Counties, CA, Prepared for Caltrans, Division of Structures.

Geomatrix Consultants, Inc., 1995, Final Report, Seismic Investigation of the Bronx Whitestone Bridge, New York, N.Y., Task 3: Seismic Environment, Prepared for Parsons Brinckerhoff Quade & Douglas, Inc./Imbsen Consulting Engineer, A Joint Venture.

Geomatrix Consultants, Inc., 2000, Final Report, Cooper River Bridges Replacement Project Ground Motion Hazard Analysis, Charleston County, S.C., Prepared for Parsons Brinckerhoff Quade & Douglas, Inc. and South Carolina Department of Transportation.

35

Housner, G.W. (Chairman), Thiel, C.C. (Editor), 1990, Competing Against Tune, Report to Governor George Deukmejian from the Governor's Board of Inquiry 1989 Loma Prieta Earthquake, Department of General Services, State of California.

Idriss, I.M., 1996, Study of seismic hazard at John F. Kennedy Airport, Prepared for the New Jersey and New York Port Authorities.

Jacob, K.S., Horton, S., Barstow, N., and Armbruster, J., (1994), Final Report, Queensboro Bridge Seismic Study, Part I: Estimates of Ground Motions on Hard Rock for Earthquakes with Expected Average Recurrence Periods of 500 and 2000 years, Prepared for National Center for Earthquake Engineering Research (NCEER) and Steinman Boynton Gronquist & Birdsall, N.Y.

Keady, K.I., Alameddine, F., and Sardo, T.E., 1999, Seismic Retrofit Technology, Chapt. 43, *Bridge Engineering Handbook*, Edited by Chen, W-F. and Duan, L., CRC Press.

Kulicki, J.M., 1999, Design Philosophies for Highway Bridges, Chapt. 16, *Bridge Engineering Handbook*, Edited by Chen, W-F., and Duan, L., CRC Press.

Leyendecker, E.V., et al., 2000, Development of Maximum Considered Earthquake Ground Motion Maps, Earthquake Spectra, Vol. 16, No. 1.

Lilhanand, K. and Tseng, W.S., 1988, Development and application of realistic earthquake time histories compatible with multiple-damping design response spectra, *Proceedings, 9th World Conference on Earthquake Engineering*, Tokyo-Kyoto, Japan.

Loh, C.H., Lee, Z.K., Wu, T.C., and Peng, S.Y., 2000, Ground Motion Characteristics of 921 Chi-Chi Earthquake, In publication,Earthquake Engineering and Structural Dynamics.

McGuire, R.K., 1993, Computations of Seismic Hazard, *Annali de Geofisica*, Vol. XXXVI, No. 3-4.

NEHRP, 1997, Recommended Provisions for Seismic Regulations for New Buildings and Other Structures, Vol. 1 (Provisions), Vol. 2 (Commentary).

Penzien, J. and Wu, C.L., 1998, Stresses in Linings of Bored Tunnels, Earthquake Engineering and Structural Dynamics, Vol. 27.

Penzien, J., 2000, Seismically induced racking of tunnel linings, Earthquake Engineering and Structural Dynamics, Vol. 29.

Priestley, M.J.N., Seible, F., and Chai, Y.H., 1991, Flexural Retrofit of Circular Reinforced Bridge Columns by Steel Jacketing, Report No. SSRP-91/05 to Caltrans Division of Structures, University of California at San Diego.

Roberts, J., and Maroney, B., 1999, Seismic Retrofit Practice, Chapt. 40, *Bridge Engineering Handbook*, Edited by Chen, W-F. and Duan, L., CRC Press.

Schnabel, P.B., Lysmer, J., and Seed, H.B., 1972, SHAKE - A Computer Program for Earthquake Response Analysis of Horizontally Layered Sites, Report No. EERC 72-12, Earthquake Engineering Research Center, University of California, Berkeley.

Seible, F., 1999, Large/Full-Scale Laboratory Validation of Seismic Bridge Design and Retrofit Concepts, Proceedings, International Workshop on Mitigation of Seismic Effects on Transportation Structures, National Center for Research on Earthquake Engineering, R.O.C.

Shin, T.C., Kuo, K.W., Lee, W.H.K., Teng, T.L., and Tsai, Y.B., 2000, A Preliminary Report on the 1999 Chi-Chi (Taiwan) Earthquake, Seismological Research Letters, Seismological Society of America.

Tseng, W.S. and Penzien, J., 1999, Soil-Foundation-Structure Interaction, Chapt. 42, *Bridge Engineering Handbook*, Edited by Chen, W-F. and Duan, L., CRC Press.

Wang, J.N., 1993, Seismic Design of Tunnels, Monograph 7, Parsons Brinckerhoff Quade & Douglas, Inc.

Weidlinger Associates, 1998, Final Report, New York City Seismic Hazard Study and Its Engineering Applications, Prepared for New York City Department of Transportation.

Werner, S.D., Taylor, C.E., Moore, J.E., and Walton, J.S., 1999, Seismic risk Analysis of Highway Systems: New Developments and Future Directions, Proceedings, ASCE, 5th U.S. Conference on Lifeline Earthquake Engineering, Seattle, WA.

Zhang, R., 1999, Seismic Isolation and Supplemental Energy Dissipation, Chapt. 41, Bridge Engineering handbook, Edited by Chen, W-F., and Duan, L., CRC Press.

Earthquake Engineering Frontiers in the New Millennium, Spencer & Hu (eds),
© 2001 Swets & Zeitlinger, ISBN 90 2651 852 8

Perspectives on successful international collaborative research

Yozo Fujino
Department of Civil Engineering, University of Tokyo, Bunkyo Tokyo 113-8656 Japan
fujino@bridge.t.u-tokyo.ac.jp

ABSTRACT: This paper discusses the long experience of past US-Japan cooperative research programs in earthquake engineering and highlights some of the keys to the success of the collaboration. The discussion includes research topics, research modes, and organizations. Although earthquake engineering encompasses nearly 50 years of experience, its importance is still increasing. Indeed, the realized and potential impact of earthquakes on human suffering and societal infrastructure still demands the attention of the best innovative and creative scientific minds. However, it may also be true that some of the problems appear old and too traditional. Thus, it is essential that students be introduced to earthquake hazard mitigation as an exciting field in need of sophisticated and creative research drawing on the latest scientific and engineering information. New and innovative advances in earthquake technology will come from engineers and scientists who are able to approach problems from a multidisciplinary perspective. Many bright students introduced to earthquake hazard mitigation framed in this challenging multidisciplinary context will choose to specialize in this field and will pursue advanced degrees in this area. Such an approach to ensuring the future success of earthquake engineering requires enhanced international collaboration. The ultimate objective of such enhanced cooperation in science and technology in the area of earthquake hazard mitigation is to reduce the impact of those disasters on the United States, China and Japan, as well as on other countries throughout the world.

1 INTRODUCTION

Earthquakes occur throughout the world on a daily basis. As global urbanization increases, the likelihood of earthquake catastrophes increases in all parts of the world. Moreover, the world is now linked. Societal disruption due to a major earthquake is no longer a local phenomenon, but rather an international one.

Last year's seismic activity in Turkey and Taiwan clearly demonstrated that there are many common engineering problems that remain to be understood and resolved. Because little intergovernmental disagreement exists regarding the importance of infrastructural issues, they would seem to be ideal vehicles for collaborative research.

This paper discusses the rich history and experience of the US-Japan cooperative research programs in earthquake engineering and highlights some of the keys to the success of this collaboration. The discussion includes research topics, research modes, and organizations. This paper emphasizes the importance of introducing bright young students into the field of earthquake hazard mitigation.

Earthquake engineering entered a new age in the 1950s under the leadership of Prof. G.W. Housner (Fig. 1). Since that time, many new and innovative

prof. G.W. Housner

Figure 1: Founder of modern earthquake engineering.

advanced technologies have been introduced, such as new materials, new sensors, and information technology, that possess high potential for significantly advancing earthquake mitigation technology. These technologies appear very attractive to young researchers and students. Proceeding in this direction of research and development will certainly enhance the future of earthquake engineering, but requires intensified international collaboration. The ultimate objective of such cooperation in science and technology in the area of earthquake hazard mitigation is to leverage limited resources to reduce the impact of those disasters on the United States, China and Japan, as well as on other countries throughout the world.

2 FIRST US–JAPAN COLLABORATIVE RESEARCH PROGRAMS ON EARTHQUAKE ENGINEERING

2.1 First US–Japan collaborative research project

Many US–Japan collaborative research programs have been conducted for the past three decades in the area of earthquake engineering (Mahin, 1998, Table 1). The seeds of the first joint research efforts, which were on seismic safety of school buildings, were planted in 1970. The principal investigators were Prof. H. Umemura (University of Tokyo) and Prof. J. Penzien (University of California, Berkeley), see Fig. 2.

Many buildings, including school buildings, suffered from severe damage due to inadequate shear capacity of RC columns during the 1968 Tokachi-oki Earthquake. This damage demonstrated the need for more coordinated research in this area. A joint USJapan seminar was held in September 1970 in Sendai, which is close to the area damaged during the 1968 Tokachi-oki earthquake. Although informal discussions of various cooperative research approaches took place at the 1970 Sendai seminar, after the close of the seminar, formal action was not taken until after the February 9, 1971 San Fernando Earthquake. Various types of urban infrastructure were heavily damaged during this earthquake; shear failure in many RC buildings was also extensively observed, clearly revealing the commonality of problems in our respective countries and the importance of the international collaboration.

The US-Japan Cooperative Research Program on Earthquake Engineering with Emphasis on School Buildings was established under the US–Japan Cooperative Science program for the period of May 1973 through October 1975. The JSPS (Japan Society for Promotion of Science) did not provide the program with research funds, but covered expenses for long-term research stays for Japanese researchers in USA. Many young Japanese researchers visited universities and institutes in USA for a year or so. During the 1970's, traveling abroad was very expensive for Japanese researchers; staying in USA at their own expense would not have been practical. These exchanges were made possible through support provided by the JSPS and by the US National Science Foundation (NSF) in the form of

Table 1: Key Events in U.S-Japan Cooperative Earthquake Research Program

May 16, 1968	Tokachi-oki Earthquake
September, 1970	1st US -Japan Joint Seminar (Sendai)
February 9, 1971	*San Fernando Earthquake*
May 1973	US-Japan Joint Seismic Research on School Buildings
September 1973	2nd US-Japan Joint Seminar (Berkeley)
June 12, 1978	*Miyagiken-oki Earthquake*
August 1979	US-Japan Cooperation Research Agreement (Reinforced Concrete Structures)
May 1981	US-Japan Cooperation Research Agreement (Steel Structures)
December 1984	US-Japan Cooperation Research Agreement (Masonry Structures)
October 17, 1989	*Loma Prieta Earthquake*
March 1991	1st US-Japan JTCC[a] meeting on Precast Seismic Structural Systems
November 1993	1st US-Japan JTCC on Composite/Hybrid Seismic Structures.
January 17, 1994	*Northridge Earthquake*
January 17, 1995	*Kobe Earthquake*
October 1998	20th Anniversary Symposium on US-Japan Cooperation
January 2000	1st US-Japan JTCC meeting on Auto-adaptive Media

a. Joint Technical Coordination Committee (JTCC).

Prof. H. Umemura   *Prof. J. Penzien*

Figure 2: Principle investigators on the first US-Japan joint research project on earthquake engineering.

postdoctoral fellowship. Researchers who took advantage of this program developed true collaborative research activities in earthquake engineering. This was a great experience for Japanese researchers in many ways, learning the US research style, becoming acquainted with many researchers in USA, publishing joint technical papers to international journals, etc. Many meetings and workshops were held during this initial phase of the US–Japan joint research program, and the technical output from these collaborations was substantial.

Indeed, the internationalization of many Japanese earthquake engineering researchers was triggered by this program, establishing a strong network between US and Japanese researchers. Those who had research experience in the US through this US–Japan program now play key roles in the area of earthquake engineering.

2.2 US–Japan large-scale testing cooperative programs

In 1979, construction of the world's largest reaction wall was completed at the Building Research Institute (BRI), Ministry of Construction, Tsukuba Science City, Japan. The US–Japan Collaborative Research Program entitled "Large-Scale Testing" was initiated in August 1979 by the STA (Science and Technology Agency) and BRI/MOC (Building Research Institute of the Ministry of Construction) in Japan and by the NSF in the USA. The principal investigators were again Profs. H. Umemura and J. Penzien. The first phase of the programs was devoted to testing R/C and steel buildings; the hybrid/composites and precast structures programs followed. US researchers spent several months or so at BRI to develop collaborative research related to large-scale testing.

At present, the US–Japan Cooperative Program in "Smart Structural Systems" is ongoing. This program deals with new intelligent/auto adaptive materials, smart damping devices, sensors, etc. The program does not necessarily require large-scale testing facilities such as the reaction wall at BRI. Although the importance of large-scale testing is still high in many fields, the present US–Japan Smart Structures Program indicates some change in the research content in the area of earthquake engineering.

3 IMPORTANT FACTORS IN INTERNATIONAL COLLABORATIVE RESEARCH PROGRAM

Observing the past US–Japan cooperative research programs, the success of an international cooperative research program should satisfy several conditions. In general, because two countries that have different historical and cultural backgrounds and are geographically distant, establishing *true* collaborative research is not easy. The program has to have something substantial to offer both sides. The followings may be important factors in the success of bilateral research collaboration:

1. Excellent initiators who proposes a timely and attractive theme for the researchers as well as stakeholders.

2. Excellent leaders who are fair and devote tremendous efforts to the research projects (see Fig. 2).
3. Excellent coordinators who are efficient managers (see Fig. 3).
4. Excellent financial supporters (see Fig. 4).
5. Balanced cooperative work among the participants.
6. More importantly, active participation of young bright researchers who make real contributions through collaborative research.

Dr. M. Watabe   *Prof. R.D. Hanson*

Figure 3: Excellent coordinators and strong supporters.

Dr. S.C. Liu

Figure 4: Strong supporter.

The US-Japan collaborative research satisfied these conditions, at least during the first phase, leading the program to success.

In a recent interview (Fujino and Abe, 1999), Professor George W. Housner stated that international cooperative research activities should: (1) be more individually-based and conducted on a voluntary basis, and (2) provide more freedom to researchers, and (3) involve a larger number of young researchers. Finally he stated that successful international collaborative research remains to be a challenging topic.

4 DIFFICULTIES IN MULTI-LATERAL INTERNATIONAL COLLABORATIVE RESEARCH

At present, a research project entitled "Development of Technology for Mitigation of Earthquake and Tsunami Disaster for Asian-Pacific Region" (EQTAP) is ongoing (Kameda, 1998). It began under the sponsorship of the Science and Technology Agency (STA) of Japan in 1998, with an intended 5 year duration. This is a multi-lateral project involving many small research projects with many different countries. The author is participating in this project, primarily conducting collaborative research with Thailand. The STA provides funding only to the research groups in Japan, with practically no funding going to their counterparts; the counterpart countries are expected to prepare funds for this collaborative research. In many developing countries, research funding is extremely limited, and furthermore researchers very often cannot afford to spend their time on research. This fact makes collaborative research very difficult.

Practically, in any multi-lateral international collaborative research program, developing countries will be involved, and these countries will receive tremendous benefit from such projects. However, it is also true that such countries may not be able to actively participate in the projects because of funding difficulties. Personally, the author feels that there is a strong need for an *international funding mechanism,* so that researchers who would like to make real contribution can freely join the project to establish a strong human network and to share advanced knowledge and experience to make a safer world.

5 OUTCOME MEASURE OF INTERNATIONAL COLLABORATIVE RESEARCH

Research projects are often evaluated directly by theories, formulae, experimental results, codi-fication, etc., obtained through the projects. This may be appropriate for certain types of research, such as problem-focused research. However, in the case of multi-lateral international collaborative research, developing countries are often involved. In such situations, one important thing is to produce young researchers who can make continued technical contribution toward the goal of earthquake engineering in the next generation. Young researchers in developing countries really need to learn the state-of-the-art in earthquake engineering, to grasp research styles from world experts, and to become acquainted with promising researchers in other countries. On the other hand, researchers in developed countries need to understand the problems in the developing

countries in order to widen their research perspective. Hence, it should be understood that exchange of researchers, in particular young researchers, is extremely important and valuable to mitigate seismic hazards throughout the world. This human exchange will have a strong impact on the young researchers on both sides and will provide a strong communication link (Neureiter and Imura, 2000).

Societal, economical, and historical backgrounds are different from country to country, and hence it is difficult to directly find solutions to local problems for which can be readily applicable in the community. Necessarily the topics for international research projects tend to be quite fundamental in nature. Of course, publishing technical articles as a result of collaborative research is an important outcome. However, the real outcome from multi-lateral international collaborative research is rather invisible; it may only be appreciated after ten years, or even more, in the case of the human network. A long-term perspective and patience are really important and necessary in multi-lateral international research projects.

Figure 5: The first US-Japan Young Researcher's Symposium, University of Tokyo, 1998.

Since 1998, the Japanese and US governments support a student summer program with a focus in earthquake engineering
(see http://www.nd.edu/quake/nhmj/).
Every summer approximately 12 US civil engineering graduate students conduct research at universities and governmental institutes in Tokyo for a period of 8 weeks. Technical trips and a small bilateral young researchers workshop are held (Fig. 3). Although the term is only two months, during the stay, they become friends with many researchers and graduate students in Japan, and learn the Japanese research style. The immediate research outcome is often not to large, but the experience and the human network developed during the stay is extremely valuable to international activities for the next generation.

6 REMODELING EARTHQUAKE ENGINEERING

Earthquake engineering is a broad and interdiscipiinary subject, including geophysical sciences, structural engineering, geotechnical engineering, transportation engineering, as well as economics and social sciences.

The goal of the earthquake engineering is to protect society from earthquake-induced hazards. Unfortunately, this goal has not yet been accomplished, as can be seen from the enormous damage in recent earthquakes.

To accomplish this goal, we need research and development, planning/design, management, administrative work, law and regulation, and even politics. There is also a strong need to train capable engineers with sound earthquake engineering backgrounds who can take the lead in these important areas. This requires many bright young students to enter the filed of earthquake engineering, again indicating the importance of development of human resources.

Past earthquake-induced damage such as was experienced during the 1906 San Francisco earthquake and the 1920 Great Kanto earthquake clearly indicated a strong necessity and societal support of this subject. Obtaining strong ground motion records such as the 1940 Imperial Valley Earthquake, California earthquake (El Centro) provided a scientific basis for earthquake engineering. Led by Professor George Housner, earthquake engineering started to come of age in the 1950s. Essential problems to be solved were identified, and the validity of numerous scientific approaches to these problems were identified. Prof. Housner's contribution to establishing earthquake engineering to be an important technological field is enormous.

Physics-based understanding of the seismic performance of infrastructures such as buildings and bridges and their aseismic design have been a primary subject in the earthquake engineering. These problems have looked challenging and attractive, and therefore many bright students came into this field. About 50 years have passed since modern earthquake engineering was initiated, and it may now be true that treating conventional materials such as concrete, steel and soil by traditional approaches looks somewhat old-fashioned in today's age of information technology. Indeed it is difficult to attract bright young students in the area of earthquake engineering in developed countries.

Recent developments of various kinds of advanced materials, control/sensor technology, and information sciences potentially changes the content of earthquake engineering; magnetorheorogical fluids, shape memory alloys, laser sensors, high damping rubber, GPS, GIS and so on, are typical examples. These new technologies may drastically reform the concept of structures; for example:

- Control rather than resist against strong ground motion
- Adaptive rather than fixed
- Realistic simulation rather than approximate formulas

It is now possible to collect detailed information on structures using advanced sensor technology and to process enormous amount of data inexpensively in a short time by advanced computer technology. Prediction of real performance of infrastructure systems based upon measured data is not impossible. There are millions and millions of infrastructure systems in our society, and determination of real performance of these structures under critical loading will be possible in the next decade; mitigation of seismic disaster has been and will continue to be dramatically improved by the advancement of computers.

Making use of these new technologies in infrastructures is to widen their hi-tech market and to bring enormous benefit to the infrastructures and our society.

These are really some of the challenging topics that can attract young students to our field.

7 RESEARCH MODE IN INTERNATIONAL COLLABORATIVE RESEARCH

In the field of civil infrastructure, large-scale testing has been a powerful and important tool to understand the true behavior of our infrastructure systems in the ultimate stage of failure and to verify various theories and formula. In the past, for international collaborative research projects, typically US–Japan collaborative research projects, the primary research mode was based on large-scale testing, with many small topics being treated under the large-scale testing.

The importance of large-scale testing still remains without question, but earthquake engineering research related to advanced technologies may take a mode different from the past. There are many types of advanced technologies, and there are many fields where these advanced technologies are applicable. Hence there are numerous research topics that deserve investigation. Integrated combination of small multidisciplinary research groups is more appropriate, and this research mode is very well fitted to the multilateral international collaborative research, since many researchers with wide variety of interests have the opportunity to participate in such research projects.

8 CONCLUDING REMARKS

The long experience of past US–Japan cooperative research programs in earthquake engineering was discussed and some of the keys to the success of the collaboration were highlighted. The discussion included research topics, research modes, and organizations. It is stressed that young students have to be introduced to the field of earthquake hazard mitigation in order to bring the best innovative and creative scientific minds into this important field. It is also indicated that such an approach to ensuring the future success of earthquake engineering requires enhanced international collaboration.

9 ACKNOWLEDGMENTS

The author would like to thank Professor S. Otani and Professor B.F. Spencer for their comments regarding international cooperative research, including the US–Japan programs. Thanks also are extended to Mr. T. Kaminosono for providing information on the US–Japan Cooperative Research Program on Earthquake Engineering.

REFERENCES

Mahin, S. 1998. Proceedings of the 20th Anniversary Symposium on The U.S.-Japan Cooperative Earthquake Research Program Utilizing Large-Scale Test Facilities, Toranomon Pastoral, Tokyo, October 9, 1998.

Neureiter, Norman P. and Imura, Hiroo (co-chairs), "An Agenda for Future US-Japan Scientific & Technical Cooperation: Report of a Joint US-Japan Dialogue Group," submitted to the Governments of the United States and Japan, May 2, 2000.

Kameda H., 1998. "Framework of STA project - Development of earthquake and tsunami disaster mitigation technologies and intergration for the Asia-Pacific region," Proc. of Multi-lateral Workshop on Development of Earthquake and Tsunami Disaster Mitigation Technologies and Their Integration for the Asia-Pacific Region, EDM-Riken/STA, Kobe, September 30-October 2, pp.110.

Fujino, Y and Abe, M. 1999, "A message from Dr. G.W. Housner," Japan Earthquake Engineering News, No. 164, January.

Earthquake Engineering Frontiers in the New Millennium, Spencer & Hu (eds),
© 2001 Swets & Zeitlinger, ISBN 90 2651 852 8

Measured ground shaking and observed damage: Do recent events confirm a direct connection?

M.A. Sozen
Purdue University, W. Lafayette, IN, USA

ABSTRACT: From the viewpoint of earthquake engineering one of the most poignant aspects of the two catastrophic earthquakes of 1999 was the measurement of two strong ground motions within three months at the same location in Duzce, Turkey. Both events caused notable structural damage in Duzce. The paper considers the simple question, "After the damage in Duzce associated with the first event (August 1999) was observed, if one were given the strong motion record for the next earthquake in Duzce, could one have anticipated the magnitude of the ensuing additional damage?" The paper contains several procedures by which this could not be done satisfactorily. In this instance, ex post facto methodologies were found to be more reliable as they have been in many other cases.

1 INTRODUCTORY REMARKS

During the second half of 1999, Duzce, a town of some 80,00 inhabitants in northwest Turkey (Fig. 1), was shaken strongly by two damaging earthquakes.

Figure 1. Location of Duzce in Turkey

The first one occurred on 17 August (the Kocaeli-Golcuk Earthquake, M_w 7.4) and the second one on 12 November (the Duzce-Bolu Earthquake M_w 7.1). The 120-km long surface trace of the strike-slip fault associated with the Kocaeli-Golcuk event reached to within 20 km of Duzce. The 40-km trace for the Duzce-Bolu event was within 9 km of the town center (Fig.2). The ground motions in Duzce were measured on both occasions by an SMA-1 strong-motion instrument near the town center.

The horizontal components of the recorded motions (http://angora.deprem.gov.tr) and the calculated linear-response acceleration and displacement

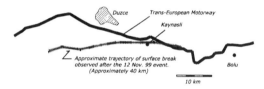

Figure 2. The 12 November 1999 Surface Trace and Duzce

spectra are in Fig. 3 through 6.

The occurrence of two strong ground motions in the same town within a three-month period is a rare event. It was unfortunate for the inhabitants of Duzce but it provided information that may help reduce vulnerability of other sites with similar construction. This paper deals with a hypothetical question of practical importance: Given the ground motion records and the damage for the August event as well as the ground motion records for the November event, could the damage in the November event have been estimated?

The question is posed with faith in the principle of proportionality, used in engineering well before mechanics. Even though it may appear to be simple and crisp, the question does suffer from epistemological flaws. Attempting to answer the question requires intelligible definitions quantifying the ground motion and the damage. And it is based explicitly on the presumption that accidents of construction can be configured with some certainty before the event.

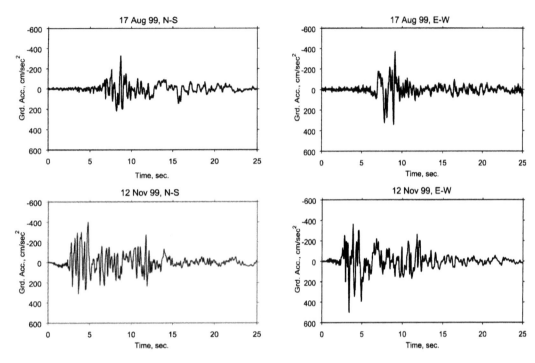

Figure 3. Measured ground acceleration in Duzce on 17 August and 12 November 1999, north-south

Figure 4. Measured ground acceleration in Duzce on 17 August and 12 November 1999, east-west

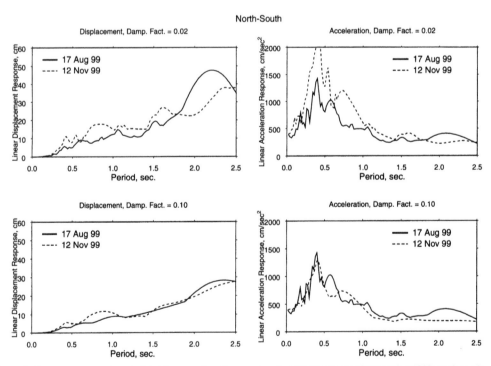

Figure 5. Liner response spectra calculated for ground motions measured on 17 August and 12 November 1999; north-south

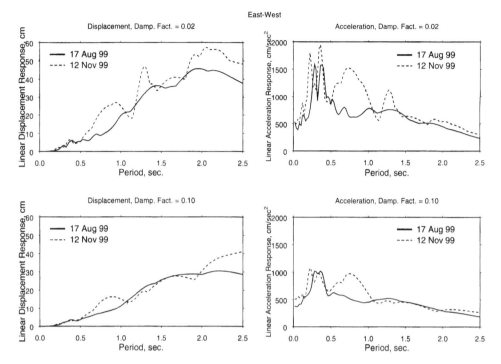

East-West

Displacement, Damp. Fact. = 0.02

— 17 Aug 99
--- 12 Nov 99

Acceleration, Damp. Fact. = 0.02

— 17 Aug 99
--- 12 Nov 99

Displacement, Damp. Fact. = 0.10

— 17 Aug 99
--- 12 Nov 99

Acceleration, Damp. Fact. = 0.10

— 17 Aug 99
--- 12 Nov 99

Figure 6. Linear response spectra calculated for ground motion measured on 17 August and 12 November 1999, east-west

2 DUZCE

A provincial center located in a pull-apart basin, Duzce is founded on deep alluvial deposits. On its southern border, the Asarsu flows on unconsolidated gravel and sand layers. The subgrade for the town includes layers of clay, silt, sand, and gravel extending to depths of at least 200 m. After both the August and November events, there were no observations of structural damage associated with ground failure. Nor was there convincing evidence of systematic variations in ground-motion amplification over the area occupied by the town buildings.

The water table is said to range from 2 to 4 m below level and has been observed to be higher in the north than in the south portion of town.

Much of the building inventory in Duzce had been completed within the last two decades. The typical building has a reinforced concrete space frame with filler walls of tile masonry. Girder spans range from 3 to 5 m. It is unusual for column centerlines to be continuous through the footprint of a building and they appear to have been dictated by architectural requirements and shape of lot.

Story height is typically 3 m. Before the two earthquakes, Duzce contained approximately 4,000 buildings. Measurements of the frame members and nonstructural walls of a sample of approximately 100 buildings (http://www.anatolianquake.org) selected randomly were made by teams of engineers from Turkey and the USA in December 1999 and June 2000. The distribution of number of stories in that sample is shown in Fig. 7. The majority of the buildings had four to six stories and the mode was five stories. All were on individual footings with grade beams.

Girder depths ranged from 0.4 to 0.6 m. Not all girder intersections in a building are supported by columns. Column sizes and shapes varied considerably, generally from 0.4 by 0.4 m to 0.3 by 1.2 m. To refer to a typical column would distort the evidence. There appeared to be a concern for orienting of the longer dimension of some rectangular columns in one direction and the longer dimension of some in the other direction so as to provide equal lateral stiffness in both orthogonal directions of a building.

Even though almost all of the construction was engineered under a prevailing seismic code, the sizes of the frame elements, the reinforcing details, and material strengths appeared to be driven by considerations and traditions related to gravity. The only indisputable evidence of sensitivity to the earthquake risk was the mixing of column orientations. Local builders attribute the use of grade beams to earthquake demand.

Concrete used ranged in compressive strength from 7 to 15 MPa. Reinforcement strengths were said to be approximately 200 MPa for plain bars and 400 MPa for deformed bars.
(http://www.anatolianquake.org)

Strong variations were observed in individual samples. Filler walls were typically made from clay tile with compressive strengths (gross section) as low as 1 MPa.

It is treason to describe the construction in a town by a limited number of parameters. Unfortunately, the need for generalization requires it. In this instance, the Hassan Index will be used (1). The Hassan Index was developed from the one proposed by Shiga et al (2) to suit the conditions in Turkey for evaluating the damage caused by the 1992 earthquake to low-rise construction in Erzincan. It assigns two attributes (pair of coordinates for each plan direction of building) to low rise buildings in relation to the cross-sectional area of the frame members and supporting walls in the lowest story. The two quantitative attributes are CI (column index) and WI (wall index) as defined below.

$$CI = \frac{\sum_1 A_c}{2 \cdot \sum_n A_f} \cdot 100 \qquad (1)$$

$$WI = \frac{\sum_1 A_{cw} + \sum_1 A_{mw}}{\sum_n A_f} \cdot 100 \qquad (2)$$

$\sum_1 A_c$: Sum of cross-sectional areas of all columns at lowest level

$\sum_n A_f$: Sum of floor areas in all n levels of the building

$\sum_a A_{cw}$: Sum of cross-sectional areas of all reinforced concrete walls at the lowest level, oriented in one plan direction of the building

$\sum_1 A_{mw}$:Sum of cross-sectional areas of all masonry walls at lowest level oriented in one plan direction of the building

The building "landscape" according to the Hassan Index (the coordinates refer to the weaker direction in each case) is shown in Fig.8. Lines I and II in the figure summarize the Erzincan experience. Buildings with defining coordinates within the area between the origin and line I were found to be extremely vulnerable to earthquake effect and those with coordinates between lines I and II were found to be susceptible to moderate damage in Erzincan in 1992. Clearly, the damage in Erzincan in 1992 was a complex function of the building quality and earthquake characteristics. It cannot be projected directly to Duzce. However, Fig. 8 does suggest that had the buildings been evaluated for susceptibility before the events of 1999, almost all of them would have been likely to have been found wanting. There is no denying the fact that from the viewpoint of earthquake resistance, the building inventory in Duzce was flawed. But this should not stop intensive study of the damage statistics from Duzce. The buildings were substandard and had a wide mix of difficult-to-define attributes. But that is the reality of countless other population centers in the world and the two successive measured ground motions provide compelling reasons for attempts to understand what happened in Duzce and why it happened. Duzce 1999 represents a rare case of incremental damage observation in an actual environment. The sequence of events resembles an earthquake simulator study. The buildings can be considered as crude and instruments with the gages "zeroed" in August.

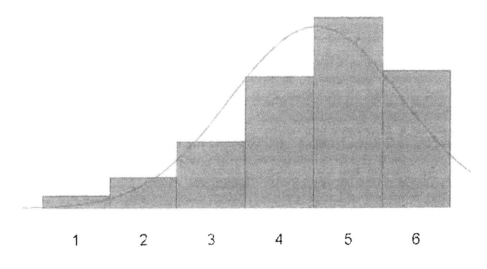

Figure 7. Distribution of number of stories in the Duzce sample

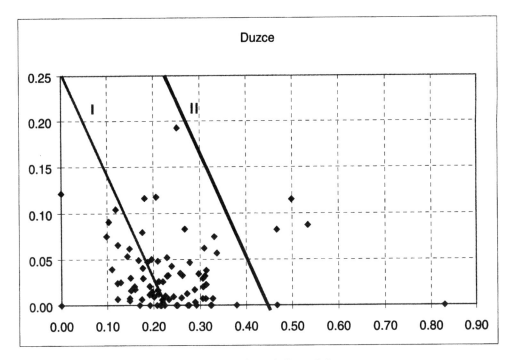

Figure 8. Distribution of building properties in Duzce according to the Hassan index

2.1 Reported damage

Information on the damage to buildings in Duzce during the August and November events was compiled by the Turkish Ministry of Public Works, Directorate for Natural Disasters and Resettlement (3). As a percentage of the total building inventory (approximately 4000 buildings) heavy damage (collapse or severe structural damage) is listed at 13 for the August and 34 for the November event. In this attempt at studying the relationship of the damage to the measured strong motion, the focus will be on the relative amount of heavy damage because errors based on incorrect observation are likely to be less for this category. No speculation will be nor need be made about what proportion of the heavily damaged buildings were moderately or lightly damaged in the first earthquake. According to direct observation there is no reason to conclude that the buildings that failed in November had been moderately damaged in August. In fact, there is observational evidence that a considerable portion of the buildings damage heavily in November had not been hurt in August. This inference is reinforced by the threefold increase in the number of casualties from the first to the second earthquake. It is known that buildings with heavy and moderate damage were not reoccupied. And it is highly unlikely that those with light damage had been fully reoccupied. (The population was forced by the oncoming winter to return reluctantly to the

buildings considered safe only days before the second event.)

Within the limits of the accuracy permitted by the precision of the data, it is plausible to accept that the ratio of the heavy damage in the November event to that in the August event is three, assuming that the same amount of damage would have occurred if only the November event had occurred.

Table 1. Damage distribution

Event	Reported Damage Intensity as Fractions of Total Building Inventory, %			
	None	Light	Medium	Heavy
17 August 1999	68	8	11	13
12 November 1999	33	21	12	34
Increment	-35	13	1	21

[a]After Sucuoglu and Gur

2.2 Direct attributes of the ground motion

It is questionable to claim that the measured ground motion was representative for the entire area occupied by Duzce. The building damage within a kilometer or two of the location where the ground motion was measured was not perceptibly different from that in other parts of town. It is thus plausible to postpone the discussion of the generality of the ground motion to another study.

47

Table 2. Ground motion parameters

	17 August 1999			
	N-S		E-W	
Maxi. Acc./G	0.31		0.36	
Max. Vel., mm/s	570		590	
Damp. Fact., %	2	10	2	10
Housner Int.[a]	271	197	194	134
Housner Int.[b]	59	41	56	38

	12 November 1999			
	E-W		N-S	
Max. Acc./G	0.52		0.41	
Max. Vel., mm/s	880		650	
Damp. Fact. %	2	10	2	10
Housner Int.[a]	330	204	209	140
Housner Int.[b]	93	57	88	53

[a] Period range from 0.1 to 2.5 sec.

[b] Period range from 0.1 to 1.0 sec.

Table 1 contains the direct quantitative attributes of the measured ground motions: peak ground acceleration (PGA) and the peak ground velocity (PGV). The larger ratio of the maximum accelerations (ratio of PGA in November to that in August) is 1.7. That observation makes it difficult to claim that PGA will explain the increase in damage by itself. Using the time of strong shaking as a modifier might make the measure more palatable but awkward. (The time of strong shaking may be said to be 7 seconds for the August and 12 second for the November event.) The calculated peak ground velocities, which are of interest on their own because they are both high relative to ground motions with comparable PGA's also fail to match the increase in damage. The ratio is 1.5. These two observations appear to preclude the possibility of a simple and direct process to estimate the damage.

The Housner Intensity (4) continues to be a convenient vehicle to describe the shaking power of a ground motion. The quantities listed in Table 1 (damping factor 2%) for option (a) do confirm that. They conform to the implications of PGA and PGV but have the advantage of being more broadly based. Considering that many of the damaged buildings in damage had initial periods less than 1 sec, a modified measure of the Housner Intensity (for velocity response over the range of periods from 0.1 to 1 sec) was also calculated. The larger ratio for this effort was 1.6.

Perhaps the most comforting comparison in the category of simple and direct comparisons is that between the range of the east-west acceleration spectra over the range of periods 0.5 to 1 sec (fig. 6). The ratio is approximately 2. But it is difficult to maintain that one would have made much of this without foreknowledge. (Under actual conditions the ground motion would not be known and if estimated it might not be trusted for such a conclusion.)

Any one of the ground-motion parameters may be shown to fit the observations by conceding a threshold for brittle response. After all, proportionality is expected but linearity is not necessary. Just as it would be reasonable to admit that even light damage might not occur up to a threshold intensity (as in the MMI scale), amount of severe damage might increase at an increasing rate with the ground-motion parameter once the brittle response threshold is reached. The brittle-threshold presumption could be related to the response of the local materials and the accidents of construction as well as to the random distribution of strength across the building landscape. Such an explanation would be reinforced by the inference that the presence of nonstructural masonry walls had a strong influence on building behavior in Duzce. But it is difficult to set this limit a priori on the basis of the available information from Duzce and then project it to other earthquakes in other towns or cities.

2.3 Indirect attributes of the ground motions

Response histories calculated for linear and nonlinear single-degree-of-freedom (SDOF) oscillators (damping factor 2%) with periods ranging from 0.5 to 1.5 sec are shown in Fig. 9 through 12. Nonlinear responses were calculated for a strength of 10% of the weight of the oscillator in all cases using a simple elasto-plastic hysteresis routine with constant slope for the linear portion loading or unloading. The comparison of the linear and nonlinear responses is not unusual except for those computed for the 12 November E-W component (Fig. 12) from approximately second 5 through second 7 of the record the relative displacement response is seen to be virtually stationary and, for T=0.5 sec, quite high with respect to the linear response. The comparison in Fig. 13 implies that the observed response may be related to the velocity history. Would this feature calculated using a SDOF oscillator with elasto-plastic hysteresis signal danger for multi-story buildings with less unrealistic force-displacement relationships?

A numerical simulation was made using an exact analysis of an approximate two-dimensional model of a five-story Duzce structure having three by four bays with the columns fixed at ground level (Fig. 14.) Assumed column and girder dimensions are shown in the figure. As indicated in the figure the column orientations were mixed. Reinforcement ratio for the columns was taken as 1%. Beam reinforcement ratio was set at 1% for the top and 0.5% for the bottom reinforcement. Story masses and column loads were based on a unit mass of 800 kg/m^2. Concrete strength was assumed to be 10 MPa and the yield stress for the reinforcement 420 MPa. Us-

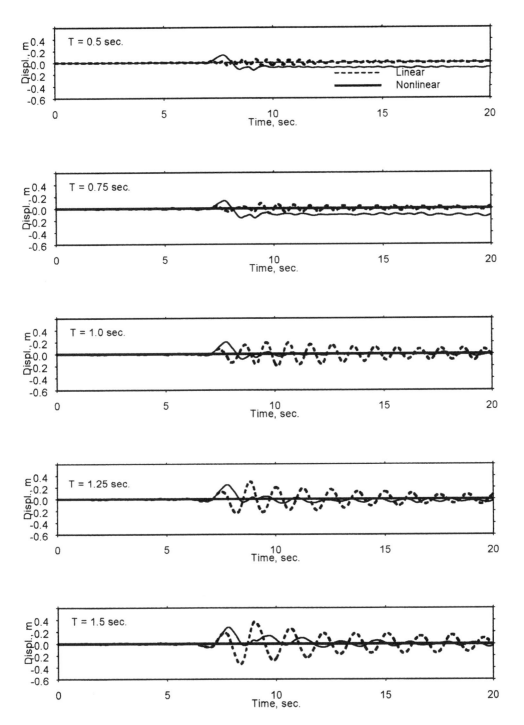

Figure 9. Linear and nonlinear response histories for SDOF systems calculated for the ground motion component measured on 17 August 1999, north-south

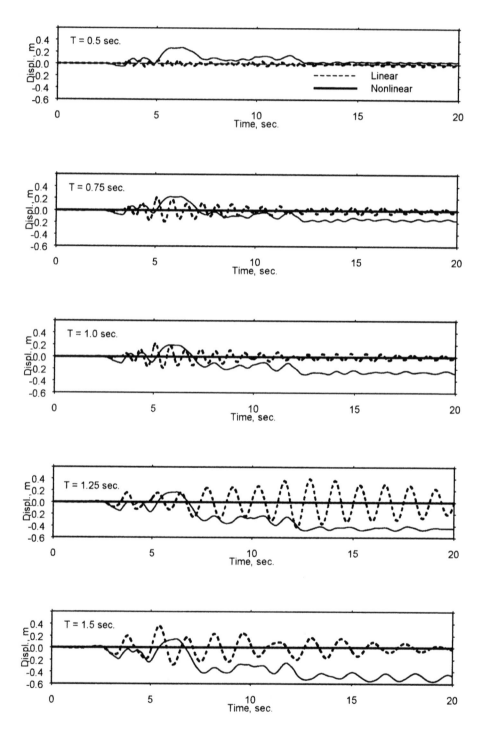

Figure 10. Linear and nonlinear response histories for SDOF systems calculated for the ground motion component measured on 17 August 1999, east-west

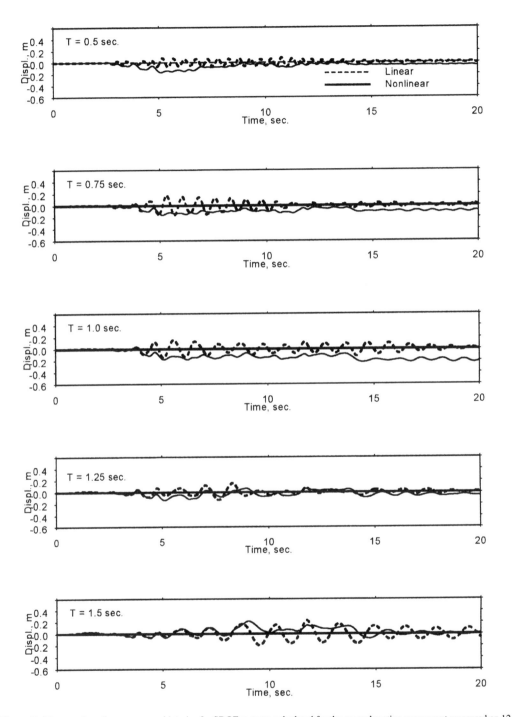

Figure 11. Linear and nonlinear response histories for SDOF systems calculated for the ground motion component measured on 12 November 1999, north-south

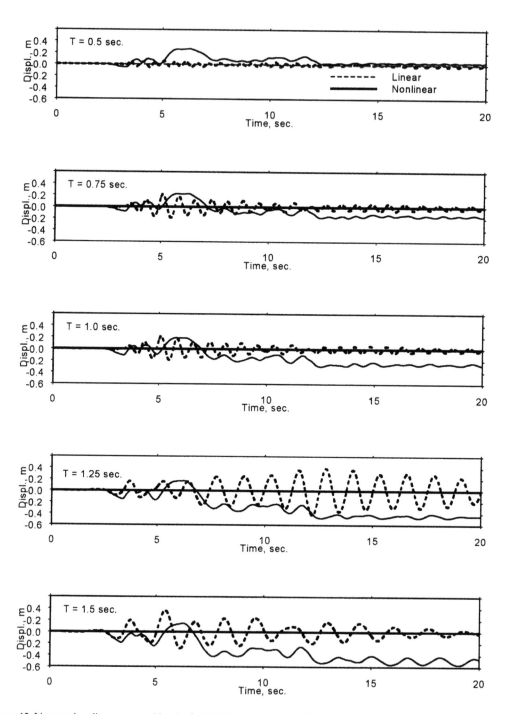

Figure 12. Linear and nonlinear response histories for SDOF systems calculated for the ground motion component measured on 12 November 1999, east-west

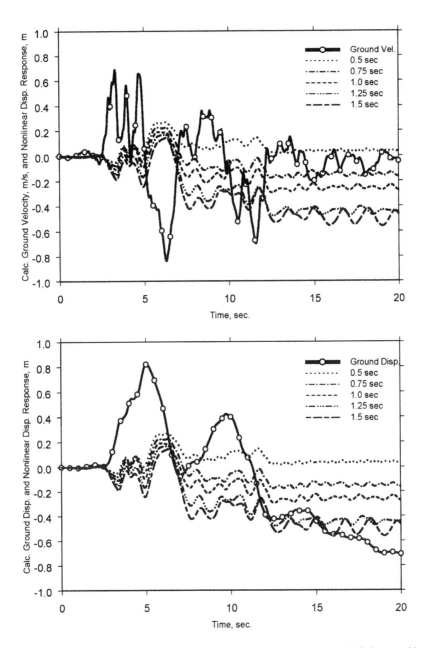

Figure 13. Comparison of calculated nonlinear SDOF response histories with ground velocity and displacement histories for the east-west component of the ground motion measured in Duzce on 12 November 1999

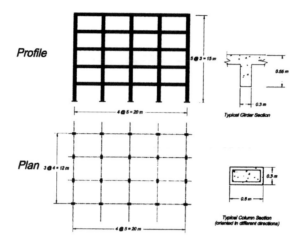

Figure 14. Footprint of five-story building frame

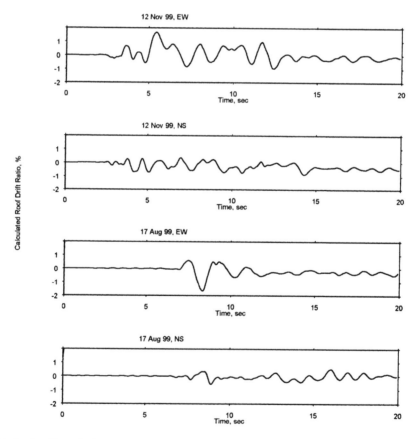

Figure 15. Calculated nonlinear displacement response, five-story building

ing LARZ, a nonlinear dynamic analysis program, the responses of the numerical model to the four considered strong motions were calculated in the long direction of the building. Computed values of the mean drift ratio (MDR), ratio of drift at roof level to the height of the roof above base, versus time are shown in Fig. 15. The values of MDR are high in general and higher in the east-west direction for both the August and November motions. Furthermore, high excitation is indicated to take place over a longer period for the November event. Admittedly, calculation of such MDR's for a frame with the details used in Duzce is a chimera. The buildings are not expected to maintain their integrity if displaced into the nonlinear range of response. But the comparison of the responses does not point to an unusual situation for the November east-west motion and the response quantities suggest that with proper details the damage in Duzce would have been tolerable.

3 CONCLUDING REMARKS

Studies made and summarized in this paper to answer the question with respect to Duzce 1999

"Given the ground motion records and the damage for the August event as well as the ground motion records for the November event, could the damage in the November event have been estimated?"

suggested that, without foreknowledge, a confident and convincing answer is not possible on the basis of direct or even indirect but simple procedures. This conclusion is clearly limited to the competence of the writer and the scope of the studies. It does not mean that the question cannot be answered or that efforts should cease to tackle the issue. On the contrary, considering the advances made in obtaining and predicting ground motion, there should be strong efforts made to connect ground motion measurement and potential damage (real and not sublimated definitions of damage).

If the question posed in this paper cannot be answered simply and directly, it is going to be difficult to have confidence in connecting potential ground motion to potential general damage in the urban environment. But this gap needs to be bridged, and crisply at that, to convince city officials to invest in preventive actions to reduce losses in earthquakes.

4 ACKNOWLEDGMENTS

The work leading to this paper was supported by National Science Foundation Grant CMS-0085270, a joint project of The Middle East Technical University (METU), Ankara, and Purdue University (PU), USA. Figure 8 was prepared by B. Koru based on the field work of many students: C. Donmez, J. Durdella, K. Engvall, Y. Firat, T. Gur, B. Ozturk, S. Pujol, J P Smith, K. Johnson, K. Tureyen (PU), S. Bayili, Z. Cagnan, A. Pay, A. Pekoz, T. Yilmaz, (METU), E. Oware (Washington University, MO), H. Sezen (University of California, Berkeley). Faculty members contributing to the project were U. Ersoy, P. Gulkan, H. Sucuoglu, G. Ozcebe, C. Yilmaz (METU), A. Kilic (Bogazici University), A. Bobet, R. Frosch, A. Johnson, J. Ramirez (PU), G. Fernandez (University of Illinois), Y. Kurama (Notre Dame University), S. McCabe (Kansas University), and S. Wood (University of Texas, Austin), A. Schultz (University of Minnesota). Acknowledgment is due Mrs. Linda Vail for her expertise in the production of this manuscript.

The writer is indebted to the Strong-Motion Study Group, Directorate for Natural Disasters and Resettlement, Ministry of Public Works, Ankara, for their promptness in making the strong ground motion records available to the public.

REFERENCES

Hassan, A. F. & Sozen, M. A. 1997. "Seismic Vulnerability of Low-Rise Buildings in Regions with Infrequent Earthquakes," *Structural Journal of the American Concrete Institute*, V. 94, No. 1, January, 31-39.
Housner, G.W. 1952. "Spectrum Intensities of Strong Motion Earthquakes, " Symposium on Earthquake and Blast Effects on Structures, EERI.
Shiga, T., Shibata, A., & Takahashi,T. 1968. "Earthquake Damage and Wall Index of Reinforced Concrete Buildings,""Proc. Tohoku District Symposium, Architectural Institute of Japan, No. 12, December, 29-32 [Japanese].
Sucuoglu, H & Gur, T. 2000. "Duzce: A Town That Experienced Two Major Earthquakes in Three Months," Engineering Report on the Marmara and Duzce Earthquakes," April, Middle East Technical University, Ankara, 163-169 [Turkish].

Earthquake Engineering Frontiers in the New Millennium, Spencer & Hu (eds),
© 2001 Swets & Zeitlinger, ISBN 90 2651 852 8

An overview and preview of China-US joint research on earthquake engineering

Xiyuan Zhou
Institute of Earthquake Engineering, China Academy of Building Research

Yuxian Hu
Institute of Geophysics, China Seismological Bureau

ABSTRACT: In this paper, the effects of earthquake engineering research in U.S. on the corresponding research and practice in China, the interaction between them and the fruitful outcomes of joint researches are overviewed in some aspects associated with the working fields of the authors. Some tentative subjects such as stochastic characteristics of strong ground motion, performance based seismic design methodologies of building and infrastructures, the applications of the high technology in earthquake engineering and real time health monitoring for essential structures and facilities are conceived of and proposed in this paper.

1 INTRODUCTION

The earthquake engineering research in China started latter than in advanced earthquake prone countries but had a commendable inception through learning from the achievements of foreign countries. Since China carried out policy of reforming and opening, the international cooperation and intercourses in science and technology have been greatly expanded. The j oint research between China and US on earthquake engineering is an outstanding example among them. Both of the partners have gained benefits from the joint research projects. In this paper, the fruitful outcomes of the joint researches are overviewed in some aspects associated with the working fields of the authors. These aspects are:

1) The principle and methodology of seismic design;
2) The characteristics of strong ground motion and attenuation law;
3) Seismic hazard and probabilistic level of seismic design criterion;
4) Seismic performance and reliability of pipeline and network;
5) The design parameters of base isolation rubber bearings.

It is pointed out that the real probabilistic reliability is far from well understanding, particularly in moderate seismic areas. In fact there are great deal of uncertainties involved in earthquake performance of structures and facilities. How to clarify the sources, properties and influences of the uncertainties is no doubt an important task in earthquake engineering, but the more essential thing is to adopt good policy to reduce the influences of uncertainties and to enhance the seismic safety of structures and facili-

ties. These two tasks might be world widely concerned topics and for searching solutions, the international cooperation could even not replaceable. Some tentative subjects such as stochastic characteristics of strong ground motion, performance based seismic design method of building and infrastructures, the applications of the high technology in earthquake engineering and real time health monitoring for essential structures and facilities are conceived of and proposed in this paper.

It is stressed that to place the performance based seismic design on a foundation of complete probabilistic reliability, the seismic input action including design spectra and acceleration time history should be considered random function, whose statistical characteristics normally being deterministic function just as peak ground acceleration (PGA) and spectral acceleration at given period being deterministic value. It is proposed that the representative parameters of the statistic characteristics of these random functions are regarded as random variables just as PGA and spectral accelerations being as independent variables in seismic hazard analysis. Working in this way, the attenuation law should be first established for these parameters to represent the statistical characteristics of ground motions. If we do that, the stochastic characteristics of the ground motion being as ran-dom process can be determined in probabilistic meaning and then the random earthquake response and reliability analyses of structures are successors. This new approach paved a way leading to performance based seismic design in probabilistic meaning. The preliminary analyses about the statistical characteristics of near source ground motion in recent 921 Chi-Chi, Taiwan Earthquake in 1999 are carried out and some results are given in this paper (Wang *et al.* 2000x, b, c). As accumulation of the instrumentation records in wide range of magnitude, fault distance and site conditions, the new approach will become realistic in near future. Finally a joint research project following this approach is proposed in this paper.

2 THE PRINCIPLE AND METHODOLOGY OF SEISMIC DESIGN

The earthquake engineering researches in China begin at mid 1950s. The Research work is carried out and almost simultaneously with development of seismic design code. The first seismic design code draft was set up in 1959. This code draft took foreign code as sample and possessed little peculiarity because the researches were just started at that time without enough products. In the early years of 1960s a new earthquake resistance design system was developed by first generation of researches whose outstanding representative is Prof Hwxian Liu. In this system the strong ground motion recordings recorded in U. S. played an important role. In consideration of the pioneer research products gained by Prof Housner, and other old generation experts and the first set of strong earthquake ground motion instrumentation records. Prof Liu offered following suggestions as principles and criteria for seismic design (Liu&Hu 1963):

1) Take real ground motion acceleration and response spectra being as design criteria rather than assumed seismic coefficient.
2) Mode superposition response spectrum analysis method should be adopted basis of seismic design.
3) Seismic loads experienced by structures can be reduced depending on ductility of the structure.
4) Design acceleration and response spectra should be modified depending on site condition.

Based on these suggestions the seismic model load is calculated using following formula

$$F_{ji} = kc\beta_j\gamma_j X_j(i)G_i \tag{1}$$

here k (in g) is effective peak ground acceleration depending upon design intensity at construction site which is comparative with MM Scale. For Intensity of 7, 8, 9, 10, k takes 0.075g, 0.15g, 0.3g and 0.6g respectively; c is structural factor 1~1/4, depending upon ductility of the structure; γ_j is jth mode participation factor, $X_j(i)$ is the value of jth mode function at point i; β_j is normalized spectral acceleration at period of jth mode; G_i is gravity load at ith story.

The normalized design acceleration spectrum is shown in Fig. 1, in which the characteristic period of T_g depends on site category. The T_g value of site category I, II, III, IV is 0.2 sec, 0.3 sec, 0.5 sec, and 0.8 sec respectively. That is so called site dependent response spectrum, which was deduced from first

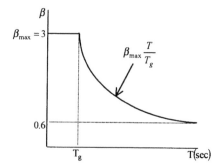

Figure 1. The normalized design spectra

batch of strong ground motion recordings obtained in U. S. We also concerned and noticed the possible influences of the site subsoil on intensity or peak ground acceleration at that time. In consideration of the following two-reversed effect of site condition on PGA (peak ground acceleration), i. e., the potential resonance of subsoil would amplify PGA but in other hand, the nonlineanty of the soft soils subject to strong earthquake motion might decrease ground surface response, hence we finally ignored the influences of the subsoil condition on PGA. However, the more direct reason of making above decision is lack of sufficient strong ground motion recordings (Zhou 1965). This decision has been generally accepted by successive seismic design codes in China until to recent revision.

Eq. (1) first appears in draft of uniformed seismic design code for buildings and other structures in 1964. This draft is not legislative but yet accepted by designer for essential structures because officially there is no seismic design code and no requirement of earthquake protection for normal structures at that time. The formal seismic design code was issued in 1974, which carried on the principles and methodology firstly put forth by Prof Liu. This code had a minor revision in 1978. The design earthquake involved in these codes is determined from deterministic intensity zoning map. In eighties the probabilistic calibration of the basic intensity shown in zoning map was carried out and the conclusion is that the return period of the intensity indicated in zoning map roughly equals 359 years in average meaning (Bao et al. 1985). In consideration of uncertainties involved in design criteria the design thought of three levels of earthquake protection was conceived sooner. The allowed exceedance probability of the basic design intensity I is 0.1 during service time of 50 years and the corresponding return period is 475 years. The following relation between intensity I and design acceleration A (in g) is adopted

$$A = 0.1 \times 2^{I-7} \tag{2}$$

A certain plastic deformation is permitted while structure subject to strong ground motion corresponding to basic design intensity. Hence, the seismic analyses of structures under action of basic intensity earthquake are not coincided with the assumptions of elastic behavior as mode superposition response spectrum analysis required. Since the basic earthquake is not frequently takes place during service period of structures, but the earthquake motion with return period of 50 year would most likely occur. So that the code defines earthquake motion with return period of 50 years first level of design criterion and requires the structural performance within elastic range of deformation. At this stage of seismic analysis is replaced to design earthquake of return period of 50 years from basic design earthquake. Owing to the severity of the consequence of earthquake, we have yet to control the plastic deformations of structures in basic intensity and to prevent collapse in extreme earthquake whose return period is normally defined 1975 years. In seismic design codes in China, usually it is only required to check the safety in most likelihood earthquake (or minor earthquake) and in extreme earthquake (or maj or earthquake).

The safety of structures in basic design intensity or moderate earthquake is ensured by corroborated structural details aiming at enhancing of ductility. This procedure is named 3-level, 2-stage design methodologies (see Fig.19), which was first accepted in seismic design code for buildings (GBJll-89) in 1989(Zhou 1992) and succeeded in its recent revision. Compared with the original Eq. (1) for calculation of seismic load, the structure factor c is removed and replaced by a reduction from basic design intensity (moderate earthquake) to most likelihood intensity (minor earthquake). Furthermore an additional adjustment coefficient γ_{RE}, of earthquake resistance capacity of structural components is introduced. So that the seismic effects S induced in each structural component should comply following equation

$$S \leq R / \gamma_{RE} \tag{3}$$

here R is earthquake resistance capacity of the component and γ_{RE}, is varying depending on types of structural components and materials but the range of variation is about 0.7~1, not as large as c in Eq.(1) and R-factor in U.S. code.

Another amendment of the new revision of seismic design code for buildings in China is going to take account the effect of site category on PGA or maximum spectral acceleration and the seismic actions of structures located on deep soft soil layers would be amplified to a certain length. This decision is based on the new acquisitions of strong ground motion recordings at comparable rock and soil sites since Loma Prieta earthquake in 1989. Although many other improvements have been made the basic principles and thoughts of seismic design laid by prof. Liu is consistently carried in successive seismic design code in China (Liu 1983,1985).

3 THE CHARACTERISTICS OF STRONG GROUND MOTION

In the mentioned design system of earthquake resistance, the strong ground motion recordings, and the dynamic characteristics and attenuation law reflected by them are of essential significance. The instrumentation observation for earthquake strong ground motion commenced in 1962 when a reservoir induced earthquake attacked Heyuan county, Guangdong province in China. Due to dispersive distribution of destructive earthquakes in China mainland, up to now still lack of near source acceleration recordings but in U. S. the obtained near source recordings expand rapidly and the maximum recorded peak ground acceleration is steadily increasing. In Cape Mendocino earthquakes of April 25-26, 1992, which included a moment magnitude 7 mainshock and two large aftershocks, the records recovered from the mainshock have some of the highest accelerations ever recorded (Dorragh 1994). Peak accelerations near 2g were recorded at the Cape Mendociono station, approximately 4 km southwest of epicenter, on hard sandstone. A peak velocity of 126 cm/sec and peak displacement near 70cm(on the vertical component) are calculated. But the duration is only about 7 seconds. Another outstanding event is magnitude 7.4 Lander earthquake of June 28, 1993. A peak acceleration of 0.9g and peak velocity of 142 cm/sec was recorded during Lander earthquake at Southern California Edison (SCE) station at Lucerne located 1km from the fault. The duration of the records obtained during Lande earthquake are about 30 seconds, 2 to 4 times longer than that of the other records including those from the magnitude 7 Loma Prieta earthquake. The large velocity impulse is recorded in 1994 Northridge M_w 6.7 earthquake with peak velocity of 177cm/s at break zone of causative fault, and the corresponding PGA and PGD is 0.85g and 50cm respectively. In 1995 Hanshin, Japan M_w, 6.9 earthquake almost the same largest velocity impulse with PGV of 176cm/sec is reported, recording station is also located at fault rupture zone (Hall et al. 1995). In China mainland only a few strong ground motion are recorded. The strongest accelerogram with peak acceleration 0.49g was recorded during a M_w 6.7 aftershock of Lancang-Gengma 1988 earthquake in Yunnan province (Wang&Li 1989).

The recording station is located on stiff soil. The period corresponding to peak spectral acceleration is about 0.5 sec, but the curve falls down rapidly beyond the peak. That means the long period components in response spectra are small. Some peak values of ground motion recordings since 1978 show in Table 1. However, the mentioned peak ground accelerations are quite large and much higher than that of considered in design codes.

Table 1. Peak near source ground motion from large earthquakes since 1978

Earthquake		Distance (km)	Acceleration (g)	Velocity (cm/s)	Displacement (cm)
Tabas (1978 Tabas, Iran, Mw 7.4)		3	0.92	125	106
El Centro (1979 Imperial Valley, USA, Mw 6.5)	Array 6	1	1.74	110	55
	Array 7	1	0.65	110	41
1987 Superstition, USA, Mw 6.6	Parachute Test Site	0	0.53	138	60
	Superstition Mountain	0	0.91	44	16
1989 Loma Prieta, USA, Mw 6.9	Los Gutos Preseutantion Center	0	0.62	102	40
	Lexingtonn Dam	5	0.44	120	40
1988, Langcang, Yunnan, China, Ms=6.7	Zhutang	1	0.44	46.1	13.1

Since earthquake ground motions were considered for the first time as random processes (Housner 1947), the concept has been accepted widely. However, because lack of sufficient strong ground motion recordings, the randomness of ground motion is far from well understanding. Fortunately, a god-given chance was provided by 921 Chi-Chi, Taiwan, earthquake, which supplied unprecedented near fault strong ground motion recordings (Lee *et al.* 1999). Prior to 1999, there were only 8 ground motion recordings worldwide for earthquakes greater than a magnitude 7.0 and at a distance of less than 20 kilometers from the causative fault, but the 921 Chi-Chi earthquake generated 65 recordings satisfied above conditions (Earthquake Engineering Research Institute, 1999, http://www.eeri.org).

Recordings of 30 near fault stations with distance to fault ranging from 3km to 20km are selected. All these stations locate on alluvium deposits of fault footwall. So the effect of site geologic conditions could be neglected. As all these recordings were recorded in same earthquake, so the influence of source mechanism is identical. According to our study (Wang et. al, 2000-a, b), in footwall side, within 3km to the causative fault, the recorded peak ground acceleration (PGA) and peak ground velocity (PGV) are much larger than those far from 3km, and the influencing of rupture directivity at hanging wall side of the thrust fault are considerable. While within distances from 3km to 20km, the PGAs and PGVs are in same order of quantity but much smaller than those within range of 3km, and the effects of fault rupture directivity and the effect of hanging wall thrust are insignificant. That means the effects of distance, fault rupture directivity and hanging wall thrust could also be neglected in this range. So the baselined corrected three dimensional component recordings of the 30 stations (Boore 1999) are regarded as an assembly of three random processes to study the randomness of near fault ground motions.

The response spectra of each component being as a random function and the autocorrelation and cross-correlation matrixes of three response random spectra are studied. The three autocorrelation matrixes and cross-correlation matrixes are shown in Fig.2. It can been seen that the autocorrelation matrixes and cross-correlation matrix of two horizontal components of response spectra are more regular than autocorrelation matrix of vertical component and cross-correlation matrixes between vertical and horizontal components. And the contour-lines of correlation coefficient in plan of autocorrelation and cross-correlation of NS and EW components roughly have diamond like shapes within period range from 0.1s to 4s. Hence the contour-lines in Fig.3 are used to depict the nonstationarity of the random response spectra.

Hence by best fit method the correlation coefficients for lower triangle can be written as (Wang *et al.* 2000b)

$$\rho(T_i, T_j) = e^{k_1 \cdot r(\frac{b\lg T_i + a\lg T_j - ab - a^2}{\lg T_i + \lg T_j - 2a} - m)^2} \cos(k_2 \omega(\frac{b\lg T_i + a\lg T_j - ab - a^2}{\lg T_i + \lg T_j - 2a} - m)) + c \quad (4)$$

Table 2. The factors involved in Eq.(4)

Comp	a	b	m	R	ω	k_1	k_2	C
NS	Lg0.065	Lg5.6	Lg5.27	-0.799	3.363	0.625	0.747	0.03
EW	Lg0.085	Lg4.5	Lg3.18	-1.323	3.157	0.642	0.736	0.105
NS-EW	Lg0.1	Lg3.5	Lg3.04	-1.287	3.291	0.585	0.669	0.045

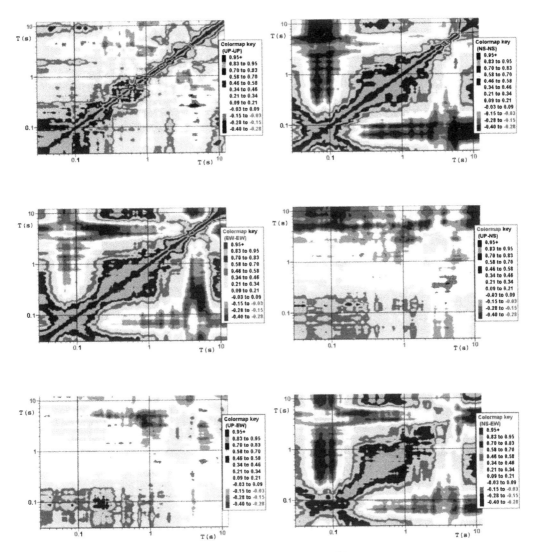

Figure 2. The autocorrelation and cross-correlation matrixes of response spectra of three components

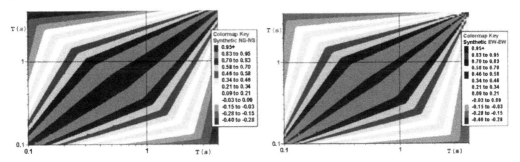

Figure 3. The synthetic autocorrelation matrixes of response spectra of two horizontal components

61

Secondly the acceleration time histories of each component are regarded as a random process and their autocorrelation matrixes and cross-correlation matrixes are studied. The distributions of the equi-value zone of correlation coefficients in plan are irregular. Apparently the correlation functions, calculated from normal algorism, are made up of two parts. One is a deterministic function; another is a random function. The deterministic process can be simulated using normal mathematic expression, for instance, exponentially decayed cosine curve. As for random part, the reasonable way is to calculate its correlation function successively. In such a case, we can call the correlation function directly from time history as the first order correlation function, and the successive correlation function from the random part of the first order one as the second order correlation function, which might be a random process as well. Using the same method, we can obtain the third, the forth and even the higher order correlation function. It is noted that the terminal correlation function could be simulated with a determinist function. This is a procedure of de-randomization. A random procedure, whose correlation function comprises random component, is called super random process (Wang *et al.* 2000c).

The mathematic expression of correlation function of the acceleration history random process can be illustrated as follows. As we known, the nonstationarity of acceleration time history displayed both in aspects of amplitude and frequency contents. Suppose $Y(t)$ is a super random process and $f(t)$ envelope function, the amplitude nonstationarity could be diminished by using stationarization treatment. And then a corresponding amplitude stationary process $X(t)=Y(t)/f(t)$ is obtained. In order to reduce the influence of the nonstationarity of frequency contents, the whole time history is divided into several segments and each segment is stationarized by the corresponding segment of the envelope function. In this section, the segment of the stationarized process from 28s to 45s is used as an example to study the characters of correlation function. The first order correlation *(FOC)* function could be written as

$$FOC = e^{-at}\cos[(\omega - \beta t)t] + XX(t)(1 - e^{-bt}) \quad (5)$$

here $e^{-at}\cos[(\omega + \beta t)t]$ is the deterministic part and $XX(t)(1-e^{-bt})$ is the random part in *FOC* . In fact $XX(t)$ is a irregularly waved long tail in correlation function. The correlation function of $XX(t)(1-e^{-bt})$ is called as the second order correlation (*SOC*) function, which could be written as

$$SOC = e^{-at}\cos[(\omega - \beta t)t] + XXX(t)(1 - e^{-bt}) \quad (6)$$

Using the same method, the mth order correlation function can be written as

$$MOC = e^{-at}\cos[(\omega - \beta t)t] + \underbrace{XXX...X}_{m+1}(t)(1 - e^{-bt}) \quad (7)$$

Finally we might obtain the terminal correlation function that only includes the deterministic part, which is expressed as

$$NOC = e^{-at}[(\cos(\omega - \beta t)t] \quad (8)$$

Eq.(8) is a typical form of correlation function but modified by a term of β_t. This modification is used to reflect the process of enlargement of the dominant period of ground motions. The first, second and third order correlation functions of each component of acceleration history are shown in Fig.4. For ease to understand the characters of cor-

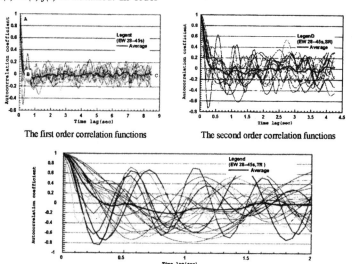

The first order correlation functions

The second order correlation functions

The third order correlation functions

Figure 4. The first, second and third order correlation functions of acceleration history random process (EW 28-45s) .

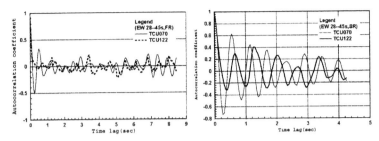

The first order correlation functions   The second order correlation functions

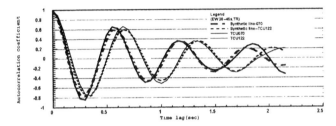

The third order correlation functions and the synthetic results

Figure 5. Map showing the first, second and third order correlation functions of the acceleration history (EW, 28-45s) recorded by station 070 and 122. The broken lines are the synthetic lines of the third order correlation functions.

relation function, the first, second and third correlation functions of the EW components of acceleration time histories (EW, 28-45s) recorded at station TCU070 and TCU 122 are shown in Fig.5 as examples. The synthetic curves of the third order correlation functions of the two samples are also plotted in Fig.5. The parameters involved in Eq. (5)-(8) are determined by best fit and the results list in Table 3.

4 THE ATTENUATION OF GROUND MOTION

In this topic we like to show two things. The first is about the definitions of distance and magnitude in ground motion attenuation law. In order to eliminate or reduce the influences of the magnitude saturation, moment magnitude MW is widely used in various attenuation equations. But in China we usually use surface wave magnitude Ms because almost all the macro (field investigation) and micro (instrument) earthquake data are modified to scale of surface wave magnitude. For moderate and strong earthquakes with magnitude 5.5~8 which

are interested in engineering circle, we have Ms ≈ Mw. This approximate relation makes the earthquake data about intensity and ground motion obtained both in China, US and other countries comparable. Different types of definitions are involved in attenuation law. However each definition actually represents a sort of seismic source model or mechanism. For example, the distance from epicenter or hypocenter corresponds to point source model, fault distance corresponds to fault rupture model. Both of the distance definitions are most popular ones. In China we usually use ellipse model of intensity or ground motion in which major and minor axes attenuation formula should be separately established. Apparently such kind of ellipse attenuation belongs to fault rupture model. Besides, we also use average distance from two ends (or focuses) of equivalent rupture on fault line. The isolines formed by attenuation law using the average distance are elliptical as well but only one attenuation equation is needed. For attenuation model with average distance, the isolines of ground motion parameter approaches circle as distance infinitively increasing. Attenuation law is a bridge which links seismic source and concerned

Table 3. The parameters involved in the correlation function used in this study

Order	a	ω rad/s	β	b
1	8	3	0	4
2	6	4	0	3
3	0.8	9.8	0.15	—

site whose seismic hazard is interested in, so that it plays an important role in hazard analysis and zoning map.

Now we have developed two kinds of zoning map, i.e., that of intensity and ground motion parameters. The problem that we confront is lack of instrumentation data in seismic hazard analysis of ground motion parameters. Even in eastern U.S.A the same problem is faced. The feasible way for solving this problem is to use the local available intensity attenuation law to amend the ground motion parameter attenuation in other region. Several schemes have been proposed in this direction. We normally use equal magnitude amendment scheme as shown in Fig.6. Suppose the lower curve in the two diagrams of Fig.6 up and down respectively is intensity attenuation law and acceleration attenuation law for a given magnitude M_1, for other region outside the concerned one. In addition, we yet know intensity attenuation law in concerned region, i.e., the upper curve in upper diagram of Fig.6. Following the routes shown in Fig.6, the target curve indicated by dotted line can be calibrated (Hu *et al.* 1982). This procedure is completed in condition of same magnitude, so equal magnitude scheme is named.

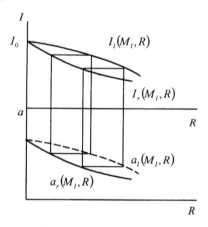

Figure 6. Modification of equ-magnitude method of ground motion

5 SEISMIC HAZARD ANALYSIS AND ZONING MAP

The randomness of earthquake event is apparent and almost no one doubt about it. The research regarding earthquake attack as random event could trace back to early 1960s when the experienced seismic intensities of a given city for example Beijing or Xian were counted one by one from earthquake catalog or historical records. By this way the occurrence frequencies for various intensities within a certain time period had been determined

for several cities in China, but without consideration of the possible migration of seismic sources and the variation of earthquake activity in time. Such kind of simple statistic method is adequate only to well civilized area where the long historical records for earthquake are available. Even in China only a few provinces have long historic earthquake records. This situation greatly constrains the application of the direct statistic method for estimating earthquake hazard. The seismic hazard analysis (Cornell 1968), which is based on identification of potential source, magnitude frequency distribution and attenuation law, and total probabilistic calculation, attracts great interests of Chinese researchers. The probabilistic approach is employed in essential engineering site and seismic zonation. Fig.7 shows the flow chart of seismic hazard analysis. The basic procedures are (Hu *et al.* 1996):

1. Identification and allocation of potential sources;
2. Determination of annual occurrence rate of earthquake, magnitude-frequency relation and magnitude upper bound for each sources;
3. Option of attenuation law of ground motion parameters adequate for the concerned region;
4. Calculating exceedance probability of different levels of ground motion parameters for various site;
5. Delineate isolines and zoning map of ground motion parameters for prescribed return period, which usually is 475 years for normal use.

The above procedures just are outlines of zoning mapping in China. In practice we use so called synthetic probability method which combine many information from various branches of sciences such as geology, particularly tectonic structures, geophysics including geomagnetism, geogravimetry, remote sensing image, seismology and archaeology, and carry out comprehensive analyses in delineating potential sources and calibrating seismicity.

Investigation of seismicity is an important procedure in hazard analysis, and the simple Poisson process is the most popular model of seismicity. Usually it is a non-memory stationary random point process. But the earthquake activities in various regions of China, there are some seismic periodicals, which consist of high and low activity episodes. In some seismic belts several earthquake circles are discovered in historical records. In order to depict the real activity of earthquakes, several non-stationary random models are investigated and used in hazard analysis, for example varying occurrence Poisson and Neyman Scott models (Zhou et al 1985), renew process, Markov model. Renew model is used to express the interval gap between two high tides of earthquake activity. In many seismic zones there is a relative quiescent period

a. Allocation of seismic sources

b. Magnitude-frequency relations

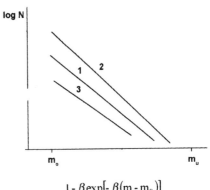

$$F(M) = \frac{1 - \beta \exp[-\beta(m - m_0)]}{1 - \exp[-\beta(m_u - m_0)]}$$

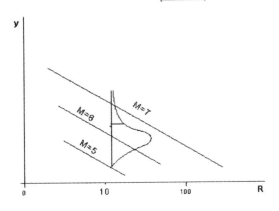

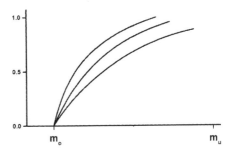

Magnitude distribution

c. Attenuation

$$P[Y > y \mid E_i] = \sum_j \sum_k p[Y > y \mid m_j . r_k] \, |F(r_k) F(m_j)$$

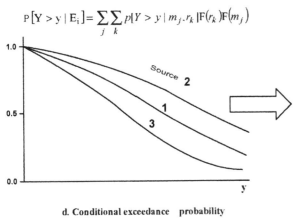

d. Conditional exceedance probability

$$P_T(Y > y) = 1 - e^{\sum P[Y > y \mid E_i, \nu, T]}$$

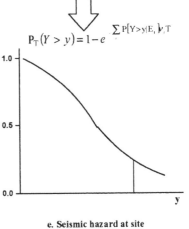

e. Seismic hazard at site

Figure 7. Flow chart of seismic hazard analysis

after a very strong earthquake for instance $M_w \geq$ 7.5 event within its surrounding area. This phenomenon is called immunity of great earthquake in near future in the surrounding area of the big event. But in other hand, another seismic belt within a certain distance to the highly activated seismic belt might enhance its occurrence probability of destructive earthquake. The immigration pattern of great earthquakes among neighboring seismic zones have been studied and the results used both in short-term or mid-term prediction and zoning mapping which is regard long-term earthquake prediction in China.

The abundant historical earthquake records are basis of seismic hazard analysis. In U.S.A the available earthquake history is brief. In order to check the feasibility of seismic hazard analysis for formulating seismic zoning map if only relatively short historical earthquake catalog is available, Cornell and McGuire closely study the seismic activity in North China. They utilized the historical earthquake records in northern five provinces of China to evaluate the adequacy of simple methods for calculating seismic hazards. Earthquake occurrences in five provinces of North China during time segments of three lengths (50, 100 and 200 years) are used as input to the seismic hazard analysis. The output is felt shaking probabilities in 62 cities within the five provinces of North China in the 50-year period following each time segment (50, 100 or 200 years). The calculated probabilities are checked by observed felt shaking frequency in the successive 50 years and the results indicate two hundred years of indiscriminately used data do not provide better estimates of probabilities of felt shaking than do 100 and 50 years of data (Mc Gwre 1979, 1981). These results are apparently related with the periodicity or non-stationary of seismicity in North China. The implication is that if seismicity is non-stationary, the best policy is to use recent 100 years database for calculation of probabilities of felt shaking in the near future. It confirms the feasibility of seismic hazard analysis for areas where only relatively short earthquake records are available even though the long-term variation rule of seismicity is unknown.

6 SEISMIC PERFORMANCE OF PIPELINE AND NETWORK

Ground surface rupture is the main reason of earthquake damage of buried oil pipeline. Twenty-five years have elapsed since N.M. Newmark and W.J. Hall (1975) first proposed seismic analysis of buried pipeline. During the time interval several modifications have been done by R.P. Kennedy (1977,1979), R.L. Leon, Wang (1981, Wang et al. 1985, Wang & Ye 1985). The improvement proposed by R.P. Kennedy includes:

1) Using a straight line and round arc to depict the deformation of buried pipeline;
2) Variable friction forces along pipe adopted;
3) Using Bilinear and Ramberg-Osgood model to express stress and strain relation of pipe material. Leon R.L. et al. developed more sophisticated analysis method in which the flexure strain and rigidity are taken into account. In this case the total deformation is the sum of lateral flexure deformation and tension deformation. They assume that the pipeline deformation model consist of two parts, i.e., transition zone and beam on elastic foundation as shown in Fig.8. As first the deformation of transition zone is assumed a round arc (Wang&Yeh 1985), but later it is replaced by a cantilever beam with elastic rotating support at one end.

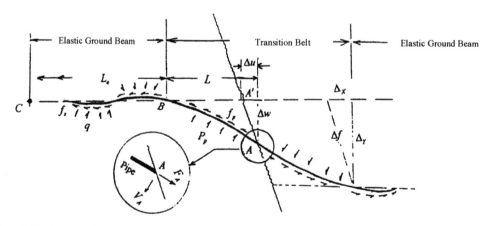

Figure 8. Displacement of buried pipeline striding fault (plane)

Recently Zhang (Zhang 1999) amended the model of elastic foundation beam at far part and cantilever beam at near part from fault, and more dedicated boundary conditions have been utilized in calculating deformation curve in stage of large deformation. Furthermore, Ramberg-Osgood model is used to consider the nonlinearity of pipeline and the effect of flexure and tension stiffness. The results of example analysis indicate that these modifications eliminate some shortcomings involved in previous models and give more reasonable deformation and interval force of pipeline.

The reliability of seismic performance of water delivery system is a common interested topic between China and U.S. researchers (Sun 1994). The methodologies for evaluating seismic performance of water delivery system have been independently developed in China and U.S. Their foundations are different. In fact, given the seismic hazard, the American researches use an empirical pipeline data, to estimate an average pipe damage rate (repairs/km) for each pipeline. In contrast, Chinese researchers estimate pipeline damage based on seismic effect analysis of pipeline subject to propagating wave and deformation capacities of pipe and joint, which are gained from experiment. The damage states of network caused by scenario earthquake are estimated both by Monta Carlo Simulation. The serviceability of the water delivery system is evaluated from the connectivity and flow analysis of seismically damaged water delivery network (Shinozuka et al. 1992; Tanaka 1996) in U. S. side but only from connectivity analysis in this case study in China side. In consideration of the randomness of seismic resistance of buried pipelines, the damage probability is determined from first order second moment method in China side. The performance of a water delivery system subject to a moment magnitude 7.0 scenario earthquake is independently evaluated using the methods developed by American and Chinese researchers. The water delivery system covers an area of approximately 10km by 20km along a river. The area contains three types of site conditions: soft, medium dense, and very dense. The epicenter distance to the study area is ranging from 36km to 58km. The damage state of each pipeline determined by American and Chinese researches is listed in reference (Hwang et al.1997). The calculated serviceability of the network subject to same scenario earthquake using American method is better than that using Chinese method. The difference is mainly caused by the different types of pipe joints used in the derivation of the damage criteria. The damage criteria used by American researches were derived mainly from seismic damage to cast iron pipes with lead caulked joints commonly found in the old pipes in U. S. and Japan, while the damage criteria used by Chinese researchers were derived based on the cast iron pipes with asbestos cement joints. If damage to pipelines is evaluated under same ground shaking severity and comparable pipe damage criteria, then the results predicted by both the methods are compatible. For instance the level of service abilities of 54.4 % of demand nodes are coincident or in same grade in total 136 nodes. The serviceability level of 37.5% of demand nodes has difference of 1 grade. The serviceability grade is divided into five grades, i.e., intact, slight damage, moderate damage, heave damage and no water. Only 8.1% of the serviceability level differs 2 grades. The joint study makes both the method more transparent and mutually well understanding. In this joint research two sides find out their advantages and shortcomings, which are helpful to further improvements. This type of joint research is mutual benefit and is worth to expand to other fields, for example, seismic design for buildings and infrastructures.

7 SEISMIC BASE ISOLATION FOR BUILDINGS AND BRIDGES

Seismic base isolation is one of relatively mature technology for greatly reducing earthquake response of structures. Normal base isolation system with rubber bearings requires the up and low supports of rubber bearings are horizontally infinitive rigid plate. In order to enhance the usable space in buildings, designers are fond of installing rubber bearing at top of basement column. In this case the isolation layer actually is a serial system composed of rubber bearings and R/C columns. If we say that such a kind of serial system can be substituted by isolation layer with up and low rigidity platforms, but it is almost inevitable in bridges. Due to these reasons calculation method of horizontal stiffness of the serial system composite of rubber bearing and column attract common attention in recent years. Imbimbo and Kelly (1997a, b) replace the column in the rubber bearing -column serial system by a elastic rotatable spring and gives a approximate formula for calculating horizontal stiffness coefficient. The closed formula of horizontal stiffness of rubber bearing and column serial system is deduced by Zhou et al. (1999). In fact the serial system can be regarded as a special case of composite isolator so that we will describe the stiffness characteristics of composite isolator firstly.

Because of the severe seismic requirements for the design of base-isolated buildings, isolators that are designed for high seismic input may be very stiff at small or moderate earthquakes. It means that the isolation system does not provide satisfactory isolation performance at low levels of seismic input and the reduction of the acceleration in the superstructure might be less than that expected. To overcome this shortcoming a composite isolator has been proposed by A.G. Tarics (1994). It-consists-of two elastomeric bearings, one on the top of the other and one much stiffer than another. This system provides a satisfactory performance both at low-level and high-level seismic inputs.

A buckling analysis of this system is shown in Fig.9, performed by Imbimbo and Kelly (1997). The boundary conditions are as follows: the top is prevented from rotating, but free to move horizontally; the bottom is fixed against both rotation and horizontal displacement. At the junction between the two isolators, continuity of the displacement and the rotation should be satisfied.

The practical composite isolator is made from two single isolators that are linked together by using rigid plate with infinitive stiffness in Fig.10. A general solution of stability and lateral stiffness of the practical composite isolator is obtained(Zhou et $al.$ 1999). The low end of the practical composite isolator is fixed and the upper one is free to move in horizontal direction but restrained against rotation. At the junction between the two isolators, the rigid plate can only rigidly rotate.

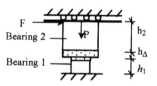

Figure 10. Composite isolator: Zhou's model

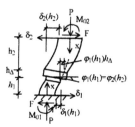

Figure 11. The deformation and reaction forces of composite isolator

Supposing a lateral force F and vertical load P are applied on top of the isolator simultaneously, the deformation patter of the practical composite isolator is shown in Fig. 11.

According to the local coordinate system, the isolated body with heigth within $0<x<h$ for each of the rubber bearing up and down can be described in Fig. 12.

Based on the equibalance and continuing conditions for various parts of the composite isolator, a closed formula for calculating horizontal stiffness and the non-dimensional horizontal stiffness is deduced (Zhou et $al.$ 1999).

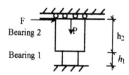

Figure 9. Composite isolator Imbimbo and Kellv's model

$$\frac{K_H(p)}{K_H(0)} = q \frac{\Psi_1 \sin p_1 \cos p_2 + \Psi_2 \sin p_2 \cos p_1 - \Psi_1 \Psi_2 \zeta_\Delta \sin p_1 \sin p_2}{2 + [\Psi_3 \sin p_1 - (1+\zeta)\Psi_2 \cos p_1]\sin p_2 - (2\cos p_1 + (1+\zeta)\Psi_1 \sin p_1)\cos p_2} \tag{9}$$

here $\quad p = \dfrac{Ph_2}{\sqrt{E_2 I_2 G_2 A_2}} \qquad\qquad \zeta = \dfrac{h_1}{h_2} \qquad\qquad \zeta_\Delta = \dfrac{h_\Delta}{h_2}$

$$\xi_b = \frac{E_1 I_1}{E_2 I_2} \qquad\qquad \xi_s = \frac{G_1 A_1}{G_2 A_2} \qquad\qquad \eta = h_2 \sqrt{\frac{G_2 A_2}{E_2 I_2}}$$

$$q = \frac{p}{\eta}\left\{ 1 + \frac{\zeta}{\xi_s} + \frac{\eta^2}{12(\zeta + \xi_b)}\left[\frac{\zeta^4}{\xi_b} + \xi_b + 4\zeta(1 + 1.5\zeta + \zeta^2) + 12\zeta\zeta_\Delta(1 + \zeta + \zeta_\Delta)\right]\right\}$$

$$p_1 = \zeta\sqrt{\frac{p}{\xi_b \xi_s}(p + \eta\xi_s)} \qquad p_2 = \sqrt{p(p+\eta)} \qquad \Psi_1 = \eta\sqrt{\frac{p\xi_s}{\xi_b(p + \eta\xi_s)}}$$

$$\Psi_2 = \eta\sqrt{\frac{p}{p+\eta}} \qquad \Psi_3 = \sqrt{\frac{\xi_s(p+\eta)}{\xi_b(p+\eta\xi_s)}} + \sqrt{\frac{\xi_b(p+\eta\xi_s)}{\xi_s(p+\eta)}} + \Psi_1\Psi_2\zeta_\Delta(1 + \zeta + \zeta_\Delta)$$

$$\frac{1}{K_H(0)} = \left(\frac{h_1}{G_1 A_1} + \frac{h_2}{G_2 A_2}\right)\left[1 + \xi_s\eta^2 \frac{\zeta^4 + \xi_b^2 + 4\zeta\xi_b(1 + 1.5\zeta + \zeta^2) + 12\zeta\xi_b\zeta_\Delta(1 + \zeta + \zeta_\Delta)}{12\xi_b(\xi_b + \zeta)(\xi_s + \zeta)}\right] \tag{10}$$

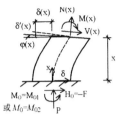

Figure 12. Izolator

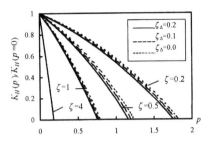

Figure 13. Relation of non-dimensional horizontal stiffness and axial pressure of composite rubber bearing ($\zeta_s = 0.5$ $\zeta_b = 0.5$ $\eta = 1.0$)

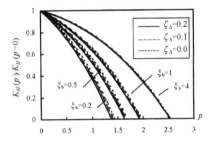

Figure 14. Relation of non-dimensional horizontal stiffness and axial pressure of composite rubber bearing ($\zeta_s = 0.5$ $\zeta = 0.5$ $\eta = 0.5$)

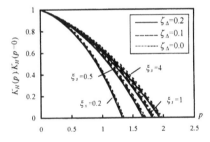

Figure 15. Relation of non-dimensional horizontal stiffness and axial pressure of composite rubber bearing ($\zeta_b = 1$ $\zeta = 0.5$ $\eta = 0.5$)

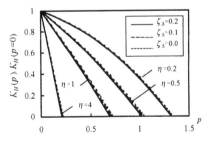

Figure 16. Relation of non-dimensional horizontal stiffness and axial pressure of composite rubber bearing ($\zeta_s = 4$ $\zeta_b = 4$ $\eta = 4$)

The relations among non-dimensional horizontal stiffness $K_H(p)/K_H(0)$ of the composite isolator and parameters p, ζ, ξ_b, ξ_s, η are given in Fig. 13-16 while $\zeta_\Delta = 0.2, 0.1, 0.0$. It can be seen from these figures that $K_H(p)/K_H(0)$ is decreased with increasing p and slightly decreased with increasing This result shows that the non-dimensional horizontal stiffness in consideration of the influence of the rigid plate is slightly lower than that of without rigid plate and as increasing thickness of the rigid plate. $K_H(p)/K_H(0)$ decreases further more. Remaining other parameters unchanged, $K_H(p)/K_H(0)$ and critical force decrease with increasing ζ as indicated by Fig. 13, increase with increasing ξ_b as show in Fig. 14. In addition, Fig. 15 and Fig. 16 show increasing and decreasing tendency of $K_H(p)/K_H(0)$ and critical force of the composite isolator respectively with increasing ξ_s and η. However the results from Fig. 13-16 indicate that the influence of the rigid plate on horizontal stiffness is insignificant if $\zeta_\Delta \leq 0.2$.

If one of bearings, for example, the lower part shown in Fig.10 is considered as column, the composite isolator becomes general model of serial system of rubber bearing and column. However the shear deformation of column usually can be neglected, horizontal stiffness of serial system of rubber bearing and bend column can be deduced from the general Eq. (9). The $K_H(p)/K_H(0)$-p curves show the relations among $K_H(p)/K_H(0)$ and p, ζ, ξ_b, η of serial system of bearing-column is similar to that of composite isolator, but the critical force is slightly increased.

Aiming at economic and safe seismic design, a kind of parallel isolation system was presented (Zhou *et al.*, 1999), that is composed of loaded rubber bearings and back up supports with a top friction sliding surface (Fig. 17). The instable failure of the rubber bearings at large horizontal displacement is averted and the shearing deformation capability of them is fully exerted. In the system the back up supports of the system should be installed after completion of superstructure construction to ensure them not to bear weight in normal state. When strong earthquake happens the loaded rubber bearings yield small vertical deformation and sinking, the superstructure moves over top friction sliding surface of the back up supports and

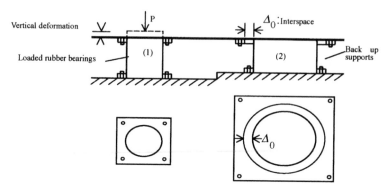

Figure 17. Parallel isolation system

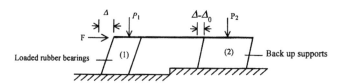

Figure 18. Horizontal displacement of parallel isolation system

then the vertical pressure gradually shifts from the loaded bearings to back up supports (Fig.18). Since the friction force on the sliding surface of the back up support will increase as its vertical pressure increasing, the back up support, who actually is a varying friction sliding load bearing device, not only prevents the loaded bearing from instable failure caused by large deformation but also provides friction damping and exerts favorable effect in aspect of reducing horizontal displacement of the isolation layer.

8 PERFORMANCE BASED SEISMIC DESIGN

The concept of performance based seismic design has partly embodied in three level two stage designs in seismic design code in China. However, the seismic design code only gives the basic performance requirements for the three earthquake protection levels. Fig. 19 shows the basic performance requirements for building earthquake protection. For normal building it is reasonable to prevent collapse during extreme earthquake to ensure life safety during basic design earthquake and to keep non-maintenance during most likelihood earthquake or minor earthquake in service period. For essential structures more strict performances are required as shown in Fig. 19. In contrast, for temporary and less important buildings, for instance non-occupancy warehouse, the seismic performance is allowed to decrease. Further effort should be focused on establishing more delicate

performance objective for different types of structures and making up quantitative performance standards. In addition, the possible worst consequences, secondary disasters, and indirect losses are worth to involve in performance design (Hu 1998).

Strictly speaking, what mentioned above is not the real performance based design, although the basic principles have permeated into seismic design code. In fact the performance based design should comprise setting up of performance objective, performance evaluating and performance control. What shown in Fig.20 is a proposed flow chart of

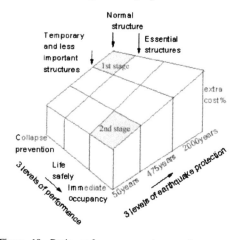

Figure 19. Basic performance requirement for structure earthquake protection

70

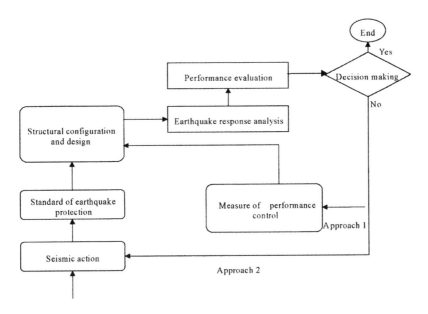

Fig.20 Flow chart of seismic performance design

seismic performance design. Distinguished with traditional seismic design, the performance evaluation and decision making are specific procedures in performance based design. Due to the uncertainty of earthquake action, performance evaluation should be put on foundation of probabilistic analysis. In probability based performance evaluation, how to select evaluation criterion is very important issue. The normally used criterion is mean annual occurrence rate of losses $\lambda\,(l \geq L)$, where L is annual losses, which may represent asset or life losses. For economic losses, L usually is expressed by proportion rate in total construction fee. According to assumption of Poisson process, $\lambda\,(l \geq L)$ approximately equals annual occurrence probability when $\lambda\,(l \geq L) < < 1$. $\lambda\,(l \geq L)$ obviously depends on seismic hazard and structural performance. Using total probability formula, $\lambda\,(l \geq L\,)$ can be written as (Cornell 2000)

$$\lambda(l \geq L) = \iint_{I\,D} \lambda(L\,|\,D)f(D\,|\,I)f(I)dDdl \qquad (11)$$

in which $\lambda(L|D)$ is annual occurrence rate of losses L under given damage state D; $f\,(D|I)$ is probability distribution of damage state D under given ground motion input I to structure; $f\,(I)$ is probability distribution of ground motion input. The damage state are divided into intact (no damage), slight damage, moderate damage, heavy damage and collapse. Input ground motion I is a seismic event, which may be presented by ground motion severity, response spectrum even time history if the corresponding mean annual rate of $f\,(I)$ is known. The

performance criterion $\lambda(l \geq L)$ can be used into seismic reliability analysis and optimization of structures. Next step is to incorporate the uncertainties of the performance of structure into the model and making the performance base design more reasonable. Furthermore, beside average annual loses, the standard deviation and propabilistic distribution are also usable in decision making and optimization analysis.

9 PREVIEW AND COMMENTS ON EARTHQUAKE ENGINEERING RESEARCH

The basic task of earthquake engineering research is to protect structures and facilities from damage and impairment at low cost. This task can be solved if the seismic hazard at the site of protected project is accurately predicted. The solutions of the task may comprise two approaches. At localities of earthquake generating belt, catastrophe seismic landslide and slump zone, debris flow route, it is not realizable to protect man made structures. The effective measures are to make structures and facilities elude from these dangerous areas. However, to identify and delineate these dangerous areas is essential prerequisite to avert seismic disasters caused by such kind of ground failure. But up to now only a few earthquake-generated faults could be found out, so that the potential danger would be existable in near future. As for protection from ground shaking there are many approaches including traditional earthquake resis-

71

tance, seismic isolation, and energy dissipation and structural control of seismic response. These techniques will be greatly progressed in next century. It can be anticipated that as the popularizing of earthquake disaster reduction to vast earthquake prone area, life loss during disastrous earthquake events will be reduced gradually but the economic losses and social disorders might be increasing in early part of the new century. So great effort should be put on earthquake protection technique of lifeline and facilities. Besides civil engineers, we should mobilize more architects and mechanical engineers to participate in earthquake engineering research, particularly in aspect of lifeline and facilities. In order to compensate the earthquake economic losses earthquake insurance will play an important role in future. In this aspect, participation of social and economic scientists is irreplaceable.

Generally speaking, seismic hazard prediction is more difficult than engineering solution for ensuring safety of the designed structure at given ground motion. The reliability of hazard analysis depends on having an adequate database of historical earthquake catalog and strong ground motion recordings in concerned region. Unfortunately such kind of database is insufficient in most of areas. For example Tangshan is located in Hebei province, North China, where the available historical records are quite long. If the hazard analysis had been done before this event we could not anticipate getting a hazard as high as Tangshan earthquake. That means the prediction for rare event with small probability is unreliable compared with that of frequently and occasionally occurred event. This situation could not be greatly improved in near future. However the consequence of the rare occurrence event is of catastrophe, if it takes place in a non-preparation area. Hence the technique of identification and zoning of great earthquake is a great task to reduce earthquake disasters particularly in moderate seismicity area. Before it is well solved, to increase the robustness of earthquake resistance structures is a practical approach. That means to design a structure insensible to uncertainties of ground motion. This approach should be a new direction in earthquake engineering research. In addition, following advances will be obtainable in next decade, these are:

1) Three dimensional base isolation of multistory building becomes reality;
2) Methodology of performance based design enters seismic design code;
3) Energy dissipation and semi-active control technique will be popularized in different types of buildings, structures and facilities;
4) Further improvement of the ductility warranted seismic design;

5) The reliability analysis, optimum design and theory of random process employed in real structures, particularly in essential structures.

10 CONCLUSION AND SUGGESTIONS

Great progress in earthquake engineering has been obtained since the China-US joint research project implemented. The Chinese researches have learned many things from American researches and their publications, particularly in following aspects: strong ground motion instrumentation recordings and analysis results, seismic hazard analysis method, geotechnical earthquake engineering, dynamic analyses of seismic response of structures, lifeline earthquake engineering, structural details for improving earthquake resistant capacity and ductility, and earthquake resistance of infrastructures. In other hand the abundant historical earthquake records in China are commonly used to study earthquake activity in time and space domain. The field testing data of liquefaction in China confirms the adequacy of evaluation of liquefaction potential. Chinese and USA researches have commonly improved the seismic performances of pipeline and network during the last two decades. The lessons and experiences from damaged and survived structures during earthquake have become common treasure and to be used to improve seismic design. The basic principles, methods and approaches of seismic design codes both in China and USA is getting close. These issues consolidate the foundation for further expanding bilateral and multilateral cooperations. The aspects listed following are suggestions for further joint researches:

1. In aspect of ground motion and design earthquake, several topics can be selected to, they are:

   a. The long period components in near and farsource sites because of the particular impairment to long period structures such as high-rise building and seismic base isolation structures;
   b. The uncertainties and reliability involved in hazard analysis, the principle and methodology for determining extreme or maximum considered earthquake for different types of structures;
   c. Study of functional relation of statistical characteristics of strong ground motion versus magnitude, distance and site conditions;
   d. The effects of site conditions and embedment on free field ground motion and input motion;
   e. Stochastic models of ground motion and response spectra;

f. The sensitivity analyses of ground motion characteristics to different styles of structures. The results are expected to develop simplified ground input model and seismic design method for different types of structures.

2. Performance based seismic design. In this direction the joint research may focus on embodiment of the design and analysis method, the suggested priority topics are:

a. The probabilistic method of performance evaluation;
b. The performance based seismic design method for typical structures;
c. The performance design criteria and requirement of building facilities including elevators, utilities and communication system;
d. Comparative study of seismic design codes through analysis and design for real structures;
e. Joint experiment study for typical and essential structure.

3. Field investigation and instrumentation observation. In this aspect following joint researches definitely are mutual benefit:

a. Fixed and mobile dense observation array of strong ground motion in high seismicity areas;
b. Post earthquake field reconnaissance and damage mechanism analyses for typical structures, lifeline and facilities;
c. The technique of data processing and interpretation of instrumentation records
d. Policy and technical measures of post earthquake repairing and retrofit.

4. Application of high technologies. The reforming of traditional earthquake resistance technology by high technology has become a new direction of earthquake engineering. In this field there are many opportunities for cooperation such as

a. Health monitoring and diagnosis of essential structures;
b. Remote sensing and its application in earthquake disaster mitigation;
c. Field identification of the dynamic property and real earthquake resistance capacity of structures;
d. Filed measuring technique of material strength and internal impairment of structural component;
e. Development of innovative, high performance and smart materials and intelligent structures;

f. Acquisition and utilization of earthquake energy for structural control and emergency treatment.

5. Seismic interaction among structures, fluids and soils. The seismic interaction problem has long been studied but still confront new challenges. The directions for further study are

a. Seismic stability of subsoils and foundation;
b. Soil structure interaction of building and infrastructure during earthquake;
c. Seismic performance of underground structures and buried pipelines;
d. Dams, bridge piers, harbor walls, petroleum and water storage tanks;
e. Offshore drilling platform and submerged structures.
f. The effect of rocking component of ground and foundation movement on seismic performance of supper tall buildings.

6. Social and economic issues. Destructive earthquakes have impact on societies and human life. These issues include pre-earthquake and post-earthquake awareness, earthquake disaster mitigation planning, emergency earthquake response of household, neighborhood and community, the countermeasures of post earthquake economic dislocation and depression, policy of evacuation, relief and urgent repairing, and education. In order to solve these issues international cooperation no doubt is able to exert important influence.

11 ACKNOWLEDGMENTS

This paper is supported by National Science Foundation of China (NSFC) under grant No.59895410. Dr. Guaoqian Wang and Suling Zhang of Institute of Geology, China Seismological Bureau, Dr. Runlin Yang, Donghw Ma and Deming Zeng of Institute of earthquake engineering, China Academy of Building Research participate in preparation of this paper. Their assistances are greatly appreciated.

REFERENCES

Bao, A. et al. 1985. Probabilistic Calibration of Basic Design Intensities of partial Regions of China, *ACTA Seismological Sinica 7(1)*.

Boore, D.M. 1999. Effect of Baseline Corrections on Response Spectra for Two Recordings of the 1999 Chi-Chi, Taiwan, Earthquake, *Open-File Report* 99-545,Version 1.0, USGS.

Cornell, C.A. 1968. Engineering Seismic Risk Analysis, *BSSA*, 58(5):1583-1606.

Cornell, C.A. 2000. Progress and Challenges in Seismic Performance Assessment, *PEER CENTER NEWS, 3(2)*.

Darragh, R.B. et al 1994. Earthquake Engineering Aspects of Strong Motion Data from Recent California Earthquakes, 5 *US National Conference on Earthquake Engineering Proceedings 3:* 99-108.

Gent A.N. 1964. Electric Stability of Rubber Compression Springs, *Journal of Mechanical Engineering Science 6(4):* 318-326.

Hall , J.F. et al. 1995. Near-source Ground Motion and its Effects on Flexible Buildings, *Earthquake Spectra* 11(4):569-605.

Haringx J.A. 1948-1949. On Highly Compressive Helical Springs and Rubber Rod and Their Applications to Free Mountings-Part I, II and III, *Philips Reports.*

Housner, G. W 1947. Characteristics of Strong Motion of Earthquakes, *BSSA* 37(1): 19-31.

Hu, YX. 1998. Application of Consequence-Based Design Criterion in Regions of Moderate Seismicity, *Workshop on Earthquake Engineering for Regions of Moderate Seismicity,* Hong Kong, PRC. Dec. 6-9.

Hu, YX., Liu, S.C., & Dong, WM. 1996. *Earthquake Engineering,* published by E & Spon, an imprint of Champan & Hall, 2-6 Boundary Row, London SEl 8HK, UK, 26-36.

Hu, Y&Zhou, X. 1999. Development Trend of Earthquake Engineering Toward New Century, *Earthquake Resistant Engineering,* No. l: 3-9(in Chinese).

Hu, Y et al. 1982. *Functional Relations of Bedrock Motion Parameters with Earthquake Magnitude and Distance,* ATCA Seismological Sinica.

Hwang H. et al. 1997. Comparison of American and Chinese Methodologies for Evaluating Seismic Performance of Water Delivery Systems, *Technical Report,* Center for Earthquake Research and Information, The University of Memphis, Memphis, TN.

Hwang H. et al. 1998. Seismic Performance Assessment of Water Delivery Systems, *Journal of Infrastructure Systems, ASCE* 4(3),118-125.

Hwang, H. et al 1999. Comparison of American and Chinese Methodologies for Evaluating Seismic Performance of Water Delivery systems, *Proceedings of the 5th U.S. Conference on Life-line Earthquake Engineering,* Seattle, Washington, August 12-14.

Imbimbo M. & Kelly J.M. 1997a. Stability of Isolators at Large Horizontal Displacement, *Earthquake Spectra* 13(3): 415-430.

Imbimbo M. & Kelly J.M. 1997b. Stability of Aspects of Elastomeric Isolators, *Earthquake Spectra 13(3):431-449*

Kelly J.M. 1996. *Earthquake Resistant Design with Rubber(* 2nd ed), London: Springer-Verlag.

Kennedy, R.P., Chow, A.W & Willwamson R.A. 1977. Fault Movement Effects on Buried Oil Pipeline, Journal of the *Transportation Engineering Division,* ASCE 103(TE5): 617-633.

Lee, W.H.K., Shin, T.C., Kuo, K. W, & Chen, K.C. 1999. CWB Free-Field Strong-Motion Data from the 921 Chi-Chi Earthquake: Volume *1. Digital Acceleration Files on CD-ROM,* Pre-Publication Version December 6, Seismology Center, Central Weather Bureau, Taipei, Taiwan.

Liu, H.X.&Hu Y. 1963. The Development Trend of Earthquake Engineering, *Chinese Science Bulletin 8(7).*

Liu, FIX 1983. Several Aspects of Earthquake Engineering in China, *International Symposium on Earthquake Engineering to Mark the Occasion of Professor K. Kub os Retirement,* Tokyo, Japan.

Liu, H.X. 1985. Some Thoughts on Earthquake Hazard Mitigation in China, *Proceedings of PRC-US Japan trilateral Symposium/Workshop on Engineering for Multiple Natural Hazard Mitigation,* Beijing, China.

McGuire, R.K. 1979. Adequacy of Simple Probability Modeles for Calculating Felt-Shaking Hazard, Using the Chinese Earthquake Catalog, *BSSA,* 69(3): 877-892.

Newmark, N.M. &Hall W.J. 1975. Pipeline Design to Resist Large Fault Displacement, *Proc.of the 1975 U.S. National Conference on Earthquake Engineering,* Ann Arbor, Michigan, pp. 416-425.

O'Rourke, M.J. &Liu X. 1999. Response of Buried Pipelines Subject to Earthquake Effects, *Monograph Series* No.3, MCEER.

Shinozuka, M., Hwang, H., & Murata, M. 1992. *Impact on Water Supply of a Seismically Damaged Water Delivery System in Lifeline Earthquake Engineering in the Central and Eastern U.S.,* ed. Ballantyne, D. B., Monograph No. 5, Technical Council on Lifeline Earthquake Engineering, American Society of Civil Engineers, New York, NY, pp.43-57.

Sun, S.P. 1994. A Review of Buried Lifeline Earthquake Engineering in China, *Proceedings of Second China-Japan-US Trilateral Symposium on Lifeline Earthquake Engineering,* Xi'an.

Tanaka, S. 1996. *Lifeline Seismic Performance Analysis for Water and Electric Power Systems,* A dissertation presented to the faculty of Waseda University for the degree of doctor of Engineering, Waseda University, Tokyo, Japan.

Tarics, A.G. 1994. Composite Seismic Isolator and Method, *International Workshop on Use of Rubber Based Bearing for Earthquake Protection of Building,* Shantou, China, May 17-19.

Toro, G.R. et al. 1992. *Analysis of Non-Stationary of Historical Seismicity in China, Selected Works on Earthquake Engineering,* Seimological Press, Beijing.

Wang, G.Q., Zhou X.Y, Ma Z.J. & Zhang P.Z. 2000a. Comparisons of Acceleration Response Spectra of the 1999 Chi-Chi, Taiwan, Earthquake with Design Response Spectra of the Building Codes, *Technical Report of Institute of Earthquake Engineering,* China Academy of Building Research.

Wang, G.Q., Zhou X.Y, Ma Z.J. & Zhang, P.Z. 2000b. A Preliminary Research on the Stochastic Characteristics of near fault response spectra of 1999 Chi-Chi, Taiwan, Earthquake, *Technical Report of Institute of Earthquake Engineering,* China Academy of Building Research.

Wang, G.Q., Zhou X.Y, Ma Z.J., & Zhang P.Z. 2000c. The Randomness of Near Fault Ground Motion of 1999 Chi-Chi, Taiwan, Earthquake, *Technical Report of Institute of Earthquake Engineering,* China Academy of Building Research.

Wang, L.R.L. 1981. Seismic Evaluation Model for Buried Lifelines, *Life Earthquake Engineering* ASCE.

Wang, L.R.L. et al 1985. *Seismic Damage Behavior of Buried Lifeline Systems During Recent Severe Earthquake in U.S., China and other Countries.*

Wang, L.R.L.&Yeh, YH. 1985. Refined Seismic Analysis and Design of Buried Pipeline for Fault Movement, *Journal of Earthquake Engineering Structural Dynamics, 13:75-96.*

Wang, Y &Li, J. 1989. *Corrected Accelerograms and Response Spectra of Lancang-Gengma Earthquake,* Seismological Press, Beijing.

Zhang, S. 1999. *Seismic Analysis of Buried Pipeline for Fault Movement,* Doctorate Dissertation of Institude of Geophysics, China Seismological Bureau.

Zhou, X. et al. 2000. The Engineering Theory and Computation Method for Designing Isolation System of Rubber Bearing and Its Composite Devices, *12WCEE,* Auckland, New Zealand.

74

Zhou, X. et al 1998. A Practical Computation Method for Horizontal Rigidity Coefficient of Seismic Isolation Rubber bearing, *Building Science,* 14(6):3-8.

Zhou, X. et al. 1999. A Simplified Calculation Method of the Horizontal Rigidity Coefficient for the Serial System of Rubber Bearing with Column, *Journal of Vibration Engineering,* 12(2): 157-165.

Zhou, X.Y. 1965. The Influences of Subsoil Conditions on The Building Endured Seismic Load, *Selected Works on Earthquake Engineering (2),* Institute of Engineering Mechanics, Chinese Academy of Sciences, Scientific Publication House, Beijing (in Chinese).

Zhou, X. et al. 1985. Seismic Hazard Analysis Based on Neyman-Scott Process, *Earthquake Engineering and Engineering Vibration, 5(2): 1-9.*

Zhou, X. Y. 1988. Intensity Attenuation Based on Earthquake-Rupture Model, *Seismological Sinica,* ACTA 1(4):85-95.

Earthquake Engineering Frontiers in the New Millennium, Spencer & Hu (eds),
© 2001 Swets & Zeitlinger, ISBN 90 2651 852 8

Research of earthquake engineering in Hong Kong: current status and future challenge

Y.L. Xu, J.M. Ko, K.T. Chau, S.S. Lam, & Y.L. Wong
The Hong Kong Polytechnic University, Hung Hom, Kowloon, Hong Kong, China

ABSTRACT: Geographically, Hong Kong is not located in a region of frequent destructive earthquakes, and historically there is no provision for seismic design of building structures in Hong Kong. Many buildings in the urban areas of Hong Kong are high-rise and close to each other and use transfer plate/beam systems to achieve maximum open space at ground level despite most of these areas are actually on reclaimed land of soil susceptible to site amplification. There is also little concern that Hong Kong is one of the major financial centers and one of the most congested cities in the world. Any interruptions to critical facilities and business operations will cause serious social and economical consequence. Such a situation has been questioned by many researchers in the recent years in consideration of recent destructive seismic events occurring in the areas of low to moderate seismicity, such as the Newcastle event in Australia and the Kobe event in Japan. Accordingly, research on earthquake engineering is gradually taking place in Hong Kong and national and international collaborations in the field are actively sought. This paper thus presents the historical review, current research status, future challenge, and research collaboration of earthquake engineering in Hong Kong.

1 HISTORICAL REVIEW

Hong Kong is located on the South China Sea Plate. To date, it has a population of about 6.8 millions living mainly in the areas of Hong Kong Island, Kowloon, and part of New Territory. It is beyond doubt that Hong Kong is one of the most congested cities in the world. The nearest active tectonic plate boundary is that of the Philippine Sea Plate and the South China Sea Plate between Taiwan and the Philippines. This active boundary is relatively far away from Hong Kong. Historical records indicate that Hong Kong has not experienced large magnitude earthquakes. However, moderate to large earthquakes did occur in the vicinity of Hong Kong. The magnitude 5.7 earthquake at Dangan Island in 1874, 30 km from Hong Kong, is the largest earthquake observed within 100 km of Hong Kong in the past 500 years. The biggest earthquake of magnitude 7.4 recorded in this century occurred in the Shantou area in 1918 which is about 300 km away from Hong Kong (GCO, 1991). Fortunately, Hong Kong was a small village at that time and no destructive damage was reported.

Probably because of the historical perception that no serious damage has been caused by earthquakes to Hong Kong and the historical fact that Hong Kong was the colony of United Kingdom while United Kingdom is situated in the area of very low seismicity, there is no provision for seismic design of building structures in Hong Kong. Another argument for the reason why there is no seismic design consideration is that the building structures in Hong Kong are designed for very strong winds, which can resist low and moderate earthquake excitation. This policy leads to many buildings constructed in Hong Kong since1970's using transfer plate/beam systems to achieve maximum open space at ground level for parking, shops, restaurants, and hotel lobby without any consideration of the seismic consequence of soft story. Because of shortage in the supply of land in the modern Hong Kong, buildings in the urban areas of Hong Kong are often built to be tall and close to each other. Some of them are actually sitting on reclaimed land of soil susceptible to site amplification. There is also little concern that Hong Kong is one of the major financial centers and the most congested cities in the world and any interruptions to critical facilities and business operations will cause serious social and economical consequence.

Such a situation has been questioned by many researchers in Hong Kong in the past ten years due to two main factors. One is several destructive seismic events recently occurring in the areas of low to moderate seismicity in other parts of the world. For instance, Newcastle, a city of low seismicity in Australia similar to Hong Kong, was attacked by a relatively small earthquake (M=5.6) in 1991 causing about 20 billion Hong Kong dollars of damage and 13 deaths (EEFIT, 1991). In 1995, Kobe, which is classified as a moderate seismic region in Japan, became the first densely populated and well-indu-strialized city to bear the full burnt of a high-mag-nitude earthquake since World War II. The other factor is that Hong Kong has been classified by the State Seismological Bureau of China as an area of earthquake intensity VII (State Seismological Bureau, 1991). According to the Chinese Building Regulations, for a city of earthquake intensity VII in Mainland China, seismic design is required (GBJ11-89). The return of Hong Kong sovereignty to China also intensifies awareness of the considerable difference in design requirement between Hong Kong and Mainland China.

In 1991, Geological Society of Hong Kong held an international conference on seismicity in Eastern Asia to address aspects of seismicity, tectonic geology, seismic hazards and earthquake countermeasures together with Guangdong Seismological Society. The Hong Kong Geotechnical Control Office also published a report entitled "Review of Earthquake Data for the Hong Kong Region" at the same year to provide data on historical earthquakes within 300 km of Hong Kong (GCO, 1991). In 1996, an inter-departmental Working Group was set up by the Buildings Department of Hong Kong to examine the likely effects of earthquakes on buildings in Hong Kong. A one day seminar on earthquake resisting structures was accordingly organized by the Hong Kong Institution of Engineers, Structural Division. More recently, an international workshop on earthquake engineering for regions of moderate seismicity was jointly organized by the three universities in Hong Kong and the Mid-America Earthquake Centre, USA, in 1998. In parallel to these activities, several special projects have been generated and undertaken by different departments in Hong Kong to investigate the seismic problems special to Hong Kong conditions. A number of research projects have been initiated and undertaken in several universities in Hong Kong to have in-depth understanding of various aspects in earthquake engineering. A large simulated earthquake-shaking table has been also set up in The Hong Kong Polytechnic University. The following is only a summary and an overview of earthquake engineering studies in Hong Kong. Although work on earthquake engineering in Hong Kong is relatively recent, the departments and researchers involved spread so widely that it is beyond our capability in a limited time to review all the relevant work in Hong Kong and to cover all relevant information in this paper.

2 CURRENT RESEARCH STATUS

2.1 Seismic hazard analysis of Hong Kong region

Since early nineties, a few researches have been carried out on seismic hazard analysis of Hong Kong

Region using the probability risk assessment methodology.

Pun & Ambraseys (1992) reviewed the earthquake records of 1000 years for a region within about 350 km of Hong Kong and obtained a b-value of 0.75 in the Gutenberg-Richter magnitude-frequency recurrence relationship given by $\log_{10} N(M_s) = a - bM_s$. Then, by using the Cornell's hazard analysis method and adopting Joyner and Boore's attenuation law (Boore & Joyner, 1991), they derived a set of probability of exceedance of peak ground accelerations (PGA) at Hong Kong for a design life of 100 years. Their results indicated that the probability of Hong Kong being subjected to strong ground shaking was quite small. However, they also reported that there was uncertainty in their prediction which came from the appropriateness of the attenuation relationship.

Scott et al. (1994) used the earthquake records for a region about 500 km of Hong Kong to determine the magnitude-frequency recurrence relationship and thereafter carried out hazard analysis using two attenuation models. The two attenuation models were based on the relationship published by Dahle et al. (1990) specially developed for bedrock sites in regions remote from plate boundaries and that by Boore and Joyner (1991) which was developed for deep soil units in the eastern part of the United States of low and moderate seismicity. They concluded that Hong Kong had a moderate seismic hazard, and for a return period of 475 years the design PGA of 0.12g at bedrock should be considered in Hong Kong. Scott et al (1994) also derived a bedrock response spectrum for Hong Kong by calculating response spectral values at a range of fundamental periods. At each fundamental period the spectral value had the same probability of being exceeded.

In 1995, the Geotechnical Engineering Office of Hong Kong (former the Hong Kong Geotechnical Control Office), the University of Hong Kong and Guangdong Seismological Bureau (GSB) examined the seismic hazard of the Hong Kong region (Lee, et al. 1998). The project team conducted a series of seismotectonic investigations in the field. The earthquake records obtained by the Royal Observatory of Hong Kong were re-evaluated, and these records, together with pertinent data from the GSB seismic monitoring network, were used to established an updated earthquake catalogue for the Hong Kong region (bounded by Latitude 22° - 23° North and Longitude 113°45' - 114°45') since 1972, for seismic events with $M_L \geq 1.8$. The analysis also considered the potential source zones and their seismological parameters. Seismic ground motion attenuation models applicable to the region were studied and compared. The peak ground acceleration and seismic intensity values were computed and analyzed for various probabilities of exceedance and return periods. The main conclusions of the study were that the peak acceleration on bedrock was in the range of 75-115 gal for the Hong Kong region, corresponding to a probability of exceedance of 10% over a return period of 50 years, and the seismic intensity of the region was rated as VII. The findings are consistent with those indicated on the seismic zoning map published by the Chinese Seismological Bureau (State Seismological Bureau, 1991).

Recently, we re-examined the completeness of earthquake records and the regional variation of the seismicity in Hong Kong (Wong et al, 1998). In our work, a reasonably large area (Latitude 18° - 25.5° North and Longitude 108° - 119° East) was considered to have better statistics. The earthquake records in this area were divided into two groups: the events between 1445-1894 and those between 1895-1995, with the minimum surface-wave magnitude of 4 ¾. The magnitudes of most events since 1895 were determined from instrumental records from worldwide sources. The magnitudes of the events between 1445-1894 were determined from the historical felt intensities by Chinese seismologists. Based on this most updated earthquake catalogue, we found that the seismicity is not evenly distributed in time and region. The regional variation of the seismicity will be a significant factor likely to affect the seismic hazard level in Hong Kong. It is recommended that the whole area considered should be divided into three major source regions. The seismicity parameters for each region are provided in Figure 1. We also found that the historical records on land since 1895 are likely to be complete for magnitude 4¾ or greater, but the number of earthquakes with magnitude 6 or greater per year in the last 100 years is much larger than that in the previous historical time (1445-1894). As a result, the models developed by us predict a considerably higher level of seismicity than those by others.

Based on the aforesaid researches, there is no doubt to conclude that the seismicity of Hong Kong is situated in a region of moderate seismicity. However, in order to come up with a widely accepted value of PGA for a given return period, further work to clarify the controversy on the attenuation models and the spatial weight of magnitude distribution in a potential source should be sought.

2.2 Effects of soil amplification in Hong Kong

Among the engineering and science community worldwide, it has been recognized that seismic ground motions on soft soil are usually much larger than a nearby rock site at a broad frequency range in which the fundamental natural frequencies of most of civil engineering structures lie in.

In early nineties, the period where dynamic properties of soil in Hong Kong were not fully understood, there was a technical note (GEO, 1991) on liquefaction of marine reclamation fills. It suggested that fill materials at the Kai Tak Airport sites were vulnerable to liquefaction when subjected to strong ground motions but extensive occurrence was very unlikely. Later on, Lee et al. (1998) examined the liquefaction potential of hydraulic sand-fills for land reclamation works in Hong Kong. Laboratory cyclic triaxial tests on local hydraulic sand were conducted to define model parameters. Site response analysis was then carried out on one site. It was concluded that liquefaction at the site was not likely to occur under the scaled-down earthquake with the maximum horizontal ground surface acceleration at 0.07g.

In 1996, an inter-departmental Working Group was set up by the Buildings Department to examine the likely effects of earthquakes on buildings in Hong Kong. The Geotechnical Engineering Office was requested by the Working Group to carry out a study on soil amplification effects (GEO, 1997). Six sites covering different types of commonly encountered ground conditions in Hong Kong were selected for the study. The fundamental periods of the sites ranged from 0.2 s to 0.8s. A peak ground accelera-

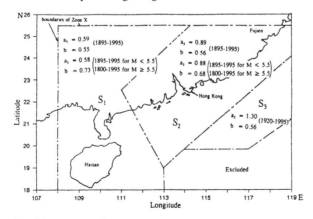

Figure 1. Seismicity models of three source regions

tion of 10% g at bedrock level was adopted from Scott et al. (1994). The one-dimensional site response analyses showed that a spectral ratio (defined as the ratio of spectral acceleration in soil to spectral acceleration in rock at the corresponding period) of up to 2.4 is possible for the selected sites at 5% of critical damping, but the ratio is generally less than 2.0 for the wide range of periods. However, they also pointed out that as local data on dynamic soil parameters were scare, the effects of soil amplification might be larger than those predicted in the study. It should be also noted that none of the sites in their study had a long fundamental period.

Many parts of the urban areas in Hong Kong are on reclaimed land which consists of soils with shear wave velocities varying between 150 and 400 m/s. The total depth of soil layers could exceed 50m. Moreover, a layer of soft marine deposits is often found below a general (sandy) fill. It is anticipated that a site with such special soil stratum might have seismic response different from that described in the recognized international/national seismic codes. Therefore, a series of field measurements, laboratory tests, and numerical simulation were conducted at The Hong Kong Polytechnic University and the major results are as follows.

2.2.1 The applicability of the Nakamura method to estimate natural frequencies of soil sites in Hong Kong

Ambient vibration measurement technique has long been used in estimating natural frequencies and damping ratios for buildings and bridges, and has also been used to estimate site natural frequency and amplification. In the past, it was considered to be more difficult to use ambient vibration measurements for soil sites than for buildings and bridges. There were too many factors influencing the characteristics of the ambient noises of a soil site apart from the soil dynamic properties. Recently, it was found that based on records of ambient vibrations from a single site, the site dynamic characteristics could be detected. This method was first proposed by Nakamura (Nakamura, 1989), and has been widely referred to as the Nakamura method.

Ambient vibration and in-situ shear wave velocity measurements were conducted in different sites in Hong Kong by Wong et al. (1998). The natural frequencies of sites determined from the Nakamura method were compared with those calculated from shear wave velocities based on simple theoretical models for wave propagation in soil layers (Zhao, 1996, 1997). It was found that that the Nakamura method worked well for soil sites in Hong Kong. The peak frequency of the Nakamura's ratio correlated extremely well with the site fundamental modal frequency estimated from soil shear wave velocities derived by a vertical seismic profiling method. The peak frequency also correlated very well with the site fundamental modal frequency estimated from borehole SPT data, within a factor of 1.1 for most of sites. Figure 2 shows the mean Nakamura's ratio S_m of a site with the first modal frequency at 1.7 Hz.

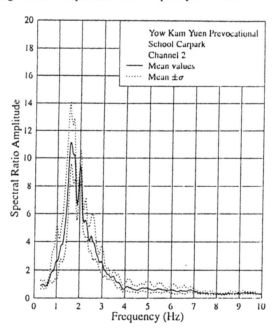

Figure 2. Mean Nakamura's ratios of 6 measurements at Yow Kam Yuen Prevocational School Carpark at Sha Tin

2.2.2 Relationship of SPT-N values and shear wave velocities

In Hong Kong, there is a rich archive of records on SPT-N values, but the field data of shear wave velocity of soil are very limited. Wong et al. 2000 measured shear wave velocities of different soil types in several sites and also collected data from one site where shear wave velocity was measured by using a suspension PS logging system (Kwong, 1998). Based on these test data, we derived a correlation relationship by least squares analysis. Wong et al. also compared the measured shear wave velocities with those predicted by Imai et al. (1982). The two correlation relationships predicted the measured shear wave velocities from SPT N-values reasonably well. The Imai's relationship was derived from a much larger data set. However, the correlation relationship derived in the Wong's study is the preferred one for Hong Kong soils, because the equipment and procedure for carrying out SPT in Japan are likely to be different from those in Hong Kong. The correlation relationship proposed by Wong et al. is suitable for marine deposits, alluvium and completely or highly decomposed rock.

2.2.3 Dynamic soil testing

Local dynamic soil testing data on shear modulus reduction curves and equivalent damping ratios are basically unavailable. These data are required for full nonlinear analysis of soil layers subject to earthquake excitations. In this respect, we carried out cyclic triaxial tests and resonant torsional column tests according to ASTM Standards (Wong at el., 2000). We derived a set of curves for alluvium sand/clay and marine deposits and then compared with the published soil models derived from a very large amount of tests on overseas soils but generically similar to the soils they tested. It is recommended

that for the marine deposits, the shear modulus reduction curve proposed by Stokoe and Lodde (1978) for the San Francisco Bay mud can be used. For the alluvium clay tested, we recommend the C2 model by Sun et al. (1988). For the alluvium clayey/silty sand tested, the modulus reduction curves can be possibly modelled by the S1 model of Seed et al. (1984).

2.2.4 Site classification and influence of soft sandwich layer on seismic response of soil sites in Hong Kong

Based on the test data and representative soil profiles of Hong Kong, we further carried out nonlinear seismic response analysis using 128 bedrock excitation of 21 earthquakes in China/overseas as input data (Wong et al 2000). We found that for reclaimed sites with a soft soil (marine deposits) layer below a general (sandy) fill, simple site classification by international/national codes (such as Chinese Seismic Code, HEHRP of US and Eurcode 8) will underestimate seismic motions at long period (Guo et al. 2000). Substantial soil amplification effects will extend to longer period (1.2 to 1.5 second), and this has not been accounted for in the Chinese seismic codes. Consequently, normalized design response spectrum for soil sites with and without a sandwich layer of soft soil has been proposed .

$$\beta = 1 + (\beta_{max} - 1)(T/T_1) \quad 0 < T < T_1$$
$$= \beta_{max} \quad T_1 < T < T_2$$
$$= \beta_{max} (T_2/T)^\gamma \quad T > T_2$$

It is suggested that for a soil site with a sandwich layer of more than 5m thick soft soil, the values of β_{max}, T_1, T_2 and γ are taken as 2.1, 0.1 s, 1.4 s and 1.0 respectively, where β_{max} is the maximum dynamic amplification coefficient. A comparison of the design response spectra with that from GBJ 89-11 is shown in Figure 3.

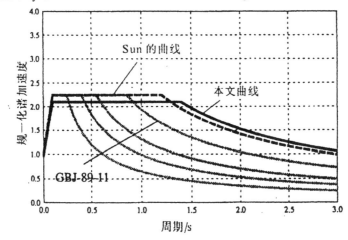

Figure 3. Comparison of suggested design spectra with that from Chinese Seismic Code

2.3 *Earthquake resistant of buildings in Hong Kong*

Studies by Lee et al. (1996) have suggested that the ground motion in Hong Kong associated with a 10% probability of exceedance in 50 years would have a peak ground acceleration around 0.1g. Seismicity level in Hong Kong could be similar to that in New York, having a seismic zoning of Zone 2A according to the American Code UBC-97. The Earthquake Intensity Zoning Map of China also indicates that Hong Kong is a region of Seismic Intensity 7, i.e. moderate seismicity. However, the seismic hazard of Hong Kong has traditionally been considered to be low, and structures can be designed with no seismic provisions. With only a few exceptions, the Mass Transit Railway Corporation and the Over Arup and Partners Hong Kong Limited for instance have issued internal guidelines to provide some consideration against seismic action.

A considerable proportion of the building stock in Hong Kong is over thirty stories. Majority of these high-rise buildings are (a) reinforced concrete structures; (b) composed of coupled shear walls and frames; (c) having transfer plates or transfer beams at a lower story to facilitate a change in the structural form; and (d) asymmetric on plan. In particular, the use of transfer beams violates the principle of "strong column weak beam" or the concept of Capacity Design, whereas a transfer plate introduces a soft story. Such structural forms are generally vulnerable to seismic action. Up to now, reinforced concrete buildings in Hong Kong are designed to either the local code of practice (based on the load factor method) or BS8110:1985 (based on the limit states concept) with no emphasis on seismic provisions. Columns are heavily reinforced with 4-5% main reinforcements. They are poorly confined with main reinforcements alternatively tied by links having 90 degrees hooks and with large spacing at 300mm. Figure 4 shows examples of some of the typical column detail. Furthermore the spacing of links when required to resist shear are not to be less than 8 times the diameter of the link, whereas seismic design codes generally limit the maximum spacing of links. Among others, Chan et. al. (1996, 1998) suggested some possible reinforcement details to be implemented in Hong Kong for seismic resistance, based on a review on the seismic provisions of the New Zealand and the American concrete design codes. Chandler et. al. (1998) recommended a revision of existing design practice. In December 1998, the Hong Kong Polytechnic University, the Hong Kong University of Science and Technology, the University of Hong Kong and the Mid-America Earthquake Center in USA have jointly organized an International Workshop on Earthquake Engineering for Regions of Moderate Seismicity in Hong Kong

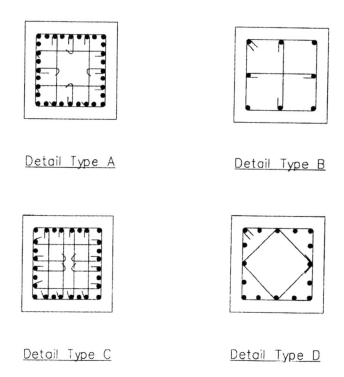

Figure 4. Typical reinforcement detail of columns in Hong Kong

(Lam and Ko 1999). It has been recommended that an acceptable approach for resistance against moderate earthquake should allow the use of "simple detailing rules to achieve limited ductility", and there should be "control of displacement and inter-story drift to minimize damages to non-structural components". There is a need to predict the response of the buildings in Hong Kong when subjected to moderate seismic action, especially for low-rise buildings where the seismic action is significantly greater than the design wind force, and for high-rise buildings incorporating considerable transfer system, e.g. a transfer plate, Scott et. al. (1994).

In what follows, we would like to limit our discussion on those research studies conducted primarily by the research team in The Hong Kong Polytechnic University. The ultimate objectives of our studies are (a) to evaluate the seismic hazards of buildings in Hong Kong and (b) to develop strategies for seismic rehabilitation. Firstly, experimental studies are conducted on key structural components including columns, lintel beams and transfer plates detailed to local practice. This will help us to quantify their seismic resistance as well as means to describe the hysteresis behavior. The studies will be integrated to provide a complete description of the seismic resistance of buildings in Hong Kong and to develop appropriate means of strengthening existing buildings.

2.3.1 Studies related to structural components

In a joint research project with the Harbin University of Civil Engineering and Architecture, 6 column specimens in 1:3 scale were tested under cyclic loading in August 2000. Figure 5 shows setup of the column tests (Lam et al. 2001). The columns were characterized by properties typical to Hong Kong including high axial load ratio (n=0.4-0.65) and low lateral confinement ratio (ρ_s=0.009-0.03). Figure 6 shows the typical modes of failure. The test data clearly indicate that the dependable displacement ductility capacity of the columns was in general poor and inadequate.

In another review study, 40 column test data were selected from the numerous cyclic lateral load tests conducted worldwide. Together with the 6 new test data, an empirical formula was suggested to estimate the ultimate drift ratio of poorly confined rectangular reinforced concrete column (Lam, 2000). Here, the ultimate drift ratio θ is defined as the ratio of the ultimate lateral deflection (corresponding to a 20% drop in the lateral load capacity) to the clear height of column. Figure 7 compares the required lateral confinement ratio against axial load ratio n based on ACT32 (1996), Wehbe (1998) and the empirical formula for a 500mm square column with normal strength concrete having 2.43% main reinforcement and ρ_s=0.002. The target ultimate drift ratios and ductility ratios are indicated in the brackets. If limited ductility is permitted (say θ=1/50), the required lateral confinement ratio could be lesser than that specified in ATC-32, and slightly over the estimates by Wehbe (1998) with target ductility equals to 2. More column tests are in preparation to refine the empirical formula so as to provide reasonable prediction suitable for design.

In September 1999, a study related to the seismic resistance of transfer plate has started. The objective is to examine the seismic response of shear wall - transfer plate – column assemblages using 1:4 scale

Figure 5. Cyclic lateral load test of a column specimen

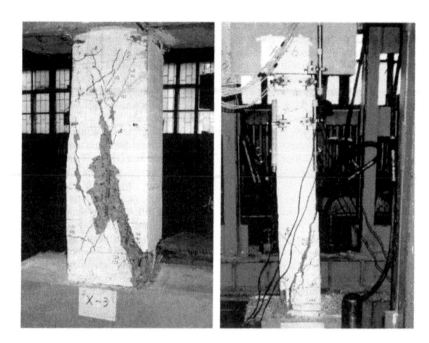

Figure 6. Typical modes of failure of column specimens

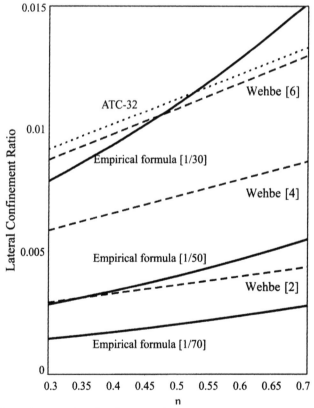

Figure 7. Comparisons of the requirement on lateral confinement ratio

experimental models. A pseudo-dynamic testing system developed based on the MTS testing equipment and modified to incorporate the capacity of sub-structuring has been completed. Verification tests on a simple cantilever as shown in Figure 8 have been reported by Chen and Lam (2000). Testing of the 1:4 scale transfer plate will be conducted in Spring 2001. Figure 9 shows set-up of the test specimen, which is essentially a shear wall - transfer plate – column assemblage. Sub-structuring technique will be implemented in the pseudo-dynamic system to enable the high-rise building to be represented by the test specimen with only two degrees-of-freedom.

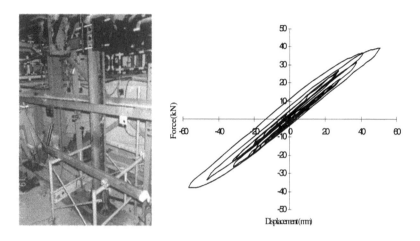

Figure 8. Verification tests of a cantilever and inelastic hysteresis response

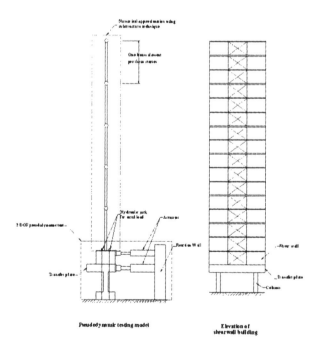

Figure 9. Setup of the pseudo-dynamic tests with sub-structuring

2.3.2 *Studies on high-rise buildings*

Response of a 7-story reinforced concrete building when subjected to earthquake action was first examined. The building was completed probably in the seventies with no seismic provisions. It is 20m high with reasonably regular vertical configuration. Structural system comprises reinforced concrete frames and traditional beam-slab floors. Deformations of the building under small and large earthquakes were estimated according to GBJ11-89. The predictions indicate that the building could be vulnerable to seismic action. In particular, inter-story drift capacities of the columns are inadequate. The use of high-yield deformed bars as transverse reinforcement and an increase in the lateral confinement ratio are recommended. Consideration has also been given to the effect due to two different soil conditions. There is significant influence of site condition on the seismic response of the building.

A research project conducted jointly by the Hong Kong Polytechnic University and the Institute of Engineering Mechanics, State Seismology Bureau in Harbin ("IEM") has been put into action with the intention to understand the seismic response of a high-rise building in Hong Kong. The building to be studied is a typical residential block with over 40 stories, and incorporating a transfer plate at the podium. Structural system comprises of a centre cores (with heavily reinforced lintel beams) throughout the elevation. Shear walls are the only vertical elements at typical floors. Below the transfer plate, the shear walls are replaced (and in fact supported by) a few heavily reinforced columns. As mentioned earlier, such structural form is typical in the high-rise buildings in Hong Kong. A 1:20 scale micro-concrete building specimen will be constructed and tested on the 25 ton, 5m x 5m shaking table in the IEM. Main objectives of the shaking table test are (a) to understand the seismic response and modes of failure of the building, and (b) to develop numerical procedures to predict the seismic response of similar buildings. Specific attention will be focused on possible damage on columns, transfer plate and lintel beams. Physical model of the building was completed in September 2000, and the shaking table tests will be conducted in January 2001. Two numerical models have been generated, namely finite element model and simplified model based on the rigid diaphragm assumption as shown in Figure 10. Inelastic response of the building will be predicted by integrating the various inelastic mathematical models developed in the course of the studies on structural components, see Section 2.3.1.

In order to maximize the use of land, buildings in Hong Kong are characterized by various degrees of irregular on plan. Structural layouts are asymmetric

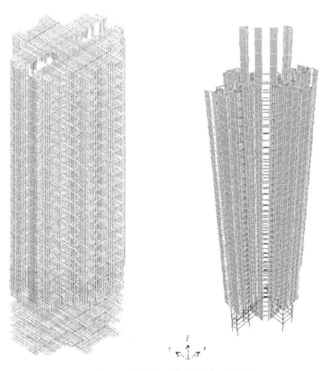

Figure 10. Finite element model (left) and simplified model (right) of a high-rise building

in the presence of appreciable torsion. As a result, various 1:3 scale one-story reinforced concrete specimens have been constructed in normal strength concrete and tested in our 3m×3m shaking table. An example is shown in Figure 11. Structural configuration and reinforcement detail of the specimens follows the local practice. To quantify the scale effect and the related similitude requirements, acrylic plastics were also used to model the one-story reinforced concrete building in smaller scales, namely 1:20 and 1:40. An example is shown in Figure 12. The contribution of torsion on the performance of a reinforced concrete frame-wall system becomes increasingly important when the structure is in a non-linear state

Figure 11. Shaking table test of a 1:3 asymmetric reinforced concrete building

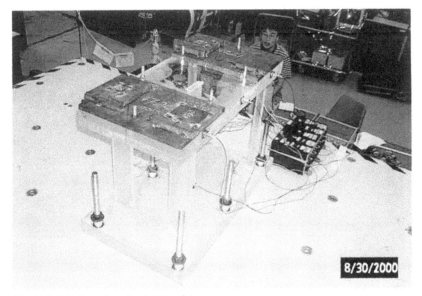

Figure 12. Shaking table test of a 1:20 acrylic plastic model

with bending cracks in columns and shear cracks in walls. Based on the assumption of rigid diaphragm and displacement-based design approach, a simplified mathematical model has been developed to predict the effect of torsion on the seismic resistance of structural components. The study will be used to predict the deformation capacity and damage on multi-story buildings when subjected to seismic action. Further to this study, all the defective concrete was repaired by epoxy grouting or external wrapping with fibre-reinforced polymer composites as shown in Figures 13 and 14 respectively. Shaking table tests were repeated after retrofitting. It was observed from the tests that both the epoxy grouting and carbon fibre wrapping are effective means of enhancing the seismic resistance of reinforced concrete columns and walls.

Figure 13. Retrofitting by epoxy grouting

Figure 14. Retrofitting by wrapping with fibre-reinforced polymer composites

2.4 *Pounding of adjacent buildings*

Pounding between adjacent structures or between parts of the same structure during major earthquakes has often been reported, and it has also been identified as one of the main causes for structural damages or for complete collapse of structures (Davis 1992, Chau & Wei 2000). For example, poundings between structures have been observed in Alaska Earthquake of 1964, San Fernando Earthquake of 1971, Mexico City Earthquake of 1985, Loma Prieta Earthquake of 1989 (Kasai & Maison 1997), Kobe Earthquake of 1995 and Taiwan Chi-Chi Earthquake of 1999. Both high-rise and low-rise inadequately separated adjacent structures are susceptible to damages induced by poundings. Therefore, pounding between adjacent buildings poses a serious seismic hazard. Hong Kong is a densely populated modern city with most buildings built closely to each other and on reclaimed land. The separation distances between adjacent buildings in some cases are very small, controlled by static wind load only without considering any seismic design requirements.

To date, the separation between adjacent buildings has been required in more recent building-code provisions (e.g. USA's UBC 1993 and NEHRP 1991, Canada's NBCC 1990, China's GBJ11-89). However, the required separation distance is still estimated based upon very primitive method, such as the "spectral difference" (SPD) method (Kasai et al. 1996) and the "complete-quadratic-combination" (CQC) method (Penzien, 1997, Hahn and Valenti,

1997). The existing building-code provisions for seismic pounding are in general too conservative, and the dynamics of pounding between adjacent buildings has not been studied comprehensively.

Pounding between adjacent structures is a highly complex phenomenon, and its accurate modeling requires great details of information of the structures in conjunction with a very reliable analytical method. In general, inelastic and plastic deformation, local crushing as well as impact-induced-fracturing may occur at the contacts during pounding. The coupled natural period of the system may only be half of that of the single stand-alone structure (Davis, 1992).

In most of the previous studies, a dashpot-spring model with either a linear stiffness or a bilinear hardening stiffness was used to model the pounding force between buildings. However, actual pounding forces between structures or between various components of a structure are highly nonlinear. In order to model the nonlinear pounding more realistically, the Hertzian impact model has been adopted by various authors (Davis 1992). The main restriction of Davis' (1992) work is that only pounding of a single-degree-of-freedom oscillator on a stationary barrier or on a barrier moving with "locked-to-ground-motion" is considered. More recently, Chau and Wei (2000) extended the analysis by Davis (1992) to model two adjacent structures as two single-degree-of-freedom oscillators and, at the same time, incorporates the Hertzian impact (Fig. 15).

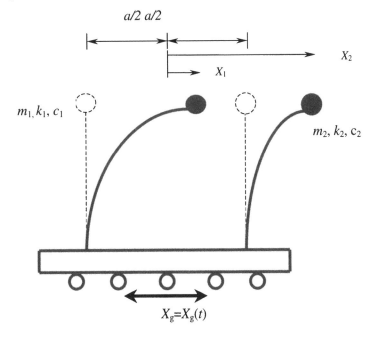

Figure 15. Pounding model of adjacent buildings

Without going into the detail, we summarize the main finding of our research here. The full detail is referred to Chau and Wei (2000). In particular, for the case of rigid impacts, a special case of our analytical solution has been given by Davis (1992) for an oscillator pounding on a stationary barrier. Our analytical predictions for rigid impacts agree qualitatively with our numerical simulations for non-rigid impacts. When the difference in natural periods between the two oscillators increases, the impact velocity also increases drastically. The impact velocity spectrum is, however, relatively insensitive to the standoff distance. The maximum relative impact velocity of the coupled system can occur at an excitation period T_n^* which is either between those of the two oscillators or less than both of them, depending on the ratios T_1/T_2 and ζ_1/ζ_2 (where T_1 and T_2, and ζ_1 and ζ_2 are the natural periods and damping ratios of the two oscillators respectively). Although the pounding force between two oscillators has been primarily modeled by the Hertz contact law, parametric studies show that the maximum relative impact velocity is not very sensitive to changes in the contact parameters.

2.5 Structural control technology

The traditional approach to seismic design of building structures has been based upon providing a combination of strength and ductility to resist earthquake loads. Significant progress has been made in the past twenty years to use structural control technology for retrofitting existing structures and enhancing the performance of prospective new structures against earthquakes (Soong & Dargush 1997). Structural control technology means regulating the structural characteristics of mass, damping, and stiffness using either passive energy dissipation devices or active and semi-active control systems. Some types of structural control have performed successfully in past earthquakes and are now being employed in China, Japan, the United States, and other countries (Housner, et. al., 1997).

Hong Kong is a densely populated modern city with a large inventory of spectacular buildings and structures clustered in a relatively small place. Most buildings in Hong Kong, however, are not designed for earthquakes and are often built closely to each other because of short supply in land and high demand for centralized service. These buildings, in most cases, are separated without any structural connection or are connected only at the ground level. Hence, earthquake resistant capacity of each building mainly depends upon itself. If the separation distances between adjacent buildings are not sufficient, mutual pounding may occur during an earthquake. A systematic investigation is thus being carried out in the Department of Civil and Structural Engineering of the Hong Kong Polytechnic University to investigate the possibility of using either passive energy dissipation devices or active and semi-active force generating devices to link adjacent buildings together for control of their earthquake responses at the same time.

2.5.1 Passive control

The first phase of the investigation is to numerically assess the effectiveness of using passive energy dissipation devices to link adjacent buildings in reducing their seismic response (Fig.16). Passive devices investigated include viscoelastic dampers and fluid dampers. The viscoelastic materials in a viscoelastic damper are typically made of polymers or glassy substances. The damping results from energy dissipation due to the shear deformation of materials. Actual viscoelastic material behavior is very complex, including frequency, amplitude, and temperature dependent properties (Soong & Dargush, 1997). A Voigt model (Sun & Lu, 1995) is employed to represent viscoelastic dampers in this investigation. The Voigt model is a combination of a linear and elastic spring and a viscous dashpot connected in parallel. Fluid dampers that operate on the principle of fluid flow through specially shaped orifices have found more and more applications to seismic mitigation of building structures. Fluid dampers are of primary interest in structural applications because they have several inherent and significant advantages. The simplest model of a fluid damper is the Maxwell model, which is a combination of a linear and elastic spring and a viscous dashpot connected in series, as a result of investigation carried out by Constantinou and Symans (1993).

Since the adjacent buildings connected by either viscoelastic dampers or fluid dampers are no longer a proportionally-damped system and the participant natural frequencies of the system may not be sparsely spaced, the traditional SRSS method may not be applicable and the traditional CQC method may be very time consuming. By considering that adjacent buildings are subjected to only small or moderate seismic events, a generalized pseudo-excitation method has been developed in the frequency domain in this study. The generalized pseudo-excitation method can naturally retain all cross-correlation terms between closely spaced modes of vibration in building response and accurately handle the non-proportional structural damping due to the installation of damping devices. The method also provides a convenient way of determining internal force response of a building. The details of the principle and algorithm of the generalized pseudo-excitation method can be found in Zhang & Xu (1999, 2000).

Two 20-story shear buildings having the same floor elevations but different natural frequencies were used to numerically assess the effectiveness of passive energy dissipation devices. The Kanai-Tajimi filtered white noise spectrum was adopted as

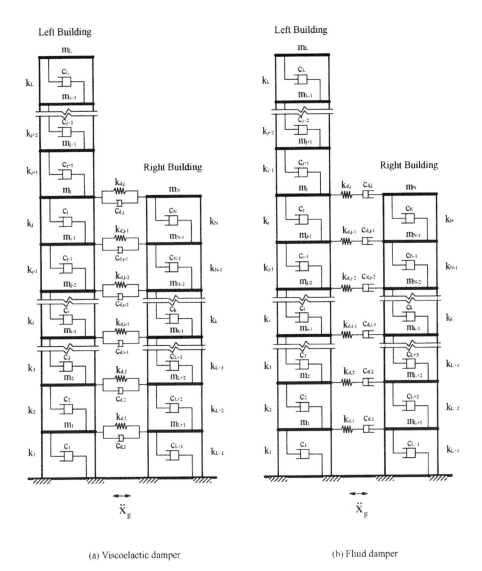

(a) Viscoelactic damper   (b) Fluid damper

Figure 16. Modelling of adjacent building linked by joint dampers

the ground acceleration spectrum in the computation. The numerical results showed that if viscoelastic damper parameters were selected appropriately, the modal frequencies of the unlinked buildings could be retained and the modal damping ratios of the system could be significantly increased. The retention of the natural frequencies of the unlinked buildings after the installation of the joint dampers is especially desirable for the existing adjacent buildings required to be strengthened. Using the viscoelastic dampers of beneficial parameters to connect the two buildings, the top floor displacement and acceleration responses and the base shear force response were reduced by more than 50% for

both buildings (Fig.17). The computation results also showed that the Maxwell model-defined fluid dampers had almost the same effectiveness as the viscoelastic dampers under the conditions that the same adjacent buildings and ground motion are considered and the parameters of each type of damper are beneficial (Zhang & Xu, 1999, 2000).

2.5.2 *Active control*
The second phase of the investigation is to numerically estimate the effectiveness of using active control devices to link adjacent buildings. A number of real implementations of active structural vibration control have been realized in many places (Spencer

91

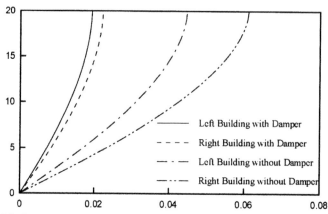

Figure 17. Variations of displacement response of adjacent buildings with height

& Sain, 1997). Correspondingly, many control algorithms have been proposed specially for civil engineering structures, such as instantaneous optimal control, sliding mode control, linear quadratic Gaussian (LQG) control, and its frequency domain analog H2 control. By using LQG controllers or H2 controllers, the estimation and control design processes can be fully separated with the design of the estimator independent of feedback to the structures. Thus, much of the stability and performance of LQG controllers can be achieved despite partial state information and noise corruption. This feature and its simplicity account for the popularity of LQG and H2 controllers over other controllers. Accordingly, LQG controllers are adopted in this investigation.

Since this investigation concerns the application of LQG controllers to connect two buildings that involve many degrees of freedom, the conventional approaches need great computational efforts to find beneficial parameters of LQG controllers and determine statistical responses of both buildings. Thus, this investigation found a general yet simple closed-form solution for the concerned problem. The derived closed from solution was then used to perform parametric studies and to assess the effectiveness of LQG controllers in reducing seismic responses of both buildings. The derivation of the closed form solution is naturally fulfilled by combining the generalized pseudo-excitation method and the residue theorem.

For application and comparison, the LQG controllers were used to link the two 20-story shear buildings discussed in the passive control. By using the closed-form solution, extensive parametric studies were carried out and the beneficial parameters of LQG controllers for achieving the maximum response reduction of both buildings were identified. The numerical results showed that if weighting matrices were selected appropriately, the modal damping ratios of the system could be significantly increased and the earthquake-induced dynamic responses of both buildings could be considerably reduced. It was also found that though the active control could perform slightly better than the passive control, its effectiveness was much dependent on the weighting matrices.

2.5.3 Semi-active control

The passive devices are relatively simple and easy to be complemented but the effectiveness of passive devices is limited due to the passive nature of devices and the random nature of earthquake events. Active devices can more effectively reduce seismic structural response than passive devices because feedback and/or feedforward control systems are used. However, the complicated control system and the large power requirement during strong earthquake hamper their implementation in practice. Therefore, different types of semi-active devices have been recently developed to equip passive control devices with actively controlled parameters forming a semi-active yet stable and low power required damping systems. Among them, MR dampers and ER dampers are two typical types of smart (semi-active) dampers under active research.

MR dampers typically consist of a hydraulic cylinder containing micron-sized magnetically polarizable particles suspended within a fluid. In the presence of strong magnetic field, the particles polarize and offer an increased resistance to flow. By varying the magnetic field, the mechanical behavior of a MR damper can be modulated. Since MR fluids can be changed from a viscous fluid to a yielding solid within milliseconds and the resulting damping force can be considerably large with a low power requirement, MR dampers are applicable to large civil engineering structures. ER dampers are essentially electric analogs of MR dampers. ER fluid contains micro-sized dielectric particles and its behavior can be controlled by subjecting the fluid into an electric field. For seismic response control, MR dampers were investigated by Dyke et al. (1996), Carlson & Spencer (1996), Spencer et al. (1997), and others while ER dampers for seismic response control were studied by Ehrgott and Masri (1992), Gavin et al. (1996a,b), Makris et al (1996), and others. With respect to control strategies of structures with MR dampers, Dyke et al. (1996) proposed a clipped-optimal force control algorithm with acceleration feedback. More recently, Ribakov and Gluck (1999) used an optimal linear passive control strategy to determine the viscous constant of the ER damper and then use active control strategy to determine control forces.

The third phase of this investigation is to use MR dampers or ER dampers link a 5 story skirt building to a 20 story main tall building, aiming to suppress the harmful whip action of the main building rigidly connected by the skirt building under earthquake excitation. This is a joint research work between Wuhan Industrial University and The Hong Kong Polytechnic University. The force-displacement relationship of a MR damper or an ER damper based on a parallel-plate model was extended to include the stiffness of brace which supports the smart damper. The optimal displacement control strategy including the controller-structure interaction was proposed for this special application. An extensive parameter study was performed in terms of the maximum yield shear stress and the Newtonian viscosity of MR or ER fluids, the brace stiffness, and the earthquake intensity. The numerical results showed that when the skirt building was rigidly connected to the main building, the seismic responses of the main building above the skirt building had a suddenly large increase due to the sudden change of stiffness, that is, the so-called whip action. When the 25 smart dampers of proper parameters were used to connect the skirt building to the main building, the whip action of the main building was eliminated. The response of the main building became more uniform along the height of the building. At the same time, the response of the skirt building was also reduced (Fig.18).

2.5.4 Experimental studies

The fourth phase of this investigation is the experimental studies, divided into two stages. In the first stage, the harmonic response of adjacent buildings connected by fluid damper was experimentally investigated using model buildings and fluid damper (Fig.19). Two building models were constructed as two three-story shear buildings of different natural frequencies. Model fluid dampers connecting the two buildings were designed as linear viscous dampers of which damping coefficient could be adjusted. The two buildings without fluid dampers connected were tested first to obtain their individual dynamic characteristics and responses to harmonic excitation. The tests were then carried out to determine modal damping ratios of the adjacent buildings connected by the fluid damper of different damping coefficients and at different locations. Optimal damper damping coefficient and location for achieving the maximum modal damping ratio were thus found. The interaction of the fluid damper with the inherent

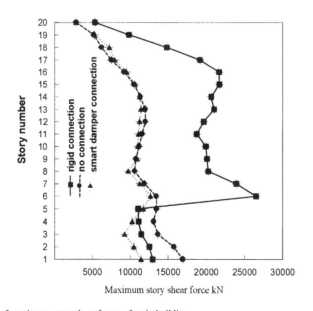

Figure 18. Comparison of maximum story shear forces of main building

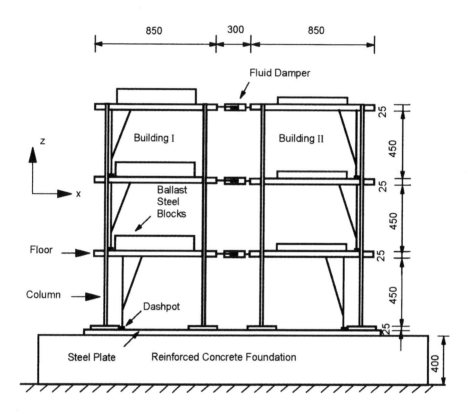

(a) Elevation (All dimensions in mm)

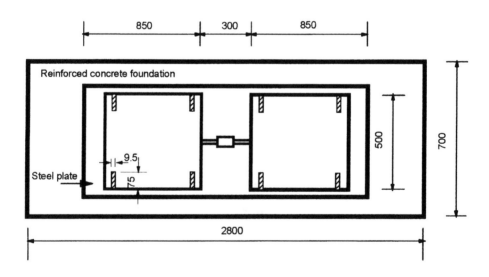

(b) Plan (All dimensions in mm)

Figure 19. Configuration of adjacent building-fluid damper system

94

properties of the buildings was observed. The experimental results showed that the overall performance of the fluid damper-building system could be significantly enhanced using the fluid damper of proper parameter to connect the adjacent buildings. The details on this work can be found in Xu et. al. (1999).

In the second stage, the simulated earthquake shaking table tests of two large-scale adjacent building models connected by fluid dampers were carried out at the Shaking Table Testing Division of the State Key Laboratory for Disaster Reduction in Civil Engineering of Tong University. This is a joint research work between Tong University and The Hong Kong Polytechnic University. Two building models were respectively constructed as a five-story and a six-story moment-resisting steel frame with a scaling factor ¼. The overall dimensions of the five-story frame model are 2.0 x 1.08m in plan and 5.0m in height while the six-story frame model has the dimensions of 1.95 x 1.9m in plan and 6.0m in height (Fig. 20). The fluid dampers connecting the frame models are the D-SERIES linear fluid dampers manufactured by Taylor Devices Inc, USA, with 375mm extended length, 101mm stroke, and 8896N peak output force.

The two frames connected by either the fluid dampers or the steel bars or without any connections were tested under one-direction horizontal excitation, two-direction horizontal excitation, and three-direction excitation, respectively. The test results from the preliminary data analysis showed that under one-direction horizontal excitation, the fluid dampers could reduce the acceleration and displacement responses of both models by 50% to 60%. Under multi-direction excitations, the fluid dampers not only could effectively reduce the horizontal responses of both models in the two horizontal directions but also had the effect on the control of torsion vibration, compared with the cases of the models connected by steel bars. The further test data analysis and the comparison with the numerical results are under way.

2.5.5 Remarks
All the studies described above show that structural vibration control technology could be used to enhance the earthquake resistant capacities of adjacent buildings by using either passive control devices or active and semi-active control devices to link them together. However, it should be reminded that if the dynamic properties of both buildings are the same or close to each other, the use of control devices linking the two buildings will not function or not function properly (Xu et al. 1999). Some other issues related to this study, such as the optimal position of dampers and the effect of non-stationary earthquake excitation need further investigation.

3 FUTURE CHALLENGE

Though the research on earthquake engineering in Hong Kong has made significant progress within a short period of time as described above, there remains some important issues that have not been satisfactorily solved or have been not yet addressed. For instance, because of few strong motion instruments installed in Hong Kong and Southern China, there are limited ground motion records. The seismic hazard levels of Hong Kong predicted by using the conventional seismic hazard analysis and by assuming ground motion attenuation relationships from so-called analogous seismic region contain some degrees of uncertainty. Also because the inherent shortcoming of the conventionally used analog accelerographs for ground motion measurements, the response spectra derived so far may fail to capture the important features of long period spectral components, which are vital for the seismic design of long period structures. Furthermore, more local data on dynamic soil parameters, in particular for hydraulic filled reclaimed land, are required in order to have reliable soil amplification factors for typical types of sites in Hong Kong. The detailed evaluation of seismic resistant performance of the existing buildings, which were not designed for earthquakes, is still on going. No definite conclusions can be drawn at this stage on their seismic resistant performance as compared with their wind resistant capacities. It is still arguable if the displacement response spectra should be used instead of the conventional acceleration response spectra for seismic design of long period structures sitting on reclaimed land in the regions of low and moderate seismicity.

Probably, a forthcoming challenge will be on how to assist the Hong Kong SAR Government in the assessment of hazard, vulnerability and risk of our building stock and infrastructures. There is also an urgent need to develop and implement remedial measures to strengthen our structures. It is our obligation to educate the public to understand the meaning of "moderate" seismic hazard and the extent of "risk" involved. Otherwise, the public could panic on the implementation of a possible future seismic code in Hong Kong in misinterpreting that the existing structures are unsafe. In recent years, many new technologies have been developed worldwide including devices on vibration control, sensor technology, smart materials, global information systems and others. Bearing in mind that structures in Hong Kong are unique, application of the new technologies will probably need some degree of modifications. Adaptation of the new technologies to improve the seismic performance of our structures is probably another challenge to be faced by the researchers and professionals in Hong Kong.

Figure 20. Shaking table test of two large scale adjacent buildings

4 RESEARCH COLLABORATION

National and international research collaboration plays a pivotal role in meeting the need of Hong Kong society for the assessment of hazard, vulnerability and risk of infrastructures and for the development of seismic design codes in a relatively short period. This is mainly because Hong Kong has a very short history of earthquake study and limited human resources in carrying out an extensive investigation of different aspects in earthquake engineering. The national and international research collaboration is also feasible because it is somewhat an inherent feature of Hong Kong society not only in seismic study but also in other aspects. Many institutions in Mainland China have been already involved in the Hong Kong earthquake research activities. In 1991, the Geological Society of Hong Kong and the Guangdong Seismological Society held a joint international conference on seismicity in Eastern Asia to address aspects of seismicity, tectonic geology, seismic hazards and earthquake countermeasures. The workshop on the draft Chinese seismic code was organized by the Hong Kong Polytechnic University in 1998, during which local researchers and practitioners had a fruitful discussion on the seismic design code with the experts from Mainland China. The University of Hong Kong and the Guangdong Seismological Society also jointly undertook a research project on the seismic hazard analysis of the Hong Kong region in 1996 (Lee, et al., 1998). The Hong Kong Polytechnic University had and continues to have a series of collaborative research work with many institutions in Mainland China, such as the Institute of Engineering Mechanics and the Institute of Geophysics of the China Seismological Bureau, the Institute of Earthquake Engineering of China Academy of Building Research of Ministry of Construction of China, Tongji University, and Harbin Industrial University. More recently, the National Natural Science Foundation of China granted its first kind ever of national important project on intelligent vibration control of civil engineering structures, in which the Hong Kong Polytechnic University is also involved.

Researchers in Hong Kong are also actively seeking international collaboration in view of its distinct advantages including the reduced duplication in research programs, the identification of new research needs, and the utilization of complementary strengths and resources. The University of California at Berkeley, USA, and the University of Turin and ENEI, Italy, have assisted the Hong Kong University of Science and Technology in completing a special investigation on hydraulic fill in Hong Kong in 1994. The Hong Kong Polytechnic University and the University of Notre Dame, USA, have coopera

tive research projects in structural vibration control. An international workshop on earthquake engineering for regions of moderate seismicity was also jointly organized by three universities in Hong Kong and Mid-America Earthquake Center, USA. It has been noticed that there are many countries, such as Singapore, Australia and Thailand or parts of some countries, such as Mid-America are all in the regions of low and moderate seismicity. It is thus necessary to have special international cooperative research on seismic hazard mitigation for low and moderate seismic regions. This is also an important objective of the China-US Millennium Symposium on Earthquake Engineering, which marks the 20th anniversary of cooperative studies in earthquake engineering and hazard mitigation between the People's Republic of China and the United States of America.

5 CONCLUSIONS

The historical review, current research status, future challenge and research collaboration of earthquake engineering in Hong Kong have been given in this paper in some detail. Significant progress has been made in Hong Kong in the field of earthquake engineering in the past ten years, but there are still other research topics and urgent tasks faced by researchers and other professionals in Hong Kong. National and international collaboration is inevitable to speed up the research progress, the development of seismic design standards for building structures, and the implementation of new technologies for seismic hazard mitigation. This can be realized through exchange of data and information, jointly organized seminars and conferences, collaborative research projects, and exchange of personnel.

6 ACKNOWLEDGMENTS

The writers are grateful for the financial supports from the Research Grants Council of Hong Kong and The Hong Kong Polytechnic University. The writers are indebted to many institutions and individuals that made direct and indirect contributions to the materials presented in this paper. In particular, the writers would like to express their sincere thanks to the following scholars for their significant contributions to the work presented in this paper. They include Prof. M.Z. Zhang, Prof. Y.F. Yuan, and Dr X. Guo from Institute of Engineering Mechanics, State Seismology Bureau; Prof. B. Wu from Harbin Industrial University; Dr W.S. Zhang from Dalian University of Technology, Prof. W.L. Qu from Wuhan Industrial University, and Prof. X.L.Lu from Tongji University.

REFERENCES

ATC-32. 1996. Improved Seismic Design Criteria for California Bridges: Provisional Recommendations, Applied Technology Council

Boore, D.M. and Joyner, W.B. 1991. Estimation of ground motion at deep soil sites in eastern North America, Bulletin of the Seismological Society of America, 81, 2167-2185.

Carlson, J.D. & Spencer, B.F. 1996. Magnetorheological fluid dampers: scalability and design issues for application to dynamic hazard mitigation, Proc. of Second Int. Workshop on Structural Control, Hong Kong, 99-109.

Chan H.C., Pan A.D.E., Pam H.J. & Kwan A.K.H. 1996. Potential problems of reinforced concrete detailing in Hong Kong under earthquake effects. Proceeding of One Day Seminar on Earthquake Resisting Structures, Hong Kong, May 3, 1996

Chan H.C., Pan A.D.E., Pam H.J. & Kwan A.K.H. 1998. Seismic detailing of reinforced concrete buildings with relevance to Hong Kong design practice, Part 1 and Part 2. Transactions of the Hong Kong Institute of Engineers, 5 (1), 6-20

Chandler A.M., Pan A.D.E., Chan H.C., Pam H.J. & Kwan A.K.H. 1998. Seismic structural design and protection issues for the Hong Kong region. Proceedings of International Workshop on Earthquake Engineering for Regions of Moderate Seismicity, December 7-8, Hong Kong, 205-215

Chau, K.T. & Wei, X.X. 2000. Pounding of structures modeled as nonlinear impacts of two oscillators. Earthquake Engineering and Structural Dynamics, (in press).

Chen A. & Lam S.S.E. 2000. Verification of a pseudo-dynamic testing system, accepted for presentation in the International Conference on Advances in Structural Dynamics, 13-15 December 2000, Hong Kong.

Constantinou, M.C. & Symans, M.D. 1993. Experimental study of seismic response of buildings with supplemental fluid dampers. The Structural Design of Tall Buildings, 2, 93-132.

Davis, R.O. 1992. Pounding of buildings modeled by an impact oscillator. Earthquake Engineering and Structural Dynamics 21, 253-274.

Dahle, A., Bungum, H. & Kvamme, L. B. 1990. Attenuation modelling inferred from intraplate earthquake recordings, Earthquake Engineering and Structural Dynamics, 19. 1125-1141.

Dyke, S.J., Spencer, B.F., Sain, M.K. & Carlson, J.D. 1996. Modelling and control of magnetorheological dampers for seismic response reduction, Smart Materials and Structures, 5, 565-575.

EEFIT 1991. The Newcastle, Australia Earthquake, Publ. Earthquake Engineering Field Investigation Team, Institution of Structural Engineers, London, UK.

Ehrgott, R.C. & Masri, S.F. 1992. Modelling the oscillatory dynamic behaviour of electrorheological materials in shear, Smart Materials and Structures, 1(4), 275-285.

Gavin, G.P, Hanson, R.D. & Filisko, F.E. 1996a. Electrorheological dampers, part I: analysis and design, Journal of Applied Mechanics, 63, 669-675.

Gavin, G.P, Hanson, R.D. & Filisko, F.E. 1996b. Electrorheological dampers, part II: testing and modelling, Journal of Applied Mechanics, 63, 676-682.

GBJ11-89. Code for Seismic Design of Buildings, National Standard, Beijing, China, 1994.

GCO 1991. Review of Earthquake Data for the Hong Kong Region, Publication No.1/91, Geotechnical Control Office (GCO) of Hong Kong.

GEO 1991. A Note on Liquefaction of Marine Reclamation Fills. Section 1, Technical Note No. TN5/91, Geotechnical Engineering Office (GEO) of Hong Kong.

GEO 1997. Pilot Study of Effects of Soil Amplification of Seismic Ground Motions in Hong Kong, Technical Note No.TN5/97, Geotechnical Engineering Office of Hong Kong.

Guo, X., Wong, Y.L., & Yuan, Y. 2000. Preliminary study on site classification for Hong Kong. World Information on Earthquake Engineering, 16(1), 51-60 (in Chinese).

Hahn, G.D. & Valenti, M.C. 1997. Correlation of seismic responses of structures. Journal of Structural Engineering, ASCE, 123, 405-413.

Housner, G.W., Bergman, L.A., Caughey, T.K., Chassiakos, A.G., Claus, R.O., Masri, S.F., Skelton, R.E., Soong, T.T., Spencer, B.F. & Yao, J.T.P. 1997. Structural control: past, present, and future, J. Engrg. Mech., ASCE, 123(9), 897-971.

Kasai, K., Jagiasi, A.R., Jeng, V. 1996. Inelastic vibration phase theory for seismic pounding mitigation. Journal of Structural Engineering, ASCE, 122, 1136-1146.

Kasai, K. & Maison, B.F. 1997. Building pounding damage during the 1989 Loma Prieta Earthquake. *Engineering Structures* 19,195-207.

Kwong, J.S.M. 1998. *Pilot Study on the Use of Suspension PS Logging*, Technical Note, TN 2/98, GEO, Hong Kong Government.

Lam S.S.E. & Ko J.M. eds 1999. *Earthquake Engineering for Regions of Moderate Seismicity*. Proceeding of an International Workshop, Hong Kong, December 1998, published in October 1999

Lam S.S.E, Wu B., Wong Y.L. & Chau K.T. 2000. Ultimate drift ratio of rectangular reinforced concrete column in Hong Kong, *accepted for presentation in the International Conference on Advances in Structural Dynamics*, 13-15 December 2000, Hong Kong

Lam S.S.E, Wu B., Wong Y.L. & Chau K.T. 2001. Behavior of rectangular columns with low lateral confinement ratio, *submitted to The International Conference on Structural Engineering, Mechanics and Computation*, Cape Town, April 2001

Lee, C.F., Ding, Y.Z., Huang, R.H., Yu, Y.B., Guo, G.A., Chen, P.L., & Huang, X.H. 1998. Seismic Hazard Analysis of the Hong Kong Region. Publication No 65, Geotechnical Engineering Office of Hong Kong.

Lee, K.M., Li, X.S., Keung, K.Y.Y., & Shen, C.K. 1998. Site response analysis of hydraulic sand-fill subjected to multi-directional earthquake loading, HKIE Transactions, 5 (2), 19-20.

Makris, N., Burton, S.A., Hill. D, Jordon, M. 1996. Analysis and design of ER damper for seismic protection of structure, J. Engrg. Mech., 122(10), 1003-1011.

Nakamura, Y. 1989. A method for dynamic characteristics estimation of subsurface using microremeter on the ground surface, Quarterly Report of Railways Technical Research Institute, 30 (1), 25-33.

Penzien, J. 1997. Evaluation of building separation distance required to prevent pounding during strong earthquakes. Earthquake Engineering and Structural Dynamics, 26, 849-858.

Pun, W.K. & Ambraseys, N.N. 1992. Earthquake data review and seismic hazard analysis for the Hong Kong region. Earthquake Engineering and Structural Dynamics, 21, 1-11.

Ribakov, Y. & Gluck, J. 1999. Active control of MDOF structures with supplemental electrorheological fluid dampers, Earthquake Engineering and Structural Dynamics, 28, 143-156.

Scott, D.M., Pappin, J.W. & Kwok, M.K.Y. 1994. Seismic design of buildings in Hong Kong. The HKIE Transactions, 1(2), 37-50.

Seed, H.B., Wong, R.T., Idriss, I.M. & Tokimatsu, K. 1984. Moduli and Damping Factors for Dynamic Analysis of Cohesionless Soils, Report No. UCB/EERC-84/14, Earthquake Engineering Research Centre, University of California, Berkeley.

Soong, T.T. & Dargush, G.F. 1997. Passive Energy Dissipation Systems in Structural Engineering, John Wiley & Sons, Chichester, England.

Spencer, B.F., Dyke, S.J., Sain, M.K., Carlson, J.D. 1997. Phenomenological model for magnetorheological dampers, J. Engrg. Mech., ASCE, 123(3), 230-238.

Spencer, B.F. & Sain, M.K. 1997. Controlling buildings: a new frontier in feedback, IEEE Control Systems Magazine, 17(6) 19-35.

State Seismological Bureau 1991. Seismic Intensity Zoning Map of China(1:10000000 scale). Seismological Press, Beijing China.

Sun, C.T. & Lu, Y.P. 1995. Vibration Damping of Structural Elements, Prentice-Hall, Englewood Cliffs, New Jersey.

Sun, J.I., Gelesrkhi, R., & Seed, H.B. 1988. Dynamic Moduli and Damping Factors for Cohesive Soil, EERC-88/15, University of California, Berkeley, USA.

Wehbe N. 1998. Confinement of rectangular bridge columns in moderate seismic areas. Earthquake Spectra, 14(2), 397-406.

Wong, Y.L., Guo, X, Yuan, Y., & Chau, K.T. 2000. Influence of soft sandwich layer on seismic response of soil sites in Hong Kong. Journal of Natural Disasters, 9(1), 109-116 (in Chinese).

Wong, Y.L., Zhao, J.X, Chau, K.T., & Lee, C.M. 1998. Assessment of seismicity model for Hong Kong region. The HKIE Transactions, 5(1), 50-62.

Wong, Y.L., Zhao, J.X., Lam, S.E.E., & Chau, K.T. 1998. Assessing seismic response of soft soil sites in Hong Kong using microtremor records. HKIE Transactions, 5(3), 70-79.

Xu, Y.L, He, Q., & Ko, J.M. 1999. Dynamic response of damper-connected adjacent buildings under earthquake excitation, Engineering Structures, 2 (21),135-148.

Xu, Y.L., Zhan, S., Ko, J.M. & Zhang, W.S. 1999. Experimental investigation of adjacent buildings connected by fluid dampers, Journal of Earthquake Engineering and Structural Dynamics, 28, 609-631.

Zhang, W.S. & Xu, Y.L. 1999. Dynamic characteristics and seismic response of adjacent buildings linked by discrete dampers, Journal of Earthquake Engineering and Structural Dynamics, 28, 1163-1185.

Zhang, W.S. & Xu, Y.L. 2000. Vibration analysis of two buildings linked by Maxwell model-defined fluid dampers, Journal of Sound and Vibration, 233(5), 775-796.

Zhao, J.X. 1996. Estimating modal parameters for a simple soft-soil site having a linear distribution of shear wave velocity with depth. Earthquake Engineering and Structural Dynamics, 25, 163-178.

Zhao, J.X. 1997. Modal analysis of soft-soil site including radiation damping. Earthquake Engineering and Structural Dynamics, 26, 93-113.

Contributed papers

Earthquake Engineering Frontiers in the New Millennium, Spencer & Hu (eds),
© 2001 Swets & Zeitlinger, ISBN 90 2651 852 8

Strengthen the link between engineering and seismology

Yuxian Hu

Institute of Geophysics, China Seismological Bureau

ABSTRACT: After a brief review of the cooperation of seismologists and engineers in the advancement of engineering seismology, some problems for further organized joint study at the linkage of seismology and engineering are discussed. Problems discussed are as follows. (1) Ground motion parameters and their estimation (data bank of strong motion records, long period motion ground motion, large velocity pulse near fault, ground motion field, attenuation laws for regions with few acceleration, and seismologist's generated time history (2) Seismic hazard assessment (uncertainties in seismicity assessment, and deterministic and probabilistic methods of earthquake hazard assessment), and (3) Design code related problems (duration of strong motion, topographical site effect, and functional or consequence design levels their definition).

1 INTRODUCTION

It has been pointed out by Professor George Housner (1984) that, at the beginning stage earthquake engineering, engineers introduced some important terms, such as intensity and distribution of intensity after a strong earthquake, which were used later mainly by seismologists before instruments were introduced to measure the earthquakes. Since the first half of the 20th Century, the engineers measured the strong earthquake ground motion, in the period range of 0.1 to 1.0 sec. especially important for the design of buildings. From the records obtained, the idea of response spectrum was introduced and used gradually over the whole world and the seismic design shifted from rough experience to a branch of science.

The engineers developed many other branches of earthquake engineering science, such as the probabilistic approach of seismic hazard assessment and seismic risk assessment and disaster mitigation, in addition to many structural problems such as base-isolation and structural control. But, at the same time, I feel that engineers should work more closely with seismologists to solve many problems in earthquake engineering field, at least in the field of engineering seismology. Gutenberg-Richter relationship, as well as the recent contribution of digital strong motion broadband instruments and records, and the recent seismic zonation map of the United States are good examples of the combined effort of seismologists and engineers. Many problems can be solved with their further joint efforts, purposely oriented to gain faster achievements.

In searching for joint effort of engineering and seismology, I have participated in three joint research projects in US-CHINA cooperative studies on earthquake engineering in the past 20 years as follows. (1) Development of probability bases for earthquake resistant design, with Prof. A.H-S. Ang of University of Illinois, from 1981 to 1986 (Ang & Hu, 1983). (2) Study on nonstationarity of historical seismicity in China, with Dr. R.K. McGuire, Prof. C.A. Cornell, and Prof. D. Veneziano, from 1988 to 1990 (McGuire et al. 1992). (3) Comparison of US and Chinese methodology for evaluating seismic performance of water delivery systems, with Prof. Howard H.M. Huang, Prof. M. Shinozuka, Prof. X.Y. Zhou, from 1993 to 1997 (Huang et al. 1998). These projects, though productive, but were on isolated problems. After 20 years cooperative research effort, I feel strongly that joint efforts among institutions and centers in USA and China, with perhaps other countries added, on important and key areas may be needed to promote faster advance of theory of design and practical application in earthquake engineering.

With this purpose in mind, this paper is directed to this goal and three related problems will be presented in the following: (1) Ground motion parameters and their estimation, (2) Seismic hazard assessment, and (3) Design code related problems.

2 GROUND MOTION PARAMETERS & THEIR ESTIMATION

2.1 *DATA bank of strong motion records*

Since accelerographs recorded acceleration time histories in the early 30's, now there are more than

5000 such accelerograph stations in the world now and with roughly 5000 strong motion records of ground motion and near 1000 records by the modern wide band digital accelerographs and seismographs. The idea of seismic response spectrum was introduced firstly in the 30's-40's on the basis of a few accelerograms and provided a reasonable basis for the seismic design of nuclear power plants and ordinary structures for more than half a century.

In the last 10 or 20 years, a much better kind accelerographs was invented and obtained records with dependable longer periods of motion than the traditional one. Those records from digital instruments obtained in Mexico, Western United States, and especially those from the recent Ji-Ji Earthquake in 1999 should provide us a very rich bank of ground motion data of a much wide band up to 20 sec. With site condition data attached, they will give us a very good chance to study the ground motion distribution, including the formation, build-up and propagation of near field strong impulses and far field of long period motions. Joint studies will help all of us to better understanding of the wide band motions from earthquakes and thus improve the design principles, especially for modern structures of long natural periods, such as high communication towels, suspension bridges, base-isolated structures, and high-rise buildings of 100 stories or more.

Important problems related to the application of the records are the site conditions and the general tectonic or geological conditions of the station, and the dependable bandwidth without possible noise polluted long period motion.

2.2 Long period motion ground motion

Because of fast urbanization, many structures of the fundamental natural period longer than 5 sec. have been built, such as bridges of more than 1000 km main span, buildings as high as 80 to 100 stories, and towers of height about 300-500 m. Their seismic design requires corresponding spectrum of long periods. The longest design spectrum specified in the current design codes is mostly only up to 3 or 5 sec. (Yu & Hu 2000) because they were derived mostly from accelerograms accurate only in that region. The seismographs used by seismologists do record motion with longer periods, but with narrow frequency band. They are used to locate the earthquake focus and the magnitude of the earthquake, with interest mainly of the earthquake itself, but not the ground motion at the station site. The records are mainly motions in terms of displacement or velocity, and accurate only in narrow frequency range near 1, 10 or 20 sec., depending upon the type of the seismograph used. That is perhaps why engineers seldom consider these records.

Based on the records available now, the author's group is trying to combine these two groups of re-

cords, the accelerograms obtained by engineers and the seismograms by seismologists, to extend the engineering design spectrum to a wider range from 0.1 to 10 or even 20 sec. In order to exclude the pollution in the records from noise, only those records from broadband digital records by seismologist are used to cover the ground motion in the long period range 2-20 sec. Fig.1 shows part of the results obtained (Yu & Hu 2000).

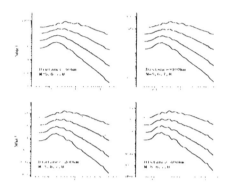

Figure 1. Spectra from both engineering and seismological data

Since strong earthquakes are rare, the accumulation of strong motion records will be a slow process. The 1999 Ji-Ji earthquake, together with its many aftershocks, provides us a huge amount of instrumental data and will certainly be well studied in the following years. However, these records all came from a given background of geology and seismicity. The conclusions derived need be checked by other strong earthquakes occurred in other places. For example, accelerograms, obtained during the Ji-Ji earthquake in the west plain, about 20-100 km away of the epicentral region, show very rich contents of long-period components around 5 – 15 sec. in the spectra.

To achieve this goal, there are two kinds of effort necessary. Firstly, it is needed to collect accelerograms obtained not only for the main shocks, but also for their aftershocks and to find out the noise-free range of the spectrum. Secondly, it is necessary to find out the site conditions and the general geologic background of the region in order to use the data effectively. Special joint studies are therefore needed.

Since strong earthquakes are rare, the accumulation of strong motion records will be a slow process. The 1999 Ji-Ji earthquake, together with its many aftershocks, provides us a huge amount of instrumental data and will certainly be well studied in the following years. However, these records all came from a given background of geology and seismicity. The conclusions derived from Ji-Ji records need to be checked by strong earthquakes in other places before they can be used in other regions. For example, accelerograms, obtained during the Ji-Ji earthquake in the west plain, about 20-100 km away of the epicenter region, show very rich contents of long-period components around 5 – 15 sec. in the spectra.

The best way to derive design spectrum of wide frequency range of 0.05-20 sec. is to collect and analyze the records of digital acceleration ground motion by the modern broadband digital accelerograms. Such records have been obtained recently near the coast of Mexico, West Coast region of USA, and the SMART-1 and 2 and the Ji-Ji earthquakes in Taiwan, China. To achieve this goal, there are at least two kinds of effort necessary. Firstly, it is needed to collect accelerograms obtained not only for the main shocks, but also for their aftershocks. Secondly, it is necessary to find out the site conditions and the general geologic background of the region. They are necessary to explain some special features, such as the especially high long-period spectra showed in the records of the ground motion on sites to the west of the epicenter of the Ji-Ji earthquake, the very large velocity impulses on sites very close to the epicenter, etc. Special joint studies are therefore needed here.

2.3 Large velocity pulse near fault

Several recent records obtained nearby or right on the fault breakage showed a very large velocity impulse, as Prof. P. Jennings and others have mentioned [Jennings, P. 2000]. This large pulse is somehow related to the dynamic offset or permanent relative displacement of the rupture surfaces, and rapidly diminished away from the records away the faulting surface. It is dangerous to structures, even ones with base-isolated foundations. A joint investigation by both seismologists and engineers may understand faster its assessment and effect on structures.

2.4 Ground motion field

For lifeline systems and other infrastructures, spatial distribution of ground motion and differential motion between nearby points are needed. With no dependable data, it is usually assumed that waves travel along the linear direction of the structure. This assumption may be too conservative. Macroseismic phenomena after strong earthquakes showed a possibility of rotation and relative displacement in a short distance, such as the twist of tom stones and the snake-shaped bending of rails. It may be the result of vibration of the structure due to ground motion simultaneously acting on a symmetrical building as the Late Prof. Newmark suggested, or due to nonlinear responses of the structure in two horizontal directions.

Spatial distribution has long been the interest of both seismologists and engineers. There were many condensed or differential arrays in Japan, United State and Taiwan of China and some results have been obtained. One of the best is from the SMART-1 array in Hualian, Taiwan. Within a radius of 2 km, there were 37 accelerographs of broadband digital instruments, which recorded strong motions from many strong earthquakes of magnitudes from 5 to 7. Careful analysis of the data available from SMART-1 and other arrays in Japan and USA with worldwide cooperation will certain help.

2.5 Attenuation laws for regions with few acceleration records

Accumulation of acceleration data recorded during strong earthquakes is quickly increased in the some part of the world; but still less or none in many other regions, and their accumulation will be slow in the coming years because strong earthquakes are rare events. Author's group has tried for some years in a special way of using the intensity attenuation curves, which may be described in the following steps. Take two sets of intensity attenuation curves, one from some region (A) with enough acceleration data and another for the region (B) with few ground motion data. Mapping these two sets of intensity attenuation curves to obtain mapping pairs of earthquake (M, R) that bring two sets coincided. Assume that the pairs will give also the same ground motion attenuation. If the mapping earthquake pair relationship is substituted in the ground motion attenuation curves of region A, the ground motion attenuation of region B will be obtained. The advantage is that not only acceleration, but also velocity, spectrum and duration can be obtained. Results obtained by this approach have been checked in some cases with enough data.

Professor Ambraseys (1998) stated recently that no significant variation of zero-period and spectral attenuation among different regions for shallow intraplate earthquakes, and he believes that the difference in intensity attenuation may be from other reasons instead of difference in attenuation itself. If it is true, money and time can be greatly saved and data obtained in the USA can be used in China mainland and continent Europe right now. Ambraseys (1998) also mentioned that collection of strong motion data is still difficult. A worldwide cooperation may be quite useful to accelerate this kind of study.

2.6 Seismologist's generated time history

Since the available real records of acceleration time history may not be suitable for an assigned specific real situation, both engineers and seismologists have tried for many years to generate mathematically ground motions. The engineers are trying to have their artificial accelerograms to patch a given spectrum and a given envelope of the time history of ground motion, with no consideration of the source properties, because they are included in the given spectrum and envelope. The seismologists start right at the faulting of the source rupture mechanism on the basis of faulting rupture model and wave propagation (Luo & Hu 1989). If there are recorded ground motion acceleration time histories obtained from smaller earthquakes occurred in the same source region considered, they may be used as the Green function, instead of the mathematical ones, and strong ground motion time histories can be generated. Though many interesting results have been obtained, such as those used in the 1997 seismic zonation of the Eastern United States as reference, there are still some things needed before used more popularly by engineers. In the following, several of them are mentioned.

(1) Near field ground motion. Near field ground motion is usually controlled by details of the fault rupture, which are full of unknown local varieties in strength, distribution, direction and sequence. Engineering-interested ground acceleration in near field is more strongly affected by these details than seismology-interested ground displacement in far field. The uncertainty of fault breakage has been simulated by random noise (Takemura & Irikura, JPE, v.36, n.3 1988) and it seems that noise dominates the high frequency motion simulated. Aki has stated in 1988: "At present, any practical simulation of high frequency acceleration requires a stochastic model because of our ignorance of the details of fault zone and rupture process".

(2) When ground motion obtained from available smaller earthquakes are used as the Green's function, the difference of the effect of boundaries of breakage has to be considered. The boundary of the source area of a small earthquake is different from that of an element breakage of a small portion of fault breakage of a large earthquake.

(3) Engineering modification. When the seismologists are trying their Green's Function approach, the engineers have worked on simulated time-history of ground motion acceleration compatible with given response spectrum and time-history envelope. Author's group has tried a modified approach by adjusting the spectrum of the Green's function on the basis of the response spectrum obtained from the engineer's result from their accelerograms of smaller earthquakes. Better results have been obtained by combining the seismologist's and the engineer's approaches.

3 SEISMIC HAZARD ASSESSMENT

3.1 Uncertainties in seismicity assessment

It is generally agreed that there are different sources and types of uncertainties in many steps of ground motion estimation, including (1) source mechanism, magnitude, and return period of possibility of occurrence, (2) propagation, refraction, reflection, super-position and attenuation of waves of different frequencies and phase angles, and (3) site effect. Some engineers believe that the largest uncertainty comes from attenuation, but I believe that the largest uncertainty comes from the source estimation, including location and magnitude in a region within a certain period of time, say 100 years. Even when the probabilistic approach of seismic hazard assessment is applied with probabilistic uncertainty considered inside the potential source zone, the large uncertainties in potential zoning, application of G-R relationship may remain.

Two questions related to this question are discussed here: the probabilistic and deterministic approaches, and the combination of various uncertainties.

(1) Probabilistic or the deterministic

Two terms or ideas have been popularly used in earthquake engineering field. One is the ground motion of probability of exceedance, which has been popularly used in seismic zonation maps in many nations and other seismic hazard assessment of important works. The other is the maximum probable or considered earthquake, which has been used in insurance business and others, such as the 1997 zonation map of United States, although this map is given primarily in terms of probability of exceedance of 10% in 50 year. The probabilistic approach considers some uncertainties such as attenuation and regional seismicity, but neglects other uncertainties such as potential source zonation; and the hazard assessment procedure is considered sometimes complicated to accept other possibilities. Some engineers wish to have one deterministic safe estimate of the hazard and adopt the term maximum possible, maximum probable, maximum capable, or maximum considered earthquake. All this kind of deterministic approach seems to me some kind of subjective selection. How to define these terms or to correlate it with a certain probability of exceedance, such as 1% in 100 years, remains a question. Cooperation and patience will perhaps solve the problem.

(2) Combination of various kinds of uncertainties

The main question I am raising here is the reasonable way of combining all these uncertainties. I was told that there was some discussion on this problem in the United States in the 70's and 80's when they were working on the seismic safety estimation of the built nuclear power plants in the Eastern United States or the EPRI's problem.

As Prof. Lind of Canada suggested, all uncertainties may be classified into three categories: namely, random, incomplete knowledge, and fuzzy (Lind 1985). The decision of EPRI and the LLNL team did not consider the fuzzy uncertainty, and suggested to use probabilistic theory for random uncertainty but to use fractile for incomplete knowledge uncertainty through logic tree. According to this approach

(Bernreuter, Savy & Mensing 1987), a ground motion will be defined by a given fractile of a given probability of exceedance. I do not know how to compare the uncertainty of a ground motion of 80% fractile of 0.5% probability of exceedance and another of 50% fractile of 0.2% probability of exceedance.

Author's group has suggested a probabilistic approach for all three types of uncertainty, by using the objective probability for random uncertainty, subjective probability for incomplete knowledge. After a conversion of the membership function in the fuzzy set into a subjective probability density function, the fuzzy uncertainty can also be handled by subjective probability approach. Cooperation will help to solve this problem.

3.2 Deterministic and probabilistic methods of earthquake hazard assessment

There are several terms used in the deterministic approach, such as scenario, maximum possible, maximum probable, maximum credible, and maximum considered earthquake. A scenario earthquake may be considered as one earthquake selected as an example. Other terms listed above are somewhat subjective and I enjoy the termed used since 1997 in the United States, the word considered is adopted to show clearly the subjective selection. Other terms with the word "maximum" show some kind of personal selection of safety level, implicitly related with a probability of exceedance usually in the range of 1/100 to 1/10000. Is it possible to specify what is the probability of exceedance attached to it?

For probabilistic approach, the consistent spectrum is usually used, which means that all ordinates of the spectrum have the same probability of exceedance, but they will not happen simultaneously. In other words, the probability of exceedance of each coordinate of the consistent spectrum is the sum of the probability of contributions from all possible future earthquakes in various potential sources. When one earthquake giving the expected spectrum ordinate at period T occurs, the spectrum at other periods will be usually less than those specified in the probability-consistent spectrum. No better and simple approach is available yet.

4 RELATED CODE SPECIFICATIONS OF ENGINEERING SEISMOLOGIC ISSUES

4.1 Duration of strong motion

Roughly 50 years ago, strong motion duration, together with peak acceleration and spectrum, was considered as one of the primary index that affected the safety of structures under earthquake action. It is known that strong motion duration has two effects on structural response, namely, resonance and inelastic response. In practical seismic design, it is not explicitly considered. For very important structures, duration is included in selection or generation of time-history accelerograms for numerical checking or shaking table testing. In random theory of seismic responses, it is considered. The China seismic design code, since 1989, duration is considered implicitly in liquefaction assessment by specifying a less number of blow counts of the same site for far distance and larger earthquake allowed for liquefaction of loose sand layers. Author has considered the effect of duration in 1985 (Hu, Tao & Zhang 1985) for low-cycle fatigue analysis of structures.

Serious and thorough joint studies are necessary to consider duration effect in practical seismic design again failure or high level functional design.

4.2 Topographical site effect

There are two kinds of site effect on ground motion; one for ground failure and another for ground motion. Only the later is considered here. With advances made in strong ground motion observation and theoretical studies, soil effect is well considered in the 1997 UBC in the United States. It considers the nonlinear effect of soft soils under strong shaking by specifying larger ratio of velocity to acceleration, and smaller acceleration to stronger motion. It may be expected that other codes will follow this trend.

On the contrary to soil effects, the topographical effects are not considered so well as the soil effect. A combination of shape and absolute scale of a hill, or some term in terms of wavelength, together with perhaps the direction of wave propagation, makes the problem much more complicated and the progress seems rather slow. No qualitative specifications are given yet in the code though it is widely accepted that the effect may be strong at least for some frequency range a hill amplifies.

4.3 Functional or consequence design levels and their definition

It is seen that functional design principle will be our next generation code principle. I have suggested one further step forward to consider a consequence design principle, which considers the effects of the loss of function inside a society. For example, the design level for an ambulance in a small city should be higher than the same type of ambulance located in a mega-city, because the consequence of loss of its functioning is quite different especially in an increasingly developed society. The value of an ambulance station is the same but the indirect loss of malfunction in modern society may be quite different.

In a two-level design such as those used in Japan and China, the design acceleration collapse level is about 5 time of that for linear behavior, which corresponds to a probability of exceedance roughly 60%

in 50 years to 2% in 50 years. The ratio was selected first and the probability was attached later to facilitate the definition. For a functional design, more levels may be used and the corresponding ground motion and probability may be assigned later on the basis of engineering experience and the specification of structures of various kinds we have now to compare with. I certainly hope that the consequences of malfunction in different societies will be considered.

There are several model codes now in the United States and we are drafting one now in China. It may be possible that we start some bilateral or multiple seminars to explore possible cooperation in this task.

REFERENCES

Ambraseys, N.N., 1998 Development and application of strong ground-motions in Europe. *Proceedings of European Conf. on Earthquake Engineering*, 45–51

Ang, A.H-S. & Hu, Y.X. 1983. Development of Probability bases for earthquake resistant design, Civil Engineering Studies, Structural Research Series No.511, UILU-ENG-83-2016

Bernreuter, D.L., Savy, J.B. & Mensing, R.W. 1987. Ground motion and engineering seismology. *Development in geo technical engineering*, 44

Housner, G.W. 1984. Historical review of earthquake engineering. *Proc. 8WCEE, Post-Conference Volume*, 25–39

Hu, Yuxian & Chen, Han-Yao 1992. Probabilistic analysis of uncertainties in seismic hazard assessment, Structural Safety, vol.11, 245–253

Hu, Y.X. & Tao, X.X. 1984. Selection and estimation of earth quake motion for critical engineering sites, Proc. Int. Workshop on Earthquake Engineering, Shanghai, v.1, pp.a-s-1-14

Hu, Y.X., Tao, X.X. & Zhang, M.Z. 1985. Joint evaluation of spectrum and duration for sites of critical structures. *Proc. 8th Int. Conf. on Structural Mech. in Reactor Tech.*, Belgium.

Huang, H.M., Zhou, X.Y., Hu, Y.X. & Shinozuka, M. 1997. Univ. of Memphis Report

Lind, N.C. 1985. Reconciling fuzzy concepts and probability in structural engineering. *Proc. NSF Workshop in Civil Engineering Application of Fuzzy Sets*

Jennings, Paul C. 2000. Ground motion pulses and structural responses. *Proc. of Earthquake Engineering Frontiers in the New Millennium*, Beijing, China.

Luo, Qi-Feng & Hu, Yuxian 1989. An improved semi-empirical method for synthesizing near-field Accelerograms of earth-quake. *Proc. Sino-Japan Conference on Seismological Re-search*, Beijing, China

McGuire, R.K. et al. 1992. Nonstationarity of historical seismicity in China. *10th WCEE*, Madrid, Spain, 287–292

Yu, Yan-Xiang & Hu, Yuxian 2000. A combine method to establish attenuation relation of long-period response spectrum. *Proc. of Earthquake Engineering Frontiers in the New Millennium*, Beijing, China.

Earthquake Engineering Frontiers in the New Millennium, Spencer & Hu (eds),
© 2001 Swets & Zeitlinger, ISBN 90 2651 852 8

The center approach in earthquake engineering research and education

G.C. Lee
State University of New York at Buffalo, USA

S.C. Liu
University of Maryland, USA

ABSTRACT: This paper briefly discusses the experience of NSF-sponsored Multidisciplinary Center for Earthquake Engineering Research (MCEER) during the past 15 years with respect to the significance and challenges of carrying out earthquake engineering research and education activities by using the Center approach to facilitate the development and implementation of knowledge and the preparation of future human resources in earthquake engineering.

Several important differences of using the center's approach by comparing with the approach of individual investigators will be discussed. These include, but not limited to, the ability to (1) shorten the time required to implement research results, (2) deliver education and outreach efforts in an organized fashion, (3) engage multidisciplinary team research efforts, and (4) leverage additional matching funds. Special emphasis will be given to the first two issues with examples.

It is recommended that center-to-center cooperative efforts be expanded to establish an internationally networked system for more effective cooperation in research and human resource development in earthquake engineering and in natural disaster reduction (International Millennium).

1 INTRODUCTION

1.1 *Background*

In 1986 the US National Science Foundation (NSF) established the National Center for Earthquake Engineering Research headquartered at SUNY Buffalo to carry out systems integrated studies in earthquake hazard mitigation (the center approach) that would yield results which could not be easily accomplished by using the individual investigators approach. The success over 10 years has resulted in an expansion of the center approach in earthquake engineering research by NSF. In 1997, NSF awarded three earthquake engineering research centers. The National Center for Earthquake Engineering Research continued its efforts with a name change to the Multidisciplinary Center for Earthquake Engineering Research (MCEER). The other two centers are Mid-America Earthquake (MAE) Center and Pacific Earthquake Engineering Research (PEER) Center. The three centers each have their own research thrusts that are coordinated through the Council of Center Directors. One of the most important efforts in pursuing the "center approach" by NCEER/MCEER has been the organized earthquake engineering educational activities, some of which would be very difficult to accomplish by isolated research and/education grants given to uncoordinated individual investigators.

1.2 *MCEER vision and its multidisciplinary team approach*

MCEER's vision is to help establish earthquake resilient communities by emphasizing the application of advanced and emerging technologies in pre-earthquake mitigation and post-earthquake response and restoration. From the viewpoint of research, NCEER/MCEER has been pioneering its multidisciplinary team approach in the past with steady progress. Cross-disciplinary fertilization and cooperation has not been popular in the past, and is viewed as an important development by many research-intensive academic institutions. MCEER has established itself as a successful pioneer in multidisciplinary research by researchers from a consortium of universities with research expertise in various engineering, architectural, and social science disciplines.

An earthquake resilient community is the result of various working professionals, government agencies at all levels, stakeholders and the public living in that community. What MCEER can contribute is to develop the knowledge and tools for those organizations and professionals concerned about earthquake hazard in a region. New knowledge components and different methods can be more or less accomplished by individual investigators. However, targeting and coordinated pursuits organized by a center can certainly enhance the efficiency. More importantly, a center has more visibility and more resources leveraged from matching commitments and partnership arrangements to transfer its research results to the users to shorten the time between research and implementation. Perhaps the most important aspect of center approach is in the area of developing future human resources in earthquake hazard mitigation and response. Cross fertilization among relevant disciplines not only benefit the research activities but also the earthquake engineering students, as they will work after graduation in a multi-professional environment involving planners, architects, decision-makers as well as the public. Since the threat of earthquake damages will always be with our society in the future, preparing for relevant future earthquake mitigation educators, researchers and implementers is a high priority item for a research center.

2 EARTHQUAKE ENGINEERING EDUCATION ACTIVITIES CARRIED OUT BY MCEER

Earthquake engineering education activities may be broadly classified into three general types. The first is the formal degree program which includes exposure, awareness and some technical knowledge in earthquakes and earthquake engineering at the undergraduate degree level, graduate degree programs with concentration in earthquake engineering, and doctoral (and post-doctoral) level of research training. The second type is the professional advancement effort using certificate programs, short courses, seminars, workshops, conferences, through "personal contact" activities and through reading and/or viewing materials including technical reports, newsletters, books, computer software, films/videos and web-based information systems. The third type is the pre-college education and general pubic awareness.

2.1 The research-education interface

It is clear that the research talents and resources in a given center are limited. All activities in a center should be established based on its vision and targeted research thrusts. MCEER has its vision to help establishing earthquake resilient communities and its targeted research programs in application of advanced and emerging technologies to pre-event mitigation of selected critical facilities and to post-event response of population centers. To illustrate the concept of research-education interface, the organized efforts of MCEER is illustrated in the following diagram:

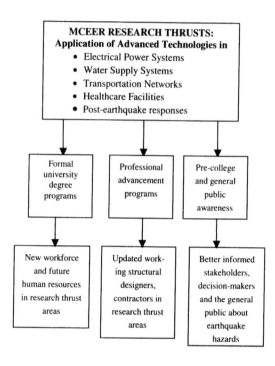

It is worth noting that although center research yields various publications documenting new knowledge discovered, existing knowledge applied and design guidelines improved and/or established, the products of new and enhanced human resources go side-by-side with the research products.

MCEER is able to deliver the three educational efforts in an organized fashion because of the multi-disciplinary and the multi-institution team approach. Furthermore, the center headquarters employed a group of professional staff in information services, publications, and industry partnerships to collectively deliver the center's knowledge transfer and outreach efforts.

2.2 From fundamental studies to implementation in compressed time

Earthquake engineering is a component of a subset in civil engineering (structural and geotechnical engineering) from the viewpoint of professional practice, because only earthquake vulnerable regions are concerned with the additional design and construction requirements to an otherwise rather standardized practice. Implementation of new research results into earthquake design guidelines would require additional time particularly in application of advanced technologies, from the already long process required for normal structural engineering practice. Civil engineering curriculum is more or less a standardized program of study, earthquake engineering components in education is more difficult to develop and implement. Many institutions do not have research activities and educational programs (or courses) in earthquake engineering. For those that have, only selected aspects of expertise exists in most of them. In order to harness the various components in research and educational programs, the multi-disciplinary and multi-institutional team approach is a desirable avenue. Experience at MCEER has shown that this "consortium efforts" in earthquake engineering can accomplish certain successful results in compressed periods of time.

3 SOME EXAMPLES

In this section, two specific examples are given to illustrate how implementation of research results and educational components are carried out through organized team efforts by using the center approach.

3.1 Multidisciplinary components in Masters of Engineering program in earthquake engineering

In 1998, encouraged and partially supported by MCEER, the Department of Civil, Structural and Environmental Engineering at the University at Buffalo developed a Master's of Engineering Program in Earthquake Engineering. The intent of this specialized degree is to provide post-graduate training for students wishing to improve their knowledge base in earthquake engineering. It is a design and practice-oriented program suitable for students planning to pursue a professional career in consulting, industry or government service. A seminar course on social and economic aspects of earthquake engineering is part of the curriculum. This seminar, offered for the first time in Spring 2000, features speakers from many center researchers with expertise in various disciplines and professions, many of whom were electronically linked to the class from remote sites. The aim of the seminar is to present earthquake mitigation issues in a problem-focused context of earthquake resilient communities as opposed to one constrained by a single discipline.

In addition to this seminar course, a separate effort to develop learning modules for certain basic courses is funded by MCEER. For example, at Notre Dame University, learning tools are being developed for the course structural dynamics in the Master's of Engineering Program.

A suite of Virtual Laboratory earthquake engineering experiments were developed using Java to provide an interactive means of instruction on structural performance under seismic conditions. Two modules were developed in 1999-2000, consistent with MCEER's emphasis on advanced technologies, to enact simulations under varying conditions. The first module, "Structural Control using Tuned Mass Dampers (TMD) and Hybrid Mass Dampers (HMD)," considers a single-degree-of-freedom building model subjected to various historical earthquake records. The second module, "Base Isolation," considers a two-degree-of-freedom building model with base isolation, which can be animated by various historical isolation earthquake records. The modules allow users to change system parameters and design TMD, HMD, and base isolators for structural response modification. Both modules were mounted on the Notre Dame web site, and hotlinked to MCEER's web site (http://mceer.buffalo.edu) to provide wide access to the modules. Each module is accompanied by technical background information in English and Japanese for the user, as well as suggested exercises and references. They are intended to provide students with a greater understanding and appreciation of simulation techniques. This first demonstration project is a success for future MCEER efforts to follow. It is especially useful for students at institutions without sophisticated experimental facilities.

These pursuits take advantages of the various expertise represented by the MCEER researchers with the purpose of collectively developing a package that may be shared by many that each individual educational institution cannot accomplish. With the rapid advances in digital and communications technologies, this effort of developing educational modules should be expanded and emphasized in the future.

3.2 Earthquake protective systems for bridges

MCEER has pursued several types of advanced technologies to explore their applications in earthquake hazard mitigation and post-earthquake responses. One of these is structural control or structural response modification (SRM) technologies. Using SRM technologies has been one of the important components in MCEER's Highway Systems Program (seismic protective system for bridges). MCEER's team approach follows the following procedure:

a) Prepare a set of design guidelines for using passive control systems for seismic protection of highway bridges based on existing knowledge before research tasks begin.

b) Consider a set of desirable design guidelines that one would like to see five years hence. It is advised by an invited group of individuals with research and design experiences.

c) Identify the necessary research tasks in order to realize (b).

d) Invite a team of experts to work on the various research tasks by allocating to them the necessary funds.

e) Invite an independent team of experts to serve as the advisory council to the research team. This council consists of a diversified group of researchers, design engineers, code officials, federal and state DOT engineers.

f) A draft for the desired guidelines would be prepared by the research team for review by the council and the design professionals at large for comments and criticisms after the research tasks are completed.

g) An interim set of guidelines will be published, by taking into considerations of the comments from (f), for trial use by the profession for a period of time before recommending final adoption.

This procedure ensures that all concerned professionals and organizations are involved from the beginning in shaping up the guideline contents and provide their collective comments and input during the time the research tasks by a team of experts are carried out. This process would require less time from research to implementation of research results. It is possible by using the center approach.

4 SUMMARY AND RECOMMENDATION

This paper briefly summarized the advantages of using the research center approach to accomplish a number of educational and knowledge implementation objectives, consistent with the research thrusts of the center. They would be difficult to achieve by using the individual investigators' approach. Special emphasis is given to the importance of developing a new generation of earthquake engineering researchers and professionals through the research-education interface activities of the center. These research center-based initiatives can transform conventional education programs into one that is more broad-based, cross-cultural, and technologically interactive than in the past.

The 1990s were recognized as the international decade for natural disaster reduction. During this decade a number of nations had made serious efforts and impressive progress in bilateral cooperative research efforts among their individual researchers funded by their governments. A follow-up effort to that of the international decade is recommended to expand the sharing and cooperative spirit in natural disaster reduction among nations for the next 10 years (International Decade II). A mechanism to accomplish this new initiative may be tried based on the experiences of the center approach in earthquake hazard mitigation.

It is recommended that a new international millennium effort in natural disaster reduction be launched in 2000. Initially, a regional effort should be established by those countries taking part in this symposium. This regional cooperation would first pursue the following objectives:

a) Center-to-center (organization-to-organization) cooperative efforts among nations (i.e. center or organization enter into a cooperative agreement). Individual investigators are affiliated with centers.

b) Encourage centers of diverse research excellence to work together to facilitate cross-fertilization and cooperation, in addition to centers of same research interest.

c) Emphasize cooperative research-education interface activities to develop future human resources in natural disaster reduction.

d) Any center (organization) can join the international team efforts of centers if it:

- Has a recognized area of research expertise and an interest to develop human resources in natural disaster reduction.
- Can commit (or obtain) center resources to international cooperative efforts.
- Agrees to develop educational modules in its area of research expertise and to provide training experience for visiting graduate students and young researchers/educators.

e) The funding agencies of the participating nations commit initial seed funding to facilitate the development of international millennium in natural disaster reduction activities. In particular, the international programs of the agencies (NSF, JSPS, CSB, NNSFC, etc.) endorse and support visiting young researchers.

f) A steering committee should be established to draft a strategic action plan for the long term establishment of an internationally networked node (centers and organizations) on natural disaster reduction), with initial focus given to the decade of 2001-2010 by first concentrating on earthquake engineering research and education activities.

Earthquake Engineering Frontiers in the New Millennium, Spencer & Hu (eds),
© *2001 Swets & Zeitlinger, ISBN 90 2651 852 8*

Improving management of urban earthquake disaster risks

Yaoxian Ye
China Building Technology Development Center, Beijing, China

Norio Okada
Disaster Prevention Research Institute, Kyoto University, Kyoto, Japan

ABSTRACT: The paper begins with a discussion of the facts of disasters, expanding the facts to include disasters and losses in the last five decades around the world and then shifting the focus from the lessons learned from recent earthquake disasters and reasons for increasing losses to urban disaster risk management requirements. A data matching earthquake loss estimation model is presented by linking together the input data, output data and the methodology of the earthquake loss estimation to address the challenge of creating urban disaster risk management tools to mitigate earthquake disasters. The paper concludes with some recommendations for improvement of earthquake loss estimation and earthquake risk management.

1 FACTS OF DISASTERS

1.1 Introduction

Technology breakthroughs are at a pace unparalleled in human history. Man can walk on the moon. Robots can land on Mars. Natural hazards may not have changed much over time. However, urban risks have actually increased sharply due to human being himself. We tend to realize that only in hindsight, when it might be already too late. We can be more hazardous to ourselves than many other hazards. Natural disasters are human-made disasters exposed by natural hazards. Earthquake disasters show that people were killed in poorly designed and constructed man-made structures, not in open fields. Mexico city is built on a very soft bowl bed known centuries to amplify seismic hazard, it has 20 million populations, it is still growing. We still build structures with inadequate disaster protection in disaster prone areas. One collapsed building could block all roads into areas under distress. People died or could not commute to work due to major highway bridge collapse. All of these indicated that we still have hard time and improving management of urban earthquake disaster risks is a critical time issue to be solved.

1.2 Disasters and losses

We did our utmost to mitigate natural disasters in the last decade through IDNDR's activities. However, both of the number of disasters and economic losses are still increasing. The number of disasters and losses and their ratio between the 1990's and other decades happened in the last five decades around the world are shown in Figure 1 and Figure 2 respectively. Figure 3 shows the direct losses from flood disasters in China in the period of 1993 to 1998.

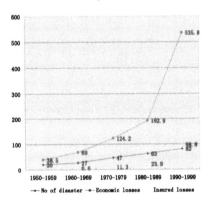

Figure 1. Number of disasters and losses in the last five decades around the world (Losses in US$ bn , 1998 value)

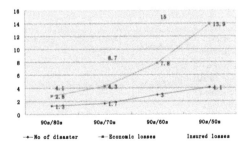

Figure 2. Increase of disasters and losses (in multiple) in the last five decades around the world

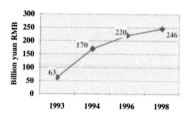

Figure 3. Direct losses from flood disasters in China

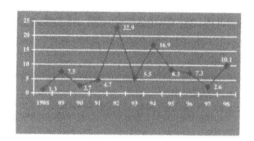

Figure 4. US insured catastrophe loss (in US$ bn)

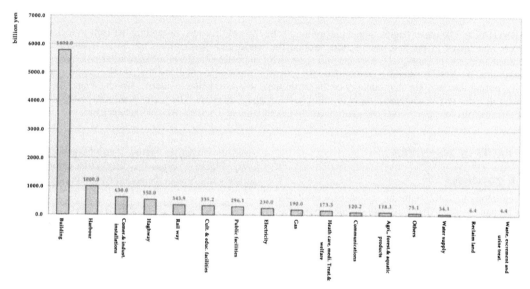

Figure 5. Property losses in 1995 Great Hashin-Awaji Earthquake, Japan

Figure 4 shows the insured catastrophe losses in the period of 1988 to 1998 in the United States. The property losses in the 1995 Great Hashin-Awaji Earthquake, Japan are indicated in Figure 5. Table 1 shows the earthquake loss estimation for three scenario earthquakes. It follows that the consequences would be too ghastly to contemplate.

Figure 6 shows the comparison of direct and indirect losses on industry in the 1995 Great Hanshin-Awaji earthquake in Japan. The indirect losses are significant in the total losses.

Table 1. Earthquake loss estimation (by Prof. Shah/RMS)

Scenario earthquake	Deaths	Serious injuries	Economic loses
Repeat of M8.3 1906 earthquake in Bay Area	2000-6000	6000-20000	$115-135B
M7.0 in Los Angles basin	2000-5000	5000-15000	$125-145B
Recurrence of M7.9 1923 Great Kanto earthquake	40000-60000	80000-100000	$800B-$1.2 trillion

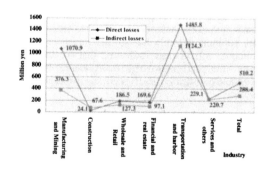

Figure 6. Direct and indirect losses in industry in 1995 Great Hanshin-Awaji earthquake, Japan

1.3 Lessons

The lessons learned from recent catastrophic earthquakes can be summarized as below:
- Environment degradation and climate change
- Population development and urbanization
- Increase of values in disaster-prone regions due to development
- Vulnerability of modern society
- Imperfect knowledge & risk awareness.

1.4 Reasons for increasing losses

The reasons for increasing losses caused by earthquakes can be listed as follows:
- Big earthquake disasters happened not only in developing countries as most of the events but also in developed countries.
- Large scale offset and ground displacement and rising caused collapse of numerous buildings, lifeline systems and dams that have not happened before, at least in the mainland of China.
- Disaster can happen even in the case of earthquakes with moderate magnitude.
- Imperfect knowledge is a big contributor of disaster. For instance, causative faults are unknown in case of Northridge earthquake in the US and Tangshan earthquake in China. Also causative faults are recognized as not very active in case of Hashin-Awaji earthquake in Japan.

2 RISK AND LOSS ESTIMATION

2.1 Risk

What does the risk mean? Risk means many things to many people. The dictionary defines risk as "exposure to the chance of injury or loss" or "possibility of meeting danger or suffering harm, loss, etc." The Standards Australia (AS/NZS 4360:1999-Risk Management defines risk as "the chance of something happening that will have an impact upon objectives. It is measured in terms of consequences and likelihood." The GeoRisk defines risk as "the convolution of exposure, hazard and vulnerability. For natural hazards we can only attempt to reduce the risk not the hazard, either by controlling exposure to hazards or their vulnerability." The insurer defines risk as "the hazard or chance of loss." Kaplan and Garrick define risk as a "the set of triplets: $R = \{(S_i, P_i, X_i)\}$, where S_i is a scenario identification or description, P_i is the probability of that scenario, and X_i is the consequence or evaluation of that scenario, i.e., the measure of damage. Although there are controversies over differing options about the proper definition for risk, there is general acceptance that risk always involves the following three characteristics:
- Uncertainty: the event that characterizes the risk may or may not happen;

- Loss: if the risk becomes reality, the losses will occur; and
- Change and choice: take measures to manage the risk that may happen.

2.2 Loss process

The earthquake loss process is shown in Figure 7.

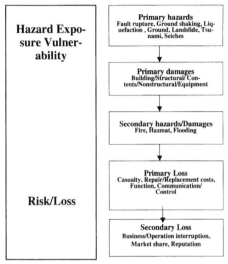

Figure 7. Earthquake loss process

2.3 Loss estimation

Estimating potential earthquake losses is a difficult but essential task to stimulate and guide earthquake disaster mitigation actions including insurance measures.

Since the 1970's many regional loss studies have been performed. In the U.S., most studies have been sponsored by either the insurance industry or government agencies. The property losses to general building stock and other economic consequences are concerned by the insurance sector. The government concerns emergency preparedness and response especially the casualties and functionality of emergency and critical facilities.

Although the early studies provided useful information as the basis for emergency response planning, they often took several years to complete. As a result, new facilities built after completion of the study were not considered in the estimation and some existing facilities were "retired" before the study completion. In order to solve this problem, the Federal Emergency Management Agency (FEMA), under a cooperative agreement with the National Institute of Building Science (NIBS), has developed a standardized, nationally applicable earthquake (and in the future wind and flood) loss estimation methodology. This methodology is implemented through PC-based GIS software called HAZUS that was studied and developed in the period of 1992 to 1994.

Some pilot studies of the methodology were conducted in early 1995 in Portland, and early 1996 in Boston and the final reports were delivered in 1997. The standardized approach for estimating earthquake losses will allow state and local governments to work more effectively with the federal agencies. By collecting and analyzing data using a consistent method, it will be possible to predict the level of resources needed more accurately and to more effectively allocate available resources.

GeoRAWS (**GeoR**isk Analysis **W**eb **S**erver) is multi-hazard multi-regional GIS-based server/client risk analysis application designed specially for use on the inter/intranet. The current scope of version 1.0 is focused on California's seismic risk, however the concept will be extended to other perils and regions of the world.

In Japan, insurance industry and government are also the main bodies concerned with the earthquake loss estimation. Many regional methods for individual disaster loss estimation including peak acceleration and intensity of ground shaking, liquefaction potential, Tsunami, land sliding, building damage by ground shaking, number of fires, building damage by fires, damage to life lines, damage to transportation systems, damage to civil engineering structures, casualties, refugees and homelessness etc. have been worked out after the Great Hanshin-Awaji earthquake. The prefecture governments select earthquake loss estimation methods by their own based on their conditions.

In China, government and insurance industry also paid attention to loss estimation. China Seismology Bureau (CSB) has developed a tool for earthquake loss estimation. Earthquake loss estimation should be made in the urban earthquake disaster mitigation plan by the requirement of Ministry of Construction.

2.4 Loss estimation process

The loss estimation process is shown in Figure 8. Exposure means "every one and every thing at risk of being exposed to seismic hazard." For homeowner, exposure = family + home.

3 DATA MATCHING METHODOLOGIES

It is recognized that the earthquake loss estimates can be made accurately, but with limited precision due to significant uncertainties. The uncertainties are inherent in almost all processes of a method, including in estimation of the ground-motion intensity and ground failure for a given earthquake scenario, in estimating damage given the intensity and ground failure, in estimating the losses given damage to the facilities, and in the process of inventorying the number of facilities in each building classification and geographic area. Therefore, the estimation re-

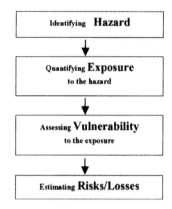

Figure 8. Earthquake loss estimation process

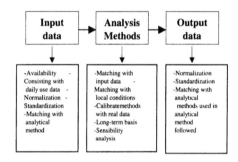

Figure 9. Data matching model for earthquake loss estimation

sults should not be looked upon as a prediction but rather as an indication of what the future may hold.

Figure 9 shows a data matching model for earthquake loss estimation. The basic requirements for the model are as follows.

- The availability of the data to be used in the loss estimation should be identified;
- The input data and the analysis methods should be matched each other;
- The input and output data should be normalized and standardized;
- The tools for earthquake loss estimation should be developed on long-term basis and consisted with that for daily use and with local conditions; and
- Calibrating with real data and sensibility analysis should be conducted for the analysis methods.

As an example, there are many formulae to assess the number of death caused by an earthquake. The more general formula is given bellow:

$$D = f (H, P, W, S, x, B, F, T, E) \qquad (1)$$

Where:

D - Number of death;
H - Number of collapsed buildings;
P - Total population in the destructive area;
W - Total number of buildings in the destructive area;

116

S - Number of severe damaged (half collapsed) buildings;

x - Building damage ratio;

B - Number of burned buildings;

F - Fire related coefficient;

T - Time occurrence related coefficient; and

E - Era related coefficient.

The following nine formulae were used to estimate number of deaths caused by the 1995 Great Hanshin-Awaji earthquake:

1. Regression Method (Ye Yaoxian)

$$Log (D/P) = 9.685 [(H + 0.5 S) / W]^{0.1} - 11.181 \quad (2)$$

2. Ohta et. Al. (1983) Method

$$D = 1.45 \times H0.93 \times F \times T \times E \quad (3)$$

3. Ohta and Gotoh (1985) Method

$$D = 0.543 \times H^{0.69} \times F^{0.29} \times T \times E \quad (4)$$

4. Tokyo DPC (Disaster Prevention Commission) (1978) Method

$$\log D = 0.9598 \times \log (H + S \times 0.5 + B) - 1.0291 \quad (5)$$

5. Saitama Prefecture (1982) Method

$$\log D = 0.3132 \times \log (H + S \times 0.5 + B) + 0.0783 \quad (6)$$

6. Ikeda and Nakabayashi Method (1996)

$$D = 0.0569 \times (H + B) \quad (7)$$

7. Osaka Prefecture Method

$$D/P(\%) = 0.000287 \ x\ 2 - 0.00780 \ x + 0.0506$$
$$(x \geq 25\%)$$
$$D/P(\%) = 0.00156 \ x - 0.00398 \quad (2.5\% \leq x < 25\%)$$
$$D/P(\%) = 0 \quad (0\% \leq x < 2.5\%)$$
$$(8)$$

8. Osaka City Method

$$D_1 = 53.3/1000 \times H$$
$$D_2 = 0.143 \times D_1$$
$$D = D_1 + D_2 \quad (9)$$

9. Shiga Prefecture Method

$$D/P(\%) = 0.0002 \ x^2 + 0.00009 \ x \quad (10)$$

The practical and estimated numbers of deaths caused by the 1995 Hanshin-Awaji earthquake in the wards of Kobe city are shown in Figure 10 and the practical (actual) and estimated total numbers of deaths caused by the 1995 Great Hanshin-Awaji earthquake in Kobe city are shown in Figure 11. The differences between the practical and the estimated numbers of deaths stem from that the input data are not matched with the analysis methods and the methods are not consisted with the local conditions.

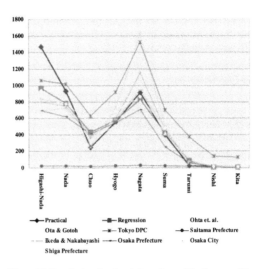

Figure 10. Practical and estimated numbers of deaths caused by the 1995 Great Hanshin-Awaji earthquake in words of Kobe city

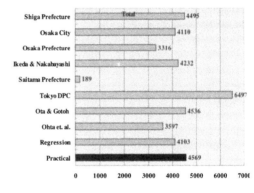

Figure 11. Practical and estimated total numbers of deaths caused by the 1995 Great Hanshin-Awaji earthquake in Kobe city

4 RISK MANAGEMENT AND RECOMMENDATIONS

- Earthquake risk management should be implemented as a comprehensive and continuous activity, not as a periodic reaction to individual disaster circumstances.
- Feasible and data matching methodologies and tools for physical and economical loss estimation, planning, continuous monitoring and review should be well developed with consideration of local conditions.
- Data to be used for earthquake loss estimation should be normalized and standardized and consisted with that of daily use.

- More attention should be paid to calibrating with real data and sensibility analysis for the earthquake loss estimation tools.

REFERENCES

Alcira Kreimer & Munasinghe (1991), Managing Natural disasters and the environment, The World Bank

GeoRisk (2000), Risk management terminology, http://www.georisk.com

John X. Wang & Marvin L. Roush (2000) Whatevery engineer should know about risk engineering and management, Marcel Dekker, Inc., New York

Kaplan, S. & B. J. Garrick (1981) On the quantitative definition of risk, Risk Analysis, Vol. 1, No.1, pp.11-27

Kobe Municipality (2000), Kobe Recovery Records for the Great Hanshin Awaji Earthquake (in Japanese)

California Universities for Research in Earthquake Engineering (CUREE) and Risk Management Software, Inc. (RMS) (1993), Assessment of the State of the Art Earthquake Loss estimation Methodologies, Task 1 Final Report (Draft)

Property and Casualty Insurance Rating Organization of Japan (PCIRO)(1998), Earthquake Loss Estimation Data Collection (in Japanese)

Panel on Earthquake Loss estimation Methodology, Committee on earthquake Engineering, Commission on Engineering and Technical Systems, and National Research Council (1989), Estimating Losses From Future Earthquakes Panel Report, National Academy Press

Robert W. Whitman and Henry J. Lagorio, The FEMA-NIBS Methodology for earthquake Loss estimation, FEMA

Toshihisa Toyoda (1997), Economic Impacts and Recovery Process in the Case of the Great Hanshin Earthquake, Fifth U.S./Japan Workshop on the Urban Earthquake Hazard Reduction, Pasadena.

Thalia Anagnos (1998), Estimation of Damage and Loss, Western States Seismic Policy Council

Robert V. Whitman, Thalia Anagnos, Henry J. Lagorio, and R. Scott Lawson (1998), Development of a National Earthquake Loss Estimation Method

Earthquake Engineering Research Institute (EERI) (1999), Research Needs Emerging from Recent Earthquakes, Recommendations from a workshop organized by the EERI for the National Science Foundation

Karen M. Clark, Robert D. Cicerone, Mark Linker (1998), Estimating the Risk: Hazard Loss Estimation

James J. Johnson, Mahmoud M. Khater, Dennis E. Kuzak, and Richard L. Clinton (1998), EQECAT Loss Modeling Methodology

William Hooke (1999), A Federal Perspective on the Economic Impacts of Natural disasters, Earthquake Quarterly-Fall 1999

UNCRD, EAROPH, RITS, UNU (1995), Innovative Urban Community Development and Disaster Management, Proceedings of the International Conference Series on Innovative Urban Community Development and Disaster Management, Kyoto, Osaka, Kobe, Japan, 1995.

Earthquake Engineering Frontiers in the New Millennium, Spencer & Hu (eds),
© 2001 Swets & Zeitlinger, ISBN 90 2651 852 8

Ground motion pulses and structural response

Paul C. Jennings
California Institute of Technology, Pasadena, California 91125, USA

ABSTRACT: The recording of large, destructive pulses in earthquake ground motions has prompted the development of simple mathematical models for the pulses and the study of the responses of simple structures to such motions. In this paper, the responses of a single degree of freedom structure to delta function pulses, to one and two-lobe sinusoidal acceleration pulses, and to one and two-lobe sinusoidal velocity pulses are reexamined and compared in the context of using the pulses to model strong ground motion. The response to pulses of strong ground motion proves to be different in significant ways from the response to longer, broadband earthquake excitation and is more complex than one might expect. For example, the maximum displacements caused by sinusoidal acceleration and velocity pulses are often not experienced by structures with the same frequencies as the pulse, but by structures with lower natural frequencies.

1 INTRODUCTION

This paper reports on work in progress on the modeling of pulses of strong earthquake motion and the response of structures to such models. Many records of strong ground shaking recorded in the near field show large impulsive motions of the ground, as opposed to the broadband, random motion seen in other strong, but usually more distantly measured records. These pulses in ground motion are typically seen most clearly in the integrated velocity and displacement traces, rather than in the accelerogram itself, because of the masking effects of the high frequency shaking which is also present in the motions. The pulses of motion can be very destructive, as they were in the Kobe and Northridge earthquakes. Consequently, they have attracted much interest in the scientific and engineering communities and several investigators have devised simple mathematical models of pulses for use in studies of structural response.

Important properties of impulsive ground motions include amplitude, duration, pulse shape, frequency content and the number of lobes the pulse contains. Another important property is which characteristic of ground motion is modeled: acceleration, velocity, or displacement. For simple pulses, these important properties are interdependent. Whichever characteristic of ground is modeled, it is the acceleration that excites the structure and models of velocity and displacement pulses have to be differentiated to study structural response.

For convenience, the response of simple structures to pulse motions can be divided into pulses of high frequency, pulses in the resonance band of structural response, and low frequency pulses. For these types of pulses, the effects of damping, the way the response increases with time, the time when maximum response occurs and other features of the response can be studied to develop an understanding of impulsive response, including how it differs from the response to other types of earthquake excitation.

2 TWO ELEMENTARY PULSES

One of the simplest representations of a pulse is a delta function, which idealizes a very short, single-lobed pulse with large amplitude. If the delta function is used to model acceleration, the area under the pulse represents the impulsive change in ground velocity, denoted by v_m. Figure 1a shows the acceleration, velocity and displacement associated with a delta function representation of an acceleration pulse; ε is an arbitrarily small positive constant. Figure 1b is similar, except in this case the delta function models a velocity pulse, so the acceleration pulse has two lobes, rather than one. Because of the two lobes of acceleration, the net change in velocity is zero, but there is an impulsive change in the ground displacement, given by d_m. The simple pulses in Figure 1 help introduce some of the features of structural response to impulsive ground motion.

The earthquake response of a single degree of freedom structure, such as used in the calculation of response spectra, is governed by the following equation of motion,

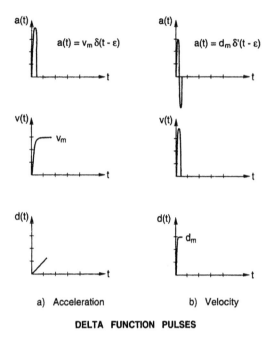

a) Acceleration   b) Velocity

DELTA FUNCTION PULSES

Figure 1. Delta function models of ground motion pulses: a) delta function ground acceleration and consequent ground velocity and displacement; b) delta function model of ground velocity with associated ground acceleration and displacement.

$$\ddot{x}(t) + 2\omega\zeta\dot{x}(t) + \omega_n^2 x(t) = -a(t) \tag{1}$$

in which $x(t)$ is the displacement of the structure with respect to the ground, $a(t)$ is the ground acceleration, ω is the undamped natural frequency of the structure and ζ is the fraction of critical damping. Denoting the displacement and velocity relative to the ground at $t = 0$ by x_0 and $\dot{x}_0$, respectively, the complete solution to Equation 1 can be written as

$$x(t) = x_0 e^{-\omega\zeta t} \cos\omega\sqrt{1-\zeta^2}\,t$$
$$+ \frac{\dot{x}_0 + \omega\zeta x_0}{\omega\sqrt{1-\zeta^2}} e^{-\omega\zeta t} \sin\omega\sqrt{1-\zeta^2}\,t - \frac{1}{\omega\sqrt{1-\zeta^2}}$$
$$\times \int_0^t e^{-\omega\zeta(t-\tau)} \sin\omega\sqrt{1-\zeta^2}\,(t-\tau)a(\tau)d\tau \tag{2}$$

With minor modifications, Equations 1 and 2 also describe the response of a mode of a multi-degree of freedom linear structure, for example, the response of the fundamental mode of a tall building. Because of the interest in the effects of very large pulses, the terms arising from the initial conditions in Equation 2 are often neglected in comparison to the response caused by the pulse.

If the acceleration in Equation 1 is the delta function from Figure 1a, and if the structure is taken to be at rest at the beginning of the pulse, then only the integral in Equation 2 remains, yielding

$$x(t) = \frac{-v_m}{\omega\sqrt{1-\zeta^2}} e^{-\omega\zeta t} \sin\omega\sqrt{1-\zeta^2}\,t \tag{3}$$

This decaying sinusoidal wave reaches its maximum absolute value the first time that

$$\omega\sqrt{1-\zeta^2}\,t = \tan^{-1}\frac{\sqrt{1-\zeta^2}}{\zeta} \tag{4}$$

For moderate damping, this angle is very close to $\pi/2$. In this vicinity, the sine wave in Equation 3 is relatively insensitive to small changes in its argument. Solving Equation 4 and evaluating the sine wave in Equation 3 shows that its value is within 2 percent of unity, even for damping ratios as high as 20 percent.

Approximating the amplitude of the sine wave by unity, the maximum damped and undamped responses can both be evaluated from Equation 3. The ratio of the peaks of damped to undamped response, r_ζ, is found to be

$$r_\zeta = \frac{e^{-\frac{\pi\zeta}{2\sqrt{1-\zeta^2}}}}{\sqrt{1-\zeta^2}} \tag{5}$$

Equation 5 shows that damping is relatively ineffective in reducing the response to the impulse. The response for five-percent damping is 93 percent of that for the undamped case and 15 percent damping is needed to reduce the response to 80 percent of that of the undamped case. This result is quite different from that of the same structure subjected to relatively long (in comparison to the natural period) segments of broadband excitation, where the effects of even small amounts of damping are quite substantial.

If the pulse is idealized as a delta function of velocity as shown in Figure 1b, then the acceleration is described by

$$a(t) = d_m \delta'(t - \varepsilon) \tag{6}$$

Again taking the structure to be initially at rest, the integral in Equation 2 can be integrated by parts to yield

$$x(t) = d_m e^{-\omega\zeta t}$$
$$\times \left[\frac{\omega\zeta}{\omega\sqrt{1-\zeta^2}} \sin\omega\sqrt{1-\zeta^2}\,t - \cos\omega\sqrt{1-\zeta^2}\,t \right] \tag{7}$$

This decaying wave has its first and largest relative maximum in absolute value at the end of the pulse, given in this representation by the value of $x(t)$ at $t = 0^+$. The maximum absolute value, d_m, is seen to be independent of the damping factor; damping does not reduce the maximum value at all for this type of pulse.

120

The above results strictly apply only when the duration of the pulse is extremely short in comparison to the natural period, $T = 2\pi/\omega$, of the structure. They are suggestive, however, of the nature of the response to more realistic models of short pulses.

3 RESPONSE TO ACCELERATION PULSES

As examples of more general models of ground motion pulses, Figure 2 depicts sine wave models of acceleration pulses. Figure 2a shows a half sine wave pulse, while Figure 2b shows a full sine wave pulse. The acceleration pulses in Figure 2 are convenient models for illustrating some of the effects of impulsive excitation, including the influence of the number of lobes, the effect of pulse duration and the reduction in response caused by damping. These pulses are familiar to many researchers. With a different initial condition for $\dot{x}_0$, Jacobsen and Ayre (1958) have examined in detail the undamped response caused by both the half sine wave and full sine wave pulses. The full sine wave pulse is also the "cycloidal type-A pulse" whose damped and undamped response has been studied by Makris and Chang (2000).

For simplicity and brevity, only the undamped response is presented here. In this case Equation 2 reduces to

$$x(t) = x_0 \cos\omega t + \frac{\dot{x}_0}{\omega} \sin\omega t$$
$$- \frac{a_m}{\omega} \int_0^t \sin\omega(t-\tau)\sin\omega_p \tau\, d\tau \tag{8}$$

Equation 8 is valid during the time of application of the pulse, which is $2t_1$ for the half sine wave pulse and $4t_1$ for the two-lobed, full sine wave pulse. Free vibrations follow after the end of the pulse in both cases.

Equation 8 can be readily evaluated, but it is useful first to think of the solution in the following form

$$\begin{Bmatrix} x(t) \\ \dot{x}(t) \end{Bmatrix} = \begin{bmatrix} A_{11} & A_{12} \\ A_{21} & A_{22} \end{bmatrix} \begin{Bmatrix} x_0 \\ \dot{x}_0 \end{Bmatrix}$$
$$+ a_m \begin{Bmatrix} B_1 \\ B_2 \end{Bmatrix} + a_m \begin{Bmatrix} C_1 \\ C_2 \end{Bmatrix} \tag{9}$$

The first term on the right side of Equation 9 represents the effects of the initial conditions and the second term represents terms attributable to the sudden application of the pulse at $t = 0$. In both cases the response is at the natural frequency of the struc-

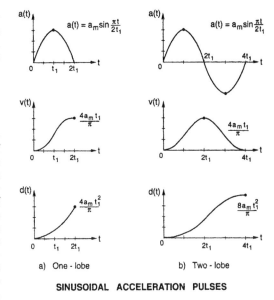

SINUSOIDAL ACCELERATION PULSES

Figure 2. Sinusoidal models of ground acceleration pulses and associated ground velocities and displacements: a) half sine wave pulse with a single lobe; b) full sine wave pulse with two lobes.

ture. The third term on the right side represents an emergent steady-state term, or resonant term if $\omega_p = \omega$, and is at the frequency of the excitation, ω_p. Evaluating the integral in Equation 8 and putting the solution in the form of Equation 9 gives, for $\omega \neq \omega_p$,

$$\begin{Bmatrix} x(t) \\ \dot{x}(t) \end{Bmatrix} = \begin{bmatrix} \cos\omega t & \frac{1}{\omega}\sin\omega t \\ -\omega\sin\omega t & \cos\omega t \end{bmatrix} \begin{Bmatrix} x_0 \\ \dot{x}_0 \end{Bmatrix}$$
$$+ a_m \begin{Bmatrix} -\dfrac{\omega_p}{\omega(\omega_p^2-\omega^2)}\sin\omega t \\[2mm] -\dfrac{\omega_p}{\omega_p^2-\omega^2}\cos\omega t \end{Bmatrix}$$
$$+ a_m \begin{Bmatrix} \dfrac{1}{\omega_p^2-\omega^2}\sin\omega_p t \\[2mm] \dfrac{\omega_p}{\omega_p^2-\omega^2}\cos\omega_p t \end{Bmatrix} \tag{10}$$

In the case in which $\omega = \omega_p$, Equation 10 is replaced by

$$\begin{Bmatrix} x(t) \\ \dot{x}(t) \end{Bmatrix} = \begin{bmatrix} \cos\omega t & \frac{1}{\omega}\sin\omega t \\ -\omega\sin\omega t & \cos\omega t \end{bmatrix} \begin{Bmatrix} x_0 \\ \dot{x}_0 \end{Bmatrix}$$

$$+ a_m \left\{ \begin{array}{c} -\dfrac{1}{2\omega^2}\sin\omega t \\[2mm] -\dfrac{1}{2\omega}\cos\omega t \end{array} \right\}$$

$$+ a_m \left\{ \begin{array}{c} \dfrac{1}{2\omega}t\cos\omega t \\[2mm] \dfrac{1}{2\omega}\cos\omega t - \dfrac{1}{2}t\sin\omega t \end{array} \right\} \qquad (11)$$

Equations 10 and 11 can be used to evaluate the response during and at the end of either the full sine wave pulse or the half sine wave pulse. With zero initial conditions, Equation 10 reduces to

$$x(t) = -\frac{a_m}{\omega_p^2 - \omega^2}\left[\frac{\omega_p}{\omega}\sin\omega t - \sin\omega_p t\right]$$

$$\dot{x}(t) = -\frac{a_m\omega_p}{\omega_p^2 - \omega^2}\left[\cos\omega t - \cos\omega_p t\right]. \qquad (12)$$

3.1 Response to a half sine wave pulse

Because it relates to the example of the delta function acceleration pulse, consider first the response of the structure to a "fast pulse", a half sine wave pulse of short duration, measured by the ratio of the pulse frequency, $\omega_p = \pi/2t_1$, to the structure's natural frequency. For a pulse of short duration $\omega_p > \omega$, so the coefficient of the first term of the displacement in Equation 12 is larger than unity. Because the velocity does not change sign, Equation 12 shows that for $\omega_p > \omega$, the maximum response always occurs after the end of the half sine wave pulse. At the end of the pulse,

$$x(2t_1) = -\frac{a_m}{\omega_p^2 - \omega^2}\left[\frac{\omega_p}{\omega}\sin 2\omega t_1\right]$$

$$\dot{x}(2t_1) = -\frac{a_m\omega_p}{\omega_p^2 - \omega^2}\left[\cos 2\omega t_1 + 1\right] \qquad (13)$$

The displacement and velocity in Equation 13 are the initial conditions for a subsequent free vibration described by the first two terms of Equation 8, with time measured from the end of the pulse. The two terms of this vibration can be combined into a single sine wave (with a non-zero phase angle) whose maximum amplitude is

$$x_m = \frac{2a_m\omega_p}{\omega(\omega_p^2 - \omega^2)}\cos\frac{\omega\pi}{2\omega_p}. \qquad (14)$$

Equation 14 gives the maximum absolute value of the response to a half sine wave pulse for $\omega_p > \omega$.

As shown in Figure 2, the half sine wave pulse of acceleration produces a maximum ground velocity of $v_m = 4a_m t_1/\pi$. Using this in Equation 14 and taking the limit for $\omega_p >> \omega$, Equation 14 yields $x_m = -v_m/\omega$, consistent with the maximum of Equation 3 when the damping is zero. This result suggests something that is, in fact, true: for very short pulses the area of the acceleration pulse determines the response, not the details of its shape. For example, if $\omega_p = 4\omega$, the response of the structure to a half sine wave acceleration pulse is the same, within 6 percent, as the response to a delta function with the same total impulse. For such high frequency pulses, the results for delta function excitation imply that damping is not very effective in reducing the maximum response.

The resonant case, $\omega_p = \omega$, is obviously of special interest. Neglecting the initial conditions, Equation 11 gives during either a half sine wave or full sine wave pulse:

$$x(t) = -\frac{a_m}{2\omega_p}\left[\frac{1}{\omega_p}\sin\omega_p t - t\cos\omega_p t\right] \qquad (15)$$

$$\dot{x}(t) = -\frac{a_m t}{2}\sin\omega_p t.$$

At the end of a half sine wave pulse at $t = 2t_1$,

$$x(2t_1) = -\frac{a_m\pi}{2\omega_p^2} \qquad (16)$$

$$\dot{x}(2t_1) = 0.$$

Equations 15 and 16 imply that for the resonant case the maximum response in absolute value occurs at the end of the pulse.

The relative hazard to structures with different frequencies, ω, of a fast half sine wave pulse of a given frequency, ω_p, can be assessed by examining the ratio Equation 14 to Equation 16 (in absolute value) as a function of ω/ω_p in the range from zero to unity. The ratio, r, is

$$r = \frac{\dfrac{4}{\pi}\cos\dfrac{\omega\pi}{2\omega_p}}{\dfrac{\omega}{\omega_p}\left[1 - \dfrac{\omega^2}{\omega_p^2}\right]} \qquad (17)$$

For frequency ratios close to unity, r approaches unity as expected, but r is always greater than unity and for small frequency ratios it increases as the inverse of the frequency ratio. This result means that the largest displacement is not induced in the resonant case by this pulse, all structures with lower frequencies respond more. The build-up of resonance is too limited by time to be as important as the fact that lower frequency, softer structures simply deflect more under comparable impulsive loads.

The response to a slow pulse, $\omega_p < \omega$, is also governed by Equation 12. In this case Equation 12 implies that the response consists of a slow wave at the frequency of the pulse plus a faster wave with smaller amplitude at the frequency of the structure. At the end of the half sine wave pulse, an analysis like that for the fast pulse shows the maximum response after the pulse to be

$$x_m = \frac{2a_m\omega_p}{\omega(\omega^2 - \omega_p^2)}\cos\frac{\omega\pi}{2\omega_p} \tag{18}$$

However, unlike Equation 14, Equation 18 can be zero for some values of the frequency ratio, e.g., $\omega = 3\omega_p$, indicating that the maximum can occur during the pulse. From examination of the velocity in Equation 12 it can be seen that there is at least one relative maximum of the displacement during the pulse. An upper bound for the maximum value of the response during the pulse can be found from Equation 12 by assuming a simultaneous peaking of the two sine waves of the displacement,

$$x_m \leq \frac{2a_m\omega_p}{\omega(\omega^2 - \omega^2)}\left[\frac{\omega + \omega_p}{2\omega_p}\right] \tag{19}$$

This upper bound is always bigger than the value achieved after the pulse.

4 RESPONSE TO A FULL SINE WAVE PULSE

If the pulse under study is the full sine wave in Figure 2b and the initial conditions are negligible, the undamped response during the pulse is governed by Equations 12 or 15, which are valid in this case until the end of the pulse at $t = 4t_1$. For the resonant case, $\omega_p = \omega$, the maximum is again at the end of the pulse and is given by

$$x_m = x(4t_1) = \frac{a_m\pi}{\omega_p^2} \tag{20}$$

a value that is exactly twice the absolute value of the maximum response to the half sine wave pulse. In the case of a fast pulse, $\omega_p > \omega$, the response to a full sine wave pulse is more complicated than was the case for a half sine wave pulse because the first relative maximum in absolute value occurs during the second half of the pulse, i. e., between $2t_1$ and $4t_1$. If the pulse is very fast, $\omega_p \gg \omega$, the time of the maximum approaches $4t_1$, the end of the pulse.

At this time, the limit of the displacement in Equation 12 approaches the negative of the maximum ground displacement, $-d_m$, consistent with Equation 7. In the general case, the fast pulse will have a maximum when the velocity in Equation 12 becomes zero, which occurs when

$$t_m = \frac{2\pi}{\omega_p + \omega} \tag{21}$$

Substituting into Equation 12 gives

$$x(t_m) = -\frac{a_m\omega_p}{\omega(\omega_p^2 - \omega^2)}$$

$$\times\left[\sin\frac{2\pi\omega}{\omega_p + \omega} - \frac{\omega}{\omega_p}\sin\frac{2\pi\omega_p}{\omega_p + \omega}\right] \tag{22}$$

Equation 22 gives the maximum absolute value of the response during the pulse, but for all values of $\omega < \omega_p$ the second maximum, which occurs after the pulse, is larger. From Equation 12, the displacement and velocity of the structure at the end of pulse at $4t_1$ are

$$x(4t_1) = -\frac{a_m\omega_p}{\omega(\omega_p^2 - \omega^2)}\sin 4\omega t_1$$

$$\dot{x}(4t_1) = -\frac{a_m\omega_p}{\omega_p^2 - \omega^2}\left[\cos 4\omega t_1 - 1\right] \tag{23}$$

Using these values as initial conditions, the subsequent maximum after the pulse has an amplitude of

$$x_m = \frac{a_m\omega_p}{\omega(\omega_p^2 - \omega^2)}2\sin\frac{\omega\pi}{\omega_p} \tag{24}$$

Forming the ratio of the maximum response for a structure with $\omega < \omega_p$, Equation 24, to that of a structure with $\omega = \omega_p$, Equation 20,

$$r = \frac{\dfrac{2}{\pi}\sin\dfrac{\pi\omega}{\omega_p}}{\dfrac{\omega}{\omega_p}\left(1 - \dfrac{\omega^2}{\omega_p^2}\right)} \tag{25}$$

This ratio approaches 2 for $\omega_p \gg \omega$, goes to unity as $\omega \rightarrow \omega_p$, and is always greater than unity in between. As in the case of the half sine wave pulse, the greatest displacement occurs for structures with periods longer than the pulse period, not for structures with the same period as the pulse. Three or more lobes of a sinusoidal pulse are needed before the resonant response is greater than the response of at least some structures with lower frequencies.

5 RESPONSE TO VELOCITY PULSES

As mentioned earlier, large pulses in the ground motion are often seen more clearly in the velocity and displacement traces than they are in the accelerogram itself. Two simple sinusoidal velocity pulses are shown in Figure 3, which is in the same format as Figure 2. The single-lobed pulse in Figure 3a can model pulses recorded close to the fault that are often associated with permanent ground displacement. The two-lobed velocity pulse in Figure 3b represents the case in which the ground displacement is an over-and back motion. This is the "cycloidal type-B pulse" whose response has been studied by Makris and Chang (2000).

In terms of acceleration, the pulse in Figure 3a consists of two anti-symmetric lobes of opposite sign. This means that the total impulse of the acceleration is zero and implies that the response of a structure to a very past pulse will resemble Equation 7, the response to the pulse in Figure 1b. The acceleration in Figure 3b consists of three pulses, with the negative one in the center twice the size of its positive neighbors. Again the total impulse of the ground acceleration is zero.

It is not possible to give here an extensive discussion of the response of a simple structure to the velocity pulses in Figure 3 and only some selected re-sults are presented. The response in the resonant case, $\omega = \omega_p$, is found to be

$$x(t) = -\frac{a_m t}{2\omega_p}\sin \omega_p t$$

$$\dot{x}(t) = -\frac{a_m}{2\omega_p}\left[\sin \omega_p t + \omega_p t\cos \omega_p t\right]$$

(26)

This result is very similar to Equation 15. The displacement is zero at the end of both the half and full sine wave pulses, with the maximum displacement occurring in the subsequent free vibrations. In both cases this maximum is larger than the relative maxima that occur during the pulse. For the half sine wave pulse, the maximum is $\pi a_m / 2\omega_p^2$, with the maximum for the full sine wave pulse exactly twice this value. With the same maximum acceleration, these results for the resonant case are the same as those for the sinusoidal acceleration pulses, Equations 16 and 20, indicating that a simple phase shift in the excitation has no effect on the peak structural response, as long as the amplitude and frequency of the sinusoidal acceleration are unchanged. However, if the same results are expressed in terms of the maximum ground velocity, the peak responses from the acceleration pulses are only half those caused by the velocity pulses. This is explained by comparing Figures 2 and 3, a comparison that shows that, depending on the phasing, an acceleration wave can either produce a maximum velocity equal to the maximum of the sinusoidal velocity wave, or a maximum twice that value.

For a fast ($\omega_p > \omega$) half sine wave velocity pulse, the displacement response is always greater than the resonant case and appears to occur always after the pulse. The response to a full sine wave velocity pulse is more complicated and may exceed that of the resonant case only over a limited range of frequencies. For very fast pulses or very flexible structures, $\omega_p \gg \omega$, both the response displacement and velocity at the end of the pulse approach zero because the ground acceleration and velocity both integrate to zero. In effect, the ground moves back and forth under the structure causing a maximum displacement response of $x_m = d_m$, which occurs during the pulse at $t = 2t_1$.

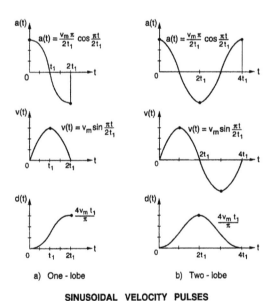

a) One - lobe   b) Two - lobe

SINUSOIDAL VELOCITY PULSES

Figure 3. Sinusoidal models of ground velocity pulses and associated ground accelerations and displacements: a) half sine wave pulse with a single lobe; b) full sine wave pulse with two lobes.

6 CONCLUSIONS

Perhaps the most significant result from this brief report is that the response to simple pulses is more complex than one might expect and that it is different in significant ways from the response to longer earth quake excitations. A case in point is the fact that a pulse may not cause the largest displacement to occur in structures with the same period as the pulse. Instead, structures with lower frequencies often can show larger responses than the resonant case. In constructing sinusoidal models of ground motion pulses, it appears that the phasing, the number of lobes used and whether the acceleration, velocity or displacement is chosen for modeling are all important factors, as well as the amplitude and frequency of the underlying sinusoid. Although damped response was not studied extensively here, the response to delta function pulses suggests that damping is not necessarily very effective in limiting structural response to large pulses in the ground motion.

REFERENCES

Jacobsen, L. S. and Robert S. Ayre (1958). *Engineering Vibrations*, McGraw-Hill Book Company, New York.

Makris, N and S-P Chang (2000). "Response of Damped Oscillators to Cycloidal Pulses", *Journal of Engineering Mechanics*, American Society of Civil Engineers, Vol. 126. No.2, pp. 123-131.

Earthquake Engineering Frontiers in the New Millennium, Spencer & Hu (eds),
© 2001 Swets & Zeitlinger, ISBN 90 2651 852 8

An overview of US-PRC cooperation on earthquake engineering at IEM and future prospects of the program

Xiaozhai Qi
Institute of Engineering Mechanics, China Seismological Bureau, Harbin 150080, China

ABSTRACT: In the 20 years since the signing of the US-PRC Protocol for Scientific and Technical Cooperation in Earthquake Studies, many research fruits have been reaped. This paper gives an overview of the program at IEM and its accomplishments, as well as possible directions for continued cooperation.

1 INTRODUCTION

Since the US-PRC Protocol for Scientific and Technical Cooperation in Earthquake Studies formally went into effect in 1980, it has provided a good opportunity for bilateral cooperation in earthquake studies between China and the United States. As a leading institute in the field of earthquake engineering in China, the Institute of Engineering Mechanics (IEM, China Seismological Bureau) has played an important role in promoting, developing and implementing the bilateral cooperation. The two important events in the history of US-PRC cooperation on earthquake engineering, which pushed the US-PRC cooperation forward, need mentioning especially here. The first one is that Prof. G. W. HOUSNER led a large American delegation (Dr. S. C. LIU was a important member of the Delegation) to visit China in 1978 and stayed at IEM about one week. Both sides discussed widely the possible cooperation prospects and directions, and suggested conducting the US-PRC cooperation study on strong ground motion firstly. The second one is the signing of the Annex 3 of the Protocol -- Cooperative Research on Earthquake Engineering and Hazards Mitigation. It was at IEM, Harbin, that the Annex 3 was discussed and drafted in 1979 under the concerted effort of Dr. R. WESSON (USGS), Dr. D. SENICH (NSF), Dr. L. JOHNSON (NSF) and Prof. LIU Huixian, the Director of IEM at the time. The event ushered in a new era in US-PRC cooperation on earthquake engineering.

During this long period — 20 years — of cooperation, both IEM and US colleagues made a number of important achievements on earthquake engineering. These accomplishments will be reviewed herein. Today, facing the new millennium, we have some new ideas for further cooperation. These ideas will be outlined in the following.

2 OVERVIEW OF US-PRC COOPERATION ON EARTHQUAKE ENGINEERING AT IEM

IEM has completed 23 cooperative projects in the past 20 years, including 12 research projects and 11 workshops/seminars, and some of them have got satisfying results and had profound effects on US-PRC cooperation. Here, some of important joint projects and symposiums/workshops are reviewed briefly.

2.1 *Joint projects*

(1) PRC-US joint project on strong ground motion studies (phase I and II)

This project was a preferential project suggested by Prof. G.W. HOUSNER when he led an American delegation to visit our institute in 1978. It was also a project with first priority in the Annex 3 of the Protocol. It ran from 1981 to 1989, through two phases, and a total 22 SMA-1 and 18 PDR-1 instruments were installed in China. Three special arrays and two networks have been deployed and a number of mobile observations have been made. In the meantime, a data processing system was established at IEM with the assistance of US experts. Over 1800 earthquake acceleration records have been obtained to date. Some of the records are rare and valuable strong ground motion observations, such as records from the 7.0 Lijiang earthquake and the 5.9 Tangshan earthquake.

Additionally, seven scientific reports, seven volumes of ground motion data and 19 papers were completed in this cooperative effort. All of the results are valuable to the study on earthquake engineering. Both sides are satisfied with the development and achievement of the project, and wish to continue the cooperation in this field.

(2) Cooperation on study of soil liquefaction (phase I and II)

This program started in 1981 and ended in 1986. It was carried out in 2 phases. Phase I was the study on soil liquefaction induced by the Tangshan Earthquake (1976), and Phase II involved site studies of liquefaction and pore pressure of soil during earthquakes. A lot of new site data on silt liquefaction, including macroscopic liquefaction data were obtained, and a method of liquefaction discrimination based on Electric Probe data was introduced. Fourteen papers were completed in the cooperation.

(3) Development of probability bases for earthquake resistant design (phase I and II)

The objective of this cooperative research was the development of probabilistic bases for the design of earthquake resistant structures. The cooperation lasted four and a half years, from March 1, 1981 to Sept. 30, 1985. During this time, the fault-rupture model for hazard assessment was evaluated and improved, the two-parameter damage criterion was developed, and a probabilistic method of earthquake damage analysis and design was developed. Twenty research reports and papers were published

(4) Risk analysis and seismic safety of existing structures

This project started in 1984. Through the cooperation, new progress in seismic hazard assessment was made, and damage probability prediction method for multi-story brick buildings was proposed. More than 30 reports were presented.

(5) Studies on seismic behavior of liquid storage tanks

A response analysis of the structure-fluid-soil dynamic interaction based on the semi-analytical element method was developed.

(6) Research on strong ground motion simulation of the San Francisco Bay

This research project was carried out from August 1998 to March 1999. In this project, GIS is used to simulate seismic ground motion in the San Francisco Bay Area. The method for calculating ground motion was given. Case studies for three scenario earthquakes are conducted to verify the method described in the methodology. The merits of the work lie in the ability to demonstrate the use of existing database and various ground motion models

(7) Studies on recorded responses for instrumented building structures

This project started officially in August 1998. The vibration properties were identified from the recorded response of an instrumented building, and a 3D finite element model was developed. Seismic response of the structure was analyzed, and reliability of current structural analysis technique was evalu-

ated by comparison between the computed responses and the recorded ones.

(8) Integrated earthquake mitigation strategies for metropolitan cities

This joint project started in 1997 and ended in 1999. The objective is to exchange knowledge and the latest experience in seismic loss estimation and hazard mitigation for metropolitan cities, and to improve the methodologies and tools recently developed in China and US. On the China side, an engineering approach was developed for vulnerability analysis and loss estimation, and implemented in the computer software DMEU. On the US side, the standardized system HAZUS was used in the study.

(9) Project on the English edition of "The Great Tangshan Earthquake"

The project began in 1992 and continued for several years. Three hundred writers and translators from more than 100 institutions all over China made great contribution. The English editions of Volumes 1 to 4 have been completed, and will be published in the US. The four volumes include three million words and two thousand photos, and describe the earthquake damage to various types of constructions in the great Tangshan Earthquake so that they provide abundant information about earthquake damage to various kinds of engineering structures and facilities. These books are considered as excellent resources and invaluable to research work in China, as well as to American researchers.

2.2 *International symposiums/workshops*

(1) The PRC-US workshop on seismic analysis and design of reinforced concrete structures

The workshop was held at the University of Michigan, USA, May 4-8, 1981. Twenty-three participants were from America and nine from China. Besides attending the workshop, the PRC delegates also visited Stanford University, the USGS in Menlo Park, University of California at Berkeley, Univ. of Texas at Austin, Univ. of Illinois and the Portland Cement Association in Skokie, Oregon.

(2) The joint PRC-US microzonation workshop

The Workshop was sponsored by the State Seismological Bureau of China and US National Science Foundation, and held in Harbin, Sept. 11-16, 1981. Thirteen participants were from America and twenty-two from China. Nine papers from the US side and 17 papers from the PRC side were presented at the seminar.

(3) PRC-US bilateral workshop on earthquake engineering

The Workshop was sponsored by the State Seismological Bureau (SSB), the Ministry of Urban and

Rural Construction and Environmental Protection of China, and the US National Science Foundation (NSF). It was held in Harbin, China, August 27-30, 1982. Fifty-one participants came from China and 21 from America. Some experts from Canada, India and Japan were also invited to the workshop. Forty-two papers were presented at the workshop

(4) PRC-US-Japan trilateral symposium /workshop on engineering for multiple natural hazard mitigation

Under the auspices of the State Seismological Bureau of China, the US National Science Foundation and the Ministry of Education of Japan, the symposium was held in Beijing, China, November 7-12, 1985. Forty-four participants were from China, 12 from America, 11 from Japan, with representation from Austria, Italy and Peru. Forty papers were presented in the symposium, which gave a broad introduction to the relevant work on natural disaster reduction in each country.

(5) PRC-US joint workshop on seismic resistance of masonry structures

Under the auspices of the State Seismological Bureau of China and the US National Science Foundation, the workshop was held in Harbin, China, May 21-24, 1986. Forty-nine participants were from China and 15 from America. Thirty-four papers were presented in the workshop. Both sides exchanged the achievements of engineering techniques and research in the field of seismic resistance of masonry structures.

(6) International seminar on seismic zonation

The seminar, sponsored by the State Seismological Bureau of China, was held in Guangzhou, December 6-10, 1987. The US National Science Foundation sent a delegation of 11 to the seminar. There were also participants from Bulgaria, Greece, Italy, Japan and Yugoslavia. In addition, 36 participants came from various parts of the host country. They were all active in the fields of geophysics, seismotectonics and earthquake engineering. Thirty-seven reports on four special topics (General, Geological, Seismological and Engineering Aspects) were presented at the seminar.

(7) Sino-American workshop on strong-motion measurement

The workshop sponsored by US National Science Foundation and the State Seismological Bureau of China was held at the California Institute of Technology, USA, December 13-15, 1988. Twelve participants from America and nine from China attended this workshop.

(8) PRC-US workshop on isolation

The workshop was held at the Earthquake Engineering Research Center, Univ. of California, May 29-30, 1990. Eighteen participants were from America and seven from China. Twenty-five reports were presented at the workshop.

(9) China-Japan-USA trilateral symposium/workshop on earthquake engineering

To bridge the cooperation between China and USA, China and Japan, as well as USA and Japan, the original bilateral seminar was expanded to a China-Japan-USA Trilateral Symposium/Workshop on Earthquake Engineering, and the workshop was held in Harbin, November 5-9, 1991. Five specialists from Shimizu Corporation of Japan, six experts from the United States, one invited guest from Korea and eighteen scholars from IEM participated in this Symposium. Twenty-six technical reports were presented at the symposium, in which a wide range of topics was covered.

(10) Workshop on prospects for PRC-US cooperation on earthquake engineering research

According to the agreement between NSF of USA and the Ministry of Construction and the SSB of China, a Joint PRC-USA Workshop was held in Guangzhou, China on April 25-28, 1992. The workshop brought together experts from both countries to identify the research strength of each country and define a joint research agenda for consideration under Annex 3 of the Protocol. National Research Council (NRC) of USA appointed an eight-member delegation of experts in earthquake engineering, engineering seismology, structural engineering, geotechnical engineering, and disaster planning and preparedness. The IEM assembled a delegation with similar experts. Twenty reports were presented.

(11) PRC-US bilateral workshop on seismic codes

The workshop was held in Guangzhou, China, December 3-7, 1996. Forty-three participants, including 10 from America, three from Hong Kong, three from Japan and about 27 from China attended the workshop.

Of the above-mentioned symposiums/workshops, all those held in China were organized and hosted by IEM including the four held in Harbin. These high-level workshops made the participants, including many famous scholars from China, America, Japan, and other countries, understand the academic thoughts and development in China and the US. These workshops have provided a good opportunity for. summarizing and exchanging widely advanced experiences and new achievements, which helped us further promote, strengthen, and enlarge the US-PRC joint effort on earthquake engineering, as well as technical exchanges all over the world.

We can see from the above introduction that through the 20-year cooperation and exchange, both of IEM and US colleagues have got fruitful results in wide fields of earthquake engineering. These cooperation and exchange lay a solid foundation for future cooperation.

3 PROSPECTS OF US-PRC COOPERATION ON EARTHQUAKE ENGINEERING AT IEM

We would like to propose the following ideas for future cooperation.

3.1 *Cooperation of center to center*

In the past 20 years, most of projects are dispersed so that the advantages or potential resources of researchers could not be fully brought into play. To improve on this situation, a new form of US-PRC cooperation, center-to-center cooperation, should be encouraged in the future by both sides. By this kind of cooperation, both sides could form a long-term, firm collaboration relationship, bring the respective expertise into full play, and also present some interesting joint projects to cooperate. This kind of cooperation will also accommodate holding symposiums/workshops periodically, which is beneficial to exchanging the information, data, staff, and sharing the research resources. This is very important in modern science. As one of earthquake engineering research centers in China, IEM would like to stress the cooperation with the three earthquake engineering research centers in US (Mid-America Earthquake Center, Pacific Earthquake Engineering Research Center and Multidisciplinary Center for Earthquake Engineering Research). Center-to-center (C2C) cooperation in fields of interest can provide a chance for scientists to learn from each other and make up deficiencies of both sides so that the both sides can make great progress on earthquake engineering research.

3.2 *Expanding multilateral cooperation*

Mitigation of earthquake damage is a complicated problem. To solve this, wide international cooperation all over the world is essential. It means that, in future cooperation on earthquake engineering, multilateral cooperation should be initiated, and IEM is very interested in that. The US-Japan cooperative research project in Urban earthquake disaster mitigation includes a lot of new fields of earthquake engineering, such as performance-based design and engineering, integrated social science and related multidisciplinary research, advanced steel structures, geotechnical engineering systems and advanced technologies. These research topics are also high-priority projects at IEM at present and in the future, so it is to IEM's advantage to join the US-Japan ef-fort, and form a US-Japan-PRC multilateral cooperation. Through this kind of cooperation, the scientists of the three countries could learn from each other, develop together and make most effective contribution to the career of earthquake disaster mitigation.

3.3 *Probable fields of cooperation for IEM and US in the future*

In the past 20 years, both China and US have achieved great successes on earthquake engineering, but damaging earthquakes all over the world still caused a great amount of injuries and deaths and heavy economic losses, especially in big cities and important infrastructures. The earthquake damage evoked worldwide repercussions in the earthquake engineering community. Summarizing the experience of earthquake damage and analyzing the observation data, the researchers have posed many new problems. Following the tracks of the developments on earthquake engineering, IEM hope to be engaged in the widespread research in the following fields with the U.S. so that both sides may achieve greater success in earthquake engineering research.

(1) Strong motion observation

Strong motion observation has been one of the subjects of interest in US-PRC cooperation. In the past 20 years, both sides were satisfied with the development and achievement of the project, and wish to continue the cooperation. Future research will continue to include establishment of strong motion array consisting of digital accelerographs, study on observation technique and data processing, and study on application of the observed data. In addition, the following studies will be considered:
  i.   Long period ground motion
  ii. Strong ground motion near active fault
  iii. Multi-component of ground motion and their coherence

(2) Strong ground motion
  i.   Strong ground motion attenuation
  ii. Relationship between ground surface rupture and earthquake source, propagation path and surface geology, and the simulating approach of the rupture
  iii. Stochastic procedure for strong ground motion simulation
  iv. Wide band Green function procedure for strong ground motion simulation
  v.   Earthquake response analysis of large sediment basin
  vi   GIS based system for active fault prospecting and evaluating

(3) Study on performance-based seismic design method of engineering structures

This project includes two parts:

i. Establishing the basic frame of performance-based seismic design. It includes seismic performance of structures and classification; study on standard of optimal performance earthquake level; multi-level check of performance-based seismic design.

ii. Presenting the basic method of performance-based seismic design. This research includes the application of new method on performance-based seismic design; the practical design solving method of equivalent single degree of freedom system; the response parameter of performance objective and the quantitative index of the response parameter of performance objective; the performance design considering about structural complication; and the performance design of nonstructural member and equipment.

(4) Structural seismic control and performance of the controlled structures

The technology of structural seismic control has been made considerable progress during the last twenty years. In present, how to use some new technologies and intelligent materials to the structural seismic control is a new direction. So we hope to be cooperative with the US in following two objectives

i. Study on the new technology of seismic isolation and its performance

ii. The intelligent materials and its application to structural control

(5) Seismic damage to major cities and critical engineering and research on earthquake disaster mitigation

i. Study on application of new technologies, such as GIS, GPS, RS etc., in the monitor and administration of earthquake damage

ii. Study on the new method for evaluating structures vulnerability, seismic economic damage and social effects

iii. Study on disaster prediction and the warning system

(6) Test technique

IEM possesses the biggest shaking table in China and plan to establish the biggest pseudo-dynamic testing device in China, which are put in a same laboratory to form a complete set, fully well-equipped laboratory of earthquake engineering. IEM hope to utilize these test devices to solve some difficult scientific problems in buildings, lifeline engineering structures, and soil etc. with US colleagues together. Both sides can also conduct study on new seismic strengthening technique using these devices

(7) Soil dynamic mechanics

In recent earthquakes, such as Turkey and Taiwan earthquakes, the foundation failure caused serious damage to structures, therefrom caused great losses. The researchers start to pay close attention to this phenomenon and study the reason. The following two aspects will be mainly considered.

i. Study on soil dynamic characteristic on the large deformation

ii. Study on ground rupture caused by fault

3.4 *International symposiums/workshops*

In the past 20 years, IEM and US colleagues held successfully 11 symposiums/workshops, which promoted and strengthened the technical exchange and cooperation among the scientists all over the world. In future cooperation, we hope to continue to organize the bilateral, trilateral, or international symposiums/workshops with the US and other countries so that good opportunities will be provided to scientists for discussing and exchanging advanced experience and new achievements on earthquake engineering.

4 CONCLUDING REMARKS

Through past cooperation activities, both Chinese and American scientists and engineers have obtained a deeper understanding of earthquakes and earthquake engineering. The cooperative projects have promoted development of earthquake engineering in each country.

Earthquake is one of the destructive natural disasters and has brought tremendous casualties and property losses to humanity. Scientists and engineers at IEM hope to develop further international exchange and cooperation in the field of earthquake engineering with scientists and engineers from the US and other countries. We hope to continue our contribution to seismic disaster reduction through various kinds of international and bilateral cooperation.

Earthquake Engineering Frontiers in the New Millennium, Spencer & Hu (eds),
© 2001 Swets & Zeitlinger, ISBN 90 2651 852 8

Catastrophe risk modeling and financial management

Weimin Dong
Risk Management Solutions, Inc. USA

ABSTRACT: The primary focus of this paper is to improve the understanding of how the technical expertise provided by engineers and other risk reduction professionals is used in the process of managing financial risk associated with natural disasters. The technical expert plays an important role in providing both the information used in decisions and in creating and using the analytical tools needed to interpret the information. This paper provides an overview of how financial risk is managed by various sectors (who bears risk, how risk can be reduced, transferred or avoided) and how engineering expertise fits into this process. The discussion is intended to be helpful for engineering professionals in gaining a better understanding of how the financial and insurance communities manage risk, and in understanding how the information they provide to various clients is part of the larger process of financial risk management.

1 ESCALATING ECONOMIC LOSS

Catastrophic economic losses associated with disasters in the U.S. and the world are increasing. In the U.S. seven of the ten most costly disasters in history, based on dollar losses, occurred between 1989 and 1994, such as Hurricane Andrew in 1992 and 1994 Northridge earthquakes. Globally there was the $120 billion earthquake in Kobe, Japan, and most recently the $30 (estimated) billion earthquake in Izmit, Turkey and $ 10 billion earthquake in Taiwan. Fig. 1 shows the potential costs of future natural disasters in USA. Over the 5-year period from 1995 to 1999, the average annual cost to The Federal Emergency Management Agency (FEMA) alone has been more than 1 billion dollars, excluding the cost of the Northridge earthquake. This is very heavy burden to the government. For Northridge earthquake, thanks to insurance social function, the most of reconstruction funds were provided by insurance (See Fig. 2). The recognition that these devastating economic losses are increasing has propelled some segments of the financial community as well as the government to carefully evaluate and identify options for managing risk. Given the enormous costs associated with recent U.S. disasters, government at all levels has been re-evaluating policies for providing disaster assistance and the need to encourage risk reduction through mitigation. FEMA is promoting a much larger focus on mitigation and local resilience through programs such as Project Impact. Governments at all levels recognize that the next major urban disaster

(earthquake or hurricane) could have economic and fiscal consequences that the public monies that have traditionally been available after such disasters may no longer exist.

Figure 1. Probable Costs of Future Natural Disasters

2 ISSUES WITH CATASTROPHE DISASTERS

There are three characters with respect to catastrophic *disasters:*

- They are uncertain. No one knows when, where and what size of disasters will occur;
- They are rare. The chance to have these kinds of events may be one of hundreds of years;
- They have tremendous impact on all walks of live and result in huge losses.

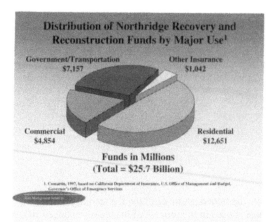

Fig. 2 Distribution of Northridge Recovery and Reconstruction Funds

Due to uncertain and rareness in nature, it is difficult to justify economically proper actions such as retrofit, enhancement of building code, etc. to be taken. Decision-makers don't have clear picture what is the consequence should a disaster occur and usually mitigation measures have been put at the end of to-do list. There is urgent need to understand how to manage these kinds of catastrophe risks.

3 UNDERSTANDING RISK MANAGEMENT

The risk management process can be defined as the process for making and carrying out decisions that will minimize the adverse effects of accidental loss, involving five steps: identifying and analyzing loss exposures (also referred to as risk assessment); examining the feasibility of alternative risk management techniques; selecting the best technique(s); implementing the techniques; and monitoring the program. Measuring the potential frequency and severity of loss must use sound scientific and engineering methods, so engineers play a large role in conducting such risk assessments or analyses. Typically these assessments include identifying the asset or property at risk; evaluating the natural hazard at the site; estimating damage; and assessing monetary loss. These analyses then form the foundation from which other decisions can be made, including the selection of alternative risk mana-gement techniques (to eliminate loss exposures, control loss exposures that are not eliminated; or finance losses that occur despite the controls).

4 TECHNOLOGICAL CHANGE

Technology is changing rapidly, and with these changes come increased opportunities for better understanding risk, and therefore better management of the risk. With the advance of science and engineering, we will be able to estimate the potential losses for future disasters with decent confidence. Many tools that have been developed in the last decade use the various loss estimation methodologies. Millions of dollars have been invested in developing these tools that allow for a more thorough understanding of expected losses associated with various events. These tools are changing how decisions are made regarding the management of financial risks associated with natural disasters. Tools such as Geographic Information Systems enable better mapping and more precise understanding of the risk. Communications changes including e-mail and the use of cellular phones are changing how business is conducted, as are changes in information technology, such as the use of the Internet to conduct financial transactions and make risk management decisions.

5 NEED FOR ENGINEERING MODEL

Estimating losses due to future catastrophe events based on actuarial (historic) data only, is inadequate for the following reasons. First, catastrophes are rare events so that the time window for which data have been collected is short compared with the return period of catastrophes. In particular, major earthquakes and extreme storms have average return period of the order of hundreds of years. The data time window is simply too short for the data to be used as the basis of an accurate estimator of the average loss. Missing an extreme loss will result in a much lower estimate of the average annual loss. Conversely, including a big one in a short time interval will overestimate the annual loss.

For illustration, Tables 1 and 2 are two tabulations of income/loss histories in California with regard to earthquakes since 1970. The tables are identical except that the Northridge event of 1994 is included in the second table but not the first. By comparing the two tables, it is obvious that a single modern event can alter the actuarial estimates significantly. In particular, according to the actuarial approach, the average loss ratio for the period 1970-1993 is 0.26, but increases to 2.07 when 1994 is included. The Northridge earthquake is only a moderate one. The loss would be much greater should an earthquake occur at the Newport Inglewood fault or should a repeat of the 1906 San Francisco earthquake take place (see Refs.1 and 2).

Another difficulty in using the actuarial approach is that, to be useful for contemporary times, historic data must be properly adjusted to account for factors that affect the ultimate loss, including:

- Inflation
- Increase in exposure
- Changes in vulnerability (updated building codes)

Table 1. Underwriting experience, 1970-1993, earthquake (in million US$) (Ref. 3)

Year	Earthquake Event	Premium	Losses
1970		5.9	0
1971	San Fernando (6.6)	4.6	.8
1972		9.0	2.1
1973		10.9	.1
1974		13.0	.4
1975	Oroville	13.8	0
1976		17.1	.1
1977		19.8	.1
1978	Santa Barbara	23.2	.4
1979	Imperial Valley (6.6)	29.0	.6
1980		38.5	3.5
1981		50.2	.5
1982		58.9	0
1983	Coalinga (6.7)	70.4	2.0
1984	Morgan Hill (6.2)	79.4	4.0
1985		132.9	1.7
1986	Southern California	180.0	16.7
1987	Whittier (5.9)	208.4	47.6
1988		277.8	31.8
1989	Loma Prieta (7.1)	333.6	433.0
1990	Southern California	384.6	180.9
1991	Northern California	427.4	73.7
1992		479.9	87.7
1993		521.0	13.2
Total		**3,389.4**	**880.9**
Average loss ratio			0.26

Table 2. Underwriting experience (1970-1994, including Northridge), earthquake (in million US$) (Ref. 3)

Year	Earthquake Event	Premium	Losses
1970		5.9	0
1971	San Fernando (6.6)	4.6	.8
1972		9.0	2.1
1973		10.9	.1
1974		13.0	.4
1975	Oroville	13.8	0
1976		17.1	.1
1977		19.8	.1
1978	Santa Barbara	23.2	.4
1979	Imperial Valley (6.6)	29.0	.6
1980		38.5	3.5
1981		50.2	.5
1982		58.9	0
1983	Coalinga (6.7)	70.4	2.0
1984	Morgan Hill (6.2)	79.5	4.0
1985		132.9	1.7
1986	Southern California	180.0	16.7
1987	Whittier (5.9)	208.4	47.6
1988		277.8	31.8
1989	Loma Prieta (7.1)	333.6	433.0
1990	Southern California	384.6	180.9
1991	Northern California	427.4	73.7
1992		479.9	87.7
1993		521.0	13.2
1994	Northridge (6.9)	619.4	7414.1
Total		4008.7	8295.0
Average loss ratio			2.07

- Deeper insurance market penetration
- Changes in policy structures.

The adjustment process is tortuous because the data reflect the time and socio-economic conditions at which they are collected. To adjust their values to modern times properly, the companion data (exposure, market penetration, policy changes) associated with each event must be collected, compiled and compared with their current counterparts. Since data on these items exist only in fragmentary form, it is difficult if not impossible to reconstruct the complete demographic and socio-economic environment in past years.

For these reasons, the use of historic data only and an empirical approach to forecast catastrophe losses is unsatisfactory; such estimates contain large uncertainties. The uncertainties can only be checked by an independent source of estimate, such as with the engineering modeling approach described below. Note that while engineering models are based on physics of the phenomenology and processes involved, they are not devoid of data. Most models are calibrated by their corresponding databases when available, e.g., building vulnerability, seismicity, attenuation functions, and so on. But the fundamental

difference is that data are used to substantiate the models for the individual processes and phenomenology, so they may reflect the state of knowledge and current conditions. The models are then combined in making the loss estimates.

In general, a complete engineering model for catastrophe impacts has four components or modules:

- A hazard module which predicts a series of events with various sizes, locations, and frequencies.
- An attenuation module which generates the local severity (ground shaking intensity or peak wind gust) from the source given the size and location of the event.
- A vulnerability module which estimates the loss given the local severity.
- A financial module which allocates loss to different owners of risk.

5.1 Hazard source

The starting point in the engineering modeling approach is the source of the disturbance. For example, for earthquake hazards, the source is the rupture of a

135

particular fault (Fig. 3). A credible scenario is postulated, which is consistent with the seismicity database on frequency, magnitude and rupture length.

Figure 3. Sample source models: major faults in California

5.2 *Attenuation*

The energy released at the source manifests itself in the form of physical hazards such as ground shaking or wind velocity. These hazards are propagated from the source region, and attenuate and disperse as they propagate. The degree of attenuation is important as it affects the severity of hazards felt at the site of interest and the extent of influence of the event. The governing physical laws are summarized in the form of attenuation functions, which very often rely on empirical data for calibration. Figure 4 is a typical attenuation relation commonly used for propagating ground shaking in rock materials, and the uncertainty bands shown are representative of the uncertainty associated with this part of the modeling.

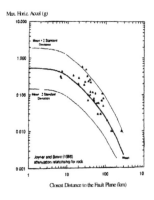

Figure 4. Attenuation of ground shaking with distance for rock sites.

5.3 *Asset vulnerability and damage*

The site hazards are applied against the vulnerability of the assets in order to determine the damage incurred in the asset. While this process entails complex physical phenomenology and engineering development, the concept can be illustrated with reference to the loss ratio curve of Fig.5 used by Steinbrugge (Ref. 4). The figure gives the extent of damage to a particular class of structures when they are subjected to various levels of ground shaking. Uncertainties in the estimate of damage, due to inherent variability in the details of the structure as well as the state of knowledge, are also noted. These curves have since been greatly extended and refined based on damage and insurance claim data collected in recent events.

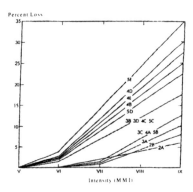

Figure 5. Loss ratio of building classes to ground shaking

5.4 *Financial module*

Financial module will allocate loss incurred to all risk bearers based on insurance or other contract.

6 FINANCIAL MANAGEMENT

In the following, we will briefly discuss how various sectors manage financial risk. For more detailed information, refer to EERI White Paper (Ref. 5).

6.1 *Public policy*

Government plays an important role in setting policies and regulations that affect risk bearing and risk transfer. Some government regulations (building codes, standards, land use regulations) and policy incentives (tax breaks, low cost loans, lower insurance costs) can reduce potential catastrophic risk. Other government policies have been used to transfer risk, such as the California Earthquake Authority. Other government policies also affect how insurers, lenders and the capital markets can develop and use new financial instruments.

6.2 Overview of the risk bearing sectors

6.2.1 The residential sector
The residential sector includes homeowners as well as renters and owners of multifamily residential buildings.

Homeowners include the owners of any residential property that could be damaged in disaster. For most people homeownership is probably the largest investment in one's life and with the investment the owner assumes the risk. The financial decisions a homeowner faces include: do nothing; risk reduction with retrofit or risk transfer through diversification of ownership, insurance or mortgage. Most of homeowners are reluctant to put up money at front to do retrofit to reduce the potential loss, which is so remote and may not happen at all. As a result, the burden of recovery after a disaster will be put on the shoulder of the government. Renters usually bear risk even more directly since they would have fewer options for mitigation and can only transfer risk via insurance.

6.2.2 The small business sector
There are several characteristics that define the way in which a small business owner manages risk. Typically small business owners are sole or majority owners. Any loans are likely personal loans or collateralized with personal assets. They rent rather than own the premise of business; thus they are not at liberty to implement any physical measures related to the structure of the premise without the landlord's consent even if they are willing to pay for the expenses. They have little discretionary income to invest in preventive measures, and no staff dedicated to risk management. They do not have the leverage in negotiating insurance coverage that big businesses have. Their diversification options are very limited. Consequently, the risk borne by small businesses is primarily indirect and related to business interruption.

6.2.3 The corporate sector
The big corporations face more serious risks from natural disasters not only their properties expose to the hazard also business interruption will bear profound consequences. Some multinational firms might cause chaos in global economy. Their financial risk management decisions have to be based on a thorough study of potential risk. Many corporations will request engineering firm to evaluate their risk and to choose from the following: strengthening the existing facilities, providing redundant system such as multi-location data centers, relocation, and risk transfer through insurance. The risk to businesses include not only the buildings but the contents or inventory, and the possibility of business interruption due to damage to a building, damage to the surrounding area or region, or damage to a sup-

plier or to clients. The possibility of business interruption is a major consideration in evaluating financial risk due to an earthquake. The impacts of ChiChi Earthquake to corporations in the Hsinzhu scientific park are a vivid example of consequence due to business interruption.

6.2.4 The lifeline sector
Lifelines are infrastructure systems, such as power, transportation, telecommunications, water and waste, gas and liquid fuel. They are vulnerable to natural and manmade hazards; their performance can be degraded; their ability to function can be disrupted due to damage to components of the system and to the support systems. They are risk-bearers, however their assets are different from a home or office building. They are massive networks consisting of myriad of components that are dispersed in space and function, yet linked to perform together as a system. They also are constrained in terms of location, i.e. they must be where they are, so certain diversification methods of risk management are not an option. Mitigation and transfer are typically the two primary methods of managing financial risk for the lifeline sector.

6.2.5 The government sector
In addition to its role as a policymaker and regulator, government at all levels is a major property owner. The federal government, for example, is the largest single property owner in the state of California, and the state government is the second largest owner. In the U.S. there are more than 78,000 different local government units, most of which own or manage property. Damage to public buildings in an disaster typically results in four types of losses to the government sector: life loss; economic loss; loss of function; and loss of cultural heritage. The three typical options for managing the risk include risk reduction (retrofit), risk transfer (typically through insurance) and risk retention. The choice of which technique to adopt is generally governed by the cost of the technique and the intended goal.

6.3 Overview of risk transfer sectors

6.3.1 Insurance
Insurance is a risk transfer mechanism that allows a property owner to transfer the risk of loss to the insurance company. Typically, insurance is based on the Law of Large Numbers (insured events are seen as independent of each other; the probability of many occurring at once is low). However, catastrophe insurance, such as earthquake or hurricane insurance, acknowledges that the loss of many properties over a large area is no longer independent. Rather, those losses are seen as correlated. This loss concentration caused 14 local insurance companies in Florida and Hawaii insolvent in 1997 as results of

hurricanes Andrew and Iniki. The less correlated the hazards, the lower the risk for a particular company. Insurance companies use the re-insurance industry to help transfer and manage their own risk. A reinsurance company will write policies for different parts of the world, spreading the risk geographically.

6.3.2 Mortgage

The mortgage sector, along with insurance and capital markets, provides a mechanism for transferring risk. Property owners, including individuals, large corporations and governments, use mortgage as mechanisms to lay off their financial risk associated with natural disasters. Different types of financial institutions offer mortgages, and each approach the financial risk associated with disasters differently. Mortgage banks typically do not retain much risk as they sell most of their loans. Depository institutions retain only the risk based on the loans they keep in their portfolios. The government-sponsored enterprises, Fannie Mae and Freddie Mac buy many of the residential mortgages, and therefore assume risk on much of their portfolios. The secondary mortgage market retains much of the risk associated with disasters, since these are the financial institutions that buy most of the mortgages.

6.3.3 The capital markets

Capital markets have a newly emerging role in catastrophic risk transfer. A basic feature of market-based economies such as found in the U.S. is the existence of a well-developed capital market system. This system allocates savings and investment capital to various economic sectors, with allocation rules based on risk and return. In practice, financial risk is packaged and transferred to investors via financial instruments, either through equities (such as common stocks), debt securities (loans, bonds, promissory notes) or derivatives (stock options, interest rate futures, foreign exchange, and commodities futures). The possibility thus exists to transfer risk from the risk-bearing sectors (the property owners broadly defined) to the capital markets.

7 CONCLUSION

This paper links together a number of disciplines by discussing how engineering is important to the various sectors of society that manage financial risk associated with natural disasters. By gaining a better understanding of how this risk is managed by each of the individual sectors the engineer should be able to more effectively play a role in the provision and analysis of technical information. Catastrophic risk management options can vary from complete retention to transfer of the risk to another sector, yet each option relies on information that comes from the engineer. Understanding the importance of this role in the context of how financial risk is managed is the subject of the paper. The interaction is iterative and evolutionary: the financial sectors pull the engineering community to provide information and develop more sophisticated tools, and the engineering community pushes the financial sectors to make decisions based on scientific and technical expertise. The paper aims to provide an overview of ALL issues, with an emphasis on those where the engineering professional has a role.

REFERENCES

RMS, "What if the 1906 Earthquake strikes again?," Topical Issues Series, May 1995, RMS, Inc. Publications.

RMS, "What if a Major Earthquake Strikes the Los Angeles Area?," Topical Issues Series, September 1995, RMS, Inc. Publications.

Richard J. Roth, Jr. & Teresa Q. Van, "California Earthquake Zoning and Probable Maximum Loss Evaluation Program," Department of California Insurance, Los Angeles, California, 1993/1994.

Algermissen, S. T. & Steinbrugge, K. V., Seismic Hazard and Risk Assessment: Some Case Studies, The Geneva Papers on Risk and Insurance, 9 (30), January 1984.

EERI, Financial Management of Earthquake Risk, Endowment Fund White Paper, EERI, May 2000

Earthquake Engineering Frontiers in the New Millennium, Spencer & Hu (eds),
© 2001 Swets & Zeitlinger, ISBN 90 2651 852 8

Some recent work in earthquake engineering at Dalian University of Technology

G.D. Cheng & X.J. Kong
Dalian University of Technology, Dalian, P.R. China

ABSTRACT: Dalian University of Technology has been engaged in teaching and research of earthquake engineering for a long time. Research activities cover numerical simulation, shaking table test and design methodology for dam, offshore platform, high-rise building, masonry structures subjected to earthquake environment and hazard mitigation. Collaboration with US and other countries contributes to our achievement a great deal. Three on-going research areas are briefly reported. Under financial support of National '211' project of Chinese government, an underwater shaking table is installed in 2000. It enhances the capacity of research on water-structure-soil at DUT. Fast computational scheme for structural random response has been carried out and extended to problems with damping and control. This approach will open new possibility to take uncertainty of earthquake into consideration of design. Performance-based structural system reliability and related structural optimization is discussed.

1 INTRODUCTION

Earthquake is one of the most serious natural disasters. Earthquake disaster mitigation is an extremely important subject for mankind. China and US are both earthquake-prone countries. Collaboration between two countries during the last twenty years has contributed a great deal to both sides and resulted many fruitful works. As a major high learning institution in the field of engineering, Dalian University of Technology has engaged in teaching and research on earthquake engineering for a long time, and taken an active part in the collaboration at various levels.

Research in earthquake engineering at Dalian University of Technology has been carried out in School of Civil Engineering and Department of Engineering Mechanics. The major interests in Department of Engineering mechanics are numerical simulations, experimental study of composite masonry multi-story building and design methodology of seismic structures. School of civil engineering covers much broad area such as shaking table test, various experimental studies, numerical analysis and design methodology of dam, offshore platform, high-rise building, masonry and reinforced concrete structures subjected to earthquake environment and hazard mitigation. This paper presents three recent research activities at DUT, which is in one way or another related to China-US collaboration.

2 UNDERWATER SHAKING TABLE

Under the financial support of National project "211" of Chinese central government for development of high education institution in order to meet the challenge of the coming 21st century, a underwater shaking table has been installed in the laboratory of seismic engineering headed by Prof. Gao LIN. It shakes in horizontal, vertical and patch direction under digital control. Its working area of quasi-ellipse has major axis of 4m and minor axis of 3m. The maximum test weight is 10 tons. It has maximum horizontal displacement ±75mm, maximum horizontal velocity±50cm/s, maximum horizontal acceleration 1.0 g, maximum horizontal acceleration 1.0g, maximum vertical displacement ±50mm, maximum vertical velocity ±35cm/s, and maximum vertical acceleration 0.7g. Its operation frequency range is 0.1-50Hz. The maximum height of gravity center is 1.0m, and maximum offset is 0.5m. The primary conceives of the research on earthquake-resistant behavior of coastal and offshore structure by means of underwater shaking table include researches on earthquake-resistant behavior of underwater pipeline, research on the failure mechanism of caisson breakwater in the earthquake, research on the failure form and stability of gravity wharf, sheet-pile wharf and caisson wharf in the earthquake, and platform-foundation-seawater dynamic interaction analysis.

3 FAST COMPUTATIONAL ALGORITHM OF RANDOM STRUCTURAL RESPONSES

It is known that the basic framework of linear random vibration theory has been well established. However, it requires a huge amount of computational efforts even for simple structures with a small number of degrees of freedom. Therefore, its applications to practical engineering have been seriously restricted. Here, a highly efficient random vibration algorithm, Pseudo Excitation Method (PEM) (Lin et al. 1992,1995,1997, Zhang 2000) will be reviewed. By using it, the multiple non-stationary seismic analysis for structures with thousands of DOF, dozens of supports, needs less than one hour on a personal computer. The PEM is an accurate CQC method because the cross-correlation terms between the participant modes and between the excitations have both been included. In addition, the PEM is easy to implement on computers because any stationary random response analysis is exactly replaced by harmonic response analysis, while any non-stationary random response analysis is replaced by a step-by-step time integration scheme. PEM is applicable to non-classical damped problems. The method has been applied to a number of practical engineering structures.

3.1 Structural response to stationary single excitation

For a linear system subjected to a zero-mean-valued stationary excitation $x(t)$, with its PSD $S_{xx}(\omega)$ has been specified, the equation of motion is

$$[M]\{\ddot{y}\}+[C]\{\dot{y}\}+[K]\{y\}=\{p\}x(t) \qquad (1)$$

If $\{u(t)\}$ and $\{v(t)\}$ are two arbitrary stationary response vectors, their PSD matrixes $[S_{uu}(\omega)]$, $[S_{vv}(\omega)]$ and $[S_{uv}(\omega)]$ can be obtained by using PEM as follows

Defining a pseudo harmonic excitation $x'(t)=\sqrt{S_{xx}}\exp(i\omega t)$, by using the traditional harmonic response analysis, the corresponding pseudo responses $\{u'(t)\}$ and $\{v'(t)\}$ can be obtained. Based on the responses, we have

$$[S_{uu}(\omega)]=\{u'\}^*\{u'\}^T ,$$

$$[S_{vv}(\omega)]=\{v'\}^*\{v'\}^T , \qquad (2)$$

$$[S_{uv}(\omega)]=\{u'\}^*\{v'\}^T$$

where the superscript * and T denote complex conjugate and transpose of the matrix respectively.

3.2 Structural response to non-stationary single excitation

The governing equation of motion for a linear system, which is subjected to a uniformly evolutionary random excitation, $f(t)=g(t)x(t)$ is

$$[M]\{\ddot{y}\}+[C]\{\dot{y}\}+[K]\{y\}=\{p\}g(t)x(t) \qquad (3)$$

in which $g(t)$ is the given slowly varying modulation function while $x(t)$ is a zero-mean-valued stationary random process whose PSD $S_{xx}(\omega)$ has also been specified. If $\{u(t)\}$ and $\{v(t)\}$ are two arbitrary non-stationary response vectors, their evolutionary PSD matrixes $[S_{uu}(\omega,t)]$, $[S_{vv}(\omega,t)]$ and $[S_{uv}(\omega,t)]$ can be obtained by using PEM as follows.

Defining a pseudo harmonic excitation $x'(t)=\sqrt{S_{xx}}g(t)\exp(i\omega t)$, by using the traditional step-by-step time integration computations, the corresponding pseudo response $\{u'(t)\}$ and $\{v'(t)\}$ can be obtained. Based on the two vectors, we have

$$[S_{uu}(\omega,t)]=\{u'\}^*\{u'\}^T ,$$

$$[S_{vv}(\omega,t)]=\{v'\}^*\{v'\}^T , \qquad (4)$$

$$[S_{uv}(\omega,t)]=\{u'\}^*\{v'\}^T .$$

In general, the step-by-step integration can be done by means of Duhamel integration formula, Wilson-θ or Newmark scheme, etc. However, the precise integration method has been proved to be extremely efficient.

The PEM also applies easily to the analysis of non-stationary random responses of structures subjected to non-uniformly modulated evolutionary random excitation.

3.3 Structural responses to stationary/non-stationary multiple excitations

Based on the fact that any PSD matrix of stationary or evolutionary random excitations can be decomposed into the sum of the PSD matrices of a limited number of mutually independent single stationary or evolutionary random excitations, the linear superposition theory can be used for such multi-excitation problems conveniently. Each of such single excitation can be converted into harmonic or time history type excitation, and is still easy to deal with.

3.4 Vibration control for wind-excited or earthquake-excited structures

In conjunction with the complex modal superposition method, the general pseudo excitation method is extended to investigate the random vibration analysis of complex structures with supplemental discrete control devices and its application to various types of vibration control problems of wind/seismic resistant buildings. By using PEM, the random loads are first transformed into deterministic pseudo loads. But for a controlled structure, the damping is assumed to be non-proportional. Therefore the responses can be solved by complex modal superposition method, and the PSD matrices can be obtained as mentioned above.

The PEM has been successfully used for many kinds of vibration control problems. The PEM is applied to find the closed form solution for wind-induced response of the building implemented by active control devices with Linear Quadratic Gaussian (LQG) controllers. The PEM is applied to investigate modal properties and seismic response of steel frames with connection dampers. Moreover, the PEM is extended and applied to vibration control of adjacent buildings under earthquakes. The visco-elastic dampers, the fluid dampers, and the active tendon devices are respectively used to link the adjacent buildings together for control of seismic responses.

4 OPTIMIZATION AND RELIABILITY OF SEISMIC STRUCTURE

The seismic capacity of a structure relies on many factors, such as structural type, structural element material, construction method and quality, initial cost, etc. How to form a rational design methodology so as to get a good balance between the structural seismic capacity and initial cost has been the key issue in earthquake engineering. The present Chinese seismic codes are believed to provide good levels of life safety protection in buildings that are properly designed and constructed under severe earthquake. However, these codes do not account for the possible economic losses due to the failure of specified structural performance under moderate and minor earthquake. With the economic development, the cost of the structural decoration, non-structural elements and information technique equipment has exceeded the cost of structure itself a lot, so the economic losses caused by the failure of specific performance may become even more serious. The Northridge earthquake and the Kobe earthquake are good examples.

The thoughts of performance based seismic design are proposed in the early 90s in USA, and have caused more and more attention in the field of seismic design recently. The basic thought of performance-based design is to design a structure so that it will perform in a specified manner when subjected to various loading scenarios. The intent is to provide owners and designers with capability to select alternative performance goals for the design of different buildings. The thoughts in performance based seismic design are essentially different from those in the present seismic codes. The former emphasize the structural "individuality" in the design process, while the latter stresses the structural "generality" and enforces minimum requirements of life safety level for the structures. In performance based seismic design, a design performance objective is an expression of the desired performance level for the building for each earthquake design level, and performance level is an expression of the maximum desired extent of damage to a building, given that a specific earthquake design level affects it.

One important aspect in structural design is the uncertainty. There are many uncertainties in external actions, structural capacity, analysis models and structural performance. Uncertainty is even more serious under earthquake hazard environment. Thus, engineers must deal with these kinds of uncertainties in the structural design process. The importance of structural reliability in performance-based design is well recognized in literature. For example, in FEMA report 283, it is stated that "The skeleton of performance based seismic design should be reliability theory". But it is also recognized that "Numerous difficulties, however, still exist with the systems approach in structural reliability.... A system approach for a performance-based seismic code, however, should not be discounted. Instead, it should be fully investigated and its long-term implementation formulated at this stage."

Although the structural system reliability is a good index that can reflect the system characteristics in concept, its application has always remained in the research field and can hardly be applied in practical design. Besides the difficulty in system reliability calculation, the reasons for such a situation can be explained from the following two aspects:

1. The failure modes of the civil engineering are very complex. Unlike the electronic system, even a simple civil engineering system will have a great number of failure modes due to its high redundancy and the large uncertainties of the external actions (especially for earthquake action). In the most of literatures of structural system reliability, the ideal and simplified analysis model (such as the frame with ideal elasto-plastic property) is employed and only the failure modes caused by the hinges are treated. This is much different from the practical case, since buckling, excess vibration, deformation, corrosion and crack may cause failure of structural performance as well. Therefore, the calculation of the structural system reliability should consider all the mechanism and phenomena that may cause the failure of structural performance, not just hinge failure modes. Nevertheless, to figure out such comprehensive system reliability is almost impossible, for the number of the structural failure modes caused by

different factors is numerous and the major failure factor may change with the design modification. This results in the inconsistency between the system reliability research and the practical situation.

2. The concept of system reliability is not very useful in the cost-benefit analysis. It is well known that different failure modes (or failure of different performance) will result in different consequence. Some are local failure modes, while the others may be the system failure modes and even make the whole structure collapse. The economic losses caused by the global failure modes are much more severe than those caused by the local failure modes. It is almost impossible to use a unique loss value to represent the complex situation of the structural system failure, and the structural system loss expectation obtained by the product of such a unique system failure probability and system loss value is even unreasonable.

In our research (Cheng et al.1998, Li 1998) the concept of performance based structural system reliability is introduced, and estimation of the structural loss expectation is based on the fact that failure of different structural performance causes different losses. With the concept of performance based structural system reliability the cost-benefit criterion based structural design consists of the following two stages. The first stage is the optimal decision of the structural performance objective based on cost-benefit criterion. Its formulation is stated as

$$
\text{Find } \left[P_{fi}\right], \qquad i = 1,2,....,n_p
$$

$$
\text{min } W\left(\left[P_{fi}\right]\right) = C_0\left(\left[P_{fi}\right]\right) + \sum_{i=1}^{n_p}\left(1 - \left[P_{fi}\right]\right)C_{fi} \tag{5}
$$

where, $[P_{fi}]$ is the design variable, which is performance based structural system reliability. C_{fi} is the corresponding failure loss value. C_0 is the initial cost, the second term in the objective function is the sum of structural loss expectation for the failure of considered performance, n_p is the number of considered performance. The second stage is the structural minimum cost design based on the optima structural performance objectives. Its formulation is as follows,

$$
\text{Find } \mathbf{X}
$$

$$
\text{min } W(\mathbf{X}) = C_0 + \sum_{i=1}^{n_p} P_{fi}(\mathbf{X})C_{fi} \tag{6}
$$

$$
\text{sub. } P_{fi} \le \left[P_{fi}\right], \quad i = 1,2,...,n_p
$$

$$
g_j(\mathbf{X}) \le 0, \quad j = 1,2,....m
$$

where, $\mathbf{X}$ is the vector of design variables, which can be the variables related to the structural topology, shape dimensions and reinforcement areas. W is the objective function, P_{fi} is structural system failure

probability related to the *i-th* performance considered, $g_j(x) \le 0$ is the deterministic constraint, such as structural strength and construction requirement.

The basic design thought in formulation (1,2) is consistent with that proposed in the *General principles on reliability for structures* (ISO 2394,1996).

To implement the idea, it is necessary to calculate the performance-based reliability efficiently. We have studied the statistical characteristics of story drift of building structures under earthquake loading using numerical simulation. We have further improved the two-point approximation of the nonlinear limit state function using intervening variables, the iterative algorithm of FOSM, and develop the reliability calculation module in general structural analysis software. Although formulation (5) is consistent with cost-benefit criterion from the view of target reliability concept, it is very difficult to solve this optimization problem. It is almost impossible to obtain the relation between the initial cost function and target reliability. To overcome the difficulty we have studied parameter programming approach and constraint relaxation approach. It was pointed out that with the aid of a number of assumptions, optimal fortification load (or optimal fortification intensity) could play the same role as the optimal performance objective does.

The work presented in this paper is under the project No.59895410 supported by the National Natural Science Foundation of China.

REFERENCES

Lin, J.H. 1992. A fast CQC algorithm of PSD matrices for random seismic responses, *Computers & Structures*, 44(3), 683-687

Lin, J.H. & Shen, W.P. Williams, F.W. 1995. A high precision direct integration scheme for non-stationary random seismic responses of non-classically damped structures, *Structural Engineering and Mechanics*, 3(3), 215-228

Lin, J.H. & Sun, D.K., Williams, F.W. 1997. Structural Responses to Non-uniformly Modulated Evolutionary Random Seismic Excitations. *Communications in Numerical Methods in Engineering*, 13:605-616

Zhang, W.S. 2000. Generalized pseudo-excitation method and its application for vibration control of wind/seismic resistant buildings. Doctoral dissertation, Hong Kong Polytechnic University

Federal Emergency Management Agency (FEMA), *Performance-based Seismic Design of Buildings*, FEMA Report 283, September, 1996

Cheng, G.D. & Li, G. 1998. Reliability-based multi-objective structural optimization under hazard load, *Proc. of Structural Engineers World Congress (SEWC'98)*, July, USA

General principles on reliability for structures (Revision of the first edition, ISO 2394:1986), 1996.

Li, G. 1998. *Reliability and performance based optimization design for seismic high-rise structures*, Ph.D. Thesis, Dalian Univ. of Tech.

Earthquake Engineering Frontiers in the New Millennium, Spencer & Hu (eds),
© 2001 Swets & Zeitlinger, ISBN 90 2651 852 8

Assessing countermeasures to emergency response in earthquakes

Yifan Yuan

Institute of Engineering Mechanics, CSB, Harbin, P.R. China

ABSTRACT: Prompt and well-planed emergency response is essential to disaster relief after earthquake occurs. Some experiences and lessons of emergency response in earthquakes of China are briefly discussed in this paper. Research topics to be needed and related to strategies and countermeasures of emergency response are proposed for further investigation.

1 INTRODUCTION

Effective emergency response plays an important role in mitigation of earthquake disaster in developing countries as well as in developed country. Especially, prompt and well-planed emergency response is an only essential way to disaster relief, when action of the earthquake is strong beyond defense level of the city, like Tangshan and Kobe earthquake.

Many studies and efforts have been made in this aspect. For example, some concepts and response activities were discussed in the International Symposium of Earthquake Countermeasures held in China, 1988, International Workshop on Earthquake Injury Epidemiology for Mitigation and Response, held in USA, 1989. Since then, valuable experiences and lessons about emergency response have been obtained in many large destructive earthquakes, such as Loma Prieta, Northridge, Kobe, Turkey and Taiwan earthquake.

Practical experiences and lessons of emergency response in Tangshan and other recent earthquake in China have been partly collected and published. Strategies and countermeasures to emergency response can be developed effectively from investigation and comparison of experiences and lessons of practice of China, USA and other countries over the world. In addition, it is necessary to conduct decision-making analysis; model study; numerical simulation; computer disaster mana-gement system and field exercise in the future to investigate effects of response activities. Based on the investigation, we may summarize principles and guide for reasonable emergency response in earthquakes, and provide useful countermeasures, which we can take before an earthquake occurs.

2 EMERGENCY RESPONSE PHASE

Obviously, main activities of each phase are different for disaster mitigation. Followings may be considered as principle activities of emergency response:

Reaction of government and public
Search and rescue
Medical aid and care
Earthquake fire and other secondary hazards
Evacuation and emergency shelters
Rush repairs of lifeline system
Urgent supply
Keep stability of community.

There is not a common definition about time-sequence phases for earthquake disaster. Lechat (1989) proposed five phases: that is, the anticipatory, pre-disaster, impact, relief and rehabilitation phase. In China, countermeasures of mitigation are briefly divided into three categories: prevention before, emergency response when, and recovery after earthquake occur. It is necessary to definite what is "when earthquake occur". In fact, the most onerous task and labor-intensive activities of emergency response have to be conducted in several days after earthquake occurs. This phase corresponds to the impact and relief phases proposed by Lechat. The duration of the phase depends on seismic intensity, casualty, population and efficiency of emergency response. Generally, 3-5 days may be the key phase of emergency response. It is necessary to emphasize timeliness of emergency response, because activities, such as search and rescue must be taken as soon as possible for disaster relief.

3 SOME EXPERIENCES AND LESSONS

3.1 Government response

Effective emergency response depends on prompt reaction of government and the public. In most case, the population is generally ill-prepared. At the first 1 or 2 days after impact, there is short disruption of local community and chaos in the public. The most urgent is to carry out well-organized and aim-determined activities of emergency response. This is duty of government. In Tangshan earthquake (also in Kobe), the most vexatious problem in initial phase was to find where is the most severe damage area and what extend of damage and casualty. Because communication system was broken down by strike of earthquake, it took 4 hours to determine meizo-seismal area, and no effective well-organized activities in whole city were carried out at the first day at least (Wang et al, 2000). Hesitate and slow emergency response caused more severe casualty in both earthquakes.

Recently, central and local governments of China have their own Preparedness Plan of Emergency Response. In the plan, organization of headquarters of disaster relief, duty of each government branch are determined in advance, so that government can start reaction immediately once earthquake occurs, in order to avoid to be in a rush.

Table.1 lists how much casualty and manpower to send to the field for disaster relief in some destructive earthquakes of China. Why did the Tangshan earthquake demand much more troops and medical teams than others did? How many troops and medical teams does it need for different earthquakes? When and where should those troops and medical teams send to? It is need to study what is a reasonable decision, which corresponding to distribution of damage, casualty and other conditions.

Table 1. Loss of some destructive earthquakes in China

Earthquake	M	I_0	Death	Injury	Loss*	Troops	Medical**
Tanghshan	7.8	11	242000	164000	1200	140000	20000***
Xingtai	7.2	10	8264	9492		20000	700
Tonghai	7.7	10	5621	5648		17000	6600
Haicheng	7.4	10	1328	4292	100		3480
Longling	7.4	10	98	451	17		332
Liyang	6.0	8	41	654	20	500	390
Heze	5.9	7	45	433	45		280
Lijiang	7.0	9	302		300	1200	400

* direct economic loss, unit: million US$.
** members of medical team
*** in addition to more than 150000 took part lifeline repair and other urgent work.

Many staffs of government were dead or were not on duty in time due to severe damage of Tangshan earthquake. Hence, it is necessary to stipulate as-sembly place for members of headquarters and to send well-trained people to the disaster area.

3.2 Search and rescue

In China, People's Liberation Army undertakes action of search and rescue for natural catastrophe. Chiefs of PLA are also members of local headquarters of disaster relief, so that government and PLA can make decision together to send search and rescue troops timely.

The principal experiences for rescue in China are that the earlier search and rescue are carried out, the more people are extricated alive. It is well-known "Golden 24 hours". Table 2 lists number and rate of extricated people alive vs. rescue time in Tangshan earthquake. The rate of extricated alive is also shown in Figure 1.

Table 2. Survival rates vs. Rescue time (Tangshan earthquake, 1976)

Time of Rescue	Extricated Number	rate(%)	Surviving Number	rate(%)	Cumulative rate (%)
0.5hour	2277	21.3	2261	99.3	29.9
1 day	5572	52.1	4513	81.0	89.6
2 days	1638	15.3	552	33.7	96.9
3 days	348	3.3	128	36.7	98.6
4 days	399	2.7	75	19.0	99.6
5 days	459	4.3	34	7.4	100.0

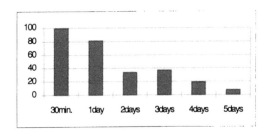

Figure 1. Survival rates (%) of extricated people vs. rescue time (Tangshan earthquake, 1976, M=7.8)

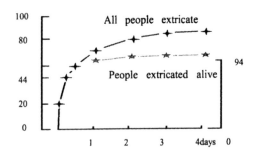

Figure 2. Total and alive cumulative rate of people extricated in an Italian earthquake (after Lechat, 1989)

144

Another similar example is Campania-Iripinia earthquake, Italy (1980). Total and alive cumulative rate of people extricated in this earthquake is shown in Fig.2 (after Lechat, 1989).

The figures indicate that self and mutual rescue as soon as possible is effective way of rescue when trapped in simple single storey houses. 15983 victims were extricated by regional troops in few days in Tangshan earthquake. It is 96% of totals extricated by PLA, although the regional troops are only 20% of total PLA to be sent to the field. It also indicates that the early extrication is a key factor.

However, it is nearly impossible to self-extricated for victims trapped in collapsed multi-storey RC buildings and in similar conditions. It takes a long time, lead to ineffective rescue. Study on collapse type of building and how to find possible voids in collapsed building is strongly suggested.

The most vexatious problem for rescue, always encountered in most earthquakes, is lack of effective rescue tools. In Tangshan and other earthquakes, people and soldiers extricated victims by bare hands, bleeding in all fingers and stripped nails. If rescue troops had effective tools or machine, more victims would be alive.

Table 3 lists some example of victims extricated alive in a long time.

3.3 Medical aid and care

Another urgent action of emergency response is first aid and medical care to injured victims. 283 medical teams including more than 20000 members took part the tremendous field medical activity in Tangshan earthquake. They were dispatched to several areas under unified arrangement. The common issues about medical care are: lack of medicine, plasma, operation appliances, heavy work on Triage, Treatment and Transfer.

There were more than 100000 of injured victims need to be treated and transferred to other region in Tangshan earthquake. This is heavy burden for emergency response beyond estimation at that time.

Corpse burial, epidemic prevention and general sanitation, including water supply and toilets setting up, is also heavy work in disaster area, especially in summer. Many experiences in practice of emergency responses in China can be found in publications (e.g. Wang et al, 2000).

3.4 Lifeline system

Normal operation of communication system is on the top in initial phase of emergency response. In Tangshan earthquake, although several telegrams about damage of Tangshan city were sent early, the central and provincial governments did not know where is the meizoseismal area until 4 hours later. Because they sent telegrams to different terminals separately. There was no fixed connection channel to transfer each other. In China, the first report about earthquake damage often comes from military communication system, but there is not a national system or organization to have responsibility for management of disaster information. It needs to assign a special agency, like FEMA in USA, to manage disaster for this purpose. In addition, it is better to set communication station in different place, in order to avoid those stations to be damaged at the same time.

Transportation is interrupted by severe damage of earthquake. Rush repair of bridges is also heavy task and takes a long time. In Tangshan earthquake, critical cross of road downtown was jammed completely by thousand vehicles and temporary houses on the second day. Some main roads were also crowded with vehicles and people and jammed seriously. The headquarters of disaster relief had to send troops of 3 companies to put it in order and control transportation from then.

Table 3. Some victims extricated alive in a long time

Name	Sex	Age	Status	Type of building	Trapped place	Supply	Trapped time*
Chen	Man	24	Soldier	Masonry	First floor	?	61
Hao	Women	24	Wife				
M.Tian	Girl	14	Pupil				
Y.Tian	Girl	12	Pupil	3-storey masonry	First floor	Urine	63
Ji	Boy	9	Pupil				
Ji	Boy	5					
Li	Man	60	Worker	1-Storey masonry		Urine, Coarse sugar	88
Z,Wang	Women	23	Nurse	4-storey hospital	First floor	Glucose Physiologic Saline	180
S.Wang	Man	27	Worker	4-storey hospital	First floor	Water, glucose	181
Lu	Women	46	House-wife	Masonry hospital	First floor	Urine, Steam bread?	302
Chen et al(5)	Man	17-55	Worker		Coal mines	Water	356

* hours

145

Air transport is particular important to emergency response of earthquake disaster. In Tangshan earthquake, air transport was real "lifeline" in initial phase of emergency response. A large part of urgent supply, as well as many injured victims, was transported by air. 2400 times of taking off and landing were carried out at Tangshan airport in safety, over the first 14 days after the earthquake occurred. In someday, average interval is only 2 minutes between adjacent taking off or landing of airplane and helicopters. The shortest interval was only 26 seconds. In Lijiang earthquake (1996, M=7.0), which occurred in mountain area, air transport also guarantee primary necessary urgent supply and transfer of injured victims.

Repair and recover of lifeline system take much time, so that alternate or spare equipment is need for emergency response.

3.5 *Urgent supply*

The most urgent supply for emergency response include:

> Machines and tools for search and rescue
> Medicine, plasma and medical appliances
> Shelters and materials of temporary houses
> Food, bedding and clothing
> Financial aid.

It is still not clear how much does urgent supply need corresponding to the extent of casualty and damage. Moreover, in many cases, some materials transported to disaster area were not urgent need, but it took much time and labor. If we have well-organized plan and arrangement beforehand, the emergency response will be more effective and smooth.

3.6 *Temporary shelter and house*

After earthquake occurs, people always concern: does another earthquake occur after main shock? Is my house and building safety for living? In practice of emergency response of recent earthquakes in China, we found that temporary house and shelter is the most difficult problem to settle residents down to live. First, materials of temporary house are not enough. Second, there is not much green land for setting up temporary house or shelter in city, so that many temporary houses are built along road and disturb transportation. Furthermore, it becomes more difficulty in raining or winter season. In Haicheng (1974, M=7.4) earthquake, which occurred in winter, fire on temporary shelter killed 341 persons and wounded 980, and 6578 were frostbitten.

4 RESEARCH NEEDS

At first, existing data about emergency response is separated, scattered and not complete. We found it is very difficult to use the raw data directly for specific research purpose, especially for a model study. It is necessary to provide models or standard for data collection and statistics.

For search and rescue, the Earthquake Injury Epidemiology is a prospective research field (Smith, 1989). Some related topics are:

Time and place of death and injury, how related to type of building? Where are victims located? What relation is it between possible void and collapse type of building? How are victims killed or injured in collapsed building? How long does a victim live in different trapped condition? How to self-rescue or extricated by external help in golden time? It needs to develop rescue technique for different building structure (RC, Masonry...), effective machine, tools, lighting equipment for search and rescue. How to organize rescue team? etc.

Based on investigation of existing data of experiences and lessons, it may proposed some special model for study effects of countermeasures or emergency response activity. For example, Earthquake casualty statistic model, Casualty predictive model, Urgent supply model, Disaster model of lifeline damage, Effects of interaction of lifeline, Spread model of fire and diffuses model of poison gas, etc.

As to social aspect, there are many interesting topics involved in emergency response. For example, How to estimate social effect of earthquake (aftershock) prediction, when we frequently encountered in disaster area and affected large city by small or distant earthquake?

China and other countries have been developing information management system of earthquake disaster, based on earthquake damage prediction and computer GIS system. It is hope that this system can provide disaster information and countermeasures quickly. Once an earthquake occurs, the system is expected to report primary estimation of casualty and economic loss, damage distribution, possible high hazard blocks in the city, and other information of emergency response. If it does work, we may establish a national-wide disaster information network. But there are some problems need to be study further, e.g. how does the result depend on database and models applied in damage prediction? How to evaluate uncertainty of damage prediction?

The larger an earthquake is, the more effective and prompt emergency response is needed. Based on investigation of mentioned above, at least, we can provide more effective guideline or guidebook of emergency response. It will be benefit to earthquake disaster relief.

REFERENCES

Comprehensive study of the great Hanshin earthquake, 1995. *UNCRO Research report series* No.12. Nagoya, Japan.

Earthquake disaster countermeasures. 1989. *Selected papers of international symposium on earthquake countermeasures.* Academic book &periodical press. Beijing.

Lechat, M.F. 1989. Corporal damage as related to building structure and design, *Proceedings of international workshop on earthquake injury epidemiology for mitigation and response,* P. 1-P.16, Baltimo, Mariland.

Smith, G.S. 1989. Research issues in epidemiology of injuries following earthquake. *Ibit.* P.61-P81, Baltimo, Mariland.

Wang, Y.Q. & Liu, Y.Q. 2000. *On-the-spot report of decision making in disaster relief of Tangshan earthquake.* (in Chinese), Earthquake press. Beijing.

Earthquake Engineering Frontiers in the New Millennium, Spencer & Hu (eds),
© *2001 Swets & Zeitlinger, ISBN 90 2651 852 8*

Development of new seismic design concepts considering moderate seismicity in Korea

Sung Pil Chang, Hyun Moo Koh, & Jae Kwan Kim
Korea Earthquake Engineering Research Center (KEERC), School of Civil, Urban & Geosystem Engineering Seoul National University, Kwanak-gu, Seoul 151-742, Korea

ABSTRACT: A significant progress has been made in R&D on earthquake engineering and applications since the Korean government has launched modern earthquake hazard reduction programs in 1990's after recognizing potential disastrous consequences of seismic events following the Northridge and Kobe earthquakes. In particular, there are currently active researches on the development of new seismic design concepts with the recognition that the design concept in low and moderate seismic regions cannot be the same as in high seismic region. This paper introduces research activities on Life-Cycle Cost and Limited Ductility Design concepts which have been developed suitable for the moderate seismicity in Korea.

1 INTRODUCTION

Korea has a long history of earthquakes. Earthquake events are well documented in the historic documents such as the Royal Chronicles of the Chosun Dynasty. Nevertheless the damage level caused by those historic and recent earthquakes was not very high, and it is believed that Korea belongs to a low to moderate seismicity zone. However, after the Northridge and Kobe earthquakes, there was a growing concern on potential devastating disruption of industrial and infrastructure systems that the economy of Korea is so much relying on. Accordingly, the Korean government has launched modern earthquake hazard reduction programs that include support to research activities on earthquake engineering and improvement of seismic performance requirements for civil and infrastructure systems.

Since the foundation of Earthquake Engineering Society of Korea (EESK) and Korea Earthquake Engineering Research Center (KEERC), the two leading organizations that execute a systematic research and collaborate program in earthquake engineering field with the support of Korea Science and Engineering Foundation (KOSEF) and the Ministry of Construction and Transportation, a significant progress has been made in applications and R&D on modern earthquake hazard reduction programs in Korea. In particular, there are currently active researches on the development of new seismic design concepts with the recognition that the design concept in low and moderate seismic regions cannot be the same as in high seismic region. Among them, two main interests are Life-Cycle Cost concept and Limited Ductility Design concept.

Optimal reliability level in the seismic design of a structure is determined as the one that involves the highest net life-cycle benefit to the society, or the minimum life-cycle cost. The life-cycle cost mainly consists of initial construction cost and damage cost estimated by failure probability over entire life cycle. A stochastic process, which can consider characteristics of moderate seismicity in Korea, is developed to evaluate the failure probability and eventually the total life-cycle cost function. Extensive study using this concept has been performed to propose design guidelines for seismic isolation of highway bridges in Korea.

In moderate seismicity regions, the gravity load may govern the design. Nevertheless many frame structures of conventional design possess considerable amount of inherent lateral strength though they may not show ductile failure mode. Since the demand is not as severe as in high seismicity regions, large inelastic deformation is not required even under the Maximum Credible Earthquake. The seismic design based on limited ductility design seems appropriate in moderate seismicity regions. The Korean research community has started to develop new design procedure based on limited ductility and reinforcement details that can supply required ductility.

2 LIFE-CYCLE COST CONCEPT FOR SEISMIC DESIGN OF ISOLATED BRIDGES

Seismic isolators are often used for bridges in low and moderate seismic region in order to reduce high cost usually caused by seismic performance requirements in such regions. However, the design codes and underlying design concept for isolators and isolated bridges usually follow those in high seismic region, which may not be appropriate to low and moderate seismic region. Optimal design procedure and design considerations for seismic isolation of bridges in such regions can be based on life-cycle cost concept. The design procedure determines optimal reliability level of isolated bridges as the one that involves the highest net life-cycle benefit to the society, or the minimum life-cycle cost. The life-cycle cost mainly consists of initial construction cost and damage cost estimated by failure probability over entire life cycle.

In the procedure input ground motion is modeled as the spectral density function compatible with design response spectrum for combinations of acceleration and site coefficients. Figure 1 describes the modeling procedure where upgrading of spectral density function is continued until the spectral density model converges to the design response spectrum. It was shown (Koh & Song 1999, Koh et al. 1999, Koh et al. 2000) that the spectral density model generated reflects properly characteristics of the acceleration and site coefficients specified in the code.

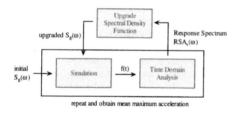

Figure 1. Generation of spectral density model compatible with response spectrum

The failure probability is calculated by crossing theory of spectrum analysis using the above spectral density model and limit states defined for super-structures, isolators and piers. Total life-cycle cost function is defined for cost optimization by considering the effects of design parameters, such as stiffness of isolators and piers and dynamic characteristics of bridges, on the failure probability. Ductility of piers and its effects on cost effectiveness are considered by stochastic linearization. Some of investigation results show that seismic isolation of bridges may be more cost-effective in low and moderate seismic region, and that optimal stiffness of isolators in such regions can be designed as more flexible than in high seismic region. Figure 2 shows an example of investigation results for cost effectiveness of isolated bridges according to acceleration coefficient and soil type. More results, such as on ductility, response modification factor and effects of intensity of input ground motion, site conditions and dynamic characteristics of isolated bridges, are pre-

sented in detail in KEERC (2000). The investigation results are being used to propose tentative design guidelines for seismic isolation of bridges in Korea.

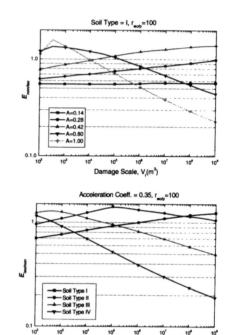

Figure 2. Cost effectiveness of isolated bridges according to acceleration coefficient and soil type

3 LIMITED DUCTILITY DESIGN CONCEPT

It is quite natural that the seismic design must be based on the dynamic response characteristics of structures under the earthquake loading. The dynamic response of structures is a function of the ground motion and mechanical properties of structures. The earthquake ground motion is local. Therefore seismic design must take into account characteristics of ground motions expected at the site. Especially the demand on the ductility capacity in the moderate seismicity region may not very large.

It has been observed in many tests that structures without seismic detailing may have significant amount of lateral resisting capacity. A 4-story reinforced concrete building (Kim and Kim 1998a) is designed according to the Korean Building Code. The model is not detailed for the lateral load such as wind and earthquake. But it has adequate amount of resistance to the design wind load. It is assumed that the walls are separated from the frames and do not contribute to the building stiffness. A capacity curve is obtained in longitudinal direction using 3-D nonlinear analysis method. It is converted into

ADRS spectra and compared with demand spectra in Figure 3. At several points on the curve PGA thresholds and maximum drift ratios are calculated. It appears that this structure may withstand the design earthquake without collapse but may fail in the event of MCE. The maximum ductility ratio is estimated to be 2.7.

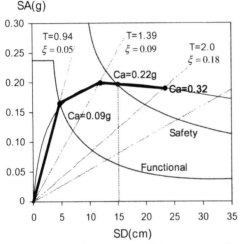

Figure 3. Load-displacement curve of 4-story concrete frame structure

The capacity of a conventionally designed bridge pier is analyzed (Kim and Kim 1998b). The pier is of solid circular section, its height 13m and diameter 2.5m. The ratio of steel is 1%. The PGA at the elastic limit is found to be 0.12g. The periods are calculated based on secant stiffness. Even to reach 1.0 % drift ratio the PGA needs exceed 0.23g.

In the above analysis, however, the longitudinal bars are assumed continuous in plastic hinge region. Very similar observations can be founded in the NCEER report on full-scale prototype and 1/3 scale model of a conventionally bridge pier (NCEER 1996a,b).

Shaking table test results of 1/3 scale model of unreinforced brick building model demonstrated that properly constructed masonry building could have considerable amount of laterally load resisting capacity. Serious cracks were initiated at PGA=0. 25G and it collapsed at the PGA=0.35G or higher (Ryu et al. 1999).

These results indicate very clearly that conventionally designed frame structures possess considerable amount of inherent lateral strength though they may not show ductile failure mode. By improving joint details, the ductility may increase to prevent premature brittle shear failure.

The inelastic behavior characteristic of the bridge pier designed conventionally has been investigated by 1/3.5 scale model test. It has lap spliced longitudinal bars in the plastic hinge zone and small amount

of transverse reinforcements with no hook. Additional model which has continuous bars instead of lap spliced one was tested together in order to investigate the influence of lap splice in the plastic hinge zone into seismic performance. Their load-displacement curves are obtained and compared together in the Figure 4. The displacement ductility is 1.5 and 4.5 in the lap-spliced model and continuous one respectively. It is worthwhile from this experimental result to note that avoiding lap splice in the plastic hinge zone can significantly enhance the seismic performance of a RC pier (Chang el al. 2000)

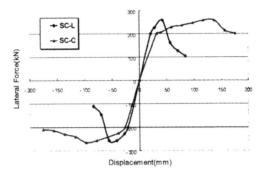

Figure 4. Comparison of load-displacement curves between lap splice model and continuous one

Similar to AASHTO requirements applicable for seismic zone 1 and 2, lap splice in plastic hinge zone is allowed implicitly in Korea Bridge Code. Large amount of a transverse reinforcement with end hook is, however, required to confine core concrete effectively so that large ductility can be achieved. This large amount and complex assembling of transverse reinforcement cause high cost in construction of a bridge. But a much simpler detailing such as continuous longitudinal bars with less amount of transverse reinforcement could be introduced as an alternative detailing if the seismic performance objective is satisfied. More extensive research on various structural type is to be conducted to develop seismic detailing appropriate in low to moderate seismicity regions.

In performance-based design, multiple performance levels are being under consideration. As pointed out by Bertero (Bertero 2000), the present design approach requires the structures undergo large amount of inelastic deformation before developing their full capacity. Even though the structure may not collapse, but the damage level will be too severe for the structure to have of any economic value after strong earthquake. Hence the present design concept based on large amount ductility may not be adequate even in the high seismicity regions.

In moderate seismicity regions, the gravity load may govern the design. It is very likely that the members proportioned for the gravity load may have significant amount of lateral load carrying capacity. If premature local failure can be prevented, the demand for inelastic deformation will not be very large even under the Maximum Credible Earthquake.

From the above observations, seismic performance levels may have to be defined based on limited damage instead of large ductility. If the required performance is to limit the damage, the natural conclusion will be seismic design based on the limited ductility (Figure 5).

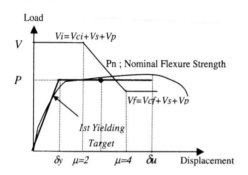

Figure 5. Concept of limited ductility design

4 CONCLUSIONS AND RECOMMENDATIONS

New developments in the seismic design concept in Korea are summarized. One approach is Life-Cycle Cost concept and the other is limited ductility design.

There are many countries that belong to moderate seismicity regions. Even in a country of strong earthquakes, there are zones of moderate seismic hazard level. The research results and experience on the new seismic design for moderate seismicity regions be of benefit to many countries. If the research toward this direction can be coordinated through participation of researchers world wide, the progress will be greatly accelerated. Development of rational and economical design procedure will help adoption of seismic design to many countries in moderate seismicity regions.

REFERENCES

Koh, H.M. & Song, J. 1999. Evaluation of economic efficiency for seismic-isolated bridges based on minimum life-cycle cost. Journal of the Korean Society of Civil Engineers, KSCE, 19(I-4): 539-550.

Koh, H.M., Ha, D.H., Song, J. & Hahm, D. 1999, Optimal design and cost effectiveness of seismically isolated bridges based on minimum life-cycle cost. In H.M. Koh (ed.), Seismic isolation, passive energy dissipation and active control of vibrations of structures; Proc. intern. Post-SMiRT conf. seminar, Cheju, Korea, 23-25 August 1999.

Koh, H.M., Song, J. & Ha. D.H. 2000, Cost effectiveness of seismic isolation for bridges in low and moderate seismic region. Proc. world conf. earthquake engineering, Auckland, New Zealand, 30 Jan.- 4 Feb. 2000: Paper No. 1100.

KEERC 2000, Improvement of seismic design and seismic performance of bridges, Final report to the Ministry of Construction and Transportation, Korea Earthquake Engineering Research Center.

Kim, J.K. & Kim, I.H., 1998a, Seismic Capacity of Reinforced Concrete Structure without Seismic Detailing and Implications to the Seismic Design in the Region of Moderate Seismicity, Proc. of Intern. Workshop on Earthquake Engineering for Regions of Moderate Seismicity, Hong Kong, October 7-8, 1998, pp 141-156.

Kim, J.K. & Kim, I.H., 1998b, Limited Ductility Seismic Design in Moderate Seismicity Regions, Proc., of Earthquake Engineering Society of Korea Conference-Fall.

NCEER, 1996a, Seismic Evaluation of a 30-Year Old Non-Ductile Highway Bridge Pier and Its Retrofit, Technical Report NCEER-96-0008, National Center for Earthquake Engineering Research, Buffalo, New York.

NCEER, 1996b, Seismic Performance of a Model Reinforced Concrete Bridge Pier Before and After Retrofit, Technical Report NCEER-96-0009, National Center for Earthquake Engineering Research, Buffalo, New York.

Ryu, H., Chae, Y.B., Jin, B.M. & Kim, J.K., 1999, Seismic response of Unreinforced Masonry Residential Building, Proc. of the First International Summer Symposium, Tokyo, Japan , August 6.

Bertero, V.V., 2000, Performance-Based Seismic Engineering: Conventional vs. Innovative Approaches, Proc. world conf. earthquake engineering, Auckland, New Zealand, 30 Jan.- 4 Feb. 2000: Paper No. 2074.

Chang, S.P., Kim, J.K., Kim I.H. & Lim, H.W., 2000, The influence of lap splice of longitudinal bars in the plastic hinge zone on the nonlinear behavior characteristics of RC piers and ne seismic detailing concept in moderate seismicity regions, Proc. of Earthquake Engineering Society of Korea, Fall 2000, pp 335-340.

Earthquake Engineering Frontiers in the New Millennium, Spencer & Hu (eds),
© 2001 Swets & Zeitlinger, ISBN 90 2651 852 8

Research on effects of seismic source on space correlation

Xing Jin, Chao Chen, Haiping Ding, & Mingyu Zhang
Institute of Engineering Mechanic, CSB

ABSTRACT: On the assumption that the seismic source is simplified as a linear fault with finite rupture length and constant rupture velocity along a single direction, this paper quantitatively analysizes the effect of seismic source on space correlation of strong ground motion based on the seismic source theory and expansion of seismic motion Fourier spectrum with the space coordinate variables, and obtains a relation of seismic motion spectra between a and its adjacent b points with interval d at a site, that is $A_b(f) = A_a(f)\exp(\alpha_1\delta\gamma + \alpha_2\delta\theta)$, where α_1 and α_2 are two complex numbers which are related to seismic source parameters, media properties, station space coordinates as well as frequency, $\delta\gamma_0$ and $\delta\theta$ are two coefficients which are mainly related to space relative positions between a and b points. According to the Taylor approximate expansion condition of Fourier spectrum with space coordinate variables, this paper also presents a theoretical formula for estimating correlated radius based on α_1 and α_2 by numerical integration. Furthermore, it is pointed out that analytical expression of α_1 and α_2 can be obtained in far-field condition. After comparing numerical integration results with those of analytical formula, this paper finds that both of them are almost same in far field. At last, this paper also discusses the problem how to design strong ground motion field using space correlation of seismic motion.

1 INTRODUCTION

With the fast development of national economy, many large scale and complex engineering structures have been built in the world, such as superhigh building, large span bridge, nuclear electricity power station and lifeline network system. The government and society specially pay attention to safety of those structures during earthquake because of their importance. The space dimension of these structures is various generally from several hundreds meters to several kilometers, which has aroused wide interests of engineers. When analyzing response of structure, engineer generally assumed that seismic motion of input foundation are same for small scale structures or variable for large scale structures, which its input satisfy propagation law of travel wave before 1980's. We did not know whether the assumption is valid because of lack of strong ground motion records.

Scientists have built a series of strong seismic motion dense arrays in high seismicity areas and obtained high quality data since 1980, this gives us an opportunity to study temporal and spatial variation of strong ground motion in details. It is well-known that temporal and spatial variation of strong ground motion generally decreases the translation response of structure, but increases rocking and torsional response of multiple support or extended foundation

(Luco and Wong, 1986). So temporal and spatial variation of strong ground motion corresponding to the dimension of engineering structures or seismic motion field is an important problem for engineers to analysize response of large scale structure, especially to long period response.

With accumulation of data, some methods and models have proposed to explain observation data recorded at dense arrays (Loh, 1985; Abrahamson et al., 1987 Jin and Liao, 1994; Hu et al 1997). By analysis of records, it was realized that temporal and spatial variation of strong ground motion has following characteristics: (1) correlation strong ground motion at adjacent points strongly depends on adjacent distance and frequency contents of seismic waves. Katayama et al. (1990) studied in details coherence function of each station pair installed at array of Chiba, Japan, and explained the phenomena. Similar results have been obtained by Abrahamson et al. (1991). (2) Spatial and temporal variation of strong ground motion is highly dependent on characteristics of seismic source. After comparing statistical atenuation properties of coherence function for four earthquakes, Harichandran and Vanmarcke (1993) found that seismic source is strongly correlated. These research results also showed that same input or travel wave assumptions applied to structure response is invalid.

This paper focuses on studying the effects of seismic source on space correlation to explain observed phenomena.

2 SPACE CORRELATION OF SEISMIC MOTION

It is well known that space correlation of seismic motion is generally studied by analysis of correlation function in time domain or power spectra density function as well as coherency function in frequency domain using strong ground motion records observed at dense array. According to the traditional engineering methods, it is easy to understand statistical characteristic of space correlation but is difficult to explore a general relationship between space correlation and seismic source parameters. It is obvious from the physical concept, the more stations is near each other, the more space correlation is strong. In other words, if we make Taylor expansion of Fourier spectrum at adjacent point relative to the referential point, it will be found that the more the station pair interval is small, the more the approximation of Taylor expansion is good. From that, it may have inner link between Taylor approximation and space correlation.

The seismic source is simplified as linear source with length L and embedded depth h as shown in Figure 1, O is the original rupture point on fault, the rupture direction is the same as the positive ζ axis, a is a point at the site, the distance from O to a is r_o, the angle between Oa and ζ axis is θ, s is the some rupture point which rupture tip reaches at velocity v_r, the distance from S to a is r. According to seismic wave spectra theory, the Fourier spectrum $A(r_o, \theta, f)$ of acceleration time history at point a is expressed by following:

$$\begin{cases} A(r_o, \theta, f) = \dfrac{C_s S(\omega)}{L} \displaystyle\int_o^L \dfrac{e^{-kr}}{r} e^{-i\omega t_s} dx \\ r = \left[r_o^2 + x^2 - 2r_o x \cos\theta \right]^{1/2} \\ t_s = \dfrac{x}{v_r} + \dfrac{r}{c} \end{cases} \quad (1)$$

with C_s a constant, ω frequency, $S(\omega)$ dislocation spectrum of point source, $k = \omega/2cQ$, quality factor, $Q = \alpha_0 f^n$, c wave velocity.

In Cartesian coordinate system $\zeta o \eta$, consider adjacent two points a and b at same site, the projections of them to $\zeta o \eta$ coordinate system are a' and b', respectively, as shown in Figure 2 and Figure 3. In these figures, Δ_o and θ_o are epicenter distance and azimuth angle, respectively. d and β are pole

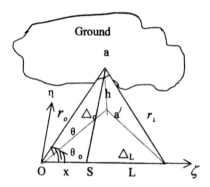

Figure 1. Seismic source model

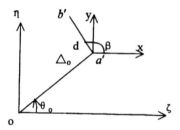

Figure 2. Geometric relation between seismic source and space point of station

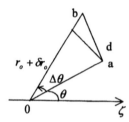

Figure 3. Geometric relation between ground points of station

coordinates of point b' in local coordinate system $\eta'a'\zeta$ with a' origin point. Leaving out the site effects, only considering spectrum difference of strong ground motion between a and b caused by the difference of space coordinates, the seismic spectrum ratio between a and b two points may be expressed as follows:

$$E(f) = \dfrac{A(r_o + \delta r_o, \theta + \delta\theta, f)}{A(r_o, \theta, f)} = 1 + \dfrac{\delta A}{A} \quad (2)$$

154

The equation (2) will be simplified into following form if the modulus $\delta A/A$ is very small.

$$E(f) = 1 + \frac{\delta A}{A} = \exp\left(\frac{\delta A}{A}\right) \tag{3}$$

Note that

$$\frac{\delta A}{A} = \frac{1}{A}\frac{\partial A}{\partial r_o}\delta r_o + \frac{1}{A}\frac{\partial A}{\partial \theta}\delta\theta \tag{4}$$

$$= a_1\delta r_o + a_2\delta\theta$$

and

$$\begin{cases} \dfrac{\partial A}{\partial r_o} = \dfrac{C_s S(\omega)}{L}\displaystyle\int_0^L \dfrac{e^{-kr}}{r}e^{-i\omega t}\left(k+\dfrac{1}{r}+\dfrac{i\omega}{c}\right)\left(\dfrac{r_o - x\cos\theta}{r}\right)dx \\[3mm] \dfrac{\partial A}{\partial \theta} = \dfrac{C_s S(\omega)}{L}\displaystyle\int_0^L \dfrac{e^{-kr}}{r}e^{-i\omega t}\left(k+\dfrac{1}{r}+\dfrac{i\omega}{c}\right)\left(\dfrac{r_o x\sin\theta}{r}\right)dx \end{cases} \tag{5}$$

Based on equations (2), (3) and (4), We may present Fourier spectra relation as follows:

$$A_b(f) = A_a(f)\exp(a_1\delta r_o + a_2\delta\theta) \tag{6}$$

Generally speaking, α_1 and α_2 are two complex numbers which are related to seismic source, propagation media, and space position of stations, and let

$$\begin{cases} a_1 = \lambda_1 + i\lambda_2 \\ a_2 = \lambda_3 + i\lambda_4 \end{cases} \tag{7}$$

where $i = \sqrt{-1}$, λ_1 and λ_2 are real and imaginary parts of a_1, λ_3 and λ_4 are real and imaginary parts of a_2. On basis of the geometrical relation shown in Fig. 2 and Fig. 3, if d/r_o is very small, it will be obtained that

$$\begin{cases} \delta r_o = \dfrac{\Delta_o}{r_o}\cos(\beta - \theta_o)d \approx \cos(\beta - \theta_o)d \\[3mm] \delta\theta = \dfrac{\sin\theta_o\sin(\beta - \theta_o)}{\left[1 - \left(\dfrac{\Delta_o}{r_o}\right)^2\cos^2\theta\right]^{1/2}}\left(\dfrac{d}{r_o}\right) \approx \sin(\beta - \theta_o)\dfrac{d}{r_o} \end{cases} \tag{8}$$

Put formula (7) and (8) into formula (6), it is easy to understand that λ_1 and λ_3/r_o describe the difference of amplitude spectra, λ_2 and λ_4/r_o describe the difference of phase spectra. We must keep in mind that the equation (6) is valid only if chosen values of λ_1, λ_3, λ_2/r_o and λ_4/r_o are so small as to satisfy the expansion condition of equation (3). Based on the idea, we can obtain following formula to estimate correlation radius of strong ground motion field where s is an acceptable fluctuation level of

$$\begin{cases} d_1 = s\left[\lambda_1^2 + (\lambda_3/r_o)^2\right]^{-1/2} \\ d_2 = s\left[\lambda_2^2 + (\lambda_4/r_o)^2\right]^{-1/2} \\ d = \min\{d_1, d_2\} \end{cases} \tag{9}$$

moduls of Fourier spectra ratio. If given site position, d will be changed with frequency.

3 APPROXIMATION OF FAR-FIELD

If far-field approximation condition $r \gg L$ is satisfied, then we will obtain analytical expression of a_1 and a_2 from formula (1) and (5) as follows:

$$\begin{cases} \alpha_1 = \lambda_1 + i\lambda_2 \\[2mm] \lambda_1 = R_e(\alpha_1) = -k - \dfrac{1}{r_o} \\[3mm] \lambda_2 = \mathrm{Im}(\alpha_1) = -\dfrac{\omega}{c} \end{cases} \tag{10}$$

$$\begin{cases} \alpha_2 = \lambda_3 + i\lambda_4 \\[3mm] \lambda_3 = R_e(\alpha_2) = \dfrac{\sin\theta\left[k^2\cos\theta - \dfrac{\omega^2}{c}\left(\dfrac{1}{v_r} - \dfrac{\cos\theta}{c}\right)\right]}{k^2\cos^2\theta + \omega^2\left(\dfrac{1}{v_r} - \dfrac{\cos\theta}{c}\right)^2} \\[5mm] \qquad - \dfrac{L\sin\theta}{1 - 2e^{-y_1}\cos y_2 + e^{-2y_1}}\left[k(1 - e^{-y_1}\cos y_2) - \dfrac{\omega}{c}e^{-y_1}\sin y_2\right] \\[5mm] \lambda_4 = \mathrm{Im}(\alpha_2) = \dfrac{\sin\theta \cdot k \cdot \dfrac{\omega}{c}}{k^2\cos^2\theta + \omega^2\left(\dfrac{1}{v_r} - \dfrac{\cos\theta}{c}\right)^2} \\[5mm] \qquad - \dfrac{L\sin\theta}{1 - 2e^{-y_1}\cos y_2 + e^{-2y_1}}\left[ke^{-y_1}\sin y_2 + \dfrac{\omega}{c}(1 - e^{-y_1}\cos y_2)\right] \end{cases}$$

$$\tag{11}$$

where $y_1 = kl\cos\theta$, $y_2 = lw(v_r^{-1} - c^{-1}\cos\theta)$. In general, k is very small in the frequency range of engineering signification. If let k=0, then

$$\begin{cases} \lambda_1 = \dfrac{-1}{r_o} \\[3mm] \lambda_2 = -\dfrac{\omega}{c} \end{cases} \tag{12}$$

$$\begin{cases} \lambda_3 = \dfrac{L}{2}\cdot\dfrac{\omega}{c}\sin\theta\, ctg\left[\dfrac{L\omega}{2}\left(\dfrac{1}{v_r} - \dfrac{\cos\theta}{c}\right)\right] - \dfrac{\dfrac{\sin\theta}{c}}{\dfrac{1}{v_r} - \dfrac{\cos\theta}{c}} \\[5mm] \lambda_4 = \dfrac{-L}{2}\cdot\dfrac{\omega}{c}\sin\theta \end{cases} \tag{13}$$

155

4 NUMERICAL ANALYSIS

In north of China, attenuation relation of Q value with frequency is $Q(f) = 300f^{0.2}$. The relation between magnitude M and rupture length L in China (shi et al, 1992) is as follows:

$$M = 2.1\log_{10} L + 3.3 \tag{14}$$

The chosen rupture velocity is 0.7c, c=3.3km/s. In the frequency range 0.1~10Hz, formula mentioned above is used to calculate values of a_1 and a_2 after using numerical filter to smooth amplitude spectra of $A(f)$ with three times. The propagator formula of chosen numerical filter is as follows:

$$\begin{cases} y_i^{(k)} = 0.25y_{i-1}^{(k-1)} + 0.5y_i^{(k-1)} + 0.25y_{i+1}^{(k+1)} \\ y_1^{(k)} = 0.75y_1^{(k-1)} + 0.25y_2^{(k-1)} \qquad i = 1, 2, \cdots n \\ y_n^{(k)} = 0.25y_{n-1}^{(k-1)} + 0.75y_n^{(k)} \end{cases}$$

$$\tag{15}$$

where k is smoothing time number. If chosen M=6.0, the values of $\lambda_i (i = 1, 2, 3, 4)$ will be calculated as shown in Figs. 4-11.

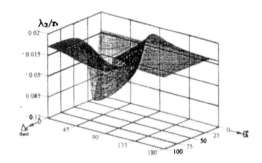

Figure 6. Variation of λ_3 / r_0 with Δ_0, ϑ_0 at f=0.1 Hz

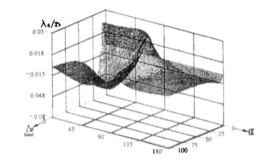

Figure 7. Variation of λ_4 / r_0 with Δ_0, ϑ_0 at f=0.1 Hz

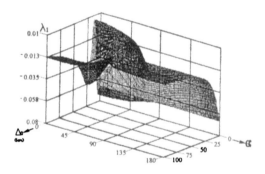

Figure 4. Variation of λ_1 with Δ_0, ϑ_0 at f=0.1 Hz

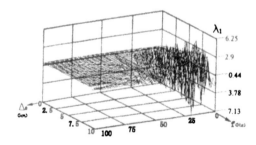

Figure 8. Variation of λ_1 with Δ_0, f at $\theta_0 = 30^0$

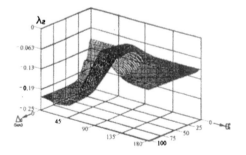

Figure 5. Variation of λ_2 with Δ_0, ϑ_0 at f=0.1 Hz

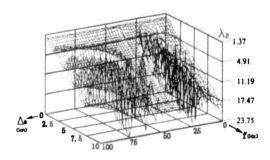

Figure 9. Variation of λ_2 with Δ_0, f at $\theta_0 = 30^0$

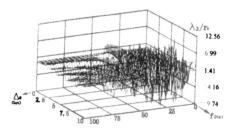

Figure 10. Variation of λ_3 / r_0 with Δ_0, f at $\theta_0 = 30^0$

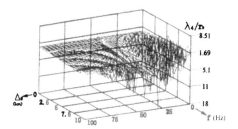

Figure 11. Variation of λ_4 / r_0 with Δ_0, f at $\theta_0 = 30^0$

In order to verify calculated results, we compare the numerical integration results with those of far field analytical formula in equations (10) and (11). We may find that both approaches results are similar, as shown in Fig. 12.

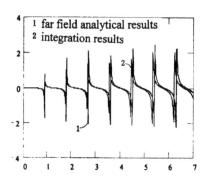

Figure 12. Comparison of far field analytical results and integration results for λ_3

It is necessary to compare results of correlation radius calculated by numerical integration method in near field as shown in Fig. 13. with those given by analytical method in far-field as shown in Fig. 14. This comparison shows that space correlation in near field is more complex than that in far field.

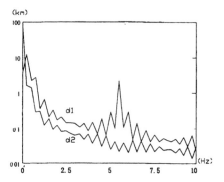

Figure 13. Variation of d_1, d_2 with f at $\theta_0 = 30^0$ and $\Delta_0 = 20km$

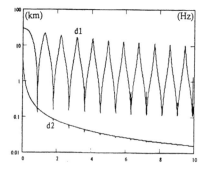

Figure 14. Variation of λ_4 / r_0 with f at $\theta_0 = 60^0$ and $\Delta_0 = 100km$

5 DISCUSSION AND CONCLUSION

This paper qualitatively studies the effects of seismic source on space correlation of strong ground motion. In contrast with the results of point source model (Jin and Liao, 1994), we can find considerable difference between the two models.

(1) In near field, both of models have different α_1 and a_2 which cause different space correlation. In the far field, they have same a_1, but different a_2.

(2) From the result of d_1 and d_2, we have found that the value of d_1 and d_2 decrease with the increase in frequencies and it has accompanied by some rinds of oscillation in near field. Furthermore, the oscillation of d_1 is more obvious than that of d_2, which is caused by the rupture mechanism. At the same time, we can also observe that the value of d_1 is larger than that of d_2 and the enveloping curve of d_2 is lower than that of d_1. This means that the am-

plitude spectra easily meet with the expanding condition within larger nterval and this is difficult for phase spectra. This further proves that the variance of phase spectra is more sensitive than that of amplitude spectra. Therefore the radius of design strong ground motion field is determined by d_2.

(3) In general, d_2 decreases drastically with the increase in frequencies and based on given radius d, the cut-off frequency $f_m(d)$ can be determined by Eq.9. By analysis of coherence function of strong ground motion, we can conclude that coherence function is close to 1 when $f < f_m$. Hence the model used in the paper is valid if $f < f_m$. Consider that there exist non-coherence waves in the design ground motion when $f > f_m$, the research on this problem should be made further.

6 ACKNOWLEDGEMENT

The research project was jointly supported by the Natural Science Foundation of China (granted 59895410) and China Seismology Foundation (granted 95-07-444).

REFERENCES

Hu Luxian, Li Xiaojun, Zhao Fengxin, 1997, Research on simulative seismic motion field with spatial correlation, *Acta Seismologica Sinica*, 19 (2):212–215.

Jin Xing, Liao Zhenpeng, 1994, Physical simulation on the random field of seismic motion, *Earthquake Engineering and Engineering Vibration*, 14(3):11–19.

Shi Zhenliang, Zhang Shaoquan, Zhao Rongguo, et al., 1992, Handbook of Earthquake Work, 237. (in Chinese)

Abrahamson, Bolt B. A., Darragh R B, et al. 1987. The SMART-1 Accelerograph Array (1980~1987). *A Review, Earthquake Spectra*, 3(2):263–287.

Abrahamson, Schneider J. E., and Steep J. C., 1991, Empirical spatial coherency functions for application to soil-structure interaction analysis, *Earthquake Spectra*, 7(1):1–27.

Abrahamsion, 1992, Generation of spatially incoherent strong motion time history, *Tenth world Conference on Earthquake Engineering*, Madrid, Spain. 845–850.

Yamazaki Y., Turke T., 1992. Spatial variation study on earthquake ground motion observed by the Chiba array, *Tenth World Conference on Earthquake Engineering*. Madrid, Spain. 651–656.

Katayama T, Yamazaki F, Nagata S, etc. al., 1990, A strong motion database for the Chiba seismometer Array and its engineering analysis. *Earthquake Engineering and Structure Dynamics*, 19:1089–1106.

Luco J.E., and Wong H.L. 1986, Response of rigid foundation to a spatially random ground motion, *E.E.S.D*, 14:891–908.

Loh C.H., 1985. Analysis of the spatial variation of seismic waves and ground movement from SMART-1 Array data, E.E.S.D,vol.13, 561–581.

Harichandran R.S., and Vanmarcke E. 1986, Stochastic variation of earthquake ground motion in space and time, *J. Engrg. Mech.*, ASCE, 112:154–174.

Earthquake Engineering Frontiers in the New Millennium, Spencer & Hu (eds),
© 2001 Swets & Zeitlinger, ISBN 90 2651 852 8

Seismic response of precast industrial buildings during the 1999 earthquakes in Turkey

S.L. Wood & M. Posada
University of Texas, Austin, Texas, USA

ABSTRACT: Precast frame buildings are used throughout Turkey for industrial facilities. These structural systems are economical to construct and provide large open areas needed for manufacturing. Many precast industrial building collapsed during the recent earthquakes in Turkey. This paper investigates the relationship between structural stiffness and the observed damage.

1 INTRODUCTION

Precast construction was first introduced in Turkey in the 1960s. Common structural systems used in Turkey today are based on systems developed in Western Europe to carry only gravity loads only (Ersoy et al. 1999). Turkish engineers modified the connection details to resist lateral loads. However, these connection details vary appreciably among the different producers of precast elements.

In June 2000, a group of researchers from the University of Texas, Kocaeli University, Boğaziçi University, Purdue University, and Middle East Technical University visited more than 50 precast industrial buildings in the epicentral region of the August 1999 Kocaeli and November 1999 Düzce earthquakes. Their observations, and the results of a parametric study to identify the causes of the observed structural damage, are summarized in this paper.

2 CHARACTERISTICS OF ONE-STORY INDUSTRIAL BUILDINGS

Although multi-story precast buildings were observed throughout the epicentral region, this paper will focus on single-story structures. Single-story structures represent the most common form of precast construction and a large portion of the structures that sustained damage during the 1999 earthquakes.

The buildings tended to be rectangular in plan with one to four bays in the transverse direction and ten to thirty bays in the longitudinal direction. Transverse bay widths ranged from 10 to 25 m, and longitudinal bay widths ranged from 6 to 8 m. Story heights also ranged from 6 and 8 m.

Figure 1 shows an elevation of a typical building with two bays in the transverse direction. The rectangular precast columns are fixed at the base. Long-span roof girders are oriented along the transverse axis of the building. The depth of these girders often varies along the length, forming the triangular shape shown in Figure 1. Beams with U-shaped cross sections are oriented along the longitudinal axis of the building. These beams function as gutters to collect water from the roof. Purlins span between the roof girders at regular intervals.

All of the precast girders, beams, and purlins were typically pinned at both ends. Dowels extended up from the supporting member. The horizontal elements were cast with vertical holes near their ends to accommodate these dowels.

Typically, lightweight roofing materials, such as metal decking or asbestos panels, were used to cover the building. Clay tile infill was used in most cases for the exterior walls, but precast wall panels were also used.

The typical, one-story industrial building depends entirely on the cantilevered columns for lateral strength and stiffness. Even when precast wall panels were used for cladding, the connection details were developed such that the wall panels did not contribute to the lateral stiffness of the building.

Three types of structural damage were frequently observed in the one-story industrial buildings: flexural hinges at the base of the columns; movement of the roof girders along their axes which led to pounding against the supporting columns or unseating of the roof girders; and out-of-plane movement of the roof girders which led to tilting or rotation off the supports.

3 BUILDING CODE PROVISIONS

The current Turkish Building Code (1998) uses the structural behavior factor, R, to convert elastic spectral accelerations to design spectral accelerations.

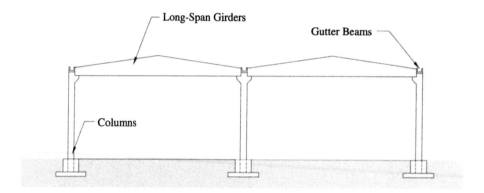

Figure 1. Elevation of typical precast industrial building

The code defines four structural systems for cast-in-place reinforced concrete buildings, four structural systems for precast concrete buildings, and four structural systems for structural steel buildings. The structural behavior factors range from 8 for reinforced concrete or structural steel moment-resisting frames to 4 for precast concrete shear walls. The structural system used for the typical one-story industrial buildings described in the previous section is assigned a structural behavior factor of 5.

The design base shear also depends on the effective peak ground acceleration, the intended use of the building, the soil characteristics at the site, and the period of the building. Elastic response spectra corresponding to the four soil categories are shown in Figure 2. The effective peak ground acceleration used to calculate these spectra corresponds to the zone of highest seismic risk in Turkey.

With the exception of very short periods, the design spectral accelerations are determined by dividing the elastic spectral accelerations shown in Figure 2 by the structural behavior factor.

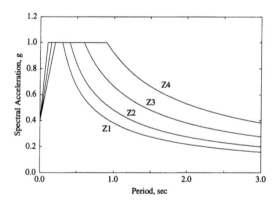

Figure 2. Elastic response spectra for seismic zone 1

Inter-story drift ratios are calculated using the lateral forces corresponding to the design spectral accelerations and must not exceed the limits given below:

$$(\Delta_i)_{max}/h_i \leq 0.0035 \tag{1}$$

$$(\Delta_i)_{max}/h_i \leq 0.02/R \tag{2}$$

where $(\Delta_i)_{max}$ - maximum inter-story displacement; h_i - height of the corresponding story, and R - structural behavior factor.

For typical one-story industrial buildings, the maximum calculated story drift ratio is controlled by Equation 1 and is limited to 0.35%.

4 OBSERVED RESPONSE

Following the Kocaeli earthquake, the Turkish Precast Concrete Association (TCPA) surveyed buildings in the epicentral region that were designed by member firms (Ataköy 1999). Less than ten percent of the 481 buildings that were included in the survey sustained serious damage. This percentage is considerably less than the percentage of all precast industrial buildings that were damaged in the earthquake. While the larger producers tend to be represented within TPCA, approximately two-thirds of the precast producers in Turkey are not members.

Most of the damaged buildings included in the TCPA survey were located in Adapazarı. Ataköy (1999) analyzed four buildings that were located in an industrial park in Adapazarı to determine the reasons of the poor performance. All four buildings collapsed during the August 1999 earthquake.

Two problems were identified in all four cases: (1) the designers had overestimated the soil properties at the site, thereby using lower spectral accelerations than intended by the building code, and (2) the buildings did not meet the lateral displacement limits given in the building code. Ataköy (1999) concluded that inadequate lateral stiffness was the primary cause of the collapses. It should be noted that flexural hinging was observed at the base of most of the cantilevered columns. No cases of foundation failures were identified, and the only observed failures of the connection between the columns and the footings occurred in buildings that were under construction where the connections had not been completed at the time of the earthquake.

5 INFLUENCE OF COLUMN STIFFNESS ON STRUCTURAL PERFORMANCE

A parametric study was conducted to relate the behavior of one-story precast industrial buildings to the column stiffness. An 80 by 200 m building in Adapazarı was selected as the prototype structure for this study. The transverse bay widths were 20 m, the longitudinal bay widths were 7.5 m, and the story height was 7 m.

A linear model of the framing system in the transverse direction was developed. Column dimensions were varied from 40 by 40 cm to 80 by 80 cm. These dimensions correspond to the smallest and largest precast columns that were observed in the epicentral region. The cross-sectional dimensions and mass of the girders, beams, purlins, roofing materials, and cladding in the prototype building were used in all analyses. The variation of the fundamental period with the column dimensions is given in Table 1.

Table 1. Scope of parametric study

Column Dimensions		Period
cm	cm	sec
40	40	1.10
40	45	1.05
45	40	0.93
45	45	0.88
50	40	0.80
50	45	0.76
50	50	0.73
50	55	0.70
55	50	0.64
55	55	0.61
60	55	0.54
60	60	0.52
65	65	0.45
70	70	0.40
80	80	0.31

Each of the buildings considered in the parametric study was subjected to fifteen ground motions that were recorded on alluvial soil deposits. Most of the recording stations were within 50 km of the epicenters of the 1999 earthquakes. Peak horizontal accelerations for most of the records were less than 0.4g, but exceeded 0.7g for two of the records.

Table 2. Ground motions considered in parametric study

Station	Component	Peak Acceleration	Epicentral Distance
		g	km
DZC	180	0.41	10**
DZC	270	0.51	10**
IZT	090	0.23	12*
IZT	180	0.17	12*
YPT	240	0.30	22*
YPT	330	0.32	22*
SKR		0.41	35*
BOL	000	0.74	42**
BOL	090	0.81	42**
GBZ	000	0.27	50*
GBZ	270	0.14	50*
ARC	000	0.21	60*
ARC	090	0.13	60*
DZC	180	0.32	110*
DZC	270	0.37	110*

*Approximate distance to epicenter of Kocaeli earthquake.
** Approximate distance to epicenter of Düzce earthquake.

The calculated drift ratios are plotted as a function of the fundamental period of the idealized build-ings in Figure 3. The shaded area indicates the range of the data, and the mean calculated drift ratio is plotted as a thick line. The mean drift ratio exceeded 1% and the maximum drift ratio exceeded 2% for periods exceeding 0.7 sec.

Each of the buildings was also analyzed using design response spectra defined in the Turkish Building Code (1998) for soil classifications Z3 and Z4. A period of 0.7 sec also corresponded to the stiffness at which the idealized building located on a site with Z4 soil conditions would satisfy the drift criteria in the building code (Equation 1). The critical period is increased to approximately 0.75 sec if the idealized building is located on a site with Z3 soil conditions. Soil conditions in Adapazarı are reported to correspond to Z4 (Ataköy 1999).

As indicated in Table 1, the fundamental period of the idealized building will be less than 0.7 sec if the column dimensions exceed 50 by 50 cm. The overwhelming majority of the buildings surveyed as part of this investigation had column dimensions that were smaller than 50 by 50 cm.

161

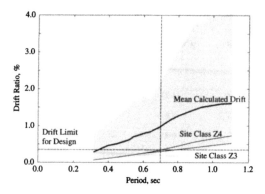

Figure 3. Calculated drift ratios for idealized buildings

6 CONCLUSIONS

This paper has focused on one aspect of the seismic behavior of one-story, precast, industrial buildings in Turkey: the flexural response of the transverse frames. Large variations in the performance of precast industrial buildings were observed in the epicentral regions of the 1999 Kocaeli and Düzce. Column dimensions and connection details are considered to have a critical influence on the performance of this type of structure. Because the structural system is not redundant, inelastic action in any column can lead to unseating of the roof girders and collapse of the roof. Drift must be controlled for this type of structural system to reliably survive future earthquakes.

7 ACKNOWLEDGMENTS

The work described in this paper was performed in conjunction with a cooperative research program sponsored by the US National Science Foundation to investigate the performance of precast buildings during the 1999 earthquakes in Turkey. The opinions expressed in this paper are not necessarily those of the sponsor. The assistance of Şevket Özden (Kocaeli University), Sami And Kılıç (Boğaziçi University), Uğur Ersoy, Güney Özcebe, and, Tuğrul Tankut (Middle East Technical University), Julio A. Ramirez and Mete A. Sözen (Purdue University), and the engineers at GÖK, Pekintaş, Set Betoya, and Yapı Merkezi, are gratefully acknowledged.

REFERENCES

Ataköy, H. 1999. 17 August Marmara earthquake and the precast concrete structures built by TPCA members. Turkish Precast Concrete Association. Ankara, Turkey. 12 p.

Code for buildings to be built in disaster areas. 1998. Ministry of Construction. Ankara, Turkey. (In Turkish).

Ersoy, U.; Tankut, T. & Özcebe, G. 1999. Damages observed in the precast framed structures in the 1998 Ceyhan earthquake and their rehabilitation. Department of Civil Engineering. Middle East Technical University. Ankara, Turkey. 11 p.

Earthquake Engineering Frontiers in the New Millennium, Spencer & Hu (eds),
© 2001 Swets & Zeitlinger, ISBN 90 2651 852 8

Outline and concept of new Japanese seismic code

Hisahiro Hiraishi, & Hiroyuki Yamanouchi
Building Research Institute, Ministry of Construction, Japan

ABSTRACT: Japanese Building Standard Law was revised in 1998. The highlight related to the seismic engineering is the instruction of the performance-based provision. This paper describes the seismic performance requirements, design response spectrum of input earthquake motions and basic verification method based on response values and limit values.

1 INTRODUCTION

Japanese standard Law was revised in 1998. The highlight related to the structural engineering is the introduction of the performance-based provisions. In the new code, the clear definitions for performance requirements and verification method based on response and limit values are specified so that the code should be applicable to any kind of materials and any type of structures such as seismic isolation systems as long as the material property is clear and the structural behavior of a building is appropriately estimated.

This paper presents the framework and concepts of performance-based new seismic provisions.

2 REQUIRED SEISMIC PERFORMANCE LEVEL'S FOR BUILDING STRUCTURES

An outline of requirements for building structures and earthquake motion levels is shown in Table 1. In the vertical column on the left hand side of the table are shown the requirements for building structures, while in the rest of the table are shown the earthquake motions to be considered and their corresponding levels for each of the requirements assigned for building structures.

As it is shown in Table 1, the requirements for building structures are classified in two categories, which are described below.

2.1 *Life safety*

The essential purpose of this requirement is the safety of human life. It should be expected that un-
der the action of earthquake motions taken into consideration, not only the building as a whole but also any story of the building should not experience any story collapse.

2.2 *Damage limitation*

The aim of this requirement is damage limitation. Under this provision, it is required first that after the action of earthquake motions taken into consideration, no structural damage which could threaten the structural safety of the building will take place. In other words, the structural safety performance required for life safety should be preserved even after the earthquake considered. Furthermore, it is required that no other kind of damage causing in the building structure a situation which does not comply with other requirements of the Building Standard Law, concerning fire safety should be experienced.

2.3 *Maximum earthquake motion level*

This level of earthquake motions corresponds to the category of requirements for life safety for building structures and is assumed to produce the maximum possible effects on the structural safety of a building to be constructed at a given site. The maximum possible earthquake motion level is determined on the basis of historical earthquake data, recorded strong ground motions in the past, seismic and geologic tectonic structures, active faults, and others.

This earthquake motion level corresponds nearly to that of highest earthquake forces used in the current seismic design practice, representing the horizontal earthquake forces induced in the building structures in case of major seismic events.

Table 1. Requirements for building structures and earthquake motion levels

Requirement	Earthquake
(a) Life Safety (to prevent failure of stories in structural frames)	Maximum Earthquake to be considered (earthq. records, seismic and geologic tectonic structures, active faults, etc.)
(b) Damage Limitation (to prevent damage to structural frames, members, interior and exterior finishing materials in order to avoid the conditions not satisfying the requirement (a) and others)	Once-in-a-lifetime Event (return period: 30-50 years)

Note: The determination of materials during the lifetime of a structure should be considered.

2.4 Once-in-a-lifetime event level

This level of earthquake motions corresponds to the category of requirements for damage limitation for building structures and is assumed to be experienced more than once during the lifetime of the building. A return period interval of 30-50 years is supposed to cover these events. This level of earthquake motion corresponds nearly to the middle level earthquake forces used in the current seismic design practice, representing the horizontal earthquake forces induced in the building structures in case of moderate earthquakes.

3 DESIGN EARTHQUAKE MOTION

The design seismic force in the previous code specified the story shear force without apparent prescription of the ground motion. Therefore, this method was so easy to pursue the design procedure. However, a contradiction existed in this method that the derived earthquake ground motions were not equal even within the same class of buildings, since the design force was derived from the response values of the building itself and the design force was prescribed uniformly with a class of buildings. Considering these situations, it was concluded that the seismic design should start with the defining the input earthquake ground motion. This also coincides with the idea of performance-based structural design aiming at more flexible design.

3.1 Design response spectrum at engineering bedrock

The ground motion is represented with its acceleration response spectrum in the new provisions. The basic ground motion is firstly defined at the engineering bedrock corresponding to the seismicity of the area.

The engineering bedrock is defined herein as follows. The engineering bedrock is underlain in the underground within the area. The geotechnical data is mostly obtained in the investigations conducted within the area. And also the considerable number of

strong motion data is obtained and the characteristics are evaluated with the recordings at the depths. This upper face of the soil layer is defined as engineering bedrock. This definition of engineering bedrock gives the idea of layers with shear velocity larger than approx. 400m/s. The amplification characteristics of the surface soil layers are to be evaluated with the geological data of the site.

The design earthquake ground motion is represented in the following equation.

$$S_A(T) = ZG_s(T)S_0(T) \qquad (1)$$

where,

$S_A(T)$ = design earthquake ground motion,
Z = regional seismicity coefficient,
$G_s(T)$ = amplification of surface soil,
$S_0(T)$ = basic spectrum at exposed engineering bedrock, and
T = period in second.

The basic spectrum is set up to be very basic. It consists of two parts, i.e., a uniform acceleration portion in shorter periods, and a uniform velocity portion in longer periods. The two intensities of uniform levels are determined with expected peak values and response factors for acceleration and velocity. The intensity level of the design motion is based on the design force for the intermediate soil class specified in the previous Building Standard Law of Japan. In addition, the relationship between the story shear force and the input motion is also taken into consideration.

The resultant design ground motion for capacity design is defined as 0.8G for the 5% damping acceleration response spectrum and as 80cm/s for the 5% damping velocity response spectrum at the exposed engineering bedrock. For the periods shorter than 1/4 of the intersection of the uniform acceleration and the velocity, the spectral values are reduced so that the acceleration amplitude at zero period equals to the peak acceleration of the input motion. The peak acceleration is set to 1/2.5 of the uniform acceleration level. The design spectrum thus defined is shown in Figure 1.

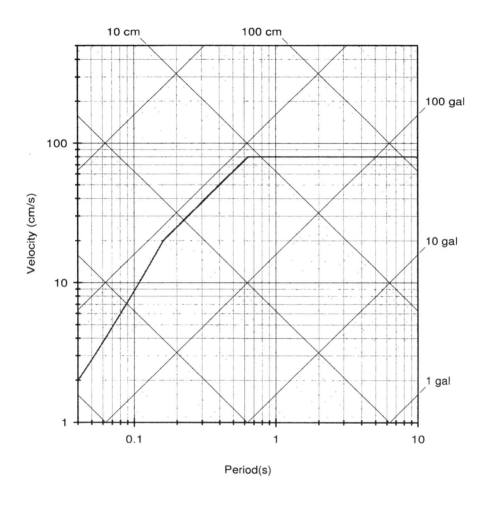

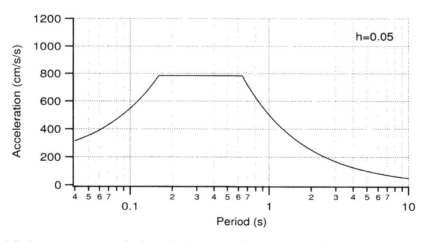

Figure 1. Design response spectrum of major earthquake motion at the exposed engineering bedrock

Table 2. Seismic performance requirements for building structures and earthquake motion levels illustrating representative response/limit values

Requirement		Earthquake	
(a) Life Safety	Maximum Earthquake to be considered	Response Value	Maximum Story shear/Displacement
		Limit Value	Limit Story shear/Displacement[*1]
(b) Damage Limitation	Once-in-a-lifetime Event	Response Value	Internal Force/Displacement taking place at each structural element
		Limit Value	Limit Strength/Displacement[*2]

Notes:
*1 - Repeating cycles effect at plastic region of response to be taken into account.
*2 - The whole building structure behaves roughly within elastic range.
1) The limit values corresponding to Maximum Event Level are determined based on the condition that equilibrium of forces and displacement compatibility in the structural system are guaranteed.
2) Displacement and acceleration related limit values, determined on the basis of the requirements for architectural, mechanical and electrical elements permanently attached to building structures, are thought to be considered in certain cases.
3) The deterioration of materials during the lifetime of a structure should be considered.

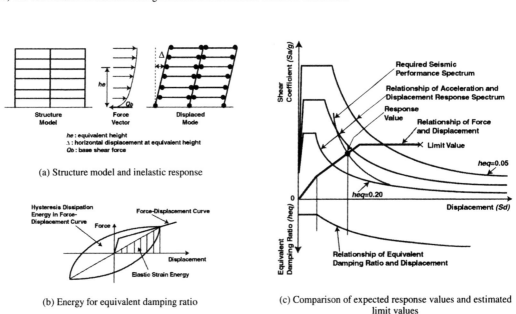

(a) Structure model and inelastic response

(b) Energy for equivalent damping ratio

(c) Comparison of expected response values and estimated limit values

Figure 2. Illustration of proposed evaluation procedures for major seismic events

166

4 BASIC EVALUATION PROCEDURES FOR MAJOR EARTHQUAKES

Various response and limit values are considered for use in the performance evaluation procedures in accordance with each of the requirements prescribed for building structures. A representative example of this arrangement is shown in Table 2. The principle of evaluation procedures is that the predicted response values due to the action of earthquake motions on building structures should not exceed the estimated limit values. In case of major earthquakes, the maximum response values of strength and displacement of a structure should be smaller than the ultimate capacity for strength and displacement.

Hereafter the focus is put on the basic evaluation procedures for major earthquakes. The analytical method to be used for predicting the structural response applies the equivalent linearization technique using an equivalent single degree of freedom (ESDOF) system and the response spectrum analysis. A flow chart of the procedures is shown in Fig. 2.

According to the procedures, the steps to be followed are:

(1) Confirm the scope of application of the evaluation procedures and the mechanical characteristics of materials and/or members to be used in a structure.

(2) Determine the response spectra to be used in the evaluation procedures.

a) For a given basic design spectrum at the engineering bedrock level, draw up the free-field site-dependent acceleration (S_a) and displacement response spectra (S_d) for different damping levels.

b) In the estimation of free-field site-dependent acceleration and displacement response (step a) above), consider the strain-dependent soil deposit characteristics.

c) In case of need, present graphically the relation of S_a-S_d for different damping levels (see Fig. 2c).

(3) Determine the hysteretic characteristic, equivalent stiffness and equivalent damping ratio of the structure.

a) Model the structure as an ESDOF system and establish its force-displacement relationships (see Fig. 2a).

b) Determine the limit strength and displacement of the structure corresponding to the ESDOF system.

c) The soil-structure interaction effects should basically be considered.

d) In case of need, determine the equivalent stiffness in accordance with the limit values.

e) Determine the equivalent damping ratio on the basis of viscous damping ratio, hysteretic dissipation energy and elastic strain energy of the structure (see Fig. 2b).

f) In case that the torsional vibration effects are predominant in the structure, these effects should be considered when establishing the force-displacement relationship of the ESDOF system.

(4) Examine the safety of the structure. In this final step, it is verified whether the response values predicted on the basis of the response spectra determined according to step 2 satisfy the condition of being smaller than the limit values estimated on the basis of step 3 (see Fig. 2c).

In order to determine the limit strength and displacement of the structure, a specific displaced mode is necessary to be assumed in advance for its inelastic response (see Fig. 2a). Basically, any predominant or possible to be experienced displaced mode of the structure subjected to earthquake motions can be applied. This implies any of the failure modes observed during major earthquakes such as beam failure, story failure or any other definite failure modes.

5 CONCLUDING REMARKS

The performance-based seismic provision was induced in revised new Japanese code. It specifies the clear definition for performance requirements and new verification method.

The earthquake motions are defined as the design acceleration response spectrum which is specified at the engineering bedrock in order to take the soil condition and soil-structure interaction effect into consideration as properly as possible. The response value is essentially calculated from this response spectrum and the structural behavior of a building considered.

The required performance shall be verified by comparing the response values with the limit values.

REFERENCES

Performance-based Building Code of Japan-Frame work of Seismic and Structural provisions (Hiraishi, Midorikawa, Teshigawara, Gojo & Okawa, 12WCEE, 2000).

Development of Seismic Performance Evaluation Procedures in Building Code of Japan (Midorikawa, Hiraishi, Okawa, Iiba, Teshigawara & Isoda, 12WCEE, 2000).

Earthquake Engineering Frontiers in the New Millennium, Spencer & Hu (eds),
© *2001 Swets & Zeitlinger, ISBN 90 2651 852 8*

The study on earthquake disaster estimation in urban areas

F.X. Zhao
Institute of Geophysics, China Seismological Bureau, Beijing 100081, China

Z.J. Han
Institute of Geology, China Seismological Bureau, Beijing 100029, China

ABSTRACT: The methodologies and main results of the cooperative research project focusing on urban earthquake risk assessment which has undertaken by China Seismological Bureau and United Nations Center for Regional Development are introduced in this paper. The earthquake risk assessment and case study in Yumin Li ward include collection of fundamental data, development of damage estimation methodology, development of GIS and case study of Yumin Li ward in Beijing city.

1 INTRODUCTION

In line with the goals of the International Decade for Natural Disaster Reduction, China Seismological Bureau and United Nations Center for Regional Development have undertaken a cooperative research project focusing on urban earthquake risk assessment. The aims of this project were to develop ideas and techniques such as the selection of scenario earthquakes, vulnerability of buildings and so on. Yumin Li ward located in the northern part of Beijing City was selected to be the area of case study for risk assessment. The area of the ward is about one square kilometer and the population of people there is about seventeen thousand. There are about 470 buildings in the ward. Most of them are residential buildings and others are hotels, museums, school classrooms, hospitals, offices, shops, cinemas and factory buildings.

The earthquake risk assessment and case study in Beijing include following contents:

1. Collection of fundamental data and database creation
2. Development of damage estimation methodology
3. Development of GIS for risk assessment
4. Case study of damage estimation of Yumin-Li Ward in Beijing City

In this paper we will introduce some of above contents.

2 CLASSIFICATION OF BUILDINGS

Seismic resistant capability of buildings is closely related to its structural type. For the research on the relation between damage of building and earthquake strength, a classification of seismic resistant capabilities of buildings must be made in accordance with their structures. It has been found that the vulnerabilities of some kinds of structures are similar. Therefore, we divide the buildings in Yumin Li ward into five types in accordance with their vulnerabilities in this study. In the seismic loss analysis, the buildings belong to same type have the same seismic damage matrix.

Type 1: steel structure and reinforced concrete, such as high-rise building of steel structures, frame shear structures and shear wall structures. The buildings in this group have the strongest seismic resistant capabilities.

Type 2: multiple story residential buildings by brick; multiple story buildings of offices and hospitals constructed by brick; multiple story buildings for school classroom constructed by brick; single story factory buildings.

Type 3: single story buildings by brick.

Type 4: single story spacious buildings, such as assembly halls, dinning halls and warehouses, etc.

Type 5: single story old brick buildings.

The numbers of buildings in Yumin Li ward responding to each building type are shown in the following table.

Type of buildings	1	2	3	4	5
Number	35	142	285	4	4

3 SEISMIC DAMAGE MATRICES OF BUILDINGS

Based on the seismic vulnerability analysis, the seismic hazard matrices corresponding to buildings of type 1 to type 4 have been calculated by Yin Zhiqian (Yin Zhiqian et al., 1991). The seismic hazard matrix of type 5 is given based on seismic damage data of previous earthquake in China. The seismic damage matrix of Type1 buildings is shown in the following table.

Seismic damage matrix of Type 1 buildings (%)

Intensity	Intact	Light damage	Moderate damage	Heavy damage	Destroyed
VI	90.00	10.00	0.00	0.00	0.00
VII	85.00	14.00	1.00	0.00	0.00
VIII	70.00	25.00	5.00	0.00	0.00
IX	50.00	31.50	14.50	3.50	0.50
X	20.00	30.00	35.00	10.50	4.50

4 SEISMIC VULNERABILITY ANALYSIS FOR HIGH-RISE REINFORCED CONCRETE BUILDINGS

Under the action of earthquake, story displacement of high-rise building is directly related to the earthquake force applied to the story. Extension rate of story displacement, which is a key parameter to characterize damage degree of the story, expresses the times that the story displacement exceeds yield displacement. The method applied here is based on the relation between story extension rate of high-rise building and its damage degree (Yin Zhiqian et al., 1992).

Based on the data obtained from model test and real building test, the statistical relation between story extension rate of high-rise building and its damage degree is given by the following table:

The relation between story extension rate and damage degree

Degree of damage	Intact	Light damage	Moderate damage	Heavy damage	Destroyed
Extension rate	$\mu \leq 1$	$1 < \mu \leq 3$	$3 < \mu \leq 6$	$6 < \mu \leq 10$	$\mu > 10$

In Yumin Li ward, there are two kinds of high-rise reinforced concrete shear wall buildings. One is of 24 stories and another is of 20 stories. The basic information corresponding to the building of 24 stories are as follows.

1. The shear wall area of every story in one direction is 25229400 mm^2.
2. The sectional area of the building is 484738100 mm^2.
3. The height of every story of the building is 2.7 m and the total height is 64.8 m.
4. For high-rise shear wall buildings in China, the relation between basic natural period and the stories is T=0.06n, where n is the total stories of the building. By the equation, we can calculate the basic natural period of the building T=1.44s.
5. The site category of the building is Type II.
6. The characteristic period T_g is 0.3s considering near-earthquake.
7. The fortification intensity of the building is VIII.

8. The concrete grate of the building is 300.
For the 20 stories building:
1. The shear wall area of every story in one direction is 22011608 mm^2.
2. The sectional area of the building is 430720222 mm^2.
3. The height of every story of the building is 2.7 m and the total height is 54 m.
4. By the equation T=0.06n, we can calculate the basic natural period of the building T=1.2s.
5. The site category of the building is Type II.
6. The characteristic period T_g is 0.3s considering near-earthquake.
7. The fortification intensity of the building is XIII.
8. The concrete grate of the building is 300.

Based on the above-mentioned deterministic method of seismic vulnerability analysis, we calculate the vulnerability corresponding to this high-rise building.

The result is shown as follows.

The vulnerability of the 24 stories buildings

Intensity	VI	VII	VIII	IX	X
Extension rate	0.23	0.44	0.86	2.58	6.54
Degree of damage	Intact	Intact	Intact	Light damage	Heavy damage

5 CASE STUDY OF DAMAGE ESTIMATION OF YUMIN LI WARD IN BEIJING CITY

5.1 Intensities due to scenario earthquakes

Through the investigation of the seismic environment around Yumin Li ward in Beijing City, we have obtained three scenario earthquakes for the case study of damage estimation. The magnitudes and distances of scenario earthquakes are (M=6.4, R=30km), (M=6.8, R=21km) and (M=7.5, R=12km) respectively.

The elliptical intensity attenuation law that is suitable to Beijing City is

Long axis $I_a = 5.720 + 1.350M - 4.075\log(R + 25)$
Short axis $I_b = 2.261 + 1.350M - 2.647\log(R + 7)$

Considering this intensity attenuation law, we can calculate the intensities of Yumin Li ward due to the three scenario earthquakes as the following table:

Scenario earthquake	Magnitude	Distance (km)	Intensity
1	6.4	30	VII
2	6.8	21	VIII
3	7.5	12	IX

Based on the intensities due to the scenario earthquakes, the damage to buildings, damage to human life and economic loss can be estimated.

5.2 Damage to building

The estimation of damage to building is an important part of seismic hazard assessment. It provides a basis for the estimation of damage to human life and economic loss. In this research, the evaluation of damage to building is based on the classification of buildings according to their vulnerabilities the determination of the three scenario earthquakes. Considering the seismic damage matrix of five types of buildings and the intensities of VII, VIII and IX caused by the scenario earthquakes, we can calculate the number of every type of buildings in different damage degrees. The results corresponding to scenario earthquake 3 is shown in the following table.

The estimation of damage to buildings for scenario earthquake 3 (Intensity=IX)

Building Type	Total number of building	Intact	Light damage	Moderate damage	Heavy damage	Destroyed
1	35	18	11	5	1	0
2	142	76	27	22	11	6
3	285	63	54	68	63	37
4	4	0	1	1	2	0
5	4	0	0	1	1	2

The map of damage assessment for buildings with exceeding probability of 50% corresponding to scenario earthquake 3 is shown in following figure.

5.3 Damage to human life

The total population in Yumin Li ward is 17,530. The damage to human life responding to scenario earthquake 1 to 3 can be calculated roughly considering the factors such as time periods, building uses and damage degrees of buildings. The results corresponding to scenario earthquake is shown in following table.

The damage to human life corresponding to scenario earthquake 3

Scenario earthquake	Time period	Deaths	Injuries
3	6:00-8:00 and 17:00-19.00	7	37
	8.00-17:00	9	49
	19:00-6:00(next day)	13	38

5.4 Damage to economics

The economic loss of buildings and the property in Yumin Li ward can be calculated considering the following factors such as area, rebuilt spending, property of each type of buildings and seismic intensity. The results corresponding to different building types and seismic intensities are shown in the following table.

Economic loss in Yumin Li ward corresponding to intensity VII,VIII and IX(Yuan,RMB)

Building type	VII	VIII	IX
1	7,860,000	19,171,000	56,087,000
2	17,577,000	34,850,000	69,388,000
3	3,963,000	8,526,000	4,092,000
4	570,000	2,047,000	4,794,000
5	182,000	284,000	374,000
Total loss	30,152,000	64,877,000	144,735,000

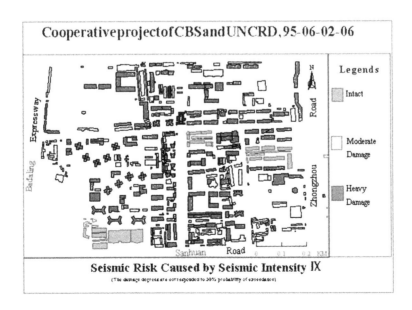

Cooperativeproject of CBS and UNCRD, 95-06-02-06

Legends

Intact

Moderate Damage

Heavy Damage

Seismic Risk Caused by Seismic Intensity IX

(The damage degrees are corresponded to 50% probability of exceedance)

REFERENCE

Yin, Z.Q. *et al*. 1985. Relation between story displacement and yield strength in multi-story framed structures and problem related controlling the displacement to prevent collapse, *Earthquake Engineering and Engineering Vibration*5(1):

Yin Z.Q. *et al*. 1990. Method for estimating seismic damages and losses, *Earthquake Engineering and Engineering Vibration*10(1):

Yin Z.Q. *et al*. 1991. Prediction of earthquake damage and earthquake damage class identification, *Earthquake Research in China*7(1):.

Research Group for Estimating Losses from Future Earthquakes. 1992. *Estimating losses from earthquakes in China in the forthcoming 50 years. Beijing: Seismological Press.*

Earthquake Engineering Frontiers in the New Millennium, Spencer & Hu (eds),
© 2001 Swets & Zeitlinger, ISBN 90 2651 852 8

Lessons on building performance from the great Hanshin-Awaji earthquake disaster in 1995

Tsuneo Okada
Shibaura Institute of Technology, Tokyo 108-0023, Japan

ABSTRACT: Lessons from the Great Hanshin-Awaji Earthquake Disaster caused by the 1995 Hyogo-ken Nambu Earthquake were reviewed and the disaster mitigation programs after the disaster in Japan on three groups of buildings; new buildings, existing buildings and damaged buildings, were summarized.

1 INTRODUCTION

The objective of this paper is to review the lessons from the Great Hanshin-Awaji Earthquake Disaster caused by the 1995 Hyogo-ken Nambu Earthquake, and to summarize the disaster mitigation programs after the disaster in Japan on three groups of buildings; new buildings, existing buildings and damaged buildings.

2 BUILDING DAMAGE DUE TO HYOGO-KEN NAMBU EARTHQUAKE

A damage statistics of buildings and houses is shown in Table-1 (Ministry of Construction 1995). About ten percents of wooden houses and six percents of reinforced concrete, steel, and steel encased reinforced concrete buildings took heavily damages. Older buildings and houses took more damage, and less damage to new buildings and houses. Table-2 is an example of relationship between grades of damage and construction years of reinforced concrete school buildings (Okada, T. et al 2000). It is clear that most of the seriously damaged buildings were constructed before 1981.

3 REVISION OF SEISMIC CODES FOR NEW BUILDINGS

Since a seismic design code was first adopted in 1924 in Japan, and revised in 1950, 1971, 1981 and 2000, the seismic performance of buildings and houses have improved in the each revision. This is one of the reasons why most buildings constructed according to 1981 seismic design code survived even to such severe ground motions in 1995 Hyogo-ken Nambu Earthquake. Other statistics showed similar trend. However, the damage grades were scattered from no damage to severe damage, and many of seriously damaged buildings were demolished even if they could prevent collapse, since the criteria of the code is only to prevent "collapse" and not to assure "reusing without heavy repairing works". Considering such seismic performance of buildings constructed after 1981, the Building Standard Law in Japan was revised in 1998 and enforced in 2000. The main points of the revision are to leave the existing seismic code with some modifications as a minimum requirement to prevent collapse and to add a new method to control damage grade based upon the required performance to buildings as an option. The optional method is to estimate a nonlinear response displacement of the building to the ground motion defined by design spectra, based upon an equivalent linear stiffness and an equivalent viscous damping. Exactly speaking, this method is not a design method but an evaluation method of the response displacement of the building, which is being designed. However, the adoption of this method will allow the structural engineer not to follow some of the specifications for structural requirements such as dimensions of members, structural detailing etc.

4 PROMOTION OF SEISMIC EVALUATION AND RETROFIT OF EXISTING BUILDINGS

4.1 *Actions before 1995*

The importance of the seismic evaluation and retrofit of existing buildings has been recognized since the early 1970's. The Evaluation Standard and Guideline for Retrofit of Existing Buildings was published in 1977 (JBDPA 1976, 1990 revised). However, it had been applied only to limited types of buildings such as reinforced concrete school buildings in limited regions such as Shizuoka Prefecture, where a possibility of an earthquake prediction is expected, and Kanto regions before 1995. A few buildings were evaluated and/or retrofitted in other regions including Hanshin-Awaji area.

Table 1. Damage Statistics of Buildings and Houses [Ministry of Construction 1995].

	Collapse And Severe Damage	Moderate Damage	Minor Damage	Total
1- to 2-storied (mostly wooden houses)	46,022 (9.4%)	42,208 (8.6%)	401,046 (82.0%)	489,276 (100%)
3-storied or more	3,081 (6.4%)	3,273 (6.7%)	42,165 (86.9%)	48,519 (100%)

4.2 Verification study

A study to verify the applicability of the evaluation standard was carried out on reinforced concrete school buildings (Okada, T. et al 2000). Figure-1 is a relationship between construction years and seismic performance indices estimated by the evaluation standard (JBDPA 1976, 1992) with degrees of damage of 104 reinforced concrete school buildings suffered from the 1995 Hyogo-ken Nambu earthquake (Okada, T. et al 2000). Figure-2 is a relationship between the damage indices and seismic performance indices of school buildings. The distributions of Is indices of reinforced concrete school buildings is shown in Figure-3. The curve 1 indicates the probabilistic density function of about two thousands school buildings in Shizuoka Prefecture approxi-

mated by lognormal function and the hatched bars indicates that of Is indices of the school buildings which took severe or moderate damage shown in Fig.2, where the probability is adjusted so that the damage ratio becomes 30%. The curve 4 indicates the probabilistic density function of the severely or moderately damaged buildings due to 1968 Tokachi-oki earthquake or 1978 Miyagiken-oki earthquake (Okada, T. et al 1989, 1999).

These results suggests that 1) seismic performance indices have increased according to the construction years, 2) less damages have occurred to buildings having higher seismic performance indices, 3) The input ground motion in 1995 Hyogo-ken Nambu earthquake might be 1.5-2 times of those of the previous two earthquakes.

4.3 Actions after 1995

It has been strongly recognized after 1995 if we had retrofitted the buildings with lower seismic indices before the earthquake, we could have prevented such serious damages. Since about sixty percents of existing building and houses in Japan were constructed before 1981, the Law for Promotion of Seismic Retrofit was enforced in 1995 to promote the seismic capacity evaluation and retrofit of such building and houses. The law places the owner of the building 1) which is constructed before 1981, 2) which is public and/or open to the public and 3) which has larger floor areas than 1,000 m^2 and higher stories than two, under obligation to evaluate the seismic performance and to retrofit, if necessary.

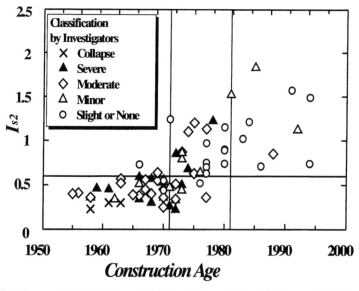

Figure 1. Seismic Indices vs. Construction Years of Reinforced Concrete School Buildings suffered from 1995 Hyogo-ken Nambu Earthquake [Okada, T. et al 2000].

174

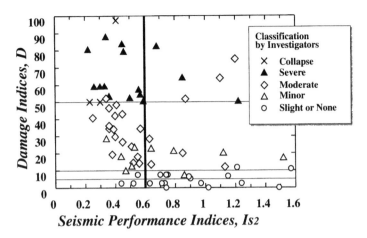

Figure 2. Seismic Indices vs. Degree of Damage of Reinforced Concrete School Buildings suffered from 1995 Hyogo-ken Nambu Earthquake [Okada, T. et al 2000].

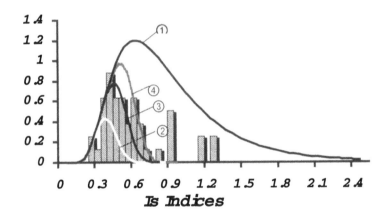

Figure 3. Distribution of Seismic Indices of Reinforced Concrete School Buildings [Okada, T. et al 1989, 1999].

The training of engineers was also needed. Therefore, the Network Committee for Seismic Evaluation and Retrofit of Existing Buildings was established consisting of the organizations related to building design and construction in Japan Building Disaster Association in 1995. Ninety-two organizations have joined in the network committee, and are carrying out the training of engineers and reviewing of the evaluation and retrofit projects.

Many public buildings such as school buildings, city halls, hospitals, etc. have been retrofitted by the financial support of the Government since 1995. However, the retrofit of wooden houses, which caused a great disaster in Hanshin-Awaji area, and private buildings have not been promoted much. Therefore, several local governments have launched recently the projects to support the evaluation and retrofit of private wooden houses financially.

Table 2. Damage Statistics of Reinforced Concrete School Buildings [Okada, T. 2000].

	Pre-1971	1971-1981	Post-1981	Total
Collapse	18(5%)	2(1%)	0	20(3%)
Severe Damage	24(7%)	9(5%)	0	33(5%)
Moderate Damage	90(27%)	39(24%)	11(8%)	140(22%)
Minor Damage	41(12%)	21(13%)	7(5%)	69(11%)
Slight/no Damage	159(48%)	95(57%)	115(87%)	369(59%)
Total	332(100%)	166(100%)	133(100%)	631(100%)

175

5 QUICK INSPECTION OF DAMAGED BUILDINGS

The quick inspection of damages is important to prevent secondary damage due to after shocks and to inform occupants of buildings and houses whether they can keep using their buildings and houses or not.

A project to make guideline for the quick inspection was launched in 1980 as a national project and the guideline to classify the damage into Unsafe (Red), Limited Entry (Yellow) or Inspected (Green) was published in 1990. Since then, it has been recommended the local governments to take actions to train engineers and establish their own quick inspection system. However, the quick inspection system had been established only in several local governments excluding the Hanshin-Awaji area until 1995. Therefore, the quick inspection of the grade of building damages was carried out by the initiation of the Government just after the Hyogo-ken Nambu Earthquake. A quick inspection team consisting of registered quick inspectors, engineers in the Government and local governments, and volunteer engineers was organized. About 40,000 buildings and 50,000 houses were inspected about in a month.

In view of the importance and effectiveness of the quick damage inspection, all local governments have established the systems to register quick inspectors and the training of engineers have been carrying out since 1995. On October 6 in 2,000, an earthquake of M=7.3 on JMA scale attacked Tottori-ken, Shimane-ken and Okayama-ken located northwest of Kobe city. Fortunately, no death toll was reported, but 138 injured, 315 buildings and houses heavily damaged, 1,649 partially damaged and 7,318 slightly or less damaged.

After the earthquake, Tottori-ken, Shimane-ken and Okayama-ken local governments initiated the quick inspection of damaged buildings and houses. 4,080 buildings and houses were inspected by 332 (man/day) registered inspectors in 2 weeks, and 443 buildings and houses were judged safe, 1,499 limited entry and 3,138 inspected.

REFERENCES

Ministry of Construction 1995. Report on Building Damage due to 1995 Hyogo-ken Nambu Earthquake. *MOC*
Okada,T. et al 2000. Improvement of Seismic Performance of Reinforced Concrete School Buildings in Japan. *12WCEE*
Japan Building Disaster Prevention Association 1976,1990. Seismic Evaluation Standard for Existing Reinforced Concrete Buildings. *JBDPA.*
Okada, T. & Nakano,Y. 1989. Reliability Analysis on Seismic Capacity of Existing Reinforced Concrete Buildings in Japan. *Bulletin of Earthquake Resistant Structure Research Center, Univ.of Tokyo. Institute of Industrial Science, University of Tokyo*
Okada,T. & Hisamatu, K. 1999. Investigation of Seismic Retrofitting of Existing RC School Buildings. *AIJ*

Earthquake Engineering Frontiers in the New Millennium, Spencer & Hu (eds),
© *2001 Swets & Zeitlinger, ISBN 90 2651 852 8*

A combine method to establish attenuation law of wide response spectrum

Yangxiang Yu & Yuxian Hu
Institute of Geophysics, China Seismological Bureau, Beijing 100081, China

ABSTRACT: In this paper a seismological-engineering approach is suggested to estimate the wide spectrum for distant earthquakes. The method combines the seismological relations of magnitude and displacement and digital records of VBB and FBA for the estimation of long period (2~20 sec) part of the spectrum with the currently available strong ground motion data from the analog accelerograph records for the short period (0.1 – 3 sec) part. Error and limitation of the results derived from the digital records are also discussed. The limited available strong ground motion data obtained by digital instruments are finally used to show the adequacy of the results obtained. It seems possible to provide dependable spectra of ground motion up to 10 to 20 sec with accuracy comparable to that from the definition of the earthquake magnitude and also comparable to the accuracy of the spectrum at shorter period part.

1 EXAMPLES OF DAMAGE OF LONG PERIOD STRUCTURES BY DISTANT EARTHQUAKE

Many examples of the selected damage of long-period structures on soft sites and at long-distant strong earthquakes have been recorded in the past. Some at hand (IAEE, 1996) are mentioned here to refresh our understanding.

1.1 *1979.03.28 Gediz, Turkey earthquake*

During this earthquake of magnitude 7 with epicentral intensity IX, the Tofas factory in Bursa City was 135 km away, in an area of intensity V, but the structures were badly damaged, with a local high intensity of VIII. The ground deposit was very soft sandy clay with a depth of a few hundred meters. Aftershock records showed a ground predominant period of 1.2 sec and the natural period of the collapsed structures about 1.25 s. The intense damage came from resonance effects of distant earthquake, soft site and long period structure (IAEE, 1996).

1.2 *Heavily damaged Mexico City during earthquakes 300-400 km away*

During several recent strong earthquakes since 1957, Mexico City, roughly 300-400 km away from the epicenters, showed abnormally high damage to tall buildings in the old city, built on very thick and soft deposits. Recent strong motion records proved the long period ground motion of 1-2 sec in the old city. This combination of very soft ground, strong distant earthquake and tall buildings was considered the cause (IAEE, 1996).

1.3 *1977.03.04 Vrancea earthquake in Romania*

Tall buildings in Sofia, 450 km away from the epicenter of this earthquake with magnitude 7.3 and with a focal depth about 100 km, suffered significant damage. At Svishtov in Bulgaria, 240 km away from the epicenter, rigid one-story brick buildings were not damaged, but buildings, tall or with a flexible first story, were damaged or even collapsed. Strong acceleration records showed that the ground motion was similar to a sinusoidal motion with long predominant periods of 1.0-2.4 sec (IAEE, 1996).

2 CURRENT CODE SPECIFICATIONS OF DESIGN SPECTRUM AT LONG PERIODS

Current codes of seismic design of structures derived their design spectra from the data available at the time of drafting the codes, mostly in the period of 1960's to 1980's. We analyzed the longest period considered in the codes listed in Regulations for Seismic Design, A World List – 1996, distributed by the International Association for Earthquake Engineering, together with some other codes at author's hand.

Some codes specify a limit of the longest period considered, for example T_{max}=3 sec in China building code GBJ 11-89, 5 sec in highway code, and 7 sec in China special structure code GB 50191-93; some give a minimum value of the design spectrum $S_a(T, \zeta)$, for example, 0.65 in Albania code, with or without limit long period; and a few codes with no limit, such as the Eurocode 8.

Out of the codes we analyzed there are 20 codes specify a minimum design value for the spectrum at long period side, 17 have a specific value of the end of the long period, 6 not clear to the author, and 2 no spectrum in the code. Among those specifying a limit of the longest period, one code limits it to 10 sec, one to 7 sec, the rests to a range of 2.25 – 5.0 with an average of roughly 3 sec. There are perhaps two main reasons for this situation. Firstly, there were not so many tall buildings as we have now. A building of 30 stories may have a fundamental period roughly 3 sec or less and the main cases considered in the ordinary codes are common buildings of lower height. That is why there are seismic design codes for tall buildings in addition to building codes, such as in China. Secondly, data of strong motion records accumulated in the world now reach a huge amount, say 10,000 traces of acceleration, but most of them are from analog accelerographs. People doubt their accuracy of periods long than 3 – 5 sec. These two topics will be discussed in the following two sections before going to the authors' suggested approach of a practical method of assessment of wider spectra.

3 CURRENT REQUIREMENT OF DESIGN SPECTRUM AT LONG PERIODS

In China, there are quite a few tall buildings of more than 50 story-high and even taller ones in planning. There are towers of height more than 100 m and bridges of main span over 1,000m. Among these tall buildings, a building of 88 stories above ground in Shanghai and a building of 68 stories in Shenzhen City were built in the nineties, both with a fundamental period of 6.2 sec. To meet the design requirement of very tall structures and other reasons in Shanghai, the mostly populated city in China had for the first time a special local seismic design code in DBJ 08-9-92 in 1992, and followed by a partially revised edition in 1996. In this local code, a special design spectrum with the longest period up to 10 sec. was specified to Shanghai region for its special seismicity environment and soft site conditions.

In addition to buildings, long suspension bridges of main span over one kilometer have built in many countries, including China; high televisions and other long-period structures are becoming popular in large cities in China.

Important long period structures should be designed with specific consideration of long period design spectrum, because large cities are usually located on large valley of soft deposit and some of them may subject to the attack of strong earthquakes at some distance. Mexico City in Mexico, Shanghai, Beijing, Taipei and Tianjin City in China are examples.

4 ACCURACY OF CURRENT DATABANK OF STRONG MOTION RECORDS

Popularly used strong ground motion data were mostly recorded by analog accelerographs, from which the design response spectra were derived. Accelerogram means the time-history of acceleration for motion in one given direction at a point on or under ground or on structure. There are now in the world about more than 10,000 accelerograms accumulated from past strong earthquakes. Roughly one third is in the USA, one third in Japan, and another one third in Taiwan of China, Europe and other places. Most of the existing accelerograms were obtained from the analog accelerographs, of which the spectrum is usually in the range of 0.1 – 3 sec. That means that those components of motion with periods out of this range, will be filtered out or strongly reduced. Corrections may be made for this band pass filtering effect, but the main point is whether the signal-to-noise ratio is still high enough to guarantee a meaningful accuracy of the results. Noise may be introduced in several steps, such as recording, processing, digitizing, scaling, and analyzing the time history of accelerograms.

5 APPLICABILITY OF DIGITAL BROAD-BAND RECORDS

Compared with the analog accelerometer, the frequency response of the digital accelerometer improved a lot, and the long-period error introduced by digitizing of analog record is avoided. So the digital records are generally considered reliable to study the long-period ground motion and some researchers hereby have calculated the long-period response spectra (e.g., Xie et al., 1990). Nevertheless, some recent researches show that when the period is longer than 10 sec, the record of digital accelerograph is unreliable due to background noise. Both the actual recording and the shaking table experiment get the same result (Chiu, 1997; Zhou et al., 1997). Fortunately, the digital seismographs used by seismologist have wide frequency response. For example, the VBB system of Chinese Digital Seismograph Network (CDSN) which is a sub-network of the Global Seismographic Network, has a flat frequency band of 8.5~0.003Hz and the FBA-23 system of CDSN has a band of 40~0.001Hz. Both systems were digitized at 24-bit resolution.

In order to investigate the applicability of VBB and FBA records of CDSN in long-period ground motion study, firstly we compared the VBB record with the FBA record to ensure the consistency of the ground motion recovered from two different records at one station. The VBB record is actually velocity

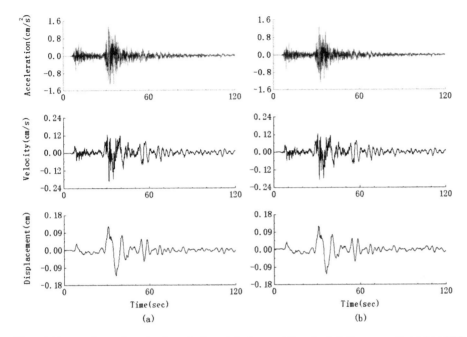

Figure 1. Recovered ground acceleration, velocity and displacement (from top to bottom, respectively) from VBB and FBA of BJI station (a) VBB (b) FBA

record. The FBA system is somewhat the same with the digital accelerograph. If the recovered ground acceleration, velocity and displacement are the same at the same station, we may say that the velocity record is also usable.

Figure 1 is the recovered ground acceleration, velocity and displacement of E-W component of VBB and FBA records in BJI station for the Zhangbei earthquake of January 10, 1998, Ms6.2. We can see from Figure 1 that the ground displacement and velocity from the two records are almost the same and no significant long-period error. This consistency may indicate that the two seismometers are reliable and the true ground motion, especially the long-period motion, can be recorded.

Another procedure to verify if the digital broad-band record can be used to study the long-period ground motion is to check the signal-to-noise ratio in long-period. The record of Zhangbei earthquake by BJI station is again used as an example to analyze the Fourier spectra of the signal and noise. After the records of VBB and FBA were corrected to acceleration, we extract the 62.5 sec of pre-event data as noise (include the instrument noise and background noise) and the following 125 sec of data as signal. The calculated Fourier amplitudes of the noise and signal are shown in Figure 2.

It can be seen from Figure 2 that the level of

noise of VBB and FBA is low enough to keep a high signal-to-noise ratio in long-period, even though the seismic signal is rather weak (the PGA is less than 2 gal).

Above studies show that the digital broad-band records are suitable for study of long-period ground motion.

6 APPLICATION OF EARTHQUAKE MAGNITUDE

The definition of earthquake magnitude has a relation with amplitude, period and distance. For example, the surface magnitude recommended by IASPEI is defined as follows

$$Ms = \log\left(\frac{A}{T}\right)_{max} + 1.66\log\Delta + 3.3 \qquad (1)$$

where A is the amplitude of horizontal component surface wave (in μm), T is the corresponding period, Δ is epicentral distance in degree. In China, when measuring the magnitude, the period T must be greater than 3 sec and $\Delta \geq 1°$. Because the magnitude is a function of ground displacement amplitude, period and distance, the ground displacement can be estimated from the corresponding period, distance and magnitude.

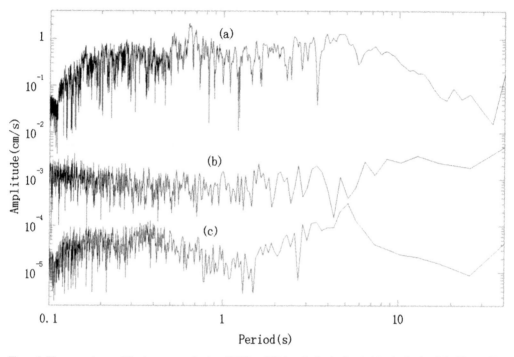

Figure 2. The comparisons of Fourier spectra of noise of VBB and FBA and seismic signals (a) seismic signal (b) FBA (c) VBB

7 A COMBINED APPROACH OF ENGINEERING DATA AND SEISMOLOGICAL DATA

The attenuation relation used by engineers and seismologists in the seismic hazard assessment or zoning map has a general form as follows

$$\log Sa(T,M,R) = C_1 + C_2M + C_4\log(R + C_5\exp(C_6M)) \quad (2)$$

where $C_i(i=1,2,\ldots6)$ are regression coefficients.

Comparing equation 1 with equation 2, we may say that they are principally the same. The engineers used to emphasize the shorter period motion, say 1 sec or less, which is controlled mostly by the near field motion, say less than 100 km; and on the other side, the seismologists paid more attention to the far field motion, say more than 100 km away, which is usually of longer periods, say more than 1 sec. It is natural then trying to combine them together with emphasis of the engineering approach on the near field and short-period motion, and the seismological approach on the far field and long-period motion, to have a better estimation of design spectrum in a wide period band, says from 0.1 to 20 sec. This is the basic idea of this paper.

Since the main part of engineering data of ground motion is obtained in the United States, the short pe-riod spectrum is then estimated on the US data. Huo (1990) had obtained such spectrum attenuation for rock sites, as given by solid lines in Figure 3 for periods smaller than T=1.7 sec and by dotted lines greater than T=1.7 sec.

To have a good estimation of the long-period spectrum two sets of data are obtained from the point of view of the seismologists. The first set is obtained directly from the definition of surface wave magnitude (Eq. 2). The displacement response spectrum value Sd is taken as Sd=2.5A, where A is the displacement amplitude in Eq. 2 and the factor 2.5 is the amplification of the ground motion to response, which is related to a damping of 0.05 for the spectrum. The spectrum value for displacement Sd is then converted to spectrum value for acceleration Sa. The second set of data is obtained from 152 horizontal components of digital broadband records in China (Wang et al., 1998; Yu et al., 2000).

With these two data sets a regression is made to obtain the long-period part (T=1.7-20 sec.) of the response spectrum. Using also the same equation 3, the results are plotted in Figure 3 in solid lines. It can be seen clearly that the spectra obtained by Huo (1990), based solely on engineering data, given in dotted lines for periods T>1.7 sec, show irregular results and not dependable in the long-period range, and that the new results presented in this paper show a much better smoothness at least.

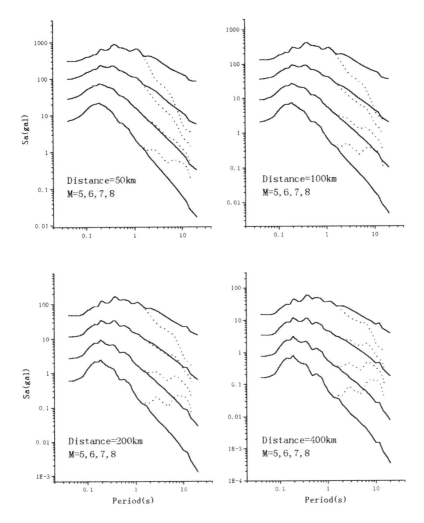

Figure 3. Acceleration response spectrum Sa(T, ξ=0.05) up to T=20 sec. Dotted line is Huo's result based on US data; solid line is the new result obtained by the suggested approach.

8 CONCLUSION AND DISCUSSION

To meet the urgent need of a wide-band design spectrum for very tall buildings and very long-span bridges, a combined approach of both the engineering acceleration data and the seismological data of magnitude definition and the broad-band data is suggested. The engineering data cover rightly the shorter period (0.05 – 1 or 2 sec) and the seismological data cover the longer period (1 – 20 sec), and the transition is smooth.

There are still something to be done in the future. For example, the amplification factor is a key problem for the combined approach. In this paper, the amplification factor is taken as 2.5. It may over-

estimate the long-period response, because the longer the period is the fewer the vibration cycles when the duration is fixed.

To expand the data set of broad-band data, other VBB or FBA records may be used, for example the TERRAscope data of Southern California. Because the VBB and FBA instruments are becoming standard instruments in seismic station, these data will accumulated quickly.

Another data set may be used. In the *Annual Chinese Earthquake Bulletin* there are displacement amplitude and corresponding period data for many stations. Although these data are used for measuring earthquake magnitude, they are also good for estimating long-period ground motion.

REFERENCES

Calvi, G.M. 1998. Performance-based approaches for seismic assessment of existing structures, *Proc. Eleventh European Conf. on Earthquake Engineering*, Paris, pp.3-19.

Chiu, H.C. 1997. Stable baseline correction of digital strong-motion data. *Bull. Seism. Soc. Am.*87(4):932-944.

Hu, Y.X., Liu, S.C. and Dong, W.M. 1996. Earthquake Engineering, E&FN SPON, Chapman & Hall, London, New York, Tokyo.

Huo, J.R. 1990. *Attenuation laws of ground motion*, Dr. Thesis, Institute of Engineering Mechanics, China Seismological Bureau. (in Chinese).

IAEE (International Association for Earthquake Engineering). 1996. A World List – 1996.

Wang, S.Y., Yu, Y.X. and Lu, H.S. 1998. Study of characteristics of long-period ground motion response spectra by using broad-band records of the Chinese Digital Seismograph Network. *Acta Seismologica Sinica*11(5):557-564.

Xie, L.L., Zhou, Y.N., Hu, C.X. *et al.* 1990. Characteristics of response spectra of long-period earthquake ground motion. *Earthquake Engineering and Engineering Vibration*10(1): 1-20. (in Chinese)

Yu, Y.X., Hu, Y.X. and Wang, S.Y. 2000. Calculation of long-period ground motion response spectrum by using broad-band digital record. *Proc. Of 12th WCEE*, Auckland, New Zealand.

Zhou, Y.N., Zhang, W.B. and Yu, H.Y. 1997. Analysis of long-period error for accelerograms recorded by digital seismographs. *Earthquake Engineering and Engineering Vibration*17(2):1-9.

Earthquake Engineering Frontiers in the New Millennium, Spencer & Hu (eds),
© 2001 Swets & Zeitlinger, ISBN 90 2651 852 8

Reliability analysis of existing frame structure based on failure correlation

Da Huo
Beijing Polytechnic University, Beijing, 100022, P.R.China

Guanghui Li, Dongwei Wang, & Fang Fu
Zhengzhou University of Technology, Zhengzhou, 450002, P.R.China

ABSTRACT: Using Monte-Carlo simulation method, through failure correlation analysis of sectional constraints under medium and small earthquakes respectively, the failure correlation rules of sectional constraints of frame structure are obtained, and they are simplified into hypothesis which can be used in reliability evaluation of existing frame structure. Moreover, the use of correlation rule in system reliability analysis of frame structure is illustrated by two examples.

1 INTRODUCTION

System reliability of structure is an important index indicting the integral safety, so it is very appropriate that system reliability is used to represent the integral bearing capacity of frame structure. However, due to the limit of research level, in current reliability appraisal practice, the bearing capacity of a member or the whole structure is determined by the bearing capacity of the most vulnerable section in a worst member. As a result, the appraisal result obtained in this way is usually far away from the factual bearing capacity of a frame structure.

The failure of a frame structure can be expressed as the failure of a series system consisting of all failure modes of the structure, and that is

$$P_f = P\{\overline{\Omega_s}\} = P\{\bigcup_{j=1}^{k} \overline{\Omega_j}\} \qquad (1)$$

Herein, $\overline{\Omega_s}$ denotes the failure domain, and $\overline{\Omega_j}, j = 1, \cdots, k$ denotes k failure modes.

While, the failure of a certain mode of frame structure system can be expressed as the failure of a parallel system composed of all failure elements of this mode, and it is as follows:

$$P\{\overline{\Omega_j}\} = P\{\bigcap_{i \in N_j} F_i\} \qquad (j = 1 \sim k) \qquad (2)$$

Herein, N_j denotes that there are totally N_j failure elements in the j_{th} failure mode, and F_i denotes the i_{th} failure element in the j_{th} failure mode.

However, there may be hundreds of thousands failure modes occurred in one factual high-rise frame structure, and there also exists complicated statistical correlation among all failure modes. So it

is very onerous and difficult to calculate the system reliability of frame structure when directly using equation (1) and equation (2), and it may be impossible at sometimes. The common practice is to introduce some hypothesis to simplify the computation of system reliability.

As to the existing frame structures, based on structural diagnose, survey and sectional check, the number of key failure sectional constraints may be limited, with the correlation between sectional constraints is taken into consideration, it is feasible to calculate the system reliability of frame structure under a certain failure criterion (Wang 1996).

2 FAILURE CORRELATION BETWEEN SECTIONAL CONSTRAINTS

2.1 *Analysis method*

The methods adopted by this paper include Monte-Carlo random simulation method and correlation checking of series. The fundamental theories of random analysis see reference (Fu 1998).

2.2 *Analysis parameters*

In this paper, random simulation test are mainly done on the RC frame structures less than 5 spans and 12 stories. In detail, the test modes can be classified into three categories, i.e. design symmetric regular frame structure, approximate symmetric regular frame structure, and non-symmetric irregular frame structure. While each kind of frame structure is also classified into two groups of structure of 6 and 12 stories respectively, and their detailed parameters can be seen in Table 1 and Table 2.

Table 1. Parameters of RC frame of 12 stories

No. of types		Span (m)	Story height (m)
A	(1)	6.0,3.6,6.0	6.0,4.5,3.6×10
	(2)	6.0,3.6,6.0	6.0,3.6×11
	(3)	6.0,3.6,6.0	6.0,3.6×10,4.5
	(4)	6.0,3.6,6.0	4.5,3.6×11
	(5)	6.0,3.6,6.0	4.5,3.6×10,4.5
	(6)	6.0,3.6,6.0	4.5,4.5,3.6×10
	(7)	3.6×5	4.5,3.6×11
	(8)	3.6×4	04.5,3.6×11

Table 2. Parameters of RC frame of 6 stories

No. of types		Span (m)	Story height (m)
A	(13)	6.0,3.3,6.0	6.0,3.6×5
	(14)	6.0,3.6,6.0	6.0,3.6×5
	(15)	6.0,3.6,6.0	4.5,3.6×5
	(16)	6.0,3.6,6.0	6.0,3.6×5
B	(17)	6.9,3.3,7.2	6.0,3.6×5
C	(18)	6.9,4.5,7.2	6.0,3.6×5
	(19)	6.0,4.8,7.2	6.0,3.6×5

A: design symmetric structure.
B: approximate symmetric structure.
C: non-symmetric structure.

2.3 Correlation rules

2.3.1 Under minor earthquakes

As to the RC frame structure designed according to seismic specification, the correlation rules among sectional constrains are very weak. The more the vertical loads taken into consideration, the weaker the failure correlation is. So according to a large amount of stochastic simulation test, we reach a conclusion that under minor earthquakes the failures of sectional constraints of RC frame structure can be regarded as independent.

2.3.2 Under medium earthquakes

1 As to the design of a symmetrical or approximately symmetrical frame structure, the failure of sectional constraints of frame beams and columns of the same type and at the symmetrical positions are fully correlated.

2 For non-design symmetrical frame structures, the failures of sectional constraints of frame beams of the same type and of the symmetrical positions are fully correlated, while the maintenance failures of sectional constraints of frame columns of the same type and of the symmetrical positions are approximately independent.

3 BASIC HYPOTHESIS AND CALCULATION METHOD

3.1 Basic hypothesis

In order to simplify the calculation of system reliability, based on the obtained failure correlation rules of sectional constraints and available earth-

quake hazard references (Liu 1985), the following hypothesis are presented.

1 There are only two state of sectional constraints taken into consideration, and they are "reliable" denoted by "1" and "failed" denoted by "0";

2 When a minor earthquake lowered than reference level occurs, all the sectional constraints of frame structure are assumed as failure independent.

3 When hit by a medium earthquake, the failure of all sectional constraints of frame columns is assumed as fully correlated, while the failure of sectional constraints of frame beams is reckoned as independent.

3.2 Simplified method of system reliability

3.2.1 Under minor earthquake

According to the stipulation of "No Damage under Minor Earthquake" in "Chinese Seismic Design Code of Building Structure" (GBJ11-89), the failure of any sectional constraint of frame structure under minor earthquake can be thought as the failure of frame structure itself. That is to say any failed sectional constraint can be regarded as a failure mode. So, the failure probability of a frame structure under minor earthquake can be written as:

$$p_f = 1 - \prod_{i=1}^{n} (1 - p_{fi}) \tag{3}$$

Herein, p_{fi} represents the failure probability of each sectional constraint.

Once the failed sectional constraints of frame structure and their failure probability are obtained through structural diagnose and analysis, the failure probability of the whole system can be obtained according equation (3).

3.2.2 Under medium earthquake

Through the analysis of reference (Li 1999), under medium earthquakes, all the sectional constraints of structure are not fully correlated, so some states in the expression of structural failure domain may not occur. In this case, fully correlated constraints in structural failure domain should be dealt with according to the hypothesis of "Weakest Constraint", and then the integral failure probability can be obtained, see reference (Li 1999).

4 EXAMPLES (LI 1999)

4.1 Under minor earthquakes

Example 1:

A six-story RC frame designed according to the seismic code is shown in Fig. 1. After serving several years under ordinary atmospheric environment, through structural inspection and sectional check, the probable failure constraints is found and num-

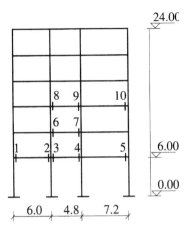

Figure 1. Sectional constraints of example 1

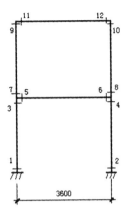

Figure 2. Sectional constraints of example 2

Table 3. Failure probability of constraints of the structure shown in Figure 1

failed constraint	1-2	3-4	5	6	7-8	9-10
failure probability	.001	.002	.001	.003	.001	.003

bered, and their places are shown in Fig. 1. Their failure probabilities are listed in Table 1. Please determine the system reliability of the structure under minor earthquakes.

When the hypothesis of "failure independence" is adopted under small earthquakes, the number of failed sectional constraints of the structure is equal to the number of the minimum failure set. So the structure in Fig. 1 has 10 minimum failure sets, and they are

$\{0_1\},\{0_2\},\{0_3\},\{0_4\},\{0_5\},\{0_6\},\{0_7\},\{0_8\},$
$\{0_9\},\{0_{10}\}.$

The failure probability of structural system with parameters shown in Table 1 will be

$$p_f = 1 - \prod_{i=1}^{10}(1 - p_{fi}) = 0.01786$$

To check the accuracy of the above result, Monte-Carlo simulation method (2000 times) was used to obtain the failure probability of the same frame structure, and the failure probability is 0.018. The error of the two methods is less than 5%.

4.2 Under medium earthquakes

Example 2:

A one-bay two-story RC frame designed according to the seismic code is shown in Fig. 2. After serving several years, through structural inspection and sectional check under medium earthquake, the reliability parameters of sectional constraints of the structure are obtained and listed in Table 2. Please determine the system reliability of the structure under medium earthquake.

1 Correlation analysis of sectional constraints

On the basis of hypothesis above, there are six pairs of constraints correlated, and they are 1 and 2, 3 and 4, 5 and 6, 7 and 8, 9 and 10, 11 and 12. Making use of the "Weakest Constraint" hypothesis we can get the representing constraint of each controlling constraint, they are shown in Fig. 2, too.

Table 4. Failure probability of constraints of the structure shown in Fig.2.

failed constraint	1	2	3	4	5	6
failure probability	0.01	0.01	0.01	0.01	0.02	0.02
representing constraint	1	1	3	3	5	5
failed constraint	7	8	9	10	11	12
failure probability	0.01	0.01	0.01	0.01	0.01	0.01
representing constraint	7	7	9	9	11	11

2 Determining the minimum failure sets and the failure field of the structure

Under the medium earthquake, "Mechanical Failure" criterion is often used to judge whether a RC frame structure is failed or not. So the fame shown in Fig. 2 has 6 failure modes as below:

$\{0_1 \cap 0_3\} \{0_7 \cap 0_9\},\{0_7 \cap 0_{11}\},\{0_1 \cap 0_5 \cap 0_7\},$
$\{0_1 \cap 0_5 \cap 0_9'\},\{0_3 \cap 0_5 \cap 0_9\}.$

The failure domain of the structure is written as:

$$\overline{\Omega}_s = \{0_1 \cap 0_3\} \cup \{0_7 \cap 0_9\} \cup \{0_7 \cap 0_{11}\}$$
$$\cup \{0_1 \cap 0_5 \cap 0_7\} \cup \{0_1 \cap 0_5 \cap 0_9\}$$
$$\cup \{0_3 \cap 0_5 \cap 0_9\}$$

3 The complete expression of the failure field

According to the "non-crossing" method (Wang 1996), the complete expression of the failure domain of the structure can be changed into

$$\overline{\Omega}_s = 0_1 0_3 + 1_1 1_3 0_7 0_9 + 1_1 1_3 1_9 0_7 0_{11}$$
$$+ 1_3 1_9 1_{11} 0_1 0_5 0_7 + 1_3 1_7 1_{11} 0_1 0_5 0_9$$
$$+ 1_1 1_7 1_{11} 0_3 0_5 0_9$$

Because the representing constraints are independent each other, we can replace the failure states (0) of the representing constraints by their failure probabilities, and the reliable state (1) by their reliable probability. The failure probability of structural system with parameters shown in Table 2 can be obtained as below:

$$P_f = 4.862 \times 10^{-4}$$

The failure probability by Monte-Carlo simulation method (10000 times) of the same frame structure system is found to be 0.0005239, and the error of the two methods is 7.75%. So, the above calculation result can basically meet the demands of application in practical engineering.

5 CONCLUSIONS

Through the analysis above, some significant conclusions and suggestions can be obtained.

1 Referring to the seismic hazard data (Liu 1985) and the analysis of this paper, we suggest that different failure correlation hypothesis should be adopted respectively considering the failure correlation of the sectional constraints under the medium and minor earthquakes;

2 Making use of the failure correlation of the sectional constraints will decrease the number of structural failure constraint and failure mode taken into consideration, and transform two kinds of correlation, namely the correlation of failure mode and the correlation of failure constraints, into one kind of correlation by "non-crossing" technique (Wang 1996);

3 For existing frame structures, the failure probability of the structural system can be obtained by the method proposed above, because the number of controlling failed sectional constraints is limited on the basis of actual diagnose and sectional check.

REFERENCES

Wang, D.W., Reliability Analysis of Building Engineering System [D]. Harbin: Harbin University of Civil Engineering and Architecture, 1996. 3–36.
Fu, F., Failure Correlation and Reliability Analysis of RC Frame Structure [D]. Zhengzhou, Zhengzhou University of Technology, 1998. 6–40.
Li, G.H., Reliability Appraisal and Maintenance & Strengthening Strategy of Existing RC Frame Structure [D]. Harbin, Harbin University of Civil Engineering and Architecture, 1999. 45–50.
Liu, H.X., The Seismic Hazard of Tangshan Earthquake (The Second Edition) [M]. Beijing: Seismological Press, 1985. 12–66.

Earthquake Engineering Frontiers in the New Millennium, Spencer & Hu(eds),
©2001 Swets & Zeitlinger, ISBN 90 265 1852 8

Dynamic model failure tests of dam structures

Gao Lin
Dalian University of Technology, Dalian 116024, China

ABSTRACT: For the earthquake safety evaluation of dam structures, it is desirable to extend the existing and to attain new knowledge on the dynamic behaviors of these structures during strong earthquakes, along with the limitation of field observation data, through laboratory experiments. Dynamic model failure tests of a number of concrete gravity dams, concrete arch dams and embankment dams have been carried out in Dalian University of Technology, China. The relevant research work is briefly introduced. The earthquake ground acceleration that induces the first tension crack at the dam body is suggested as the basic index for the safety assessment of concrete dams and is predicted through dynamic model failure tests. A rather simple similarity rule for the dynamic model failure tests of embankment dams is developed; it can be used for the quantitative assessment of strong earthquake response of fill type structures.

1 INTRODUCTION

A great number of high dams will be built in highly seismic areas, the safety evaluation of these structures during strong earthquakes are of great concern, since tremendous material damage and loss of human lives may be caused by the failure of these structures. In the last two or three decades, our ability to analyze mathematical models of dam structures subjected to earthquake ground motions has improved dramatically. Sophisticated computers programs have been developed and used for numerical nonlinear analysis of the concrete as well as embankment dam structures. In spite of the advances in the theory and numerical procedures, these analyses have failed to predict the response of real structures particularly at their ultimate states. As a consequences of this and also because the lack of reliable models to predict the failure model of real structures, a considerable degree of uncertainty still exists in the seismic design of dam structures. The needs to extend the existing and to attain new knowledge on the dynamic behaviors of dam structures motivate, along with the limitation of field data, laboratory experiments on small-scale models.

A number of dynamic failure model tests of several real concrete gravity dams, arch dams and embankment dams, which are going to be constructed in the near future in China, have been carried out on a earthquake simulating shaking table installed in the Dalian University of Technology. The relevant research work is briefly introduced.

2 DYNAMIC SIMILARITY RULE

The equation of motion of the structure subjected to earthquake excitation is expressed in the following form:

$$[M]\{\ddot{U}\}+[C]\{\dot{U}\}+[K]\{U\}=-[M][E]\{\ddot{U}_g\} \qquad (1)$$

where [M], [C], [K] are the mass, damping and stiffness matrixes of the structure respectively; $U,\dot{U},\ddot{U}$ are the displacement, velocity and acceleration of the structure respectively; and $\ddot{U}_g$ is the earthquake ground acceleration; and [E] is the transformation matrix.

It is seen that the dynamic response of the structure is affected by the inertia forces, the damping forces, the elastic restoring forces and the earthquake exciting forces. It leads to the similarity relationship as expressed below. we call it the similarity rule of elasticity (Lin 1958).

$$\tau = \lambda \lambda_\rho^{1/2} \lambda_E^{-1/2} \qquad (2)$$

where λ -the geometric scale of the model; τ, λ_ρ and λ_E are the scale factors(ratios of prototype to model) of time, mass density and modulus of elasticity respectively.

If viscous damping is assumed, the critical damping ratio should be made equal for the real structure and the model. And the scale factor of earthquake ground acceleration equals $\lambda\tau^{-2}$.

In case of full reservoir, it is necessary to keep the similarity of earthquake induced dynamic water pressure of the model and the prototype. That is, the ratio of density of the model and prototype reservoir liquid should be equal to the ratio of the mass density of the model and dam material (Lin 1958).

$$\gamma_m/\gamma_p = \rho_m/\rho_p \qquad (3)$$

where γ_m, γ_p -density of the liquid in the reservoir of model and prototype respectively; ρ_m, ρ_p are the mass density of dam material in the model and prototype respectively.

To study the dynamic behaviors of dam structures at near failure stage and to study the arch dam response including joint opening effect, the role of hydrostatic pressure and the gravity force should be taken into consideration, we arrive at the additional similarity requirement (Lin 2000) as follows, we call it the similarity rule of gravity.

$$\tau = \lambda^{1/2} \qquad (4)$$

To design a model that satisfies both the relationship Eq. (3) and Eq. (4) as shown in Eq. (5) is exceptionally difficult.

$$\tau = \lambda\lambda_\rho^{1/2}\lambda_E^{-1/2} = \lambda^{1/2} \text{ or } \lambda\lambda_\rho\lambda_E^{-1} = 1 \qquad (5)$$

Because once the model material is assigned, the scale factors λ_ρ and λ_E are fixed, the geometric scale of the model λ cannot be chosen freely, unless we turn to the centrifuge modeling, where the scale factor of gravity acceleration may be adjusted.

In addition, at near failure stage, the restoring force becomes nonlinear, we need further to ensure similarity of the stress-strain curve for the model and real dam material.

For concrete dam to keep similarity between model and the prototype, several alternating approach as shown in Fig.1 may be used. It brings some convenience in the model design.

Experience of Dalian University of Technology in dealing with dynamic model failure tests of dam structures is briefly introduced.

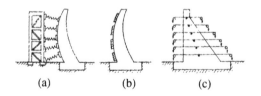

(a)   (b)   (c)

Figure 1. Simulation of hydrostatic pressure (a), hydrodynamic pressure (b) and self-weight (c) in the model to satisfy similarity requirement

3 SIMILARITY RULE FOR DYNAMIC MODEL FAILURE TESTS OF CONCRETE DAMS

The results of model experiments and the historical cases of concrete gravity dams as well as buttress dams, such as the Hsin Fen Jiang dam in China in 1962, the Koyna dam in India in 1967 (Chopra et al. 1972) and the Sefid Rud dam in Iran in 1990 (Ahmadi 1992) indicate that tension cracks are the significant overload responses of concrete dams during strong earthquakes. So, we choose the earthquake ground acceleration that induces the first tension crack at the dam body as the basic index for the safety assessment on this type of dam. It has been found that the most vulnerable part of such dams is the head of gravity and buttress dam, or the top central part of the arch dams.

In order to obtain data similar to real structures, it is hoped that the basic mechanical properties of the model material satisfy the following requirements to be greatest extent. First, it is concrete-like; second, its modulus of elasticity keeps as low as possible, so the primary frequencies of vibration of model structures lie in the effective range of the shaking table; and, third, its tensile strength is maintained sufficiently low which allows the model to be damaged under the nominal exciting force of the shaking table. The mixes of our laboratory are selected to consist of cement, barite sand, stone powder and water. It gains strength with age, and its mechanical properties are very similar to concrete. By adjusting the material age in performing the failure test in accordance with the room temperature, a model dam with proper modulus of elasticity and tensile strength is achieved. The density of the material may be kept almost the same as concrete.

The geometrical scale is assigned such that the model dam is nearly 1.5m to 2m high. The exciting acceleration is gradually increased. The possible damages inflicted on the structure under various intensities of earthquake ground motion are identified. Fig. 2 shows the typical crack formation of the model of a 195m high rolled-concrete gravity dam corresponds to various levels of excitation (Lin et al. 1993).

The problem is how to infer the seismic ground acceleration at the dam site that will produce the first tension crack at the real dam, based on the failure test results of model dams, taking into consideration that in general the damage condition in the model can correctly match that of the actual scenario in the real dam. So far as the dam behaves mainly in the elastic range before and even after the appearance of the first tension crack, the law of elastic similarity

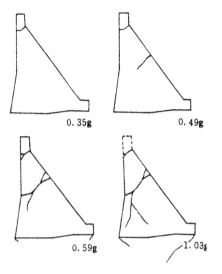

0. 35g   0. 49g

0. 59g   1. 03g

Figure 2. Crack formation corresponding to various levels of excitation

may be applied to determine the relationship between the physical values of model and prototype with respect to their proportion and material properties.

Based on the discussion stated above, the earthquake input acceleration at the dam base that induces the first tension crack at the dam body A_c is determined as follows (Lin et al. 1993)

$$A_c = a_c \cdot \frac{1}{\lambda} \frac{\beta_m \rho_m R_p}{\beta_p \rho_p R_m} \qquad (6)$$

where a_c-input acceleration at the model base that induces first tension crack at the model dam; R_p, R_m are dynamic tensile strength of the prototype and the model material respectively; β_p, β_m are dynamic amplification factors corresponds to damping ratios of real dam and model material respectively.

Some correction factors may be introduced to take into account of the boundary conditions that affect the dynamic response of the structure and that are different in the model in comparison with the prototype (Lin et al. 1993), such as the loading rate and the size effect of the fracture.

Table 1.Test results of the model of a 195 m high concrete gravity dam

No.	Material density (kN/m³)	Compressive strength (MPa)	Tensile strength (MPa)	a_c (g)	A_C (g)
2	23.7	0.624	0.0851	0.311	0.541
3	24.7	0.696	0.1060	0.346	0.503
4	24.3	1.004	0.1356	0.421	0.414
5	24.8	0.070	0.0960	0.386	0.543

Table 1 and 2 illustrate the test results correspond to a 195m high rolled-concrete gravity dam and a 90.5m high concrete gravity dam constructed of poor concrete before the founding of our new People's Republic.

Table 2. Test results of the model of a 90.5 m high concrete gravity dam

No.	Dam type	Wave form	Material density (kN/m³)	Tensile strength (MPa)	a_c (g)	A_C (g)
A-1	Re*	S*	26.3	0.0589	0.58	0.220
A-2	Re*	S*	26.0	0.0503	0.36	0.158
A-3	Re*	S*	26.5	0.0311	0.37	0.249
A-4	Re*	R*	26.6	0.0306	0.48	0.177
A-6	Re*	R*	26.6	0.0367	0.56	0.172
B-1	Ov*	S*	25.8	0.0729	0.51	0.153
B-2	Ov*	S*	25.4	0.0574	0.48	0.181
B-3	Ov*	S*	25.6	0.0669	0.75	0.244
B-4	Ov*	S*	26.4	0.0608	0.67	0.247
B-5	Ov*	S*	25.3	0.0591	0.51	0.185
B-6	Ov*	R*	25.8	0.0655	0.87	0.178
B-7	Ov*	R*	26.6	0.0455	0.75	0.186

*Re-Retaining dam; *Ov-Overflow dam;
*S-Sinusoidal; *R-Random

All the experiments are performed under the condition of empty reservoir. Because the stress distribution over the weak section of the dam calculated by finite element analysis in case of full reservoir resembles that in case of empty reservoir closely as shown in Fig. 3, the response of the dam in case of the (full reservoir may be obtained from that of the empty reservoir by introducing a correction factor.

Dynamic model failure tests of a 292 m high arch dam in case of empty reservoir has also been carried out. The damage pattern of the model is schematically shown in Fig. 4 and the test results are summarized in Table 3. The standard strength of concrete of the real dam is assumed to be C30. It is seen, that the arch dam is more safe to withstand the earthquake shock than that of gravity dam.

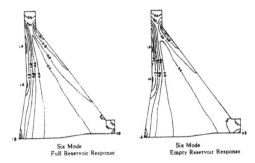

Six Mode
Full Reservoir Response

Six Mode
Empty Reservoir Response

Figure 3. Isolines of seismic principle stresses of the dam in case of empty reservoir and that in case of full reservoir

Table 3. Test results of the model of a 292 m high arch dam

No.	Material Density (KN/m³)	Compressive Strength (MPa)	Tensile Strength (MPa)	a_c (g)	A_C (g)
B-1	25.5	0.59	0.0862	0.83	1.142
B-2	26.5	0.714	0.0856	0.73	1.051
A-3	26.3	0.696	0.0780	0.65	1.019
A-4	24.8	0.217	0.0297	0.36	1.398
A-5	26.5	0.484	0.0436	0.35	0.989
B-3	26.6	0.416	0.0515	0.387	0.929

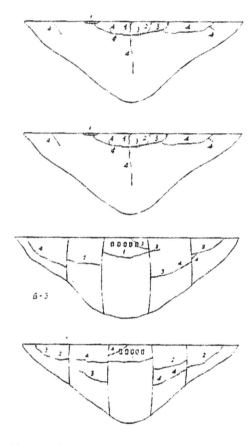

Figure 4. Damage pattern of the model of a 292 m high arch dam

simple similarity relationships between model and real structure of geotechnical structures.

We start with the discussion of earthquake response of fill-material structures.

4.1 Model experiment

According to the dynamic experimental results performed at our laboratory for a model dam 1.06m high (Fig. 5), it is revealed, that the distribution of acceleration response along the dam height varies with the intensity of base excitation. When exciting acceleration is very low, the dam behaves like an elastic continuum, the crest amplification is about 2.0. By increasing the level of excitation, the distribution of acceleration gradually becomes more or less uniform with the depth, and finally, the dynamic amplification approaches 1.0 for the whole dam. (Fig. 6)

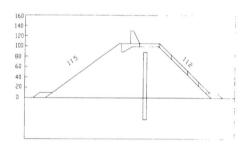

Figure 5. Shape of model dam

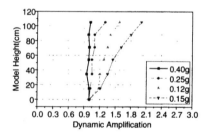

Figure 6. Distribution of acceleration response due to different level of excitation

4.2 Numerical analyses

Numerical analyses of a concrete-faced rock-fill dam 120m height has been conducted by us with equivalent linear method. The calculated distribution of acceleration at the instant when the crest displacement reaches its maximum value is demonstrated in Fig. 7. It is seen that the crest amplification varies from 3.3 for input acceleration 0.1g to nearly 1.1 for input acceleration increased to 0.5 g.

4 SIMILARITY RULE FOR DYNAMIC MODEL FAILURE TESTS OF EMBANKMENT DAMS

Since soils and rocks are characterized by their highly nonlinear behaviors, most experimental investigation of fill type dam in the literature was carried out qualitatively. Based on fairly long time practice and experience of performing dynamic model tests, we get some idea to establish rather

190

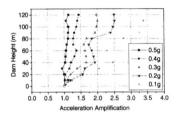

Figure 7. Calculated acceleration amplification of a 120m height dam

duce as the size of the earthquake increases. The size of the earthquakes is characterized by the amount of crest settlement referring to prototype terms, which is shown in parentheses in the figure. The fundamental natural frequency of the embankments appears to be about 6–7 Hz. For the smallest earthquake the average amplification reaches 3.6 and for the largest earthquake (CIV), it reduces to 1.16. The second natural frequency of the embankments ~15 Hz appears to be damped out when excited by the largest earthquake.

4.3 Field measurement

Similar results can be found from the literature. Field measurements were performed at several earth and rock fill dams during the Loma Prieta earthquake as well as previous events (Harder 1991). The peak transverse accelerations recorded at both the base and crest of these dams are depicted in Fig. 8. As may be observed, the points indicate that at low accelerations, the amplification through embankment dams is relatively large. However, as the peak base acceleration becomes larger, the amount of amplification is relatively low, the average amplification factor is about 1.1 for exciting acceleration 0.5 g. Further reduction of amplification for more intensive excitation could be expected.

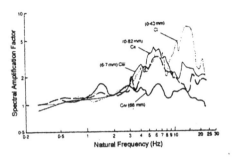

Figure 9. Spectral amplification factors of the embankments

4.5 Inelastic analysis

Gazatas (1987) computed the seismic response of a 40m high earth fill dam with Layered Inelastic Shear Beam LISB) method. The typical results are given in Fig.10 and Fig.11. The dam property corresponds to a stiff silty-clay material of an actual modern earth dam. The. excited earthquakes are five historic recorded motions, all scaled to three different peak accelerations: 0.2 g (moderate), 0.4 g (strong) and 0.7 g (very strong). Notice the diminishing amplification of peak crest accelerations from about 2.1 to 1.6 and finally to 1.0, as the peak ground acceleration increases from 0.2 g and 0.4 g to 0.7 g.

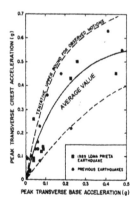

Figure 8. Comparison of peak base and crest accelerations measured at earth dams

4.4 Centrifuge test

Kutter and James (1989) carried out experiments of five clay embankment models subjected to 23 simulated earthquakes on the Cambridge geotechnical centrifuge. The influence of earthquake intensity on dynamic magnification was studied. It is seen in Fig. 9 that the spectral amplification factors, i.e. ratio of spectral acceleration at crest to that at the base, re-

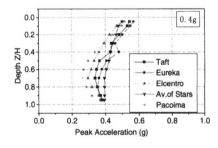

Figure 10. Typical results of LISB analysis for a modern earth-fill dam section subjected to different level of earthquake excitation

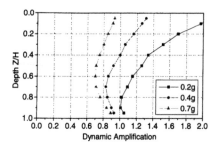

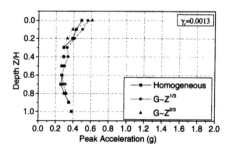

Figure 11. Average amplification for different level of earthquake excitation

Figure 12. Peak acceleration distribution for different constant values characterizing nonlinearity, $\gamma_r = 0.0030$ and $\gamma_r = 0.0013$

Gazatas (1987) also studied the effect of inhomogeneity and degree of nonlinearity of the material on the dam response with the LISB method. The dam is a hypothetical one, 120m in height and having an average low-strain S-wave velocity $V_{max}=360$ m/s, and subjected to the Taft 1952 NE component of recorded motion, scaled at peak acceleration 0.40 g. It is modeled as a nonlinear shear beam, the initial shear modulus (G_{max}) is taken to increase in proportion to z^m, where z is the depth from the crest, while m is parametrically varied from o to 1/3 and to 2/3. The average stress-average strain relationship is assumed to be hyperbolic:

$$\tau(z,\gamma) = G_{max}(z) \cdot \frac{\gamma}{1 + \dfrac{\gamma}{\gamma_r}} \qquad (7)$$

where $G_{max}(z)$-the law-strain elastic modulus and γ_r-the reference strain, which characterizes the degree of nonlinearity. Fig.12 plots the distribution of peak response with depth for three values of m and two values of γ_r. It is seen that the acceleration distribution becomes more uniform, when the material varies from moderately nonlinear ($\gamma_r = 0.003$) to strongly nonlinear ($\gamma_r = 0.0013$), in despite of the variation of the stiffness (inhomogeneity) along the depth (m=0,1/3,2/3).

4.6 Dynamic behavior of fill-type structures

From the above investigation several points are worthily note: (1) Subjected to low levels of earthquake excitation, fill type dams behave more or less like an elastic structure, dynamic magnification appears noticeable. (2) Under strong earthquake shocks, the dynamic amplication becomes substandically reduced and the distribution of acceleration response tends to be uniform along the depth. The higher the nonlinearity the dam material exhibits, the more evident this phenomenon.

4.7 Similarity rule for failure model tests of fill type structures

Noticing the fact, that during strong earthquakes, the dynamic amplification along the height of the dam becomes relatively low with the increased level of excitation and at near failure stage the averaged value approaches 1.0, possibly due to the higher damping and yielding of soil materials. We arrive at the conclusion, that under this condition, the distribution of dynamic acceleration over the dam body does not sensitive to the stiffness variation of the dam material. In other words, the similarity requirement of the restoring force, that depends on the stiffness of the structure and that determines the dynamic amplification corresponding to the different wave form of earthquake ground nation, could be relaxed. It results in the rather simple similarity relationship expressed by Eq. (4), while the similarity relationship expressed by Eq. (2) (for the embankment structures, it is usually to replace E in Eq. (2) by the shear modulus of elasticity G) may be ignored. In other words, the similarity of the elastic property G between the model and the prototype is not required, only the proportion of G_i among every parts of the structure in the model and in the prototype should be maintained. This simplifies the design of models to a great extent. In case the geometric scale of the model and the model materials λ, λ_ρ have been selected, the scale factors of other physical values can be determined with the aid of Eq. (4) straightforwardly. In this way, the model experiment of geotechnical structures to study its performance subjected to strong earthquake excitation at near failure stage could be performed quantitatively. The accuracy may be considered sufficient for engineering design purpose. And the higher the nonlinear behavior the dam material exhibits, the better results may be expected.

It is need to point out, that at near failure stage the restoring force is no longer a linear function of the displacement, it is expressed as {R (G,U)}instead of [K]{U}.To keep similarity of the nonlinear constitutive relationship between the model and the prototype material is also important. In order to study the failure mechanism of the structure, the similarity of the resistance is required. In general, the Mohr-Column principle is used as the strength criterion of the soil media, then the following expression should be satisfied for the frictional angle and the cohesion of the prototype and model material

$$\lambda_\phi = 1 \text{ and } \lambda_c = \lambda \cdot \lambda_\rho \qquad (8)$$

where λ_ϕ, λ_c - the scale factors of frictional angle and the cohesion of the material respectively.

It is seen from Eq. (8), the friction of the prototype and model material should be equal, while in comparison with the real dam material the cohesion of model material needs to be reduced nearly in proportion with the geometric scale, because in most cases λ_ρ is probably equal to 1.0.

A number of dynamic failure model tests of coastal structures which protect the intake tower of cooling water for nuclear power plant were also carried out in our laboratory. Safety of these structures to withstand design earthquake shocks is assessed.

5 CONCLUSION

A number of dynamic model failure tests of concrete gravity dams, concrete arch dams and embankment dams have been carried out in Dalian University of Technology, China.

The earthquake ground acceleration that induces the first tension crack at the most vulnerable part of concrete dams is selected as the basic index for the safety assessment for this type dams. Some tests results of gravity as well as arch dam were introduced.

Based on the investigation of the results through shaking table and centrifuge experiments on small models, numerical analyses with equivalent linear method and layered inelastic shear beam method for typical dams, and field observation data for several dams measured during Loma Prieta earthquake and previous events, a rather simple similarity rule for dynamic model experiments to study the performance of earth and rock- fill dams subjected to strong earthquake excitation at near failure stage is developed. In this way, the experiment may be carried out quantitatively. The accuracy is sufficient for engineering design purpose. The higher the nonlinear behaviors the dam material exhibits, the better results could be expected.

REFERENCES

Ahamadi, M.T., 1992, Sefidrud dam's dynamic response to the large near-field earthquake of June 1990, *Dam Engineering*, III-2.

Chopra, A.K.& Chakrabart, P., 1972, The earthquake experience at Koyna dam and stresses in concrete gravity dams, *Earthquake Eng. Struct. Dyn.*, 1,151-164.

Gazetas, G., 1987, Seismic response of earth dams: some recent developments, *Soil Dyn. & Earthq. Engrg.*, 6-1, 1-47.

Harder, L.F., 1991, Performance of earth dams during the Loma Prieta earthquake, *Proc. Second Intern. Conf. on Recent Adr. In Geo. Earthq. Engrg. and Soil Dyn.*, St.Louis, MO, 1613-1629.

Kutter, B.L. & James, R.G., 1989, Dynamic centrifuge model tests on clay embankments, *Geotechnique*, 39-1, 91-106.

Lin, G., 1958, Similarity rule for the study of vibration of arch dams, *Journal of Hydraulic Engineering China*, 1,80-104.

Lin, G., Zhou, J. & Fan, C. 1993, Dynamic model rapture test and safety evaluation of concrete gravity dam, *Dam Engineering*, IV-3, 173-186.

Lin, G., Zhu, T. & Lin, B., 2000, Similarity technique for dynamic structural model test, *Journal of Dalian University of Technology*, 40-1, 1-8.

Modeling earthquake impact on urban lifeline systems: Advances and integration in loss estimation

S.E. Chang
University of Washington

A.Z. Rose
Pennsylvania State University

M. Shinozuka
University of Southern California

K.J. Tierney
University of Delaware

ABSTRACT: This paper summarizes the development and application of an advanced, integrated earthquake loss estimation methodology for urban lifeline systems. The methodology, which evaluates direct and indirect economic losses from lifeline failures, provides a means for assessing both expected losses from future earthquakes and potential loss reduction from mitigation alternatives. This effort builds on and coordinates contributions from lifeline earthquake engineering, geography, sociology, and economics by researchers of the Multidisciplinary Center for Earthquake Engineering Research (MCEER). The methodology combines Monte Carlo simulations, geographic information systems, business resiliency questionnaire surveys, and economic computable general equilibrium modeling. It is applied to the water delivery system of Memphis, Tennessee, the major city in the New Madrid Earthquake Zone.

1 INTRODUCTION

In recent years, the massive losses caused by major urban earthquakes, combined with the increasing capabilities of new tools such as geographic information systems (GIS), have led researchers and practitioners to focus on computerized methodologies for estimating expected losses from future earthquake disasters. At the same time, new empirical data on the physical and economic effects of recent earthquakes provide opportunities for reevaluating, refining, and calibrating loss estimation models (e.g., National Research Council, 1999; and *Earthquake Spectra*, 1997, special issue).

The MCEER loss estimation research effort focuses on economic losses associated with earthquake-induced failures of critical urban lifeline systems, particularly water and electric power systems. It uses a multi-disciplinary, coordinated approach to address some of the most salient challenges in the state-of-the-art of earthquake loss estimation: capturing the systems response of engineering and economic systems; calibrating models to limited empirical data; improving our understanding of loss mechanisms, risk factors, and indirect economic loss; acknowledging loss estimate uncertainties; and, perhaps most fundamentally, integrating engineering and socio-economic models in a balanced, insightful, and productive way.

This paper describes the methodological approach and initial results of an application to the Memphis Light, Gas and Water (MLGW) water delivery system of Shelby County, Tennessee. Other ongoing efforts pertain to the MLGW electric power and the Los Angeles Department of Water and Power systems. Our ultimate objective is to demonstrate how the methodology can assist end-users to assess and compare the potential benefits of pre- and post-earthquake loss reduction strategies.

2 TECHNICAL SUMMARY

2.1 *MLGW case*

Memphis and Shelby County (pop. 900,000) are at risk from earthquakes originating in the New Madrid Seismic Zone (NMSZ) in the Central U.S. The NMSZ produced the largest earthquakes in the recorded history of the U.S. in the winter of 1811-12, including at least three events with magnitude 8.0 or greater. This paper presents initial findings for 3 scenario earthquakes (magnitudes 6.5, 7.0, and 7.5, respectively) with epicenter at Marked Tree, Arkansas, 55 km northwest of downtown Memphis.

MLGW is the primary supplier of water for Shelby County, with the exception of a few unincorporated municipalities. The water source is an underground aquifer accessed by wells. The water delivery system consists of a large low-pressure system and several high-pressure systems located on the outskirts of Memphis city. The network includes about 1370 km of buried pipes and a number of pumping stations, elevated tanks, and booster pumps.

2.2 *Methodological approach*

Previous research under the National Center for Earthquake Engineering Research (NCEER) had developed a lifeline loss estimation methodology and demonstrated its application to the MLGW electric power and water systems (Shinozuka et al., eds., 1997; Chang et al., 1996; Rose et al., 1997). While that effort provides a foundation for the current work, numerous significant refinements have been made. Highlights of current advances include:

- Model integration - "Seamless" integration of engineering and direct economic loss models;
- Multiple earthquake scenarios - Modeling of losses for potential earthquakes, and incorporation into a probabilistic risk framework;
- Spatial modeling - Refinement of the spatial resolution of analysis through advanced GIS and more effectively use digital spatial data;
- Temporal modeling - Development of a lifeline restoration model that incorporates post-disaster response parameters;
- Empirical data - Use of data on business losses in the Northridge Earthquake to refine direct and indirect loss models;
- Indirect loss modeling - Computable general equilibrium (CGE) approaches for estimating indirect losses and business resiliency effects;
- Uncertainty - Uncertainties deriving from both engineering and economic models.

Figure 1 illustrates the framework for implementing the loss estimation methodology for the MLGW water delivery system. Individual earthquake scenarios are evaluated for events with different magnitudes and locations. For each scenario earthquake, an integrated model (shaded box) is used to estimate damage and loss. Its core consists of an engineering model previously developed by Shinozuka et al. (1994) to simulate the expected damage and water service outage from major earthquakes. Here, the model is expanded to incorporate a direct economic loss component.

The new model uses a Monte Carlo simulation approach to evaluate both physical damage and economic loss. The model simulates damage to water system components, water leakage, and initial water outage throughout the system. It then simulates how the situation is improved on a weekly basis as repairs are made. Direct economic loss from business interruption is evaluated probabilistically at weekly intervals and summed over time. This enables economic loss to be estimated probabilistically in a manner consistent with engineering damage estimation. Direct loss estimates, water outage results, and industry resiliency data are input into a CGE model to calibrate and assess indirect economic loss.

Results can be summarized in "economic fragility curves" that indicate the exceedance probability of different levels of economic loss associated with different levels of earthquake severity. The economic fragility curves can then be combined with information on the occurrence probabilities of earthquakes of various severity levels. The ultimate result consists of an estimate of the expected annual loss -- in terms of repair costs and economic output loss -- from water disruption in future earthquakes (see, e.g., Shinozuka and Eguchi, 1997).

Note that the model is also capable of evaluating the effects of various loss reduction strategies. These include both pre-disaster mitigation measures such as seismically upgrading vulnerable components or increasing network redundancy, as well as post-disaster mitigations such as implementing mutual aid repair crew agreements or restoring damaged facilities according to an optimal restoration pattern. The model can therefore evaluate losses "with" and "without" mitigations to estimate the loss reduction benefits of these actions.

2.2.1 *Damage, outage, and restoration*

The MLGW water delivery system is represented in the new model by roughly 960 demand and supply nodes and 1300 links. The model estimates damage probabilistically using a Monte Carlo simulation approach (Shinozuka et al., 1994). In particular, the damageability of system components such as pipes is represented in terms of fragility curves that indicate the probability of failure for a given level of ground motion input, local soil condition, and component characteristics (e.g., pipe material and diameter). Fragility curves for pumping stations were updated using results in Hwang et al. (1998). For each earthquake scenario, 100 simulation cases of system-wide damage were produced. Each simulation case represents a deterministic damage pattern that is a possible outcome of the earthquake; collectively, they represent the probabilistic damage outcome of the event. Only damage from ground shaking is considered here. For each simulation case of scenario damage, the model evaluates the loss of water service by undertaking a complex system flow analysis. The analysis translates pipe damage into water leakage and solves for a new state of system equilibrium for the damaged state. The ratio of water flow in the damaged versus the intact state, evaluated at each of the demand nodes, provides the basic indicator of water outage that is used in the economic loss model.

Economic loss depends upon not only the water outage that occurs immediately after the earthquake, but how and over what time period water service is restored to the community. Restoration is modeled here using a resource constraint approach, which specifies the number of repairs that can be made in any time period according to the number of repair personnel available. Damage to large pipes require more worker-days to repair than damage to smaller pipes. Model parameters are based on a survey of current lifeline restoration models and data from the Kobe earthquake (Chang et al., 1999). In addition, the restoration model specifies a sequence of restoration based on engineering priorities and observations in the Kobe and Northridge disasters. Specifically, any damage to large pipes is restored first, in order to bring the transmission "backbone" of the system online before repairing service pipes. The restoration of smaller pipes then occurs in spatial sequence, from census tracts with the lowest damage density (number of pipe breaks per square kilometer) to those with the highest. Initial damage patterns are updated on a weekly basis until repairs have been completed. At each weekly interval, system flow analysis is conducted on the updated network. (Note that alternative restoration patterns can be simulated, see Rose et al., 1997.)

2.2.2 Analysis zones

The extent of economic loss depends not only on water outage, but also on how outage corresponds with the location of economic activity. Loss estimation methodologies have typically used census tracts, of which there are 133 in Shelby County, as the unit of analysis. In contrast, the current analysis defines approximately 3,400 Analysis Zones using GIS overlay and intersection operations integrating two input datasets. The first data set is Service Zones, a contiguous set of 971 polygons for the county, each representing the inferred service area of one node on the water network. The Service Zone polygons were defined by performing a GIS Thiessen polygon operation on the point distribution of water delivery nodes, implementing an assumption that a particular business location will be served by the closest node on the water delivery network. The second dataset pertains to employment, distributed into 515 Traffic Analysis Zone (TAZ) polygons. Defining Analysis Zones in this way reconciles point data on water outage with polygon data on employment. It therefore allows implementation of the loss model at the same spatial resolution as that of the input data and maximizes use of information on the spatial variability of factors contributing to loss.

2.2.3 Business resiliency

The economic impact of the water loss to businesses is mitigated by their resiliency, or their ability to withstand temporary water disruption. A 50 percent loss of water does not, in other words, necessarily reduce economic output by 50 percent. Until recently, empirical calibration of resiliency has been severely limited by lack of appropriate data, and many loss estimation methodologies use expert opinion data for this purpose.

In the current study, new empirical data from a business survey of the Northridge Earthquake provide an important source for more accurately calibrating business resiliency (Tierney, 1997; Tierney and Dahlhamer, 1998). The principal difficulty in using data from an actual disaster in this way, however, is that business loss would have been influenced by many sources of disruption besides loss of water. Isolating the effects of water outage is critical. A second difficulty consisted of limited sample size. Although the Northridge dataset contained responses from 1,110 businesses, less than one-fifth of them (207) had actually lost water service in Northridge. Moreover, only 6 businesses indicated that loss of water was the main reason for closing temporarily after the earthquake. These data limitations were addressed by dividing the analysis into two steps: a) looking at how disruptive water outage was to businesses in different industries, and b) evaluating how disruption from all sources impacted business activity.

Resiliency factors were developed for 16 industries comprising the private sector economy and for 3 water outage durations (less than 1 week, 1 week, and 2 or more weeks). Particularly relevant survey data included information on whether or not a business lost water, the duration of this outage, the disruptiveness of this outage, the disruptiveness of numerous other types of damage (e.g., building, transportation), whether or not the business closed for any length of time, and the most important reasons for this closure.

2.2.4 Direct economic loss

Direct economic loss—defined here as business interruption loss caused by water outage at the site of production—results from a combination of outage pattern, its spatial coincidence with economic activity, and business resiliency to water loss (for details, see Rose et al., 1997, and Chang et al., 1996). A key innovation made here in modeling direct economic loss consisted of probabilistically simulating loss in a Monte Carlo approach rather than estimating it deterministically. Specifically, resiliency factors were applied as a parameter indicating the probability that a business or group of businesses would close if it lost water service. In this simulation, each specified industry in each Analysis Zone was modeled as either closed or operating normally in each time period. (Partial closures were also allowed for cases with partial water availability.) This probabilistic approach enables consistent treatment of engineering and economic portions of the overall model.

2.2.5 Indirect economic loss

Indirect economic loss is here defined as business interruption deriving from interactions between businesses through a chain reaction of changes in input demands or output sales. This study applies Computable General Equilibrium (CGE) methods to assess indirect loss. CGE analysis is the state-of-the-art in regional economic modeling for impact and policy analysis. It is defined as a multi-market simulation model based on the simultaneous optimizing behavior of individual consumers and firms, subject to economic account balances and resource constraints (see, e.g., Shoven and Whalley, 1992). The CGE formulation incorporates many of the best features of other popular model forms, but without many of their limitations. The basic CGE model represents an excellent framework for analyzing hazard impacts and policy responses.

The methodology begins by recalibrating the parameters of the CGE model for earthquake simulations to yield direct business disruption losses consistent with empirical results of other MCEER researchers, as referenced above. The parameter adjustments are linked to specific real world examples of business resiliency (e.g., conservation, use of back-up supplies, increased substitutability). Different resiliency measures may result in the same direct output reduction level, but have different implications for indirect impacts, however.

The production side of the CGE model used in this paper is composed of a multi-layered, or multi-tiered, constant elasticity of substitution (CES) production function for each sector. We explicitly separate water out as a major aggregate in the top tier of the production function. Adjustment types are linked to the appropriate production function layer and parameters and the recovery/reconstruction stage (time period) to which each is applicable. Resiliency adjustments are incorporated by altering the parameters and variables in the sectoral production functions of the CGE model.

Results of the questionnaire survey for the Northridge Earthquake are adapted to specify the resiliency to water disruption in Memphis. Information on what actions businesses took—if any—to remedy the loss of water provided the basis for calibrating parameters pertaining to adjustment types. The proportion of total explained resiliency from the Memphis economy was apportioned to major adjustments in the following manner:

Conservation of water	=	41.9%
Increased substitutability	=	47.7%
Back-up supplies/costless	=	8.1%
Back-up supplies/cost-incurring	=	2.3%
Total of all adjustments	=	100.0%

Overall, the estimation of indirect losses involves a multi-step procedure: (1) Extract the sectoral production functions from the CGE model and adjust parameters and variables in them one at a time to match direct loss estimates; (2) Reinsert the recalibrated sectoral production functions into the CGE model, reduce water supply to a level consistent with simulation estimates, and compute total regional losses; and (3) Subtract direct losses from total losses to determine indirect losses.

2.3 Findings

Initial results from applying the above methodology to the MLGW water system have been encouraging and insightful. Figure 3 summarizes direct economic loss results for the magnitude 6.5, 7.0, and 7.5 Marked Tree earthquakes, respectively. Loss values are averaged over 100 Monte Carlo simulations.

As expected, losses increase exponentially with magnitude since the three events have identical epicentral locations. The table provides additional insights. First, loss standard deviations are very high relative to mean values, indicating substantial variation from one simulation case to the next for a given scenario. Recall that this variation derives from both engineering and economic model uncertainty. In the case of the M=7.5 event, results are limited by the outcome that in the first 4 weeks of system repairs, damage is sufficiently great that the system flow model fails to solve, indicating that water outage is likely to be complete throughout the county. In this case, maximum economic loss (defined by resiliency parameters) is assumed. Damage (number of pipe breaks) and restoration times also increase exponentially with magnitude.

However, closer inspection of the variability in the individual simulation results showed that simulation cases with high dollar loss often had similar numbers of pipe breaks as those with low loss. This insight suggests that, in the MLGW case, pumping station retrofits would be more effective at reducing losses in moderate earthquakes than wholesale pipe upgrades.

Results on indirect economic loss are summarized in Table 1. The initial indirect loss results shown here are based on one scenario characterized by a 7.0 magnitude earthquake (average damage and direct loss over 100 simulations), a one-week outage, and resiliency adjustments involving only the substitution of other inputs for water utility services.

In Table 2, the sector labels on the left-hand side refer to the economic producing units of the Memphis CGE model. Direct water disruption for each sector is presented in column 1, and sums to a 20.6 percent decline in water available in Week 1. As noted in column 2, this constraint is binding on all but two sectors as the general equilibrium adjustments work themselves out. Baseline output is presented in Column 3 and reflects the relative prominence of sectors in Shelby County economy and

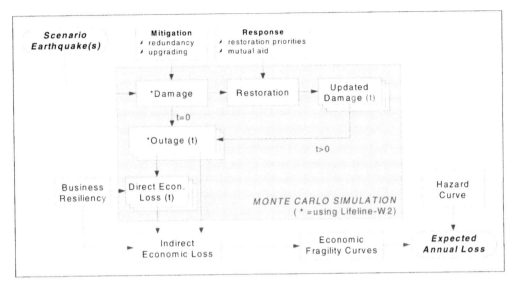

Figure 1. Implementation Framework

Table 1. Partial and general equilibrium changes in output following a water outage in Shelby county (1-week outage following a 7.0 magnitude earthquake)

	Sector	Water Input Direct Disruption	Water Input Total Unused	Output Baseline (10^6 1995)	Output Change from Water Outage Our Direct w/o Adjust[a]	Output Change from Water Outage Chang's Direct	Output Change from Water Outage Indirect	Output Change from Water Outage Total[b]	Elasticity (σ_u) Without Adjustment	Elasticity (σ_u) With Adjustment
1	Agriculture	-26.7%	-26.7%	128	-21.0%	-11.6%	-11.5%	-23.1%	0.01300	0.01325
2	Mining	-16.5%	-16.5%	20	-9.1%	-8.5%	0.2%	-8.7%	0.01240	0.01241
3	Construction	-25.2%	-33.0%	2782	-18.3%	-8.3%	-24.6%	-32.9%	0.01300	0.01322
4	Food Processing	-19.5%	-19.5%	1999	-17.0%	-15.8%	-0.3%	-16.1%	0.01006	0.01008
5	Manufacturing	-20.1%	-20.1%	8151	-14.5%	-14.0%	-3.0%	-17.0%	0.01200	0.01201
6	Petroleum Refining	-19.5%	-19.5%	511	-17.4%	-15.8%	-0.7%	-16.5%	0.00945	0.00948
7	Transportation	-10.1%	-10.1%	6235	-9.1%	-7.8%	0.2%	-7.6%	0.01150	0.01153
8	Communication	-22.2%	-22.2%	654	-14.7%	-10.8%	-5.5%	-16.3%	0.01300	0.01308
9	Electricity Services	-22.2%	-22.2%	530	-18.5%	-10.8%	-9.5%	-20.3%	0.01200	0.01228
10	Gas Distribution	-22.2%	-22.2%	21	-12.9%	-10.8%	-3.7%	-14.5%	0.01300	0.01304
11	Water & Sanitation	-22.2%	-22.2%	57	-19.1%	-10.8%	-9.2%	-20.0%	0.01300	0.01343
12	Wholesale Trade	-19.7%	-19.7%	4581	-12.7%	-12.7%	-1.7%	-14.4%	0.01300	0.01300
13	Retail Trade	-28.7%	-28.7%	4148	-23.1%	-14.1%	-10.1%	-24.2%	0.01300	0.01324
14	Finance, Insurance & Real Estate	-24.7%	-24.7%	6931	-18.8%	-16.5%	-3.2%	-19.7%	0.01300	0.01306
15	Personal Services	-23.9%	-23.9%	1546	-19.0%	-13.9%	-6.8%	-20.7%	0.01300	0.01316
16	Business & Professional Services	-22.2%	-22.2%	3917	-14.9%	-14.1%	-3.0%	-17.1%	0.01300	0.01302
17	Entertainment Services	-30.8%	-30.8%	293	-25.5%	-15.3%	-11.8%	-27.1%	0.01300	0.01329
18	Health Services	-18.9%	-18.9%	3143	-22.9%	-22.6%	6.8%	-15.8%	0.01050	0.01051
19	Education	-29.1%	-54.0%	1324	-20.7%	-16.1%	-37.9%	-54.0%	0.01300	0.01308
20	Government	-26.9%	-26.9%	3291	-21.6%	-15.7%	-9.1%	-24.8%	0.01200	0.01215
	Total	-20.6%	-20.8%	50263	-16.8%	-13.9%	-5.1%	-19.0%		

[a] From partial equilibrium analysis
[b] Following CGE simulation

serves as a reference point for the impact simulations. Note that Water sector (11) gross output represents only 0.113 percent of the regional economy.

Column 4 presents the direct output losses that are estimated in the CGE model before any resiliency adjustment. Chang's estimates of direct output losses, which incorporate the extent of resiliency, are presented in Column 5. Our direct loss estimates are based on input substitution elasticities (the lowest possible values that yielded an equilibrium solution) presented in the next to the last column in Table 1. Note that the CGE direct loss estimates exceed those of Chang in every sector because they omit all resiliency options except normal input substitution. The final column shows the elasticities necessary to incorporate resiliency measures consistent with the Chang estimates.

Our estimates of the indirect and total regional economic impacts of the water lifeline disruption are presented in Columns 6 and 7. Overall, they yield a 5.1 percent indirect reduction in regional gross output and an 19.0 percent total reduction in regional gross output for the week. The former represents $49.3 million and the latter $184.6 million.

Some interesting aspects of indirect losses are indicated by Table 1. First, they are slightly more than one-third the size of direct losses. In the context of an input-output (I-O) model, this would be a multiplier of about 1.37. The Shelby County overall output multiplier is larger than this, but the CGE model incorporates many other factors that mute the unidirectional and linear nature of the pure interdependence effect of the I-O model. For example, it is able to capture price declines due to decreased intermediate goods demand pressure, various substitutions aside from those relating to water, and various income and spending considerations on the consumer side.

Overall, our results appear to be reasonable for the economy as a whole, for individual sectors, and for individual impact stages (direct and indirect). The analysis is one of the few CGE simulations on any subject that has distinguished direct and indirect effects in detail. Moreover, we have developed a methodology that enables CGE users to recalibrate their models to make use of empirical data on individual parameters and direct impacts, such as those associated with responses to water lifeline disruptions following an earthquake.

3 CONCLUSIONS

The MLGW water system application demonstrates numerous methodological advances in earthquake loss estimation for lifeline systems, ranging from integrated probabilistic modeling of engineering and economic loss, to refinements based on Northridge data, to development of new Computable General Equilibrium models for indirect loss evaluation. Priorities for continuing work include sensitivity analysis, modeling other New Madrid Seismic Zone earthquakes, and evaluating the effects of pre- and post-disaster loss reduction measures. Ultimately, losses will be summarized in the form of economic fragility curves that indicate the likelihood of exceeding different levels of economic loss, with and without loss reduction actions. These results will be combined with probabilistic hazard information to estimate the expected benefits of mitigations which can then be compared with their costs. This information can assist lifeline agencies in determining which mitigation options and how much mitigation to undertake. The methodological approach can be implemented for other lifelines, such as electric power systems, and to other seismically vulnerable urban areas.

4 ACKNOWLEDGEMENTS

The research summarized in this paper was supported by the U.S. National Science Foundation through grants from the Multidisciplinary Center for Earthquake Engineering Research (MCEER), Buffalo, New York. Research assistance was provided by Walter Svekla, Gauri Guha, Chunsheng Shang, and Shu-Yi Liao.

REFERENCES

Chang, S. E., Seligson, H. A., & Eguchi, R. T., 1996. *Estimation of the Economic Impact of Multiple Lifeline Disruption: Memphis Light, Gas and Water Division Case Study.* Technical Report NCEER-96-0011. Buffalo, NY: National Center for Earthquake Engineering Research.

Chang, S. E., Shinozuka, M., & Moore, J. E. II., Probabilistic earthquake scenarios: extending risk analysis methodologies to spatially distributed systems, *Earthquake Spectra*, forthcoming.

Chang, S. E., Shinozuka, M., & Svekla, W., 1999. Modeling post-disaster urban lifeline restoration. In W.M. Elliott & P. McDonough (eds.), *Optimizing Post-Earthquake Lifeline System Reliability: Proc. of the 5th U.S. Conference on Lifeline Earthquake Engineering,* ASCE Technical Council on Lifeline Earthquake Engineering Monograph No.16.

Earthquake Spectra, 1997, Special Theme Issue on Earthquake Loss Estimation (13) 4.

Hwang, H. H. M., Lin, H. & Shinozuka, M., 1998. Seismic performance assessment of water delivery systems, *Journal of Infrastructure Systems* 4 (3): 118-125.

National Research Council. 1999. *The Impacts of Natural Disasters: A Framework for Loss Estimation.* Washington, DC: National Academy Press.

Rose, A., Benavides, J., Chang, S., Szczesniak, P., & Lim, D. 1997. The regional economic impact of an earthquake: direct and indirect effects or electricity lifeline disruptions, *Journal of Regional Science* 37: 437-58.

Shinozuka, M., & Eguchi, R. 1997 "Seismic Risk Analysis of Liquid Fuel Systems: A Conceptual and Procedural Framework for Guidelines Development," National Institute of Standards and Technology Report No. GCR 97-719, Gaithersburg, Maryland.

Shinozuka, M., Rose, A. & Eguchi, R. (eds). 1998. *Engineering and Socioeconomic Impacts of Earthquakes: An Analysis of Electricity Lifeline Disruptions in the New Madrid Area*, Buffalo, NY: Multidisciplinary Center for Earthquake Engineering Research.

Shinozuka, M., Tanaka, S., & Koiwa, H. 1994. Interaction of Lifeline Systems under Earthquake Conditions, *Proc. 2nd China-U.S.-Japan Trilateral Symposium on Lifeline Earthquake Engineering*, pp. 43-52.

Shoven, J. B. & Whalley, J., 1992. *Applying General Equilibrium*. Cambridge, UK: Cambridge University Press.

Tierney, K. J., 1997. Business impacts of the Northridge Earthquake. *Journal of Contingencies and Crisis Management* 5: 87-97.

Tierney, K. J., & Dahlhamer, J. M. 1998. Business Disruption, Preparedness, and Recovery: Lessons from the Northridge Earthquake," in *Proceedings of the NEHRP Conference and Workshop on Research on the Northridge, California Earthquake of January 17, 1994*, Vol. IV, Richmond, California: California Universities for Research in Earthquake Engineering.

Earthquake Engineering Frontiers in the New Millennium, Spencer & Hu (eds),
© 2001 Swets & Zeitlinger, ISBN 90 2651 852 8

Earthquake behavior of arch dams

Chuhan Zhang, Yanjie Xu, Guanglun Wang, & Feng Jin
Department of Hydraulic Engineering, Tsinghua University, Beijing, China

ABSTRACT: The earthquake behavior of arch dams is studied including dam-canyon interaction and nonlinear response of dams due to contraction joint opening and joint reinforcements. Significant effects of canyon radiation damping, non-uniform free-field excitations and nonlinear joint behavior on dam response have been confirmed.

1 GENERAL INTRODUCTION

Starting with the new millennium, construction of a series of high arch dams up to 250-300m in height is being planned or conducted (Table 1).

Table 1. Project data and Design PVA

Project	Dam height (m)	Reservoir capacity ($10^9 \, m^3$)	Power capacity (MW)	Design PVA* (g)
Ertan	240	5.8	3300	0.20
Laxiwa	250	1.0	3720	0.22
Xiaowan	292	15.3	4200	0.308
Xuluodu	278	13.0	14400	0.32
Jinping	300	7.7	6000	0.20
Miaojiaba	306		920	
Goupitan	240		2000	

* PVA: Peak value of acceleration

Most of these projects are located in seismically active regions in southwest China. For seismic resistance design of these high dams, it is necessary to study some important effects of arch dams during strong earthquakes. These include but not limited to: interaction of dam-foundation-reservoir system and seismic input mechanism-especially non-uniform free-field motions of excitation; nonlinear behavior of contraction joints and artificial peripheral joints, and dynamic fracture propagation of dams; strengthening and control measures such as joint reinforcements, post-tensioned cables and joint dampers; dynamic stability of dam abutments and foundation interface; and dynamic strength and constitutive behavior of mass concrete.

Since early 1980s the Tsinghua Group of Earthquake Resistance on Dams has carried out a systematic research on the aforementioned topics. In 1986, the China-US Joint Workshop on Earthquake Behavior of Arch Dams was held at Tsinghua and chaired by Professor Ray W. Clough and Professor Zhang.

Guangdou. Most study on the topics was accomplished during the past 15 years. This presentation only focuses on some important aspects, i.e., dam-canyon interaction and nonlinear behavior of contraction joints.

2 SIMULATION OF INFINITE FOUNDATION AND MODELING OF INTERACTION BETWEEN ARCH DAMS AND ROCK CANYONS

First, a novel element-infinite boundary element (IBE) was developed and coupled with the normal boundary element (BE) to model the irregular dam canyon (Zhang & Song, 1986). Further, a time domain procedure for three-dimensional (3D) coupling of finite elements (FEs), BEs and IBEs was presented by Zhang et al. (1995) for numerical modeling of seismic interaction between arch dams and rock canyons. In this method, the impedance functions on the dam-canyon interface in the frequency domain can be obtained by descritizing the infinite canyon into BE-IBE coupling. The impedance functions are then transformed approximately into frequency independent discrete parameters by a curve-fitting technique. Finally, these discrete parameters are coupled with the dam structure which is discretized by finite elements, thus allowing the response of the arch dam-canyon system to be evaluated in time domain from which the nonlinear behavior of dams can be taken into account (Figure 1). Two important points may be noted in the interaction analysis by using this method: (1) Earthquake excitations are defined as the free-field motions acting on the dam-canyon interface, thus making spatially varying input motions possible. These free-field motions taking into account the topographical effects of the canyon can be obtained from solving the wave scat-

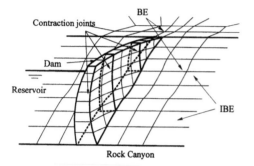

Figure 1. Schematic layout of dam-canyon system with contraction joints

tering equation for free-field canyons; (2) The procedure has the capability of incorporating the dam-canyon interaction analysis with non-linear behavior of structures due to the frequency independence of the impedances.

3 EFFECTS OF RADIATION DAMPING OF INFINITE CANYON ON DAM RESPONSE

The problem was studied by Chopra et al. (1992) and Dominguez et al. (1992) using procedures in frequency domain. Herein, the proposed procedure in time domain was also employed in seismic analysis of several high arch dams listed in Table 1, such as Laxiwa, Xiaowan and Xuluodu Dams. In the computation, infinite canyon and Westergaard reservoir are included. A comparison of the frequency response functions of Laxiwa arch dam (250m high)

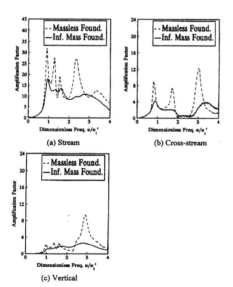

Figure 2. Comparison of response functions between the massless and infinite mass foundation

between the interaction model and the standard massless model is shown in Figure 2 (Zhang et al. 1992).

Also, the comparison of maximum dynamic stresses of Xiaowan arch dam (292m high) between the two models under the design earthquake is listed in Table 2. Significant reduction in dam response in the entire frequency range for interaction model is observed. The reduction in maximum stresses for the interaction model may range from 20-30% when compared with that of the massless foundation.

Table 2. Comparison of max. arch and cantilever stresses (MPa) between the interaction and massless model for Xiaowan dam (PVA=0.308g, design response spectrum, full reservoir)

Earthquake Direction	Model	U/S* arch	U/S canti.	D/S** arch	D/S canti.
Stream -wise	Interaction	10.5	5.4	7.2	4.4
	Massless	13.2	6.4	9.4	5.8
	Reduction	20%	16%	23%	25%
Cross Stream	Interaction	3.5	2.8	3.3	2.7
	Massless	4.9	5.4	4.2	4.3
	Reduction	29%	49%	21%	43%
Vertical	Interaction	2.0	1.9	2.2	2.0
	Massless	3.7	3.6	4.2	3.7
	Reduction	45%	48%	48%	46%

* U/S: Upstream; ** D/S: Downstream

4 EFFECTS OF NONUNIFORM FREE-FIELD INPUT ON DAM RESPONSE

In order to evaluate the effects of non-uniform free-field motions on dam response and to compare with the results by uniform input, it is assumed that an identical earthquake wave is traveling upward through point A and B as shown in Figure 3. Point A is located below the canyon and point B is underneath the half-space. Two equivalent input mechanisms for uniform and non-uniform input respectively are compared:

1 Uniform free-field motions obtained at the half space surface under the earthquake wave incidence, but acting along the canyon surface uniformly;

2 Non-uniform free-field motions obtained directly at the canyon surface due to the same earthquake incidence. The BE-IBE coupling methods employed to obtain the non-uniform free-field mo-

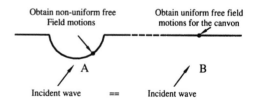

Figure 3. Definition of earthquake input mechanism

tions for SH, SV, P waves at different angle of incidence while the uniform free-field motions are computed by analytical solutions. The results are summarized in Table 3.

Table 3. Comparison of maximum stresses between uniform and non-uniform free-field motions (MPa) PVA=0.308g

Wave type Angle of incidence	Case	U/S Arch	U/S Canti.	D/S Arch	D/S Canti.
SH/0°	non-uniform	7.5	4.5	6.8	4.8
	uniform	8.6	5.7	9.0	4.5
	difference	-14%	-21%	-25%	+5%
SH/30°	non-uniform	6.2	3.9	6.0	3.7
	uniform	8.8	5.7	9.0	4.5
	difference	-29%	-32%	-33%	-19%
P/30°	non-uniform	2.1	1.5	1.5	1.6
	uniform	1.8	1.2	1.4	1.3
	difference	+10%	+25%	+7%	+18%
SV/30°	non-uniform	5.2	4.0	4.2	3.9
	uniform	4.2	2.9	3.4	3.1
	difference	+22%	+38%	+22%	+27%

It is shown that the effects of non-uniform input of earthquake are also significant but different depending on wave components of the input. A further decrease of the response is observed for non-uniform SH wave excitation in stream direction. However, a significant increase of stresses is also evident for inclined non-uniform input of SV and P waves which are equivalent to cross-stream and vertical excitations. The overall effects of radiation damping and non-uniform input deserve further study.

5 NONLINEAR BEHAVIOR OF ARCH DAMS DUE TO CONTRACTION JOINT OPENING AND JOINT REINFORCEMENTS

5.1 Modeling of joints and reinforcements

The most important nonlinearity of arch dam response is initiated by the contraction joint opening during strong earthquakes. This phenomenon often occurs in the upper portion of a dam where the largest tensile stresses up to 5-6 MPa are expected to occur in the arch direction during moderately strong earthquakes. Thus, the opening of contraction joints is inevitable, leading to a substantial reduction of tensile stresses in arch direction, while, on the other hand, a significant increase of cantilever stresses. In addition, a large opening of joints may damage waterstops between joints and cause leakage. It may be necessary to strengthen the dam integrity with joint reinforcements.

The problem of contraction joint opening was raised by Clough (1980) and studied by Fenves et al. (1989). The objective of this research is to study the effects of joint opening and joint reinforcements on nonlinear response of the Xiaowan dam. In reality, 48 joints are designed for dam construction with an approximately equal spacing of 20m. For computa-

tional economy, fewer joints may need to be simulated in the analysis. Herein, several cases of joint number simulation are compared, i.e. 1, 3, 5, 7, 9, 21 joints with approximately equal spacing of 40m, and 25 and 35 joints with 20m and 40m spacing for central and side joints respectively. The contraction joint layout and its FE discretization of the dam (case of 21 joints) are shown in Figure 4. The constitutive relationships for contraction joints and reinforcements are shown in Figure 5.

Herein, a 3-D non-linear joint element shown in Figure 5(a) used for contraction joint simulation by Fenves (1988) is employed. The modeling of contraction joints with reinforcements is shown in Figure 5(b). In Figure 5, q_n is the resisting stresses in the normal direction, k_n the stiffness of the joint in compression, v_n the relative displacement, q_{on} the specified tensile strength of the joint. In addition in Figure 5b, k_{ns1}, k_{ns2} represent the different average stiffness of the joint with different reinforcement ratios. Assuming that the relative displacement in normal direction n only produces stresses in that direction, the stress-displacement relationship of the joint with reinforce ments before the first opening can be expressed as

$$q_n = \begin{cases} k_{cn}v_n, & v_n \leq q_{on}/k_{cn} \\ k_{ns}v_n, & v_n > q_{on}/k_{cn} \end{cases} \quad (1)$$

After the first opening of the joint and the strength q_{on} is reached, subsequent tensile stress is $k_{ns}v_n$, i.e.

By using the virtual work principle, the stiffness

$$q_n = \begin{cases} k_{cn}v_n, & v_n \leq 0 \\ k_{ns}v_n, & v_n > 0 \end{cases} \quad (2)$$

of the joint with reinforcements can easily be obtained.

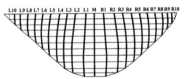

Figure 4. Simulation of contraction joints for Xiaowan arch dam (case of 21 joints)

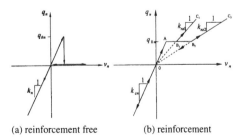

(a) reinforcement free   (b) reinforcement

Figure 5. Stress-displacement of joint element

The dam response with joint opening and the effectiveness of the control measure by reinforcements are studied. Strength and flexibility requirements of reinforcements for joint opening control are established (Zhang et al., 2000). To produce the required deformation for the expected joint opening under control, a sufficient length of cohesion-free sections for joint reinforcement in concrete can be evaluated. The reinforcement ratios and layout for joints are determined from the maximum tensile stresses in arch components under control. 20000-40000 ton of reinforcements are designed for different alternatives. One of the reinforcement layout alternatives is shown in Figure 6. Detailed formulations and dam parameters can be found in Zhang et al. (2000).

5.2 Results with 9 joints simulation

From the comparison of different cases of joint number simulation, it is concluded that starting from the case of nine joint simulation, all cases of larger number of joint simulation show little difference in stresses. However, it shows significant difference in maximum joint opening. This difference mainly depends on the spacing and layout of the joint rather than the number of joints simulated. Herein, for saving the computational efforts, only the case of 9 joints with spacing of 40 m are calculated for joint reinforcement analysis. Following conclusions are summarized.

Table 4 shows the comparison of maximum joint opening between conditions of reinforced and reinforcement-free joints. It is concluded that the lowest reservoir is the most critical case for joint opening control, although the tensile stresses for full reservoir is also needed to be considered for safety evaluation. On the other hand, the effectiveness of joint reinforcements for reducing the joint opening has been confirmed. This reduction of joint opening is a benefit to improve the dam integrity and to prevent damage to waterstops.

Table 4. Comparison of maximum joint opening (mm) between conditions of joint reinforcement (case of 9 joints)

Reservoir elevation	Design basis earthquake PVA=0.308g		El-Centro 1940 PVA=0.308g	
	Reinforcement-free	Reinforcement	Reinforcement-free	Reinforcement
Lowest	25.2	19.0	50.0	33.0
Full	7.1	/	19.7	/

Figure 7 shows the maximum joint opening at the instance of maximum response of the dam. It is evident that the upper-middle portion of the joints experiences complete opening from upstream to downstream faces. At least one-quarter of the joint height from the dam crest is subject to significant opening.

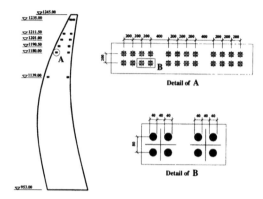

Figure 6. Layout of joint reinforcements (elevation: m; detail: mm)

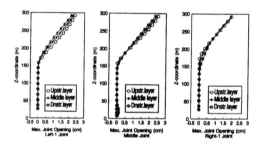

Figure 7. Max. joint opening of Xiaowan dam under lowest reservoir and design basis earthquake (PVA=0.308g, case of 9 joints)

Figure 8 shows the envelopes of maximum stresses of the arch and cantilever components for joint closed, joint opening and joint reinforced cases under the EL Centro earthquake. It is evident that due to joint opening, the maximum arch stresses have released dramatically from 5-9 to 0-1 MPa. On the other hand, the increase of the cantilever tension is also substantial from 1-2 to 5-6 MPa.

By comparing the stresses for joint opening with that of the joint reinforced, the arch tensions for the latter resume to 2-3MPa locally while the cantilever tensions have only a minor reduction from 5-6 to 4-5MPa, indicating the stress redistribution due to reinforcement control is not significant.

5.3 Further study on joint number simulation

As mentioned previously, more cases of larger number of joint simulation are also studied to disclose the effects of the joint number simulation on the response of the dam, including modeling of 21, 25 and 35 joints with the latter being closed to the real situation. The spacing of joints are different for the three cases as mentioned in section 5.1. The comparison of the results is shown in Figure 9.

206

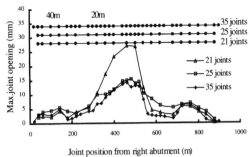

Figure 9. Distribution of max. joint opening along the dam crest

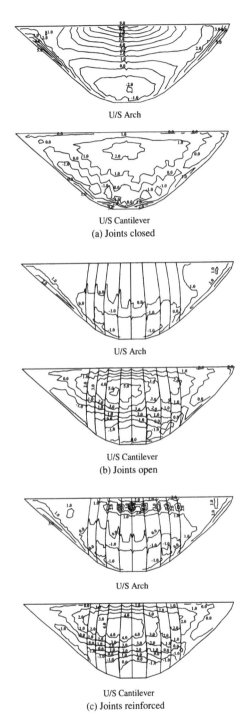

U/S Arch

U/S Cantilever

(a) Joints closed

U/S Arch

U/S Cantilever

(b) Joints open

U/S Arch

U/S Cantilever

(c) Joints reinforced

Figure 8. Envelopes of Max. stresses (Mpa) due to El-Centro earthquake (PVA=0.308g, case of 9 joints).

By comparison of stresses among the three cases (not shown for space limitation) and with Figure 8 where only 9 joint are simulated the difference is trivial, indicating the simulation with 9 joints is sufficient to obtain accurate results in stress distribution;

By comparison of joint opening distribution at top arch, much differences are observed between the case of 21 joints and the other two cases. It is evident that the case of 21 joints with 40m spacing does not present accurate joint opening distribution for the dam although it does provide excellent accuracy in stresses. By comparison of the results between the cases of 25 and 35 joints, it is seen that where the spacing of joints is the same where the values of joint opening are similar, significant differences in joint opening will occur in the region where the spacing of joints is different. For accurately modeling the real behavior of joint opening for a specific arch dam, the designed spacing and layout of the dam must be simulated at least for the central portion of the dam. The simulation of joints near the abutments may be simplified.

6 CONCLUSIONS

1 Effects of radiation damping of infinite mass canyon on arch dam response are significant in comparing with the massless foundation. A reduction of 20-30% in dam response is observed when this effect is taken into account.

2 Effects of non-uniform free-field seismic input on dam response are also evident. It may reduce or increase the dam response depending on the wave components and the direction of incidence. The overall effects due to radiation damping and non-uniform input deserve further study.

3 The opening of contraction joint in arch dams releases the arch tensions dramatically, but the in-

207

crease in cantilever tensions is also substantial. Joint reinforcements can effectively reduce the joint opening for improving dam integrity, although it has only minor influence on stress redistribution between arch and cantilever components.

4 For obtaining accurate stresses, 9 joints simulation is sufficient for Xiaowan dam. However, for obtaining accurate joint opening distribution along the dam, the real design spacing and layout of joints for the central portion of the dam must be modeled accurately.

REFERENCES

Clough, R.W. 1980. Nonlinear mechanisms in the seismic response of arch dams. International Research Conference on Earthquake Engineering. Skopje, Yugoslavia.

Chopra, A.K. & Tan, H. 1992. Modeling of dam-foundation interaction in analysis of arch dams. Proc. 10th World Conf. Earthquake Eng., Madrid 8, 4623-4626.

Dominguez, J. & Maeso, O. 1992. Model for the seismic analysis of arch dams including interaction effects. Proc. 10th World Conf. Earthquake Eng., Madrid 8, 4601-4606.

Fenves, G.L., Mojtahedi, S. & Reimer, R.B. 1989. ADAP-88: A computer program for nonlinear earthquake analysis of concrete arch dams. Report No. EERC 89-12, EERC, University of California, Berkeley, CA.

Zhang Chuhan & Song Chongmin 1986. Boundary element technique in infinite and semi-infmite plane domain, Boundary Elements, Du Qinghua ed. Pergamon Press.

Zhang Chuhan, Jin Feng & Pekau, O.A. 1995. Time domain procedure of FE-BE-IBE coupling for seismic interaction of arch dams and canyons, Earthquake eng. struct. dyn. 24, 1651-1666.

Zhang Chuhan, Jin Feng & Wang Guanglun 1996. Seismic interaction between arch dam and rock canyon. Proc. 11th World Conf. Earthquake Eng., Paper No.595, Acapulco, Mexico.

Zhang Chuhan, Xu Yanjie, Wang Guanglun & Jin Feng 2000. Nonlinear seismic response of arch dams with contraction joint opening and joint reinforcements. Earthquake eng. struct. dyn. 29, 1547-1566.

Earthquake Engineering Frontiers in the New Millennium, Spencer & Hu (eds),
© 2001 Swets & Zeitlinger, ISBN 90 2651 852 8

Seismic reliability of lifeline systems

Yang Han
He Nan Building Research Institute, Zhengzhou, China

Shaoping Sun
Beijing Municipal Engineering Research Institute, Beijing, China

ABSTRACT: In this paper the seismic reliability of lifeline systems is presented by means of graphic theory and topological classification method. Meanwhile an improved probabilistic prediction model to seismic damage of urban underground lifeline is also presented.

1 INTRODUCTION

Lifelines are those services vital to the health and safety of community and the functioning of an urban and industrialized society. Lifelines include the systems for water, gas, sewer, electric power, transportation, communication and so on. Lifelines are critical for emergency response, such as earthquake[1]. In this paper, only water supply network is covered. These large and complex network systems are consisted of different kind of components connected in different ways.

The pipelines of water supply are divided into two kinds, i.e., continuous and segment. We often use the segment pipe in China, so the seismic reliability analysis of the network is mainly based on the displacement of pipe joints and site conditions.

According to the seismic damage state of buried pipelines we established the probabilistic prediction model, and the program of the model has been put into work. The application of the graphic theory, simplified upper bound algorithm for assessing reliability for a large and complicated network are presented, with these method a general program has been made to predict the seismic damage of the pipeline network and its feasibility and effectiveness have been verified in practice.

2 THE BURIED PIPELINS RESPONSE TO TRAVELLING SEIMIC WAVE

The buried pipelines response to traveling seismic wave is not only related to wave velocity, wave length and amplitude, but also to the displacement of the pipe and its transmission coefficient, as following expression:

$$D_p = \beta \cdot D_S(x,t) \qquad (1)$$

where D_p: the longitudinal displacement of pipe,

$D_S(x,t)$: the displacement of free field, β: the transmission coefficient of longitudinal displacement

$$\beta = \cfrac{1}{1 + \cfrac{EF}{K}\left(\cfrac{2\pi}{L}\right)^2 \cos^2\alpha\left(\cfrac{D}{D_0}\right)^r} \qquad (2)$$

where D: the diameter of the pipe, F: the pipe cross section, K: Modulus of foundation, D_0: transmission diameter of pipe, r: constant.

From past earthquake the displacement of pipe and joint we can define the seismic damage of buried pipeline into three states, described as follows:

Undamaged: No damage on pipe body, deformation at rigid joints is in elasticity and its relative displacement S is less than the fissure limit R_1. Joints may have very small fissures. There might be very little leakage at joints.

Slight damage: Deformation at rigid joints is in the stage of elasticity to plasticity, and the relative displacement S exceeds the fissure limit R_1. There is slippage between rubber ring and pipe body at flexible joints. Water pressure in pipelines may be reduced.

Serious damage: The relative displacement S of the joint exceeds the leaking limit R_2, and the filler is loosened with serious leakage even out of operation.

3 THE PROBABILISTIC PREDICTION MODEL

Let the relative displacement of the joint be S as the earthquake effect, and the allowable displacement be R as the structural resistance of the pipeline, then the effective function Z of the variables S and R can be obtained:

$$Z = f(S,R) = R - S \qquad (3)$$

S and R are both random variables, put them into Equation (2), then the failure probability is

$$P_F = P(Z < 0) \tag{4}$$

Assume that R and S follow the random variables of the two normal distributions, $N(\mu_R, \sigma_R)$ and $N(\mu_S, \sigma_S)$ respectively, then Z is a normal random variable, and follows the normal distribution $N(\mu_Z, \sigma_Z)$, in which

$$\mu_Z = \mu_R - \mu_S, \sigma_Z = \sqrt{\sigma_R^2 + \sigma_S^2}.$$

The $\mu_R, \sigma_R, \mu_S, \sigma_S, \mu_Z, \sigma_Z$ are the expected values and standard deviations of the random variables R, S, and Z respectively.

So, the probabilistic prediction model of seismic damage of the pipeline is:

1 The basic perfect state

$$P_{F1} = P(Z_1 > 0) = \phi(\frac{\mu_1}{\sigma_1}) \tag{5}$$

2 The serious damaged state

$$P_{F3} = P(Z_2 \le 0) = \phi(-\frac{\mu_2}{\sigma_2}) \tag{6}$$

3 The moderate damaged state

$$P_{F2} = 1 - P_{F1} - P_{F3} \tag{7}$$

in which $\mu_1 = \mu_{R1} - \mu_S$, $\sigma_1 = \sqrt{\sigma_{R1}^2 + \sigma_S^2}$, $\mu_2 = \mu_{R2} - \mu_S$, $\sigma_2 = \sqrt{\sigma_{R2}^2 + \sigma_S^2}$.

By means of these formulas and test statistical data, seismic risk analysis, microzonation data and pipeline data, the probability can be determined.

4 EXACT ALGORITHM FOR RELIABILITY OF NETWORK

4.1 Graph model of the network

According to the graphic theory, a network is made up by some nodes and edges linked them, its mathematical definition is:

Suppose that the node-set V, edge-set E, and weight-set P of edges are $V = \{v_1, v_2, \cdots, v_n\}$, $E = \{e_1, e_2, \cdots, e_m\}$, and $P = \{p_1, p_2, \cdots, p_m\}$, respectively, and they meet to the non-empty set V, each edge e_i of the set E is unordered element pair $\{v_s, v_t\}$ of the set V, and every element of the set P has one-to-one correspondence with that of the set E.

The one-to-one correspondence can be also established between the network of water supply system and the elements of the weighted digraph G: the water supply plant corresponds to the source, the inter section of the pipelines to the node, the end-user to the sink, the pipeline to the edge, and the weight of the edge corresponds to the probability that the pipeline can be operated during earthquake. A weighted digraph G as an instance is given in Fig.1.

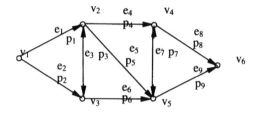

Figure 1. Weighted digraph G

4.2 The terminal-pair reliability problem

A basic problem in reliability analysis is to find the probability that there exist at least one path from a specified source node s to another specified terminal node t along which all nodes and edges are good.

A simple series network with m edges, its reliability $R(G)$ is:

$$R(G) = \prod_{i=1}^{m} p_i \tag{8}$$

A simple parallel network with m edges its reliability $R(G)$ is:

$$R(G) = 1 - \prod_{i=1}^{m} (1 - p_i) \tag{9}$$

To a complex network that cannot be simplified into a series or parallel network, the practical methods of exact reliability analysis are path or cut-set enumeration methods. If a network contains m paths, its terminal-pair reliability is:

$$R(G) = p_r \{\bigcup_{i=1}^{m} P_i\} \tag{10}$$

where P_i is the ith path of G. To evaluate the above expression, the following principle of inclusion-exclusion for the probability of the union of m events is often used.

$$p_r\{\bigcup_{i=1}^{m} P_i\} = \sum_{i=1}^{m} p_r\{P_i\} - \sum\sum_{1 \le i_1 < i_2 \le m} p_r\{P_{i_1} P_{i_2}\}$$

$$+ \sum\sum\sum_{1 \le i < i_{2l} < i_3 \le m} p_r\{P_{i_1} P_{i_2} P_{i_3}\}$$

$$- \cdots + (-1)^{m-1} p_r\{P_1 P_2 \cdots P_m\} \tag{11}$$

The number of terms in the expression is $2^m - 1$. The computation using Eq. (11) is clear complexity (e.g., if m=20, then the number would be nearly 10^6).

The fundamental concept of disjoint sum algorithm is connecting set of some shortest paths to disjoint events set, thus logic operation becomes arithmetic operation.

In general cases:

$$p_r\{\bigcup_{i=1}^{m}P_i\} \neq \sum_{i=1}^{m}p_r\{P_i\} \qquad (12)$$

If we express the event of network reliability as a sum of disjoint events, then we can easily obtain the reliability as follow:

$$R(G) = \sum_{j=1}^{k}p_r\{T_j\} \qquad (13)$$

We adopt following process to get above formula (13). At first, due to

$$\bigcup_{i=1}^{m}P_i = P_1 + \overline{P_1}\bigcup_{i=2}^{m}P_i \qquad (14)$$

where $\overline{P_1}$ is the complementary set of P_1. If it is evident that the two items on the right are disjoint, let P_1 as first disjoint item T_1, then simplify the second item by following basic rules:

$$AB \subset A, \ A\overline{A} = \Phi, \ A\bigcup AB = A,$$

$$\overline{A\bigcup B} = \overline{A}\overline{B}, \ \overline{AB} = \overline{A}\bigcup\overline{B},$$

in which A & B are arbitrary sets; after simplifying,

$$\overline{P_1}\bigcup_{i=2}^{m}P_i = \bigcup_{l=1}^{n}P_l \qquad (15)$$

Repeating the right of Eq. (15), $\bigcup_{i=1}^{m}P_i$ becomes a disjoint sum after limited steps.

In this paper, the logic comprehension and the multidimensional calculation methods are applied in above disjoint process.

5 A TOPOLOGICAL CLASSIFICATION METHOD FOR NETWORK ANALYSIS

A lifeline network is usually very large in size and complicated, often having hundreds of nodes and edges. The traditional graphic analysis method can't meet the need in practice. On other hand, we are not quite sure about the seismic damage mechanism in different parts of the lifeline network, and there are uncertain and inaccuracy factors in resistance and computational model, so it is practical to develop approximate algorithm which should be simple, time-saving, and can deal with the large and complicated network, and can meet certain precision.

For the above reason the topology methodology

and classification method are used to get the reliability of upper bound in a large and complicated network. This method is simple and practical, and can work out the reliability of the lower bound by dual theory (Han 2000).

5.1 *Topological classification of node set*

The basic idea of the classification method is as following: according to topology of node set we classify the network into ordered subsets, then calculate the reliability of every node from the source to the sink step by step.

The following theorem is used in the network division: any no-circle directed graph has at least one node whose out going degree is zero and one node whose in coming degree is zero.

We can set $G(V,E)$ as no-circle diagram, in which V is the node set, and E is the edge set. If N_0 is the node subset of G, whose coming-in degree is zero and it is nonempty, that is:

$$N_0 = \{ v_i \in V \mid \Gamma^-(v_i) = \Phi\}$$

where $\Gamma^-(v_i)$ is the coming-in neighbor set of v_i and Φ is empty set.

For $G_0 = \{V - N_0, E - \Gamma^+(N_0)\}$, $\Gamma^+(N_0)$ is the going-out neighbor set of N_0 and it can fit the definition of no-circle directed diagram. if G_0 has nodes, it has at least one whose coming-in degree is zero.

These nodes make up a new set that is another nonempty set and can be noted as N_1

$$N_1 = \{ v_i \in (V - N_0) \mid \Gamma^-(v_i) \subseteq N_0\}$$

For the same reason, we can get

$$N_2 = \{v_i \in [V - N_0 \bigcup N_1] \mid \Gamma^-(v_i) \subseteq (N_0 \bigcup N_1)\},$$

$$N_n = \{v_i \in [V - \bigcup_{k=0}^{n-1}N_k] \mid \Gamma^-(v_i) \subseteq \bigcup_{k=0}^{n-1}N_k\}.$$

The parts $N_0, N_1, \cdots, N_n$ of the node set are called the topological classification of the node set, as shown in Fig. 2.

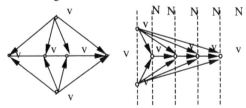

(a) no-circle directed graph   (b) newly-arranged node after the topology metho dology

Figure 2. The topological classification of the node set.

5.2 Topological classification method

From topological classification of node set (the former section) we define k is the order of v_i to $v_i \in N_k$. We can learn that the coming-in neighbor set of node belong to N_k only exist in those nodes whose order are lower than k. If we want to get the reliability of node belong to N_k, we have to take the contribution of these nodes whose order are lower than k into consideration. Let the initial value of the source node $R_1 = 1$, because we can get the upper bound of system reliability by using 1 minus the probability of system out of operation, so we can use the following formula as the basis of this calculation, and get the reliability of every node one by one in the order of the nodes.

$$R_{ij} = 1 - \prod_{m=1}^{n(i,j)} (1 - \prod_{(k,l) \in T_{ij}^m} p_{kl}) \qquad (16)$$

where R_{ij}: the reliability of the path from the node i to node j. T_{ij}^m: the edge set of the mth shortest path from i to j, p_{kl}: the reliability of edge (k,l).

The following two cases sketch the reliability of every node how to work out by the classification method.

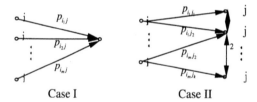

Case I   Case II

Figure 3. Two cases of the classification method

In case I, the node j has no like node and has m forward nodes, its reliability is:

$$\overline{R}_j = 1 - \prod_{k=1}^{m} (1 - \overline{R}_{ik} p_{ikj}) \qquad (17)$$

in which $\overline{R}_j$ and $\overline{R}_{ik}$ are the upper bound of the reliability of the node j and i_k, p_{ikj} is the reliability of the edge (i,j).

In case II, the node j has the like nodes, then the question is that how to solve the undirected edge(s). In this case the calculation for related path is needed in the sub-graph made by like nodes. Take a node j_i, along the undirected edge and use the depth first search (DFS) to find a like node j_k until it fits to case I. After that, apply Eq. (16) to work out temporary reliability of the point j_k. Repeating above steps for all undirected edges, the point j_i can be approached to the case I.

Now we have adopted the computational program YU3 in the calculation. In the project for the city lifeline networks in cities Anyang, Zhengzhou and Xuchang, validity of the method has been proven (Han 1990, Sun 1997).

6 APPLICATION

Fig. 4 shows a water supply network in a city. The whole network can be simulated as a directed graph. It has 142 nodes; among them 4 are source nodes, 22 sink nodes, and 194 edges. This simulation can be simplified as a directed graph only with one source and one sink in imagination. The direction of the edges can be simplified according to the following principles: the main pipeline near the source and the branches to the end are set as directed edges. The others can be made according to such factors as flowing from big pipes to small one, distance to the source, height of nodes, etc. Some factors of pipeline are uncertain can be taken as undirected edges.

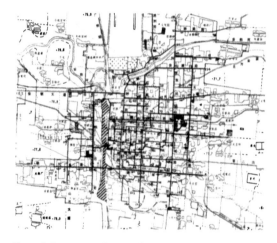

Figure 4. A water supply network

By the program YU3 which we made (Sun & Han 1990), we can work out 21 categories of network nodes.

Table 1 shows the calculated results of some nodes and the comparison with the Monte Carlo simulation, from which we can see that its accuracy is satisfactory.

Table 1. Reliability analysis of a water supply network

Node Sort	Classification method	Monte Carlo simulation times		
		1000	2000	3000
5 (N_8)	0.984	0.983	0.984	0.984
9 (N_{16})	0.849	0.845	0.843	0.845
10 (N_{16})	0.876	0.862	0.853	0.850
29 (N_6)	0.978	0.983	0.980	0.979
39 (N_7)	0.975	0.971	0.976	0.976
53 (N_{17})	0.775	0.721	0.723	0.725
55 (N_{19})	0.496	0.463	0.474	0.479
79 (N_{13})	0.825	0.828	0.822	0.823
131 (N_5)	0.645	0.645	0.645	0.645
139 (N_{17})	0.545	0.543	0.541	0.543

7 CONCLUSIONS

1 In this paper the probabilistic prediction model of pipeline is improved, its feasibility has been verified in practical engineering.
2 It is easy to get the upper bound value of reliability in a large and complex network by means of the classification method; meanwhile it may be found the lower bound by dual theory.
3 The logic comprehension and multidimensional calculation methods improve the shortest path of network of indisjoint process, and it also can be used to solve the main pipeline network precisely.

REFERENCES

Wang L.R.L.etc.,1985, Seismic Damage Behavior of Buried lifeline Systems During Recent Server Earthquake in US.,China and Other Countries, *Tech. Report DULEE-02,Old Dom-inion University*, U.SA.

Shinoguka. M, Kaike. T, 1979 Estimation of structural strains in underground lifeline pipes, *Technical Report No.NSF-PER-78-15049*,U.S.A

Sun Shaoping, 1983 Pipeline Damage and its Relationship with Joints *Earthquake Behavior and Safety of Oil and Gas Storage Facilities, Buried Pipelines and Equuipment". PVP.Vol.77. ASTM, Portland*, U.S.A.

Sun Shaoping, Han Yang, 1990 prediction to Seismic Damage of Urban Buried Pipeline Network *proceedings of China-Japan Symposium on Lifeline Earthquake Engineering, Beijing*, China

Han Yang 1990, Seismic Risk Analysis of Life-line Network, *The Proceedings of 3rd national Earthquake Engineering Conference, Dalian*, China(in China)

Sun Shaoping, 1997, The Aseismic measures for water supply systems, *Proceedings of the 4th International symposium on water pipe systems, Kobe*, Japan

Owens. F.C., 1999 Seismic Analysis of Buried Pipelines, *Proceedings of the 5th U.S. conference on Lifeline Earthquake Engineering, Seattle*, U.S.A

Han Yang, 2000, Research on the lower bound of reliability of network by dual theory. *3rd Journal of Northern Hydraulic and Electric University*, China.

Earthquake Engineering Frontiers in the New Millennium, Spencer & Hu (eds),
© 2001 Swets & Zeitlinger, ISBN 90 2651 852 8

A cell-phone based wireless information system and its application

G.Z. Qi
InfraTech Inc. Silver Spring, Maryland, USA

ABSTRACT: This paper presents a cell-phone based wireless information system and its application for intelligent prevention of ignition of fires following earthquakes. Besides the severe ground shaking, fires induced by earthquakes are usually the major causes of human casualties and properties losses. Ground information facilities may be damaged during earthquakes. So, application of wireless information technology becomes more important for earthquake hazard mitigation. Satellite, TCP/IP network-interface based Internet and cell-phone are existing technologies of telecommunications. Results showed that the cell-phone based wireless information system developed by InfraTech, Inc (CPBWI) is an applicable, effective and costless system, which can be widely used for earthquake hazard data transmission.

1 INTRODUCTION

It has been our strategy to utilize advanced technologies for civilian application including hazard mitigation. In past decade, this strategy has led to significant change in state-of-the-art of earthquake engineering, particularly, in structural control, sensing technology, structural monitoring and smart material and structures (Johnson, et al. 1999. Niwa 1999, Soong 1999, Wang 1997, Tamura, 1999). Most advanced technologies used in air space and military are usually very expensive. So cost-effectiveness is one important concern for their civilian application in addition to their applicability and effectiveness.

Ground information facilities may be damaged during earthquakes. So, application of wireless information technology becomes more important for earthquake hazard mitigation. Satellite, TCP/IP network-interface based Internet and cell-phone are existing technologies of telecommunications. Upon extensive study of their applicability, effectiveness and cost, a more reliable and cost-effective cell-phone based wireless information system (CPWLI) has been developed for earthquake hazard data transmission.

Besides the severe ground shaking, fires induced by earthquakes are usually the major causes of human casualties and properties losses (Chung 1995). As an example, an application of CPWLI system for intelligent prevention of ignition of fires following earthquakes is presented in this paper.

2 CPWLI SYSTEM

2.1 *System*

The cell-phone based wireless information system includes two units, a wireless information unit and a central process unit. As shown in Figure 1, the wireless information unit consists of a microcontroller and a cell-phone. It takes the data from sensors, encodes and transfers the data into remote central process unit. The central process unit consists of the PC, cell-phone and intelligent signal processor (ISP) software. The cell-phone receives the data from the remote sites and input them into PC. The ISP includes a code classification window to encode the data based on the tasks' requests, a rule and knowledge window to build up the system rules, an analysis window to analyze the coded data based on the rules, and a displaying window to show the events' information on geographic map by using GIs technology (see Figure 3). ISP is an intelligent platform tool and can be used for any disaster data treatment. The central process unit also wirelessly sends necessary information and commands to the remote sites if necessary.

2.2 *Cost analysis*

Cost is an important concern in application of advanced technologies for civilian in addition to their applicability and effectiveness. Wireless information technologies may be classified in to three categories, satellite based, Internet based and cell-phone based

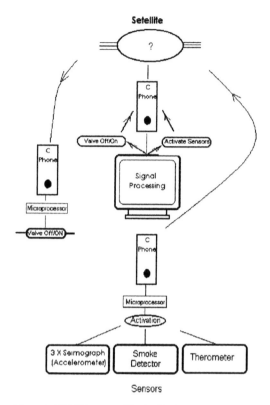

Figure 1. CPBW System and its Application.

3.1 *Auto-off/on valve for gas pipe*

Breakage of gas distribution lines has been identified as the major factor leading to the ignition of fires following earthquakes. Therefore, a law requiring an approved auto-shut-off valve to be installed on all gas-serviced buildings has been adopted in California. Existing valves used in California for prevention of fires were designed by using a moveable ball, which is sensitive to the ground vibration (Figure 2a). In addition to its false shutoff, the major disadvantage is that the valve has to be reset by hand after shutoff. In a new valve developed by InfraTech, Inc shown in Figure 2b, the moveable iron-ball is controlled by the iron-core surrounded with electric coil, which is connected with a 12 volt DC battery through a relay. The relay is remotely controlled by CPWLI system. If the relay is in on-position, the iron-core should be magnetized and the movable ball will be pulled up to close the gas flow; if it is in off-position, the ball will fall down to resume the gas flow. Thus, the new valve can be set on off-position or on-position automatically by the predetermined criteria in the intelligent signal processing unit (ISP) of CPWI system.

systems. Table1 shows a comparison of their effectiveness and costs. Of course, the costs and effectiveness are case dependent. The fix costs are for required equipments of the systems, such as receiver, antenna and encode & decode device for satellite, the TCP/IP interface device & software and small PC for Internet, and cell-phone and microcontroller for CPWLI. Note that the cell-phone based wireless information technology is most cost-effective and can be affordable for individual users.

Table 1. Cost and Effectiveness of wireless technologies.

System	Cost ($)		Effectiveness
	Fix	Lease	
Satellite	5,000 - 45,000	3,200/m	Effective
Internet	2,700	22/m	Less Effective
Cell-Ph	150	20/m+ fee	Effective

3 CPWLI APPLICATION

An example of CPWLI applications for in protection of earthquake-induced urban fires is presented here.

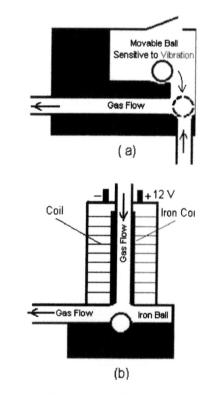

Figure 2. Auto-Off (a) and Auto-Off/On Valves (b).

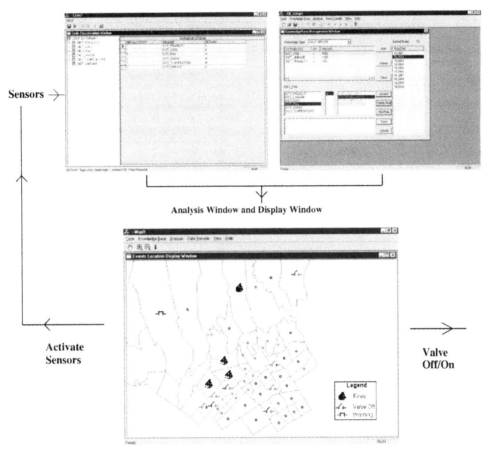

Figure 3. The Intelligent Signal Processing (ISP) Unit and its Functions.

3.2 *ISP unit*

The ISP is a necessary unit of CPWI system. ISP is a platform of disaster data analysis software consisting of the code classification window, rule and knowledge window, analysis window and display window. As shown in Figure 3, the code classification window receives the data from sensors and encodes them into different categories. For protection of earthquake-induced fires, the sensors include thermometer, smoke detector and seismometers (Figure 1). The rule and knowledge window is to build up and encode the rules, which will be the bases of the decision-making of ISP. For instance, the fire case should include rules of the fire potential and spreading and valves' shut-off/on. The analysis window is to determine whether and where there will be fires' potential, if fire occurs how the fire will spread, and whether and where the valves need to shut-off/on based on the coded data and coded rules. The display window geographically shows the events occurred, such as the locations of the fires, valves' shut-off/on and warnings.

3.3 *System test*

The effectiveness of the system has been tested by using a set-up shown in Figure 4. There are a lamp and a gas-stove in the set-up, which are connected with a relay switch. The relay-on/off is controlled by the CPWLI system. Test showed that the lamp' light and stove' fire could be automatically on/off in terms of the commands sent by the ISP unit via CPWL.

4 ACKNOWLEDGEMENT

This project is sponsored by National Science Foundation, USA under the SBIR grant DMI-9801436. The author is grateful to Dr. Patrick Johnson, the Program Director of NSF for his support and valuable advices.

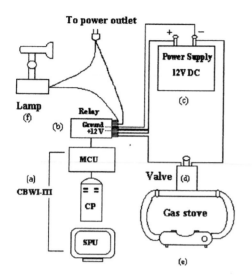

To power outlet

Power Supply
12V DC
(c)

Lamp
(f)

Relay
(b)   Ground
       +12 V

MCU

Valve (d)

(a)
CBWI-III

CP

Gas stove

SPU

(e)

Figure 4. Test Set-Up for Shut-Off/On of Gas Valve Controlled by CPWL System.

REFERENCES

Chung R. M. et al Editors, 1995. "Proc. of Post-Earthquake Fire and Lifeline Workshop," Long Beach California, January 30-31, 1995, NTIS Special Publication 889.

Wang, M.L., Satpathi, D. and Heo, G. 1997. "Damage Detection of a Model Bridge Using Modal Testing." Structural Health Monitoring, September 18-20, 1997: 589-595.

Soong, T. T., 1999. "Structural Control: Impact on Structural arch in General." Proceedings Of The SReseecond World Conference On Structural Control, Volume 1: 5-14.

Johnson, E.A., Ramallo, J.C., Spencer Jr., B.F. and Sain M.K., 1999. "Intelligent Base Isolation Systems." Proceedings Of The Second World Conference On Structural Control, Volume 1: 367-376.

Niwa, N., Kobori, T., Takahashi M., Matsunaga, Y., Kurata N., and Mizuno T., 1999. "Application of Semi-active Damper System to an Actual Building." Proceedings Of The Second World Conference On Structural Control, Volume 1: 815-824.

Tamura M., Yamamoto S., Sone A., and Masuda A., 1999. "Measuring of Dynamic Bending Strain of Structure Subjected to Dynamic Load by Using a Couple of Vibratory Gyroscope." Proceedings Of The Second World Conference On Structural Control, Volume 2: 1213-1220.

Earthquake Engineering Frontiers in the New Millennium, Spencer & Hu (eds),
© 2001 Swets & Zeitlinger, ISBN 90 2651 852 8

Effects of sediment on the dynamic pressure on rigid dams

Xiuli Du & Jinting Wang
China Institute of Water Resources and Hydropower Research, Beijing 100044

Tin-Kan Hung
University of Pittsburgh, PA 15261

ABSTRACT: After a reservoir having run a period of time, a sediment layer will be deposited before its dam. The studies of the effects of sediment layer on the hydrodynamic pressures of impounded water and the seismic responses of dam are less, and the factor is not almost considered in the seismic design of dams to be constructed and the assessment of earthquake safety of existing dams. A common action of sediment layer and foundation is modeled as a partial energy absorption boundary while considering the compressibility of impounded water in the less research materials. In the paper, it is revealed that sediment layer not only has the energy absorption effect, but also has the inertia amplification effect for the responses of dynamic pressures of water and sediment induced by vertical ground motion on dams through studying for a series of theoretically models and verifying of the test on indoor shaking table. With increasing of the thickness and porosity of sediment, the frequency corresponding to the inertia amplification effect appears to vary gradually. The comprehensive idea in the paper is important for further improving the analytical model and method of seismic responses of dams.

1 INTRODUCTION

Rational and reliable seismic design theories and analytical methods are substantial to the seismic design of dams to be constructed and the assessment of earthquake safety of existing dams. Many researches on the seismic design theories, analytical methods and seismic tests of dams have been extensively done during last several decades, and many improvements have been obtained. The researches concentrated mainly on the seismic design theories, main influence factors on seismic responses of dams, analytical methods of dams and verification tests for the analytical methods. The many investigations stated above show that a rigorous seismic analytical model of dams should include the factors as follows: the dam-foundation dynamic interaction, the action pattern of earthquake on dams, the effects of closing-opening of contracted joints of dams, the impounded water-dam- sediment-foundation dynamic interaction. Some comprehensive understandings and regular conclusions have been obtained for the influences of the first three factors. But, the effects of impounded water on the seismic responses of dams are not yet considered rationally, in which there are the two focus problems: the one is how to assess the effects of compressibility of impounded water, the another is the dynamic interaction among water, sediment and foundation.

The hydrodynamic pressure responses of compressible impounded water on the face of rigid dams with vertical upstream face were firstly investigated by Westergaard (1933). Von. Karman (1933), A.K. Chopra (1967), Chwang and Housner (1978), Chakrabati and Chopra (1973a, 1973b) investigated also the hydrodynamic pressure responses of compressible water on dams sequentially. These researches show that when the stimulating frequencies of external actions exerted are near the natural frequencies of impounded water, the hydrodynamic pressures generate a great increase (usually called the resonance phenomena). Tadashi (1958) carried out the stimulating tests for the two prototype gravity dams to examine the hydrodynamic pressure responses of impounded water but, did not observe the resonance phenomena stated above. Tadashi recognized that the outgoing pressure waves from impounded water impinging upon the reservoir bottoms are not wholly reflected back into impounded water because of the compressibility of reservoir bottoms, the part is reflected by the reservoir bottoms, the another part is absorbed by the reservoir bottoms, it is the reason that the resonance responses of hydrodynamic pressure of impounded water were not observed. To verify the view, an indoor model test was designed and carried out by Tadashi (1966), in which the energy absorption effect of sediment reduced the resonance responses of compressible water greatly made the test results accord approximately with the results obtained by the analytical method without considering the compressibility of water. Some investigators, however, doubted this conclusion and recognized that the energy absorption effect of sediment may not be so remarkable, the dynamic behavior of the tank used in the model test may have the some important influences on the test results. No matter how to say, this is still a creative and typical test reflecting the energy absorption effect of sediment. Rosenbluth (1968) proposed a reservoir bottom absorbing boundary which reflects the partial energy absorption effect of foundation flexibility on hydrodynamic pressure waves, and obtained the smaller hydrodynamic pressures exerted on dams. Chopra and Gupta (1981) investigated the dynamic interaction of gravity dam –foundation – compressible water system. The influences of foundation on the responses of impounded water was modeled as a damping absorbing boundary by Hall and Chopra (1982) through introducing the conception of reflection coefficient of pressure waves. Fenves and Chopra (1983) studied the effects of reservoir bottom absorbing boundary on the hydrodynamic pressure responses of impounded water. Fenves and Chopra (1983) considered also the effects of the bottom absorbing boundary including the action of sediment upon foundation on the hydrodynamic pressure responses. Cheng (1986) and Medina (1990) investigated theoretically the damping absorbing boundary under the common action of sediment and foundation. Yan (1989) studied the eva-luation of pressure wave reflection coefficient by an indoor test. With the investigation results stated abo-ve, the energy absorption effect of sediment and foundation has been extensively accepted by engineers and investigators. Sino-American investigators cooperated to carry out a series of prototype tests and compared the test results with the analytical results given by the method with the assumption of added mass model ignoring the compressibility of impounded water and by the method considering the compressibility of impounded water and the reservoir bottom absorbing boundary (Clough et al. 1984, Clough et al. 1984). It is shown that the test results do not accord with the results by any one of the two analytical methods, on the other hand, the basis for which model is more rational can not be obtained. For simplicity, the added mass model has been extensively used in the present seismic analysis of dams.

Sediment layer was simulated by Chen and Hung (1993) as a poroelastic media for studying the effects of sediment layer on the hydrodynamic pressure responses of impounded water in the case of rigid gravity dam and rigid foundation, in which the hydrodynamic pressure responses at the initial stage were solved by assuming seismic ground motion to be a constant acceleration and the amplification action of hydrodynamic pressure induced by sediment layer was observed. Although there are the some problems to be discussed, such as the solutions with incompressible water is steady, and the solutions with compressible water is transient, the comparison

between the two can not explain the problem better, yet the work has the very important academic values which reveal the aforementioned energy absorption effect of sediment may not be comprehensive. Recently sediment layer was simulated by Dominguez, Gallego and Japon (1997) as a poroelastic media fluid-filled and the effects of foundation were considered for calculating the top displacement of gravity dams under the action of harmonic seismic ground motion, in which the cases of saturation and partially saturation sediment were considered, the results show that the top peak displacement at the low frequency stage varies with the thickness of sediment layer and the saturation level of sediment, the reason generating the peak displacement at the low frequency and action mechanism of sediment had not be explained in their paper. In this paper, sediment was simulated as a fluid saturated poroelastic media for analyzing the influences of sediment on the hydrodynamic pressure responses of impounded water. The action effect of sediment bellow the first natural frequency of impound water was mainly studied, and the theoretical results were verified by using the model test on indoor shaking table.

2 TEST STUDY

2.1 The description of test model and measurement system

2.1.1 Water tank
The water tank with 50cm length, 35cm width, and 150cm height used in the test was made from fine alloy steel plate. The thickness of alloy steel plate is 15mm. The displacement of hydrodynamic pressure sensors A1, A2, A3, A4, A5 far from the level surface of water are 15 cm, 40 cm, 80 cm, 110 cm and 120 cm respectively. The water depth in the tank is 130 cm. The two tri-direction acceleration sensors B1, B2 are located at the shaking table and the top of water tank respectively.

2.1.2 Sediment layer
The sediment layer in the tank was made by using the sinking method. Before every shaking test, sand layer has consolidated for over 24 hours under gravity condition. The fine sand, with average diameter about 2mm, was used in the test.

2.1.3 Shaking table and measurement instruments
The electromagnetic double-direction shaking table, 614L, made in Japanese was used for the test. Its maximum load-bearing capacity is 500 kg, frequency respond range 0~1200 Hz, crest amplitude 300 mm, maximum acceleration 1.0 g, table size 1×1 m². The hydrodynamic pressure sensors used are the type of XY-05 made in China, its measurement

range is ±0.15 kg/cm², frequency respond range 0.2–400 Hz, the acceleration sensors used are the type of YD49A made in China, its measurement range is 0.005–10 g, frequency response range 0.2–1000 Hz. Moreover, the porous water pressure sensor FTS and piezoelectric hydrodynamic pressure sensor BJ1YY2-2 were equipped in the tank, due to the greater measurement range of the both type of sensors, their results were for reference only.

2.2 Test process

2.2.1 Measurement of fundamental parameters of shaking table

The comparison of the output acceleration amplitudes of shaking table and the acceleration values measured by the tri-direction accelerator fixed on the table was performed firstly. The output vertical acceleration amplitudes with harmonic vibration are 0.1 g and 0.3 g, respectively. The observed frequency range is 3–300 Hz. The measurement results show that the output values of shaking table agree well with the values recorded by the acceleration sensor equipped on the shaking table, and the traverse effect of shaking table size is ignorable.

2.2.2 Dynamic response measurement of the water tank

The acceleration responses at the top of water tank without water and with impound water in the tank were measured. The results show that, the acceleration responses at the top of water tank are very accordance with the acceleration response at the shaking table at the range of frequency below 35Hz in the both cases aforementioned while the vertical harmonic vibration were exerted.

2.2.3 Measurement of hydrodynamic pressure responses

The hydrodynamic pressure responses of water in the tank caused by vertical harmonic vibration of shaking table were measured in the two cases with sediment and without sediment.

2.3 Analysis of the test results

2.3.1 Hydrodynamic pressures at 10 Hz

The hydrodynamic pressure distributions along the depth of tank induced by vertical vibrating of the table with the acceleration amplitude of 0.3 g are shown in Fig. 1(a) for the thickness of sediment 0 cm, 10 cm and 30 cm respectively. It can been seen that hydrodynamic pressure distributions are upside-down triangular approximately, which agree well with the results of theoretical calculation ignoring the compressibility of watér. We can also see that the influences of sediment on the hydrodynamic pressure responses are small, the average has a slight increase.

2.3.2 Hydrodynamic pressures at 40 Hz

According to the aforementioned test study, the dynamic behavior and traverse effects of the water tank are notable over 40 Hz, the tank was no longer satisfied with the assumption of rigid tank. At this moment, even if the sediment layer changes the dynamic behavior of water tank, the real action mechanism of sediment effects on the hydrodynamic pressures cannot be distinguished. In fact, 10 cm sediment layer equipped on the bottom of tank hardly changes the dynamic behavior of tank, which has been verified by the measurement. The change of hydrodynamic pressures, in the two cases without sediment and with 10 cm sediment layer, can reveal the action mechanism of sediment. Fig. 1(b) shows the hydrodynamic pressure distributions along the depth of tank when the amplitude of vertical vibration acceleration is 0.3 g and the stimulating frequency is 40 Hz. The sediment layer causes an obvious increase of hydrodynamic pressure. This shows the amplification effect of sediment.

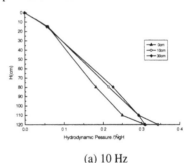

(a) 10 Hz

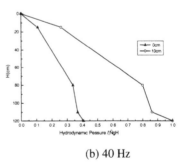

(b) 40 Hz

Figure 1. Hydrodynamic pressure distributions along depth

3 THEORETICAL ANALYSIS

3.1 Sediment media model

In this paper, the poroelastic theory advanced by Biot (1956) was used to model sediment media. According to the Biot's dynamic theory, the wave equations of fluid saturated porous media can be expressed as follows:

221

$$N\nabla^2\boldsymbol{u}+\nabla[(A+N)e+Q\varepsilon]=\frac{\partial^2}{\partial t^2}(\rho_{11}\boldsymbol{u}+\rho_{12}\boldsymbol{U})+b\frac{\partial}{\partial t}(\boldsymbol{u}-\boldsymbol{U})$$

$$(1)$$

$$\nabla[Qe+R\varepsilon]=\frac{\partial^2}{\partial t^2}(\rho_{12}\boldsymbol{u}+\rho_{22}\boldsymbol{U})-b\frac{\partial}{\partial t}(\boldsymbol{u}-\boldsymbol{U})\qquad(2)$$

where $\boldsymbol{u}$ is the displacement vector of solid skeleton, $\boldsymbol{U}$ is the displacement vector of porous fluid; $e=\nabla\cdot\boldsymbol{u}$; $\varepsilon=\nabla\cdot\boldsymbol{U}$; N and A are the parameters related to drained Lame constant μ, λ; R and Q are the material constant, which are measured by test (see Biot and Willis (1957)), $\rho_{11}=(1-n)\rho_s+\rho_a$, $\rho_{22}=n\rho_f+\rho_a$, $\rho_{12}=-\rho_a$,; ρ_s, ρ_f, ρ_a are the solid skeleton density, the porous fluid density and added apparent density (in general $\rho_a=0$) respectively, n is the porosity; The dissipation constant $b=\nu n^2/k$, k is the permeability coefficient, ν is the viscosity of fluid (when $\nu=0$, the equations (1) and (2) became elastic wave equations without dissipation).

3.2 Calculation of the rigid water tank model

3.2.1 Material parameters
To compare the results between the theory calculation model and the test measurement, the calculation model is the same as the test model. In calculating, the sides of tank were assumed as rigid. The impounded water has a density $\rho_L'=1000kg/m^3$ and a compression modulus $K_L=2.0\times10^9$ pa. The bulk modulus of the pore water $\beta_L'=1.0\times10^8$ pa (corresponding to a degree of saturation 98.6%). The material parameters of poroelastic media used in the computation are listed in Table 1.

3.2.2 Calculating results and analyses
Hydrodynamic pressure distributions along the depth of tank are shown in Fig. 2, when the parameters of sediment are the set I, the set II, the set III, and the set IV respectively, and the thickness of sediment are 0 cm, 10 cm and 30 cm, respectively, and the vibration frequency is 40 Hz with the acceleration amplitude of 1.0 g. Fig. 2 shows that, in the case without sediment, the hydrodynamic pressure distribution along the depth of tank is upside down triangular, which agrees well with the results of test. At the frequency 40 Hz, sediment obviously changes

hydrodynamic pressures. At this case, ignoring the effect of sediment is unreasonable. Qualitatively, the amplification effect of sediment is identical with that observed in the test.

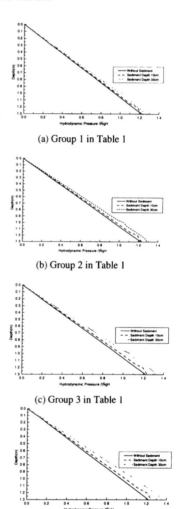

(a) Group 1 in Table 1

(b) Group 2 in Table 1

(c) Group 3 in Table 1

(d) Group 4 in Table 1

Figure 2. The effects of sediment thickness on the hydrodynamic pressures (40 Hz)

Table 1. Sediment (fluid-saturated two phase media) material parameters (SI units)

Set	ρ_S	ρ_L	n	b ($\times10^3$)	A ($\times10^8$)	N ($\times10^7$)	R ($\times10^8$)	Q ($\times10^8$)
I			0.3	0.9	1.716	1.250	0.3	0.7
II			0.4	1.6	0.972	1.068	0.4	0.6
III	2100		0.5	2.5	0.561	0.914	0.5	0.5
IV			0.6	3.6	0.319	0.782	0.6	0.4
V	2301							
VI	2501	1000						
VII	2701		0.3	0.9	1.716	1.250	0.3	0.7

3.3 Computation of the model with rigid vertical dam and rigid foundation

3.3.1 Description of the calculation model

An example was examined for analysis of the influences of sediment on the dynamic pressure responses of sediment and water exerting on the upstream face of dams, in which a dam with height of 170 m and water depth of 150 m and foundation were assumed as rigid and the material parameters of sediment are seen in Table 1. The incident displacement amplitude is 1.0.

3.3.2 Influences of the thickness of sediment on the dynamic pressures

The integration values of dynamic pressures along the depth of dam are shown in Fig. 3 (a, b, c, d) while the thickness of sediment are 0 m, 5 m, 10 m, 20 m, 40 m, 70 m and 100 m respectively and the material parameters of sediment are the set I, the set II, the set III, the set IV, respectively, in which a non-dimensional form was used, the vertical axis represents the ratio of the integration value of dynamic response values along the depth of dam to the integration of hydrostatic pressures along the depth of dam and the horizontal axis represents a dimensionless frequency $a_0 = \omega H / C_w$, in which C_w is the wave velocity of water and H is the depth of sediment and water. It can been seen that the existence of sediment makes the resonance peak corresponding to the first natural frequency of impounded water without sediment disappear, which shows a energy absorption effect of sediment, but, the action of sediment generates some new resonance peaks. With increasing of the thickness of sediment, the frequency of the new resonance response decreases gradually. For the parameters in this paper, the dynamic pressures exerting on the dam below the first natural frequency of impounded water without sediment increase due to the new resonance response. Therefore, the conventional opinion of energy absorption effect of sediment is not comprehensive.

3.3.3 Influence of the porosity of sediment on the dynamic pressures

The influences of the different porosities of sediment (see Table 1) on the dynamic pressure on dam are shown in Fig. 4 (a, b) while the thickness of sediment layer are 20 m and 100 m respectively. We can see that while the thickness of sediment layer is thin, the influences of the porosity of sediment on the dynamic pressures are large roughly. Increasing the porosity of sediment reduces the amplitude and frequency of the first peak response of dynamic pressures remarkably.

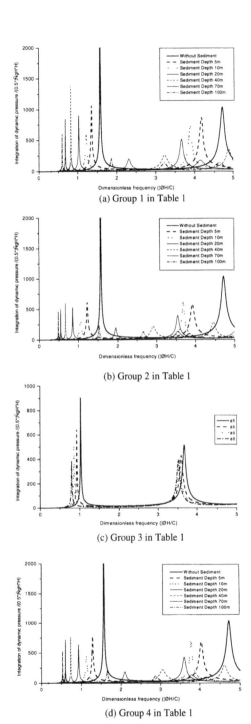

(a) Group 1 in Table 1

(b) Group 2 in Table 1

(c) Group 3 in Table 1

(d) Group 4 in Table 1

Figure 3. The effects of the sediment thickness on the dynamic pressures

223

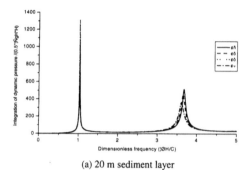

(a) 20 m sediment layer

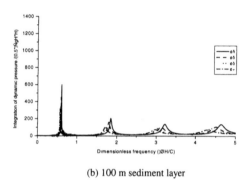

(b) 100 m sediment layer

Figure 4. The effects of the porosity of sediment on the dy-

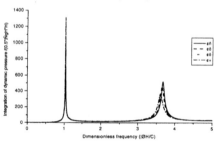

(a) 20 m sediment layer

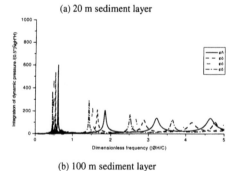

(b) 100 m sediment layer

namic pressures

Figure 5. The effects of the densities of solid skeleton on the dynamic pressures.

4 CONCLUSIONS

Some conclusions can been drawn through the analyses of the test and numerical calculations in the paper as follows:

1 Sediment layer has the energy absorption effect which reduces the amplitudes of dynamic pressures of impounded water with compressibility and sediment on dams greatly. In addition, sediment layer has the resonance amplification effect for the dynamic pressures of water and sediment induced by vertical ground motion on dams. With increasing of the thickness of sediment layer, the frequency of the resonance phenomena appears to vary gradually.

2 The porosity of sediment has an important influence on the dynamic pressures of sediment and water on dams. With increasing of the porosity, the amplification effect of sediment layer vary gradually, in the meanwhile, the frequency corresponding to the resonance response appears to vary too.

3 The influences of the density effect of solid skeleton of sediment on the hydrodynamic pressures on dam are not remarkable.

The effects of the changes of sediment parameters, the flexibility of dam and the flexibility of foundation on the dynamic pressures on dams will be analyzed in other papers.

5 ACKNOWLEDGMENTS

This research has been supported by CNSF Grant (No. 59739180 and 59409001), and the last author was supported by NSF Grant CMS-9319341.

REFERENCES

Westergaard, H.M., Water pressures on dams during earthquakes, Trans. ASCE, Vol. 98 (1933), 418-433.

Von Kármán, T., Discussion of water pressures on dams during earthquakes, by Westergaard, H.M., Trans. ASCE, Vol.98 (1933), 434-436.

Chopra, A.K., Hydrodynamic pressures on dams during earthquake, J. Eng. Mech. ASCE, Vol.93 (1967), 205-233.

Chwang, A.T. and Housner, G.W., Hydrodynamic pressures on sloping dams during earthquake, Part 1:Momentum Method, J.Fluid Mech., Vol.87 (1978), 335-341.

Chakrabarti, P. and Chopra, A.K., Hydrodynamic pressures and response of gravity dams to vertical earthquake component, Earthq. Eng. and Struct. Dyn., Vol.1 (1973), 325-335.

Chakrabarti, P. and Chopra, A.K., Earthquake analysis of gravity dams including hydrodynamic interaction, Earthq.Eng. and Struct.Dyn., Vol.2 (1973), 143-160.

Tadashi Hatano, Tadashi Takahashi and Hajime Tsutsumi, Vibration tests and their studies on Tsukabaru gravity dam. Transactions of the Japan Society Civil Engineers, No.59 (1958), 8-16. (in Japanese).

Tadashi Hatano, An examination on the resonance of hydrodynamic pressure during earthquakes due to elasticity of water. Transactions of the Japan Society Civil Engineers, No.129 (1966), 1-5. (in Japanese).

Rosenbluth, E., presión hidródinámica en presas debida a la acceleractión vertical conrefraccion en elfondo., 2nd Congreso Nacional de Ingenieria Sismica, Veracruz, Mexico, May (1968).

Chopra, A.K., and Gupta, S., Hydrodynamic and foundation interaction effects in earthquake response of a concrete gravity dam. J.Struct.Div., ASCE, Vol.107 (1981), 1399-1412.

Hall, J.F., and Chopra, A.K., Two-dimensional dynamic analysis of concrete gravity and embankment dams including hydrodynamic effects. Earthq.Eng.and Struct.Dyn., Vol.10 (1982), 305-332.

Fenves, G., and Chopra, A.K., Effects of reservoir bottom absorption on earthquake response of concrete gravity dams. Earthq.Eng. and Struct. Dyn., Vol.11 (1983), 802-829.

Fenves, G., and Chopra, A.K., Earthquake analysis of concrete gravity dams including reservoir bottom absorption and dam-water-foundation rock interaction. Earthq.Eng. and Struct.Dyn., Vol.12 (1984), 663-608.

Cheng, A.H.D., Effects of sediment on earthquake-induced reservoir hydrodynamic response. J.Eng.Mech., ASCE, Vol.111 (1986), 654-663.

Medina, F., Dominguez, J. and Tassoulas, J.L., Response of dams to earthquake including effects of sediment. J. Struct. Eng., ASCE, Vol.116, No.1, (1990), 3108-3121.

Chen, B.F. and Hung, T.K., Hydrodynamic pressure of water and sediment on rigid dam. J.Eng.Mech., ASCE, Vol.119, No.7 (1993), 1411-1433.

Dominguez, J, Gallego, R. and Japon, B.R., Effects of porous sediments on seismic response of concrete gravity dams, J. Eng. Mech., Vol.123, No.4 (1997), 302-311.

Yan Chengda, effect study of bottom sediment on reflection behavior of fluid pressure wave and hydrodynamic pressure on concrete dam. Ph.D. thesis, Tsinghua University. (in Chinese).

Clough, R.W., Chang, K.T., Chen, H.Q., Stephen, R.M., Wang, G.L. and Ghanaat, Y., Dynamic response behavior of Xiang Hong Dian dam. Earthquake Engineering Research Center, University of California, Berkeley, Report No. UCB/EERC-84102, 1984.

Clough, R.W., Chang, K.T., Chen, H.Q., Stephen, R.M., Ghanaat, Y., and Qi, J.H., Dynamic response behavior of Quan Shui dam. Earthquake Engineering Research Center, University of California, Berkeley, Report No. UCB/EERC-84120, 1984.

Earthquake Engineering Frontiers in the New Millennium, Spencer & Hu (eds),
© 2001 Swets & Zeitlinger, ISBN 90 2651 852 8

Seismic response characteristics of multispan cable-stayed bridges with stabilizing cables

J.M. Ko & Y.Q. Ni
The Hong Kong Polytechnic University, Hung Hom, Kowloon, Hong Kong

G.J. He
The Hong Kong Polytechnic University, Hung Hom, Kowloon, Hong Kong;
Central South University, Changsha 410075, P. R. China

ABSTRACT: A recent trend in the design of long-span bridges is the multispan cable-stayed bridges with three or more towers. A critical problem of multispan cable-stayed bridges is the stabilization of the central tower(s), which has resulted in increasing application of stabilizing cables. The Ting Kau Bridge (TKB) in Hong Kong is one of the few multispan cable-stayed bridges adopting stabilizing cables ever built. In this paper, the seismic response characteristics of multispan cable-stayed bridges with stabilizing cables and the effect of stabilizing cables on the seismic response are studied by referring to the Ting Kau Bridge.

1 INTRODUCTION

The multispan cable-stayed bridges with three or more towers have been a recent design trend of long-span bridges (Virlogeux 1999). For a conventional three-span cable-stayed bridge with two towers, each of the towers is connected through outermost stay cables to the fixed anchorage or anchor pier, which provides effective support to the towers. However, in a multispan cable-stayed bridge, the beneficial effect of the fixed anchorage or anchor piers diminishes for the central tower(s). As a result, an important design concern of multispan cable-stayed bridges is the stabilization of the central tower(s) under unbalanced loading. This has resulted in increasing application of stabilizing cables when the bridge does not have very stiff towers (Tang 1995).

The Ting Kau Bridge (TKB) in Hong Kong is one of the multi-span cable-stayed bridges using stabilizing cables in practice. The bridge, as shown in Figure 1, is a three-tower cable-stayed bridge with two main spans of 448m and 475m respectively, and two side spans of 127m each (Bergermann et al. 1995, Bergermann and Schlaich 1996). In this bridge, the deck is separated into two carriageways with width of 18.8m each, between them being slen-

der single-leg towers. These two carriageways, with a 5.2m gap, are linked at 13.5m intervals by main crossgirders. Eight longitudinal stabilizing cables with lengths up to 465m are installed to diagonally connect the top of the central tower to the deck near the side towers. They provide restraint to the central tower from traffic-induced unbalanced loading and vertical wind loading. Both the central and side towers are stiffened by a total of sixty-four transverse stabilizing cables in the lateral direction. They provide restraint to the towers under horizontal wind loading (Schlaich and Bergermann 1998).

Because only a few multispan cable-stayed bridges with stabilizing cables have been built and the stabilizing cables were designed mainly for wind resistance, it is desirable to understand the performance of this type of bridges subject to earthquake excitation. In this paper, by referring to the TKB, we analyze the seismic response characteristics of multispan cable-stayed bridges and the effect of stabilizing cables on the seismic response behavior. Based on a precise and validated three-dimensional finite element model, seismic responses of the TKB under the 1979 El Centro, the 1994 Northridge and the 1995 Kobe earthquakes are analyzed by using the modal superposition method in the assumption of

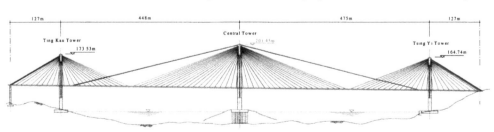

Figure 1. Ting Kau Bridge

synchronous excitation. The analysis is first made with and without accounting for cable local vibration modes in the response prediction. Then a comparative study is conducted for three situations: (a) with transverse and longitudinal stabilizing cables; (b) without transverse and longitudinal stabilizing cables; and (c) with only transverse but no longitudinal stabilizing cables.

2 MODAL PROPERTIES

A precise three-dimensional finite element model of the TKB has been developed and validated by use of the measured modal parameters (Ni et al. 2000). A hybrid-element cable system is adopted in the model in which all the eight longitudinal stabilizing cables are modeled by multi-element cable system taking into account the sag and tension effect (geometric stiffness) while other cables are modeled by single-element system. The resulting bridge model contains 2901 nodes and 5581 elements (totally 16806 degrees of freedom).

With the developed finite element model, the modal analysis of the TKB has been carried out. The first 125 modes were computed with the highest frequency 1.046Hz. This indicates that the bridge has a lot of closely spaced vibration modes with low frequencies. These vibration modes can be classified into two categories: global modes and local modes. Here the global modes are defined as those in which the vibration of the deck and/or tower(s) participates, whereas the local modes are those in which vibration occurs only in the cables. Although only the local vibrations of longitudinal stabilizing cables are allowed in the model, among the 125 modes 78 modes are cable local modes.

Modeling the longitudinal stabilizing cables as multi-element system not only produces a lot of cable local modes, but also reveals a number of special globally coupled vibration modes with strong modal interaction among the deck, towers and cables. Figure 2 shows the first mode of the TKB. This is the first vertical bending global mode with the natural frequency of 0.163 Hz. In this mode, besides both the deck and the central tower participate greatly in the modal motion, the four longitudinal stabilizing cables on one main span vibrate in phase with large amplitude. The modal shape of the cables in this

global mode is identical with the first mode of the separated cables with fixed ends, but the frequency is far less than the lowest frequency of the individual cables. The modal interaction in this mode is not due to resonance between the cables and the deck/tower that may occur in conventional cable-stayed bridges, because the natural frequencies of the individual cables are away from those of the bridge superstructures. This type of coupled modes is also observed in higher global modes, and has been verified using the field measurement data from a long-term monitoring system with accelerometers installed simultaneously at the deck, towers and stabilizing cables of the bridge (*Report No. TKBD-1* 2000).

The first vertical bending modal frequency and the first torsion modal frequency of the TKB are 0.163Hz and 0.514Hz respectively, which are lower than the corresponding 0.199Hz and 0.569Hz of the world's longest cable-stayed Tatara Bridge (Endo et al. 1991). In this sense, the TKB is one of the most flexible cable-stayed bridges in the world.

3 EFFECT OF CABLE LOCAL MODES

The seismic response characteristics of the TKB are studied by using three typical strong ground motion records as earthquake excitations. They are the 1979 Imperial Valley earthquake (El-Centro horizontal N-S record), the 1994 Northridge earthquake (horizontal N-S record), and the 1995 Hyogoken-Nanbu earthquake (Kobe horizontal N-S record). The peak accelerations of these three records have been scaled to $0.1g$ in accordance with the earthquake intensity in Hong Kong region (Pun et al. 1998). In this preliminary study, a synchronous earthquake excitation is assumed. Each earthquake record acts on the bridge in longitudinal, lateral and vertical directions respectively.

The first 200 modes of the TKB have been included in the seismic response analysis using the modal superposition method. Among the first 200 modes (in the frequency range of 0 to 1.46Hz) 69 are global vibration modes (including globally coupled modes) and 131 are cable local vibration modes. In order to verify the influence of the cable local modes on the bridge seismic response, the response prediction is conducted using all the 200 modes (including both global and local modes) and using only the 69

Figure 2. First vibration mode of Ting Kau Bridge (f = 0.163Hz)

global modes. It should be noted that here the response prediction results even disregarding the cable local modes are different from those obtained by directly modeling all the cables as single-element system, because in the latter case the globally coupled modes between the cables and the deck/towers cannot be revealed.

3.1 Structural displacement response

Tables 1 to 3 list the maximum displacement responses of the three bridge towers under different earthquake excitations, where the quantity with superscript 't' denotes the response obtained by taking into account the total first 200 modes including both global and local modes, and the quantity with superscript 'g' denotes the response obtained by only accounting for the global modes. Here x, y, and z are respectively the bridge longitudinal (axis), vertical and lateral (transverse) directions. The following observations on the tower response characteristics are made: (a) Under longitudinal or vertical earthquake, the longitudinal response of the towers is dominant. Under transverse earthquake, the lateral response of the towers is dominant; (b) The longitudinal displacement response of the towers under longitudinal earthquake is basically the same amplitude as the lateral displacement response of the towers under transverse earthquake; (c) Under longitudinal (or transverse) earthquake, the maximum longitudinal (or lateral) displacement response of the central tower is greater than the maximum longitudinal (or

lateral) displacement response of the side towers; (d) Under vertical earthquake, the longitudinal displacement response of the central tower is much less than the longitudinal displacement response of the side towers. This means that the longitudinal stabilizing cables are effective for vertical earthquake resistance; (e) Under longitudinal earthquake, the maximum longitudinal displacement response of the central tower does not occur at the tower top, instead at the section of approximately 72% tower height. The displacement at the tower top is only 75% to 85% of the maximum displacement. The maximum longitudinal displacement response of the side towers occurs at the tower top; (f) Under transverse earthquake, the maximum lateral displacement response of both the central tower and the side towers occurs at the tower top; (g) Under vertical earthquake, the maximum longitudinal displacement response of all the three towers does not occur at the tower top. This means that the transverse stabilizing cables provide resistance to vertical earthquake.

The local vibration modes mainly affect the longitudinal displacement response of the central tower under longitudinal earthquake. Under lateral or vertical earthquake excitation, the predicted displacement response for each of the three towers is almost identical when using the total 200 modes and when using only the 69 global modes. Under longitudinal earthquake excitation, the local vibration modes still have no influence on the displacement response of the two side towers, while the longitudinal displacement response of the central tower obtained without accounting for the cable local modes (using only the 69 global modes) is greatly less than that obtained with taking into account the cable local modes (using the total 200 modes). The difference of the predicted tower maximum displacement with and without considering the local modes comes to 22% to 39%. The tower top displacement predicted without considering the local modes is only 57% to

Table 1. Tower displacement under El-Centro earthquake

Excitation direction	Ting Kau tower	Central tower	Tsing Yi tower
Longitudinal (x)	$u_x^t=0.072$m $u_x^g=0.069$m	$u_x^t=0.097$m $u_x^g=0.059$m	$u_x^t=0.069$m $u_x^g=0.058$m
Lateral (z)	$u_z^t=0.069$m $u_z^g=0.067$m	$u_z^t=0.091$m $u_z^g=0.091$m	$u_z^t=0.062$m $u_z^g=0.063$m
Vertical (y)	$u_x^t=0.045$m $u_x^g=0.045$m	$u_x^t=0.014$m $u_x^g=0.014$m	$u_x^t=0.047$m $u_x^g=0.047$m

Table 2. Tower displacement under Northridge earthquake

Excitation direction	Ting Kau tower	Central tower	Tsing Yi tower
Longitudinal (x)	$u_x^t=0.099$m $u_x^g=0.077$m	$u_x^t=0.109$m $u_x^g=0.072$m	$u_x^t=0.085$m $u_x^g=0.072$m
Lateral (z)	$u_z^t=0.041$m $u_z^g=0.044$m	$u_z^t=0.088$m $u_z^g=0.088$m	$u_z^t=0.070$m $u_z^g=0.067$m
Vertical (y)	$u_x^t=0.078$m $u_x^g=0.078$m	$u_x^t=0.011$m $u_x^g=0.010$m	$u_x^t=0.054$m $u_x^g=0.054$m

Table 3. Tower displacement under Kobe earthquake

Excitation direction	Ting Kau tower	Central tower	Tsing Yi tower
Longitudinal (x)	$u_x^t=0.045$m $u_x^g=0.039$m	$u_x^t=0.049$m $u_x^g=0.038$m	$u_x^t=0.040$m $u_x^g=0.030$m
Lateral (z)	$u_z^t=0.042$m $u_z^g=0.044$m	$u_z^t=0.048$m $u_z^g=0.048$m	$u_z^t=0.051$m $u_z^g=0.051$m
Vertical (y)	$u_x^t=0.048$m $u_x^g=0.048$m	$u_x^t=0.015$m $u_x^g=0.015$m	$u_x^t=0.044$m $u_x^g=0.044$m

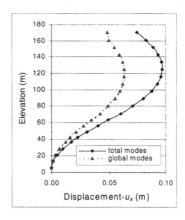

Figure 3. Displacement response envelope of central tower under longitudinal El-Centro earthquake

Table 4. Deck displacement under different earthquakes

Excitation direction	El-Centro earthquake	Northridge earthquake	Kobe earthquake
Longitudinal (x)	u_y^l=0.115m	u_y^l=0.117m	u_y^l=0.074m
Vertical (y)	u_y^l=0.157m	u_y^l=0.133m	u_y^l=0.092m
Lateral (z)	u_z^l=0.114m	u_z^l=0.056m	u_z^l=0.046m

67% of that predicted with considering the local modes. Figure 3 illustrates the longitudinal displacement response envelope of the central tower under longitudinal El-Centro earthquake excitation.

Table 4 lists the maximum displacement responses of the bridge deck under the three earthquake excitations. It is seen that the maximum vertical displacement of the deck under vertical earthquake is slightly greater than the maximum vertical displacement of the deck under longitudinal earthquake, but the maximum lateral displacement of the deck under transverse earthquake is much less. The cable local modes have no effect on the deck vertical displacement response and have a little effect on the deck lateral displacement response.

3.2 Structural internal force response

Due to the limited space, only the maximum shear forces and bending moments of the three towers under the Northridge earthquake are given in Tables 5 and 6. The following observations on the tower shear force response are made: (a) Under longitudinal or vertical earthquake, the shear force of the towers in the longitudinal direction, Q_x, is dominant. Under transverse earthquake, the shear force of the towers in the lateral direction, Q_z, is dominant; (b) Under longitudinal earthquake, the maximum shear force (Q_x) of the central tower is much greater than the corresponding values of the two side towers. The maximum shear force of the central tower is 3 to 6 times the maximum shear force of the Ting Kau and

Tsing Yi towers; (c) Under lateral earthquake, the maximum shear force (Q_z) of the central tower is also greater than the corresponding values of the two side towers; (d) Under vertical earthquake, the maximum shear force (Q_x) of the central tower in the longitudinal direction is always less than the corresponding values of the two side towers.

On the tower bending moment response, the following observations are obtained: (a) Under longitudinal or vertical earthquake, the tower bending moment about the lateral axis, M_z, is dominant. Under transverse earthquake, the tower bending moment about the longitudinal axis, M_x, is dominant; (b) Under longitudinal earthquake, the maximum bending moment (M_z) of the central tower is much greater than the corresponding values of the two side towers. The ratio of maximum bending moment of the central tower to the side towers is between 3 and 6; (c) Under lateral earthquake, the maximum bending

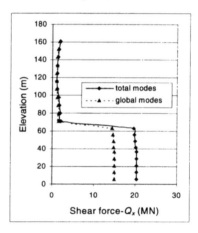

Figure 4. Shear force envelope of central tower under longitudinal Northridge earthquake

Table 5. Tower shear force under Northridge earthquake

Excitation direction	Ting Kau tower	Central tower	Tsing Yi tower
Longitudinal (x)	Q_x^l=3.68MN Q_x^g=3.67MN	Q_x^l=20.5MN Q_x^g=14.9MN	Q_x^l=3.81MN Q_x^g=3.77MN
Lateral (z)	Q_z^l=3.39MN Q_z^g=3.28MN	Q_z^l=8.17MN Q_z^g=7.21MN	Q_z^l=3.74MN Q_z^g=3.57MN
Vertical (y)	Q_x^l=5.48MN Q_x^g=5.46MN	Q_x^l=2.81MN Q_x^g=2.58MN	Q_x^l=4.01MN Q_x^g=4.01MN

Table 6. Tower bending moment under Northridge earthquake (Unit: MN-m)

Excitation direction	Ting Kau tower	Central tower	Tsing Yi tower
Longitudinal (x)	M_z^l=223 M_z^g=204	M_z^l=1115 M_z^g=807	M_z^l=236 M_z^g=224
Lateral (z)	M_x^l=171 M_x^g=166	M_x^l=443 M_x^g=409	M_x^l=180 M_x^g=169
Vertical (y)	M_z^l=308 M_z^g=308	M_z^l=137 M_z^g=124	M_z^l=224 M_z^g=224

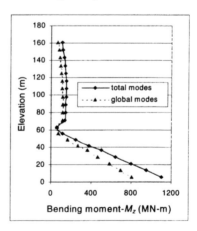

Figure 5. Bending moment envelope of central tower under longitudinal Northridge earthquake

moment (M_x) of the central tower is also greater than the corresponding values of the two side towers; (d) Under vertical earthquake, the maximum bending moment (M_z) of the central tower is always less than the corresponding values of the two side towers.

The local vibration modes mainly affect the internal force response of the central tower under longitudinal earthquake. Under longitudinal earthquake excitation, the maximum shear force (Q_x) of the central tower obtained without accounting for the cable local modes (using only the 69 global modes) is only 72% to 75% of that obtained with taking into account the cable local modes (using the total 200 modes), and the maximum bending moment (M_z) of the central tower obtained without considering the local modes is only 70% to 73% of that predicted with considering the local modes. Figures 4 and 5 show respectively the shear force (Q_x) and bending moment (M_z) envelopes of the central tower under longitudinal Northridge earthquake excitation.

4 EFFECT OF STABILIZING CABLES

Although the stabilizing cables were designed for the purpose of providing restraint to the towers, modal analysis showed that introduction of the longitudinal stabilizing cables also influenced the global dynamic properties of the bridge (Chan et al. 2000). For example, the fundamental natural frequency of the TKB increases from 0.141Hz to 0.163Hz after installing the eight longitudinal stabilizing cables. In order to evaluate the effect of the stabilizing cables on the bridge seismic response, a comparative study has been conducted for the three situations: (a) *with-SCB*: the TKB has transverse and longitudinal stabilizing cables; (b) *without-SCB*: the TKB does not have transverse and longitudinal stabilizing cables; and (c) *without-LSCB*: the TKB has only transverse

but no longitudinal stabilizing cables. For each situation, the bridge static (equilibrium) configuration is achieved through a nonlinear iteration scheme.

4.1 Structural displacement response

On the bridge tower displacement response, the following observations are made: (a) Under longitudinal or vertical earthquake, the tower maximum longitudinal displacement response in the situation with stabilizing cables (with-SCB) is almost same as that in the situations without stabilizing cables (without-SCB or without-LSCB); (b) Under transverse earthquake, the tower maximum lateral displacement response obtained with both transverse and longitudinal stabilizing cables (with-SCB) is almost identical with that obtained with only transverse stabilizing cables (without-LSCB), but can be significantly different from that obtained without stabilizing cables (without-SCB). Figure 6 shows the lateral displacement response envelope of the central tower under

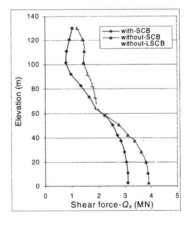

Figure 7. Shear force envelope of Tsing Yi tower under longitudinal El-Centro earthquake

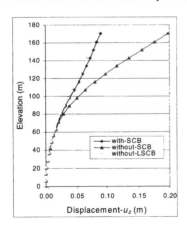

Figure 6. Displacement response envelope of central tower under transverse El-Centro earthquake

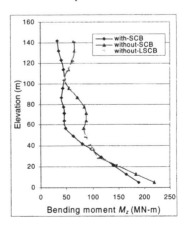

Figure 8. Bending moment envelope of Ting Kau tower under longitudinal El-Centro earthquake

transverse El-Centro earthquake excitation for the three situations.

Both the transverse and longitudinal stabilizing cables have almost no influence on the deck vertical displacement response subject to longitudinal or vertical earthquake. Under transverse earthquake, the transverse stabilizing cables can reduce the deck lateral displacement response amplitude.

4.2 Structural internal force response

The stabilizing cables have a more significant influence on the tower internal force than the tower displacement response. The following observations are made on the tower internal force response: (a) Under longitudinal earthquake, using stabilizing cables can greatly reduce the tower maximum shear force in the longitudinal direction (Q_x) and the tower maximum bending moment about the lateral axis (M_z), especially for the central tower; (b) Under transverse earthquake, the tower maximum shear force in the lateral direction (Q_z) obtained without stabilizing cables (without-SCB) is less than that obtained with both transverse and longitudinal stabilizing cables (with-SCB) or with only transverse stabilizing cables (without-LSCB), but the maximum bending moment of the central tower about the longitudinal axis (M_x) is still larger in the case without stabilizing cables than in the case with stabilizing cables; (c) Under vertical earthquake, the stabilizing cables have ignorable effect on the tower shear force Q_x and have slight effect on the tower bending moment M_z. Figures 7 and 8 show respectively the shear force (Q_x) envelope of the Tsing Yi side tower and the bending moment (M_z) envelope of the Ting Kau side tower under longitudinal El-Centro earthquake excitation for the three situations.

5 CONCLUDING REMARKS

As a typical modern cable-stayed bridge with extremely low natural frequencies, the TKB is deemed to be a good candidate selected for studying the active/semiactive seismic response control of cable-stayed bridges. In this paper, the effects of the cable local modes and the stabilizing cables on seismic response of the TKB have been studied. This preliminary study is a prelude to developing a mode-based reduced-order model for the TKB suitable for the accomplishment of control algorithms.

The extremely long longitudinal stabilizing cables of the TKB have been modeled as multi-element system in the seismic response analysis. Modeling the longitudinal stabilizing cables as multi-element system not only invokes cable local modes, but also reveals globally coupled modes among the deck, towers and cables. The local modes have negligible effect on the bridge deck response, while the glob-

ally coupled modes participate significantly in the bridge seismic response. In addition, some local modes greatly affect the seismic response of bridge towers. These results attain to two consequences: (1) we should consider the specific type of modal coupling among the deck, towers and cables in designing the control strategy for the TKB; (2) one or a few cable local modes which greatly affect tower response should be included as dominant modes in the reduced-order model for control purpose.

6 ACKNOWLEDGEMENT

This study was supported partially by a grant from the Research Grants Council of the Hong Kong Special Administrative Region, China (Project No. PolyU 5045/00E) and partially by a grant from the Hong Kong Polytechnic University (Project No. G-YW29). The authors wish also to thank the Highways Department of the Hong Kong SAR Government for providing all the supports to the research.

REFERENCES

Bergermann, R. & Schlaich, M. 1996. Ting Kau Bridge, Hong Kong. Structural Engineering International 6: 152-154.

Bergermann, R., Schlaich, M., Holmes, D. & Arnold, D.C. 1995. The design of the Ting Kau cable-stayed bridge in Hong Kong. In Bridge into 21st Century: Proceedings of the International Conference, The Hong Kong Institution of Engineers, Hong Kong, 171-178.

Chan, T.H.T., Wang, J.Y., Ni, Y.Q. & Ko, J.M. 2000. Importance of longitudinal stabilizing cables in dynamic characteristics of Ting Kau Bridge. In J.M. Ko & Y.L. Xu (eds.), Advances in Structural Dynamics, Vol. I, 467-474. Oxford: Elsevier.

Endo, T., Iijima, T., Okukawa, A. & Ito, M. 1991. The technical challenge of a long cable-stayed bridge–Tatara Bridge. In M. Ito, Y. Fujino, T. Miyata & N. Narita (eds.), Cable-Stayed Bridges: Recent Developments and their Future, 417-436. Oxford: Elsevier.

Ni, Y.Q., Wang, J.Y. & Ko, J.M. 2000. Modal interaction in cable-stayed Ting Kau Bridge. In J.M. Ko & Y.L. Xu (eds.), Advances in Structural Dynamics, Vol. I, 537-544. Oxford: Elsevier.

Pun, W.K., Lee, C.F. & Ding, Y.Z. 1998. Seismic risk in Hong Kong. In E.S.S. Lam & J.M. Ko (eds.), Proceedings of the International Workshop on Earthquake Engineering for Regions of Moderate Seismicity, Hong Kong, 19-24.

Schlaich, M. & Bergermann, R. 1998. Monoleg towers with transverse stabilising cables. Structural Engineering International 8: 252-155.

Tang, M.-C. 1995. Multispan cable-stayed bridges. In M. Sanayei (ed.), Restructuring: America and Beyond, ASCE, New York, Vol. 1, 455-458.

Virlogeux, M. 1999. Recent evolution of cable-stayed bridges. Engineering Structures 21: 737-755.

Report No. TKBD-1 2000. Spectral analysis and modal identification of the Ting Kau Bridge. Department of Civil and Structural Engineering, The Hong Kong Polytechnic University, Hong Kong (http://www.cse.polyu.edu.hk/~dynamics).

Earthquake Engineering Frontiers in the New Millennium, Spencer & Hu (eds),
© 2001 Swets & Zeitlinger, ISBN 90 2651 852 8

Earthquake input and reliability analysis of water delivery systems

Xiaojun Li
Institute of Geophysics, China Seismological Bureau, Beijing, China

ABSTRACT: The earthquake input should be a ground motion field for the response analyses of the lifeline systems and some important structures with large-span. The main problems on an earthquake ground field are the studies of the earthquake ground motion at point and the spatial correlation of the motions at different points nearby. An empirical model of simulation of earthquake ground motion field is introduced for earthquake input of seismic performance analysis of lifeline system in this paper. In this model, the distance between spatial points is introduced only as a parameter controlling the uncertainty effects, and the source-to-site distance and its variation are used to reflect the correlation of motions at points. A method of the reliability analysis of the seismic performance of water delivery systems under a scenario earthquake is proposed based on the empirical model of earthquake input. In the method, the soil-pipe interaction and uncertainty of mechanical parameters of pipes are considered. As an application, a case study is shown for an urban water delivery system.

1 INTRODUCTION

The lifeline systems and some important engineering structures are spatially extended and distributed on a large scale. The earthquake responses for these systems and structures are spatially correlative. Damage of one unit or part does not mean that another similar unit or part in the system or structure is damaged to the same level. The earthquake inputs are different for the foundations of different parts of a system or structure. Therefore, the earthquake input for the response analysis should be a ground motion field. Furthermore, the damage of some units or parts does not mean the damage of the functions of the system. The damage of the functions of the system is related to spatial distribution of damaged units in the system. Therefore, the key problem is the determination of earthquake ground field. The earthquake ground motion field. is to describe the properties of the earthquake ground motion at points and the spatial correlation of the motions at different points nearby. The observational data of the earthquake ground motions show that earthquake ground motion fields are random fields that are neither completely correlated nor completely independent.

A method is introduced here for the reliability analysis of water delivery systems under a scenario earthquake. Several problems are considered in the method, which include simulation of ground motion field due to a scenario earthquake, response analysis of pipelines and performance analysis of a system.

2 SIMULATION OF EARTHQUAKE GROUND MOTION FIELD

The researches of the spatial correlation of earthquake ground motions began in the late 70's (Matsushima, 1975; Feng, et al., 1981). Most of spatial correlation models are expressed in the cross power spectral density function or the coherency function between the ground motions at two points (Hao, 1989; Abrahamson, 1992). The distance between points is used as the key parameter describing the spatial correlation in the models. Recently, author and cooperators have done some relative researches (Li, et al., 1997). A model of ground motion field is proposed for describing the spatial variation of ground motions due to a scenario earthquake (Li, et al., 1998). The ground motion field is directly expressed by spatially correlative Fourier amplitude and phase spectra. In the suggested model, the earthquake magnitude (M), the source distance (R) and the distance between spatial points (d) are introduced to reflect the effects of the source, the wave travelling paths, the relative orientation between the source and engineering site, and random factors. The model is described in detail as follows:

1 The ground motion field $u(x,y,z,t)$ is described in a discrete form, that is, $u(x,y,z,t)$ is constituted by the ground motions $u_P(t)$ at the spatially discrete points $P(x_P,y_P,z_P)$.

2 $u_P(t)$ is described in Fourier amplitude spectrum $U_P(f)$ and Fourier phase spectrum $\varphi_P(f)$.

$$u_P(t) = \frac{1}{\sqrt{2\pi}} \int_{-\infty}^{\infty} U_P(f) e^{-i(2\pi ft + \varphi_P(f))} 2\pi df \qquad (1)$$

3 $\varphi_P(f)$ are decomposed into two parts, a large-scale part for a large-scale spatial variation and a small-scale part for a small-scale spatial variation as following relations

$$U_P(f) = U_l(M, R_P, f) + U_s(M, R_O, r_P, f) + \\ \varepsilon_{U_l}(M, R_P, f) + \varepsilon_{U_s}(M, R_O, d_P, f) \qquad (2)$$

$$\phi_p(f) = \phi_p(f) + \int_{f_p}^{f} \Delta\phi_p(f) df \qquad (3)$$

$$\Delta\phi_P(f) = \Delta\phi_l(M, R_P, f) + \Delta\phi_s(M, R_O, r_P, f) + \\ \varepsilon_{\Delta\varepsilon_l}(M, R_P, f) + \varepsilon_{\Delta\phi_s}(M, R_O, d_P, f) \qquad (4)$$

In above equations, f is the engineering frequency; the subscript "O" indicates the reference point in the site; R_O and R_P are the source distances at a reference point O and an arbitrary point P; r_P is the differential source distance at point P and point O; d_P is the distance between point P and point O; U_l and $\Delta\phi_l$ (where 'l' means 'large-scale') are the relations of Fourier amplitude spectrum and Fourier phase difference spectrum of ground motion field in the case of large-scale spatial variation; U_s and $\Delta\phi_s$ (where 's' means 'small-scale') are the relations in the case of small-scale spatial variation; ε_{U_l}, $\varepsilon_{\Delta\phi_l}$, ε_{U_s} and $\varepsilon_{\Delta\phi_s}$ represent the effects of the random factors; $\phi_P(f_0)$ is the phase value corresponding to the reference frequency f_0 for the ground motion at point P.

In the model, the large-scale part and the small-scale part respectively represent the spatial variation of the Fourier spectra in the cases of regarding the source and point O as the reference point. In actual fact, the small-scale part is supplementary to the large-scale part of Fourier spectrum, which increases the accuracy of simulating the ground motion field, especially, simulating the difference between the ground motions in an engineering site. In the method, $\Delta\phi_s$ is a key quantity to describe the spatial correlation of the ground motions.

All of the function relations relevant to the model are expressed in empirical relations. These empirical relations can be obtained based on statistical analyses of earthquake records from seismic stations and arrays (Li, et al., 1998).

3 RESPONSE ANALYSIS OF PIPELINES

The buried pipeline network is a structural system buried in foundation. The relation is weak between the earthquake responses of its different parts. The response analyses of the network can be regarded as the response analyses of each pipeline of the network. The earthquake input on the response analyses

is the motion of the foundation of the pipeline. For a scenario earthquake, the motion of the foundation of each pipeline can be simulated based on the ground motion field model introduced above, and they are expressed in spatially discrete form.

Earthquake damages and research results show that the axial motion of pipelines is the main reason to cause the damages of pipelines. Therefore, the axial responses of pipelines are studied here, and the pipelines are simply considered as straight line. For each segment of a pipeline, the dynamic equation is

$$m\frac{\partial^2 u_P(x,t)}{\partial t^2} - EA\frac{\partial^2 u_P(x,t)}{\partial x^2} + K_s(u_P(x,t) - u_S(x,t)) = 0 \quad (5)$$

where m is the mass of per meter pipe; K_S is the stiffness coefficient of foundation in the axial direction of pipeline; E is elasticity modulus of pipe material; A is the sectional area of pipe; $u_P(x,t)$ and $u_S(x,t)$ are the displacement motions of pipe and foundation. Following dynamic equation is obtained by differentiation of equation (5) with x

$$m\frac{\partial^2 \varepsilon_P(x,t)}{\partial t^2} - EA\frac{\partial^2 \varepsilon_P(x,t)}{\partial x^2} + K_s(\varepsilon_P(x,t) - \varepsilon_S(x,t)) = 0 \quad (6)$$

where $\varepsilon_P(x,t)$ and $\varepsilon_S(x,t)$ are the strain responses of pipe and foundation.

For a short segment of a straight pipeline, the displacement motions of pipe and foundation could be approximately considered as a hyperbola function of x. That is

$$\varepsilon_P(x,t) = \varepsilon_P(0,t) + (\varepsilon_P(L_e,t) - \varepsilon_P(0,t))\frac{x}{L_e} \qquad (7)$$

$$\varepsilon_S(x,t) = \frac{1}{L_e}(4u_S(0.5L_e,t) - 3u_S(0,t) - u_S(L_e,t)) \\ + \frac{4x}{L_e^2}(u_S(0,t) + u_S(L_e,t) - 2u_S(0.5L_e,t)) \qquad (8)$$

In equations (7), (8), L_e is the length of the short segment of pipeline; $u_S(0,t)$, $u_S(0.5L_e,t)$ and $u_S(L_e,t)$ are the spatially discrete values of ground motion field. $u_S(0,t)$, $u_S(0.5L_e,t)$ and $u_S(L_e,t)$ for a scenario earthquake can be obtained by the ground motion model introduced in this paper. Based on above consideration, equation (6) is simplified as

$$m\frac{\partial^2 \varepsilon_P(x,t)}{\partial t^2} + K_s\varepsilon_P(x,t) = K_P\varepsilon_S(x,t) \qquad (9)$$

If considering the ealstoplastic connection and relative slip between pipe and foundation, the equation (9) becomes

$$m\frac{\partial^2 \varepsilon_P(x,t)}{\partial t^2} + \tau_R(x,t) = 0 \qquad (10)$$

$$\tau_R(x,t) = \begin{cases} K_S(\varepsilon_P(x,t) - \varepsilon_S(x,t)) & , \\ \tau_{RC} Sign(\varepsilon_P(x,t) - \varepsilon_S(x,t)) & , \\ |\varepsilon_P(x,t) - \varepsilon_S(x,t)| < \tau_{RC}/K_S \\ \text{else} \end{cases} \quad (11)$$

where τ_{RC} is the critical shear stress of relative slop between pipe and foundation.

The center difference method is used to solve equation (10), and $\varepsilon_P(x,t)$ for $x=0$, $0.5L_e$ and L_e, that is $\varepsilon_P(0,t)$, $\varepsilon_P(0.5L_e,t)$ and $\varepsilon_P(L_e,t)$, are calculated by

$$\varepsilon_P(x,t+\Delta t) = -\frac{\Delta t^2}{m}\tau_R(x,t) + 2\varepsilon_P(x,t) - \varepsilon_P(x,t-\Delta t) \quad (12)$$

$\varepsilon_P(x,t)$ for any x can be calculated from equation (7).

4 CRITERIA OF PIPE DAMAGE AND PROBABILITY ANALYSIS OF PIPE DAMAGE

In the processes of earthquake responses of pipelines, the pipe segment will be damaged if the response strain of a pipe segment exceeds a critical value. The damage degrees are affected by the response strains and the parameters relevant to the pipe, such as pipe material, pipe joint. Figure 1 shows the relation between pipe damage degree and response strain adopted in this paper. In figure 1, ε_{NC} is the critical strain of pipe material or pipe joint corresponding to the $N\%$ damage degree, ε_Y and ε_U are the yielding stain and ultimate strain of pipe material or pipe joint.

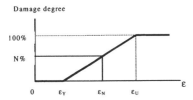

Figure 1. Relation between pipe damage and response strain

Experimental results show that the critical strains of material are not fixed values but the uncertainty values. Therefore, in the case of given tension strain of pipe (ε_{TP}), compression strain of pipe (ε_{CP}) and tension strain of pipe joint (ε_{TJ}), the probability of pipe damage with damage degree less than $N\%$ is calculated by

$$P_{N\%} = \frac{1}{\sqrt{(2\pi)^3}\sigma_{TPN}\sigma_{TJN}\sigma_{CPN}} \int_{\varepsilon_{TP}}^{\infty} e^{\frac{(x-\varepsilon_{TPN})^2}{2\sigma_{TPN}^2}} dx$$
$$\int_{\varepsilon_{TJ}}^{\infty} e^{\frac{(x-\varepsilon_{TJN})^2}{2\sigma_{TJN}^2}} dx \int_{\varepsilon_{CP}}^{\infty} e^{\frac{(x-\varepsilon_{CPN})^2}{2\sigma_{CPN}^2}} dx \quad (13)$$

5 PREFORMANCE-BASED RELIABILITY ANALYSES

The performance-based reliability analyses of water delivery systems include the connectivity analysis and water supply analysis. Because the damage degrees of pipes are expressed with probability, a Monte Carlo simulation technique is used to simulate the performance of systems.

5.1 Modeling of a damaged pipeline

In the reliability analyses of water supply systems, the damaged pipeline segment is modeled by adding an orifice node with an opening area A_0 at the middle of the pipeline (Hwang, et al., 1997). The opening area A_0 is determined based on the damage degree of pipe. The discharge through the orifice node can be determined using the following equation (Gupta, 1989)

$$Q_o = \begin{cases} A_o(2g(H_o - G_o))^{0.5} & , \quad H_o > G_o \\ 0 & , \quad H_o \le G_o \end{cases} \quad (14)$$

where g is the gravity acceleration; H_0 is the water pressure elevation; G_o is the elevation of the orifice node.

5.2 Dynamic equilibrium equation of water flow

For a water-demanded node i, the dynamic equilibrium equation is

$$\sum_j q_{ij} + Q_{ri} + Q_{li} = 0 \quad (15)$$

For an orifice node i, the dynamic equilibrium equation is

$$\sum_j q_{ij} + Q_{oi} = 0 \quad (16)$$

where j is the node connecting with node i; q_{ij} is the water flow in pipeline ij; Q_{ri} is the supply water flow at node i; Q_{li} is the leaky water flow at node i; Q_{oi} is the leaky water flow at orifice node i. Following equations are also needed

$$q_{ij} = R_{ij}(H_i - H_j)^{\alpha_1} \quad (17)$$

$$Q_{ri} = \begin{cases} Q_{di} & , \quad H_i \ge H_{di} \\ Q_{di}\left(\frac{H_i - G_i}{H_{di} - G_i}\right)^{0.5} & , \quad G_i < H_i < H_{di} \\ 0 & , \quad H_i \le G_i \end{cases} \quad (18)$$

$$Q_{li} = \begin{cases} C_i(H_i - G_i)^{\alpha_2} & , \quad H_i > G_i \\ 0 & , \quad H_i \le G_i \end{cases} \quad (19)$$

235

where H_i and H_j are the water pressure elevations at nodes i and j; Q_{di} is design supply water flow at node i; H_{di} is design minimum water pressure elevation at node i, G_i is the elevation at node i; R_{ij}, α_1, C_i and α_2 are the parameters relevant to pipelines.

5.3 Reliability of water pressure and flow

The dynamic equilibrium equations of water flow are solved by numerical method. For each sample of simulated network with damaged pipelines by the Monte Carlo technique, the elevation of the water pressure H_{ik} and the water flow Q_{rik} can be calculated for node i and kth sample of simulated networks. Here, $k = 1,2,...,N$; N is the number of simulation of the damaged network by the Monte Carlo technique.

The probability (P_{HS} or P_{QS}) of satisfying a given condition that the water pressure rate or water flow rate is not less than D_S can be calculated by following formulas.

$$P_{HS} = \sum_{k=1}^{N} V_{HSK} \bigg/ N \tag{20}$$

$$P_{QS} = \sum_{k=1}^{N} V_{QSk} \bigg/ N \tag{21}$$

$$V_{HSk} \quad or \quad V_{QSK} = \begin{cases} 1 & , \quad D_{HC} > D_S \\ 0 & , \quad D_{HC} \leq D_S \end{cases} \tag{22}$$

$$D_{HC} = \begin{cases} 1 & , \quad H_{ik} \geq H_{di} \\ \dfrac{H_{ik} - G_i}{H_{di} - G_i} & , \quad H_{di} > H_{ik} > G_i \\ 0 & , \quad H_{ik} \leq G_i \end{cases} \tag{23}$$

$$D_{QC} = Q_{rik} / Q_{di} \tag{24}$$

6 CASE STUDY

The performance-based reliability analysis method is systematically introduced above. As an application, the method is used to analyze the earthquake reliability of an urban water delivery network. The urban water delivery network is shown in figure 2. There are 4 pumping stations, 140 water-demanded nodes and 199 pipelines in the water delivery network. There are 4 soil types in the area of the foundation of the network. The pipe material is cast iron, and the material of the pipe joint is backfilling concrete. The stiffness coefficient of foundation in the axial direction of pipe K_S is given by

$$K_S = \pi D K_0 \tag{25}$$

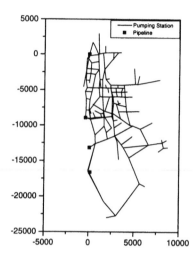

Figure 2. Water delivery pipeline network for a case study

where D is the diameter of a pipe and K_0 are separately 6000000, 9050000, 3310000 and 1110000 N/m³ for the 4 soil types. The critical stresses of relative slop between pipe and foundation are separately 29400, 44000, 16000 and 5390 N/m² for the 4 soil types. The density of pipe material is 7170 kg/m³. The average value and variance are 0.0003 and 0.00006 for the tension yielding strain of pipe material, and 0.002 and 0.0004 for the tension ultimate strain of pipe material. The average value and variance are 0.000033 and 0.0000067 for the tension yielding strain of pipe joint material, and 0.00033 and 0.000067 for the tension ultimate strain of pipe joint material.

In the earthquake reliability analyses, it is assumed that a scenario earthquake with magnitude $M_S = 7.0$ and source depth $H = 10$ km occurs at the west of the network with epicentral distance $R=50km$. Some of the calculating results are shown in Figs. 3–5.

7 CONCLUSION

In this paper, a performance-based reliability analysis method is systematically introduced for water delivery systems under a scenario earthquake. The earthquake input is assumed as a ground motion field under a scenario earthquake, and the interaction and relative slop between foundation and pipe are considered in the method. The damage degrees of pipelines and connectivity and serviceability are expressed as probabilities because the uncertainties of mechanical parameters of pipeline material are considered. A case study is given to show the application of the introduced method.

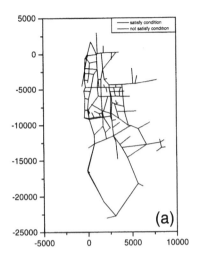

(a) Probability of no damage is bigger than 0.75

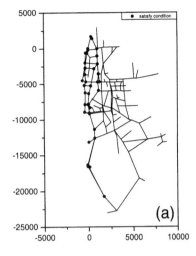

(a) Probability of no damage for at least one connecting path is bigger than 0.75

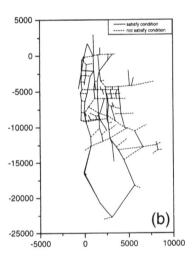

(b) Probability of no damage is bigger than 0.50

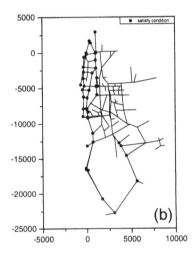

(b) Probability of no damage for at least one connecting path is bigger than 0.50

Figure 3. Damage state of pipelines in the network

Figure 4. Connectivity state of nodes in the network

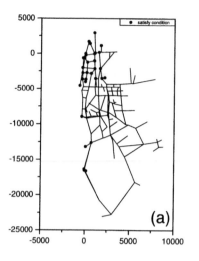
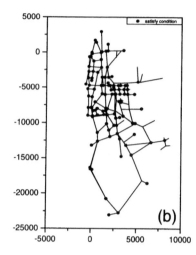

(a) Probability of over 75% water flow rate is bigger than 0.75

(b) Probability of over 50% water flow rate is bigger than 0.75

Figure 5. Water flow state of nodes in the network

REFERENCES

Abrahamson, N.A., 1992. Generation of Spatially Incoherent Strong Motion Time History. *Tenth World Conference on Earthquake Engineering*, 845–850.

Feng, Q.M. and Hu, Y.X., 1981. Spatial correlation model of ground motion. *Earthquake Engineering and Engineering Vibration*, 1(2):1–8 (In Chinese).

Gutpa, R.S., 1989. *Hydrology Hydraulic System*. Prentice Hall, Englewood Cliffs, NJ.

Hao, H. and Oliveira, C.S. and Penzien, J.,1989. Multiples-Station Ground Motion Processing and Simulation Based on Smart-1 Array Data. *Nuclear Engineering and Design*, 111:293–310.

Hwang, H., Lin, H., 1997. GIS-Based Evaluation of Seismic Performance of Water Delivery Systems, Technical Report, Center for Earthquake Research and Information, The University of Memphis, Memphis, TN.

Li Xiaojun, Zhao Fengxin and Hu Yuxian, 1997. Study on simulation of spatially correlative earthquake ground motion field. *ACTA Seismologica Sinica*, 10(2):267–271.

Li Xiaojun, Hu Yuxian and Zhao Fengxin, 1998. Earthquake Ground Motion Field for Seismic Safety Analysis of Lifeline. *Proceedings of Third China-Japan-Us Trilateral Symposium on Lifeline Earthquake Engineering*, Kunming, China, 105–113.

Matsushima, Y., 1975. Spectra of Spatially Variant Ground Motion and Associate Transfer Function of Soil Foundation System. *Proceedings of 4th Japan Earthquake Engineering Symposium*.

Earthquake Engineering Frontiers in the New Millennium, Spencer & Hu (eds),
© *2001 Swets & Zeitlinger, ISBN 90 2651 852 8*

Seismic reliability analysis of electric power network system

Jie Li
Department of Building Engineering, Tongji University, Shanghai, 200092, China

Jun He
Department of Building Engineering, Tongji University, Shanghai, 200092, China

ABSTRACT: This paper presents an analytic solution to solve problems of fake mini-paths and space complexity in evaluation of node-weight network reliability, therefore, it is especially suitable to analyze seismic reliability of large scale electric power network system. A case study is presented and the analysis results show the validity of the suggested method.

1 INTRODUCTION

Electric power system is an important lifeline engineering system that has much to do with the national economy and the people's livelihood. With social progress, electric power system is having characteristics of high pressure, remoteness and high capacity. Therefore, reliability and safety of electric power system become more and more important.

Earthquake resistant analysis of electric power system stems from post-seismic inspection made by Ang A.H.S in early 1970's. However, electric power system didn't be analyzed as an integrated system until 1990's. Up to now, methods most in use includes PNET method, Mini-paths method and Monte Carlo simulation method. Because NP-hard problem exists in evaluation of system reliability, most of above analytic methods become invalid for reliability analysis of large electric power system. On the other hand, although Monte Carlo simulation method has capability to analyze seismic reliability of large electric power system, it can't estimate error bound of analysis results and its computation time is nonlinear increase with system dimension.

An analytic method is presented in this paper to evaluate seismic reliability of larger scale electric power system. Using disjoint-decomposition technique, the method can directly get disjoint mini-paths and mini-cuts of system. So the problems of space complexity and fake mini-paths introduced by traditional analytic algorithm can be solved.

2 RELIABILITY ALGORITHM OF LARGE NODE-WEIGHT NETWORKS

If analysis aim is to assess importance of power plans and electric substations in electric power network, damage of transmission lines can be neglected. Therefore, the electric power system may be regarded as node-weight network systems which nodes weight denote seismic reliabilities of power plans and electric substations. In reliability computation of larger scale node-weight networks, there are two problems including space complexity and time complexity. Following analytic solution can effectively solve the problem of space complexity, therefore, analysis of the problem becomes much more easy.

2.1 Assumptions

1 Nodes failure are s-independent;
2 System and its nodes have two states: operative or failure.

2.2 Fundamentals

Definite an arbitrary smallest minimal path from the source to the terminal of a network system (shown by graph G) as $L_0 = \{s_1 s_2 \cdots s_{|s_0|}\}$, here $s_i, i = 1,2,\cdots,|S_0|$ are nodes or edges of the system, $|S_0|$ is the number of components making up S_0. Let system structure function $\Phi(G)$, then according to the absorption law, there is

$$\Phi(G) = L_0 + \bar{L}_0 \Phi(G) \tag{1}$$

According to De.Morgen law, there exists

$$\begin{aligned}\Phi(G) = L_0 &+ \{\bar{s}_1\}\Phi(G_1) + \{s_1 \bar{s}_2\}\Phi(G_2) + \cdots \\ &+ \{s_1 s_2 \cdots \bar{s}_i\}\Phi(G_i) + \cdots \\ &+ \{s_1 s_2 \cdots s_i \cdots s_{|s_0|}\}\Phi(G_{|s_0|})\end{aligned} \tag{2}$$

where G_i is a sub-graph received through deleting component $s_i \in L_0$ from original graph G.

If sub-graph $G_i, i = 1,2,\cdots,|S_1|$ in Eq. (3) still has minimal path $S_i, i = 1,2,\cdots,|S_1|$ from the source to the terminal, and let c_i be the responding coefficient term in front of $\Phi(G_i)$ in Eq. (3) and name it the decomposition factor, then Eq. (3) can be transformed as

$$\Phi(G) = L_0 + \sum_{i=1}^{|S_i|} c_i S_i + \sum_{i=1}^{|S_i|} c_i \overline{S}_i \Phi(G_i) \tag{3}$$

where $\overline{S}_i$ is decomposed according to Eq. (2).

Then there are $|S_0|-|S_1|$ sub-graphs G_j, $j=1,2,\cdots,|S_0|-|S_1|$ in Eq. (3) that do not have minimal paths from the source to the terminal, let F_j be the responding coefficient term in front of $\Phi(G_i)$ in Eq. (3), according to complementation

$$\begin{aligned} \Phi'(G) &= 1 - \Phi(G) \\ &= \sum_{j=1}^{|S_0|-|S_1|} F_j + Q \end{aligned} \tag{4}$$

where $\Phi'(G)$ is failure function of the system; Q is the remaining term.

According to above principle, Eq. (3) can be continually recursively decomposed until all generated sub-graphs do not exist any minimal paths form the source to the terminal, then there exist

$$\begin{aligned} \Phi(G) &= L_0 + \sum_{i=1}^{|S_1|} c_i S_i + \sum_{i=|S_1|+1}^{|S_1|+|S_2|} c_i S_i + \cdots + \sum_{i=|S_1|+\cdots+|S_{N-1}|+1}^{|S_1|+\cdots+|S_N|} c_i S_i \\ &= \sum_{i=0}^{N} L_i \end{aligned} \tag{5}$$

where N is total number of disjoin minimal paths, $L_i = c_i S_i$ is the i-th disjoin minimal path of original system G.

At the meantime, the integrated form of Eq. (4) can be obtained as follows:

$$\Phi'(G) = \sum_{j=1}^{M} F_j \tag{6}$$

where F_j is the disjoin minimal cut of the system G, M is total number of disjoin minimal cuts.

According to Eq. (6), reliability of system is

$$R(G) = \sum_{i=1}^{N} p_r\{L_i\} \tag{7}$$

$$p_r\{L_i\} = \prod_{i=1}^{N_i}(1-p_i) \cdot \prod_{i=N_i+1}^{N_i+K_i} p_i \tag{8}$$

where p_i is the reliability of i-th component in the network system G, N_i is the number of failure components in L_i, K_i is the number of operative components in L_i.

Analogous to Eqs. (8) and (9), the failure reliability of system is as follows:

$$\overline{R}(G) = \sum_{j=1}^{M} p_r\{F_j\} \tag{9}$$

$$p_r\{F_j\} = \prod_{j=1}^{M_j}(1-p_j) \cdot \prod_{j=N_j+1}^{M_j+K_j} p_j \tag{10}$$

where M_j is the number of failure components in F_j, K_j is the number of operative components in F_j.

2.3 *Multiple-source system reliability*

In the seismic reliability analysis of power electric system, we usually evaluate system reliability from K sources to a terminal, i.e. evaluate K-terminals system reliability. On this condition, a subjunctive node directing into K sources can be set. The subjunctive node and those directed edges only have operative states (i.e. their reliability all equal to 1.0). During computation process they only take the effect of connection and do not take place in recursive computation. Having been treated by this way, the K-terminals reliability problem would be transformed into two-terminal reliability problem from the subjective to a certain terminal.

Take Fig. 1 node weight network system as an example. There are two sources S_1, S_2, and one terminal T in the system. S is the subjective node directing into S_1 and S_2.

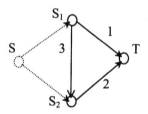

Figure 1. Node weight network system

When the subjunctive node S has not been set, for the source S_1 the set of minimal paths includes $\{1\}$ and $\{32\}$, and for the source S_2 the set of minimal paths includes $\{2\}$, then

$$\Phi(G) = \bigcup_{i=1}^{3} L_i = 1 \bigcup 32 \bigcup 2 = 1 \bigcup 2 = 1 + \overline{1}2 \tag{11}$$

However, when the subjunctive node S is set, utilize the above-suggested analytic algorithm, there is

$$\Phi(G) = 1 + \overline{1}\Phi(G_{-1}) = 1 + \overline{1}(2 + \overline{2}\Phi(G_{-1-2})) = 1 + \overline{1}2 \tag{12}$$

Comparing Eq. (11) with Eq. (12), it is known that the subjunctive node not only ensure the accuracy of calculation result but also simple the computation process of multiple-sources problem.

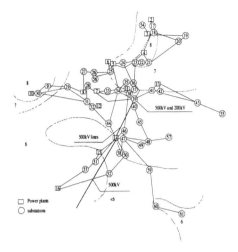

Figure 2. Network graph of a district power electric system

Table 1. Given information and evaluation results

Series node of substations	Seismic reliability of substation	System seismic reliability	Numbers of disjoin minimal paths
17	0.9994	0.999279	3
18	0.6981	0.698100	3
19	0.9994	0.697681	3
20	0.9994	0.699784	3
21	0.8415	0.740255	132
22	0.8415	0.774093	133
23	0.8415	0.841497	45
24	0.9994	0.999398	50
25	0.9994	0.999400	3
26	0.9999	0.999900	5
27	0.9999	0.999900	3
28	0.9999	0.999900	11
29	0.9999	0.999900	13
30	0.7745	0.774500	14
31	0.9999	0.999900	38
32	0.9999	0.999900	28
33	0.9999	0.999900	340
34	0.9999	0.999800	340
35	0.9999	0.990400	126
36	0.9904	0.990033	343
37	0.9994	0.999398	577
38	0.9998	0.999800	346
39	0.9998	0.999800	145
40	0.9999	0.999900	147
41	0.9994	0.999400	74
42	0.9994	0.999400	74
43	0.9999	0.999900	74
44	0.9999	0.999900	15
45	0.9999	0.999900	149
46	0.9999	0.999900	145
47	0.9998	0.999800	145
48	0.9999	0.999900	147
49	0.9999	0.999900	147
50	0.9999	0.999899	2
51	0.9999	0.999900	2
52	0.9928	0.992800	2
53	0.9999	0.999900	2

3 CASE STUDY

Fig. 2 is a network graph of a district provincial power system in china. It is a system of node failure under seismic excitations with 60 nodes and 160 edges. There are 15 power plants from node 2 to node 16. Seismic reliabilities under given earthquake excitations of power plants and substations have been evaluated and shown in table 1. Using above analytic algorithm, exact values of system seismic reliabilities from 15 power plants to each substation can be computed within 5 seconds. the results are shown in Table 1.

4 CONCLUSIONS

The analytic method presented in the paper is, in fact, a recursive decomposition algorithm for system seismic reliability estimation of electric power system with the aim at solving space complexity in analysis of system reliability. Because categoricalness of structure function is not destroyed during computation and avoid getting fake elementary paths during calculation of node weight system seismic reliability, accurate value of system seismic reliability can be evaluated faster than ever.

REFERENCES

Duke, C.M. & Moran, D.F (1975). 'Guidelines for evolution of lifeline earthquake engineering'. Proc. of the U.S. National Conference of Earthquake Engineering, Michigan, pp. 367–376.

Ang, A.H.S., Pires, J., Villarerde, R., Schireinger, R. (1992). 'Seismic probabilistic assessment of electric power transmission systems'. Proc. of 10th Word Conference of Earthquake Engineering. Madrid, Spain.

Li Guiqing. Theory of Structure Dynamic Reliability and Its Application (in china). Beijing, Earthquake Publishing House, 1993. pp. 335–340.

Li Jie, Li Guoqiang. Introduction to Earthquake Engineering. Beijing (in china), Earthquake Publishing House, 1992. pp. 22–31.

Shinozuka, M., Tanaka, S. (1996). 'Effects of lifelines interaction under seismic conditions'. Proc. of 11th Word Conference of Earthquake Engineering. Mexico.

Vanzi, I. (1996). 'Seismic reliability of electric power networks: methodology and application'. Structural Safety. pp. 311–327.

Giannini, R. & Vanzi, I. (2000). 'Seismic reliability of electric power networks and interaction with other damage indications'. Proc. of 11th Word Conference of Earthquake Engineering.

Li Jie & He Jun. 'Reliability analysis of power engineering system with identical seismic risk'. Proc. of International Conference on Safety, Risk and Reliability. Malta.

He Jun, Li Jie. Algorithm complexity of large lifeline system seismic reliability and its simplification (in china). World Earthquake Engineering. 2000, Dec.

Earthquake Engineering Frontiers in the New Millennium, Spencer & Hu (eds),
© *2001 Swets & Zeitlinger, ISBN 90 2651 852 8*

Seismic behavior of underground structures and site response

Y.M.A. Hashash
University of Illinois at Urbana-Champaign, Urbana, IL, USA

ABSTRACT: The past 20 years have seen major progress in understanding of seismic response of underground structures. Seismic design loads for underground structures are characterized in terms of deformations and strains imposed by the surrounding ground. In contrast, above-ground structures are designed for inertial forces imposed by ground accelerations. Design methods have been developed to estimate seismic loads on underground structures. These methods include pseudo-static analysis whereby free-field motion is imposed on the underground structure as well as dynamic soil-structure interaction analysis. Free-field ground deformations and velocities due to a seismic event are estimated using site specific response analysis that accounts for local geology. Historically, underground facilities have experienced a lower rate of damage than above-ground structures. However, recent failures of underground structures during recent earthquakes were reminders of the need to include seismic considerations in design details of underground structures.

1 INTRODUCTION

Underground facilities are an integral part of the infrastructure in an increasingly urban world. Underground space is used for subways, highways, material storage and water and sewage transport. Underground facilities built in areas subject to earthquake activity must withstand both seismic and static loading. Historically underground facilities have experienced a lower rate of damage than have above-ground structures. Nevertheless, some underground structures have experienced significant damage in recent large earthquakes.

Underground structures have features that make their seismic behavior distinct from most aboveground structures, most notably (1) their complete enclosure in soil or rock and (2) their significant length (i.e., tunnels). This paper focuses on relatively large underground facilities commonly used in urban areas. This paper does not discuss pipelines or sewer lines that are conventionally labeled lifelines. Large underground structures can be grouped into three broad categories, each having distinct design features and construction methods: (1) bored or mined tunnels, (2) cut-and-cover tunnels, and (3) immersed tube tunnels (Power, et. al, 1996). These structures are commonly used for metro structures, highway tunnels, and large water and sewage transportation ducts.

2 FRAMEWORK FOR CHARACTERIZING SEISMIC BEHAVIOR OF UNDERGROUND STRUCTURES

Earthquake effects on underground structures can be grouped into two categories: (1) ground shaking and (2) ground failure such as liquefaction, fault displacement, and slope instability. Ground shaking, which is the primary focus of this paper, refers to deformation of the ground produced by seismic waves propagating through the earth's crust. The major factors influencing shaking damage include: (1) the shape, dimensions, and depth of the structure; (2) the properties of the surrounding soil or rock; (3) the properties of the structure; and (4) the severity of the ground shaking (Dowding and Rozen, 1978, St. John and Zahrah, 1987).

Seismic design of underground structures is unique in several ways. For most underground structures, the inertia of the surrounding soil is large relative to the inertia of the structure. Measurements made by Okamato, et al. (1973) of the seismic response of an immersed tube tunnel during several earthquakes show that the response of a tunnel is dominated by the surrounding ground response and not the inertial properties of the tunnel structure itself. The focus of underground seismic design, therefore, is on the free-field deformation of the ground and its interaction with the structure. This concentration on displacement is in stark contrast to the design of above-ground structures, which focuses on inertial effects of the structure itself.

Three types of deformations (Owen and Scholl, 1981) express the response of underground structures to seismic motions: (1) axial extension and compression, (2) longitudinal bending, and (3) ovaling/racking (Figure 1.). Axial deformations in tunnels are generated by the components of seismic waves that produce motions parallel to the axis of the tunnel and cause alternating compression and tension. Bending deformations are caused by the components of seismic waves producing particle motions perpendicular to the longitudinal axis. The design of a tunnel lining to accommodate axial and bending deformations generally concentrates on the direction along the tunnel axis (Wang, 1993).

Ovaling or racking deformation in a tunnel structure develops when shear waves propagate normal or nearly normal to the tunnel axis, resulting in a distortion of the cross-sectional shape of the tunnel lining. Design considerations for this type of deformation are in the transverse direction. The general behavior of the lining may be simulated as a buried structure subject to ground deformations under a two-dimensional plane-strain condition.

Diagonally propagating waves subject different parts of the structure to out-of-phase displacements (Figure 1.), resulting in a longitudinal compression-rarefaction wave traveling along the structure. In general, larger displacement amplitudes are associated with longer wavelengths, while maximum curvatures are produced by shorter wavelengths with relatively small displacement amplitudes (Kuesel, 1969).

3 SITE RESPONSE AND FREE FIELD DEFORMATIONS

Deformations and velocities in the surrounding ground dominate the estimate of seismic demand on an underground structure. A site-specific response analysis is needed to estimate expected free field ground deformations and velocities. It has long been recognized by researchers that local soils modify the ground motion.

One-dimensional site response analysis is used to solve the problem of vertical propagation of horizontal shear waves (SH waves) through a horizontally layered soil deposit. Seed, Idriss and co-workers introduced the equivalent linear approximation method to capture non-linear cyclic response of soil. For a given ground motion time series (T.S., also referred to as time history) and an initial estimate of modulus and damping values, an effective shear strain (equal to about 65% of peak strain) is computed for a given soil layer. Modulus degradation and damping curves are then used to obtain revised values of shear

modulus and damping. Solution of wave propagation equations is performed in the frequency domain. An iterative scheme is required to arrive at a converged solution. This approach has provided good results compared with field measurements and is widely used in engineering practice (e.g. SHAKE, Schnabel et. al., 1972, FLUSH, Lysmer et. al., 1975). More recently Sugito (1995) and Dominic et al. (2000) extended the equivalent linear approach to include frequency and pressure dependence of soil properties.

The equivalent linear approach is computationally easy to use and implement. However, it does not capture the full range of cyclic behavior of soil, including modulus degradation due to number of loading cycles, permanent (residual) straining of soil and excess pore pressure generation. Non-linear analysis is used to capture these important aspects of soil behavior. In this approach equations of motion and equilibrium are solved in discrete time increments in time domain. A constitutive model is used to represent the cyclic behavior of soil.

The earliest constitutive relations use a simple model relating shear stress to shear strain, whereby the backbone curve is represented by a hyperbolic function. Lee and Finn (1978) developed one-dimensional seismic response analysis program using the hyperbolic model. Matasovic and Vucetic (1995) further extended the model with modification of the hyperbolic equation. Plasticity models have also been used to represent cyclic soil behavior. Borja et al. (1999) used a bounding surface plasticity model to represent cyclic soil response at Lotung Site in Taiwan.

Hashash and Park (2000) introduce an extension of the model by Matasovic to account for the influence of confining pressure on soil modulus and damping properties. Hashash and Park (2001) show that confining pressure has an important impact in computed ground motion in a site response analysis. They show that significant portions of high frequency components of ground motion are propagated through deep soil deposits and that Propagation of seismic waves through very deep deposits result in the development of long period ground motion. Spectral amplitudes of propagated ground motions are higher than what would be obtained using conventional wave propagation analyses.

One dimensional site response analysis provides data useful in the analysis of racking deformations in an underground structures and is usually sufficient for analysis of short structures such as subway station. Three dimensional wave propagation analysis is required to develop ground deformations along the length of a long tunnel to properly account for variability in ground conditions and it influence on ground motion incoherency, phase shift and arrival delay times. These type of analyses are not commonly performed due to their relative complexity.

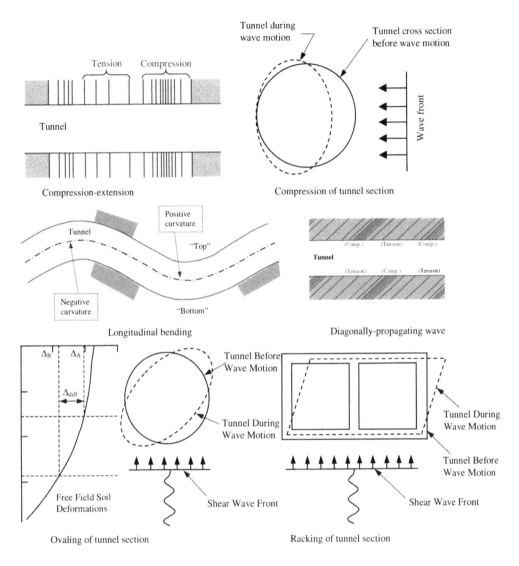

Figure 1. Deformation modes of tunnels due to seismic shaking

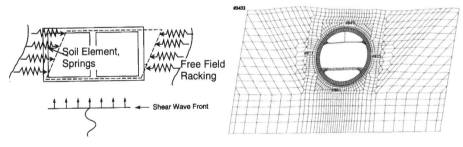

a) Frame analysis with equivalent soil spring

b) Racking analysis for Alameda Tubes, 12 m diameter

Figure 2. Pseudo static analysis approach

4 DESIGN AND ANALYSIS METHODS

Extensive reviews of seismic design procedures of underground structures are presented in St. John & Zahra (1987) and Wang (1993). The study by Wang (1993) is an unpublished monograph by Parsons Brinckerhoff used by many designers. A report under preparation by Hashash et. al. (2000) presents an extensive review of seismic design approaches for underground structures.

4.1 *Free field deformation method*

Early solutions developed to compute tunnel deformations due to seismic wave propagation assume that the tunnel conforms to free field ground deformations. This may lead to an over estimate of structure deformations especially in soft soils where the tunnel maybe stiffer than the surrounding soil.

4.2 *Pseudo-static soil-structure interaction method*

A pseudo-static analysis method is commonly used to account for soil-structure interaction effects. In a simplified procedure described by St. John and Zahrah (1987) the tunnel is simplified as an elastic beam on an elastic foundation representing the surrounding soil. Closed form solutions have been developed to account for lining stiffness effect on ovaling deformations of circular tunnels. These solutions use the concept of flexibility ratio that relates elastic soil stiffness to elastic tunnel lining stiffness (Einstein and Schwartz, 1979, and Peck et. al, 1972). Wu and Penzin (1997) present closed-form solutions for circular tunnels. For rectangular sections the structural racking deformations are computed by applying free field racking deformations in a frame analysis. Soil is represented by elastic springs. Deformations can be applied as boundary displacements in a finite element or finite difference analysis as shown in Figure 2. Figure 3. illustrates the use of the pseudo static approach in the longitudinal analysis of a tunnel. In this analysis the forcing function is the free field displacement time history. Tunnel soil interaction is represented as the interaction between longitudinal beam elements and soil spring.

The pseudo static approach may be valid for weak ground motion where deformations are elastic. Ground response is highly non-linear for strong ground motion and motions with "fling effect".

4.3 *Dynamic soil-structure interaction method*

There is an increasing use of dynamic soil structure analyses for underground structues whereby the soils as well as the underground structures are represented using finite elements. Wang (1993) used the program FLUSH to analyze rectangular tunnel sections. An unpublished analytical study was performed for the stations for the future East Side extension of the LA Metro Red Line using time domain non-linear seismic soil structure interaction analysis. Figure 4 shows a typical mesh used in the analysis. The analyses show that structural racking displacements exceeded the free field racking displacements (Figure 4). That result was surprising and contrary to conventional design approaches. The analyses imply that the ground displacements were amplified due to the presence of the station.

4.3.1 *Wave scattering solutions*

A class of solutions has been developed to solve the problem of scattering and diffraction of elastic plane waves by an arbitrarily shaped opening below ground surface in an elastic half space. Manoogian (1998) analyzes deep and shallow tunnels of circular, rectangular and elliptical cross sections using this approach. Analyses show that for shallow tunnels, and for higher wave frequency components, surface displacement amplitudes are amplified by a factor of up to 3.5 within a distance of 3 tunnel radii from tunnel centerline. The study by Manoogian has important ramifications, it implies that placement of tunnels amplifies anticipated ground motion, and adversely affects adjacent structures.

5 OBSERVED SEISMIC RESPONSE OF UNDERGROUND STRUCTURES

Dowding and Rozen (1978), Owen and Scholl (1981) and Sharma and Judd (1991) present summaries of case histories of damage to underground facilities. Damage to tunnels is greatly reduced with increased overburden, but damage is greater in soils than in competent rock. More recently several large earthquakes resulted in damage to modern underground structures in major urban centers.

5.1 *Underground structures in the US*

Bay Area Rapid Transit System, San Francisco, CA: The BART system was one of the first underground structures to be designed with seismic considerations (Kuesel, 1969). On the San Francisco side, the system consists of below ground stations and tunnels in fill and soft Bay Mud deposits, and is connected to Oakland via the transbay tube. During the 1989 Loma Prieta Earthquake flexible joints connecting the transbay tube to the San Francisco and Oakland vent buildings displaced few inches (PB 1991). Peak ground accelerations (PGA) experienced did not exceed 0.3 g. The BART system is now slated for seismic retrofit design for anticipated earthquake events with PGA~0.8g and considerable near field effects such as high velocity pulses.

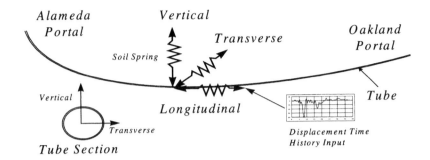

Figure 3. Simplified three-dimensional model for analysis of the global response of an immersed tube tunnel (Hashash et al. 1998)

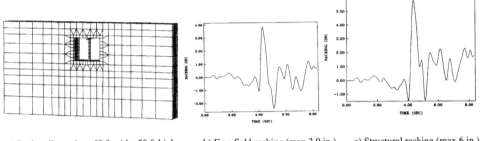

a) Station dimensions 60 ft wide, 50 ft high   b) Free field racking (max 3.9 in.)   c) Structural racking (max 6 in.)

Figure 4. Dynamic analysis of little Tokyo subway station, LA Metro

Alameda Tubes, Oakland-Alameda, CA: The Alameda tubes connect Alameda Island to Oakland through a pair of immersed tube tunnels built in 1930's-1950's. The tubes, and vent buildings experienced some structural cracking that caused an increased water leakage into the tunnels during the Loma Prieta Earthquake.

L.A. Metro, Los Angeles, CA: The Los Angeles Metro is being constructed in several phases, some of which were complete and operational during the 1994 Northridge Earthquake. PGA near the affected sections was around 0.3 g. Concrete lining of bored tunnels experienced some cracking.

5.2 Underground structures in Kobe, Japan, 1995 Hyogoken-Nambu Earthquake

The 1995 Hyogoken-Nambu earthquake caused a severe damage to Rokko, Bantaki and Naruo-Mikage Tunnels and resulted in a major collapse of Daikai subway station in Kobe, Japan (Nakamura, et. al, 1996). The collapse experienced by center columns of the station was accompanied by collapse of the ceiling slab and settlement of soil cover by more than 2.5 m. The station design in 1962 did not include specific seismic provisions.

5.3 Other tunnels

Tunnel performance in Taiwan, September 21, 1999 Chi-Chi earthquake: Most tunnels in Central Taiwan survived the severe shaking intact. Limited spalling and cracking was observed. The main damage was limited to slope failure at tunnel portals which resulted in blockage of tunnels.

Tunnel performance in Turkey during the 1999 Koceli earthquake: A 100-m section of twin shotcrete supported tunnels collapsed during the earthquake. The tunnels were under construction. The cause of the collapse is currently being investigated (Mair, 2000).

6 CURRENT CHALLENGES AND AREAS OF FUTURE RESEARCH

Underground structures in the US have experienced limited damage during Loma Prieta and Northridge earthquakes, but shaking levels were much lower than maximum anticipated events. More recent large earthquakes in Kobe, Japan and Koceli, Turkey have resulted in severe damage to some tunnel structures. This may point to limitations in our understanding of seismic behavior of

247

underground structures. There is a need to enhance our knowledge to design earthquake resistant underground structures for high levels of shaking.

As we look into the future, development plans in urban areas in the USA, China as well as other countries include the construction of major underground subway/metro systems. Seismic consideration will have to be included in the design of such systems. There are still many issues that are not well understood as they relate to the behavior of underground structures. Research in the future will have to answer some important questions in our quest to reduce earthquake hazard. The following are some proposed areas that will contribute to reduction of uncertainty and increase in our knowledge:

1. Instrumentation of tunnels and underground structures to measure their response during ground shaking: These instruments would include measurement of vertical and lateral deformations along the length of the tunnel. This will be useful to understand the effect of spatial incoherencey and directivity of the ground motion on tunnel response as well as structural deformations.

2. Study of the influence of high vertical accelerations on the generation of large compressive loads in tunnel linings and subway station columns: Large vertical forces may have been a factor in the collapse of the Daikai Subway station as well as other tunnel structures in Japan and Turkey.

3. Development of improved numerical models to simulate the dynamic soil structure interaction problem of tunnels, as well as portal and subway structures: These models will be useful in studying the effect of high velocity pulses (fling effect) generated near fault sources on underground structures.

4. Ground Motion Incoherence: Evaluation of the significance of ground motion incoherence on the development of differential movement along the length of a tunnel (Power et al., 1996). Ground motion incoherence is particularly important in soft soils and shallow tunnels where the potential for slippage between the tunnel and soil is high.

5. Effect of the underground structure on the local ground motion: Evaluation of the influence of underground structures on local amplification or attenuation of propagated ground motion. Change in ground motion due to shielding and amplification effects.

6. Tunnel Components: Research into the application of non-conventional lining, bolting, and water insulation materials that can be used for seismic joints to enhance seismic performance of a tunnel.

7. Site Response: Improved resolution of local site response will be important for estimating ground deformation around an underground structure.

REFERENCES

Borja, R.I., Chao, H-Y, Montans, F.J., & Lin, C-H. 1999. Nonlinear ground response at Lotung LSST site. *J. Geotechnical and Geoenvironmental Engineering*, Vol. 125, No. 3, 187-197.

Dowding, C. H. & Rozen, A. 1978. Damage to rock tunnels from earthquake shaking. *Journal of the Geotechnical Engineering Division*, ASCE, Vol. 104, GT2, 175-191.

Dominic, A., Kausel, E. & Whittle, A. 2000. A model for dynamic shear modulus and damping for granular soils , *Journal of Geotechnical and Geoenvironmental Engineering, ASCE*, Nov. 2000.

Einstein, H. H. & Schwartz, C. W. 1979. Simplified analysis for tunnel supports. *Journal of the Geotechnical Engineering Division*, ASCE, Vol. 105, GT4.

Hashash, Y. M. A., Tseng, W. S., & Krimotat, A. 1998. Seismic soil-structure interaction analysis for immersed tube tunnels retrofit. *Geotechnical Earthquake Engineering and Soil Mechanics III*, ASCE Geotechnical Special Publication No. 75, Vol. 2, 1380-1391.

Hashash, Y.M.A. & Park, D., 2001. Non-linear one-dimensional seismic ground motion propagation in the Mississippi Embayment, to be published, *Engineering Geology*.

Kuesel, T. R. 1969. Earthquake Design Criteria for Subways. *Journal of the Structural Division*, ASCE, ST6, 1213-1231.

Lee, M.K.W. & Finn, W.D.L. 1978. DESRA-2, Dynamic effective stress response analysis of soil deposits with energy transmitting boundary including assessment of liquefaction potential. *Soil Mecahnics Series No. 36, Department of Civil Engineering*, University of British Columbia, Vancouver, Canada.

Lee, V.W. & Trifunac, M.D. 1979. Response of tunnels to incident SH-waves, *Engineering Mechanics Division*, ASCE, Vol. 105, 643-659.

Lysmer, J., Udaka, T., Tsai, C. F., & Seed, H. B. 1975. FLUSH: A computer program for approximate 3-D analysis of soil-structure interaction problems. Report No. EERC 75-30, *Earthquake Engineering Research Center*.

Mair, R. Personal communications, 2000.

Manoogian, M.E. 1998. Surface motion above an arbitrarily shaped tunnel due to elastic SH waves, ASCE, Geotechnical Special Publication No. 75, *Geotechnical Earthquake Engineering and Soil Dynamics III*, 754-765, Seattle.

Matasovic, N. & Vucetic, M. 1995. Seismic response of soil deposits composed of fully-saturated clay and sand layers. Proc. *1st International Conference on Earthquake Geotechnical Engineering*, JGS, Vol. 1, 611-616, Tokyo, Japan.

Okamoto, S., Tamura, C., Kato, K., & Hamada, M. 1973. Behaviors of submerged tunnels during earthquakes. Proceedings of the *Fifth World Conference on Earthquake Engineering*, Rome, Vol. 1, 544-553.

Owen, G. N. & Scholl, R. E. 1981. Earthquake engineering of large underground structures. Report No. FHWA/RD-80/195, *Federal Highway Administration and National Science Foundation*.

PB, 1991. Trans-bay tube seismic joints post earthquake evaluation, Bay Area Rapid Transit District, report prepared by *Parsons Brinckerhoff Quade & Douglas*, Nov.

Peck, R. B., Hendron, A. J., & Mohraz, B. 1972. State of the art in soft ground tunneling. Proceedings of the *Rapid Excavation and Tunneling Conference*, American Institute of Mining, Metallurgical, and Petroleum Engineers, New York, 259-286.

Power, M. S., Rosidi, D., & Kaneshiro, J. 1996. Volume III Strawman: Screening, evaluation, and retrofit design of tunnels. Report Draft, *National Center for Earthquake Engineering Research*, Buffalo, New York.

Sharma, S. & Judd, W.R., 1991. Underground Opening damage from earthquakes, *Engineering Geology*, 30.

Schnabel, P. B., Lysmer, J., and Seed, B. H. 1972. SHAKE – A computer program for earthquake response analysis of horizontally layered sites. Report *No. EERC 72-12*, University of California at Berkeley, California.

St. John, C. M. & Zahrah, T. F. 1987. Aseismic design of underground structures. *Tunneling and Underground Space Technology*, Vol. 2, No. 2, 165-197.

Sugito, M., 1995. Frequency-dependent equi-linearized technique for seismic response analysis of multi-layered ground. *Doboku Gakkai Rombun-Hokokusho/Proceedings of the Japan Society of Civil Engineers*, No. 493, 3-2, 49-58.

Wang, J.-N. 1993. *Seismic Design of Tunnels: A State-of-the-Art Approach*. Monograph 7, Parsons Brinckerhoff Quade & Douglas, Inc., New York.

Wu, C. L. & Penzien, J. 1997. Stress analysis and design of tunnel linings. *Proceedings of the 1997 Rapid Excavation Tunneling Conference*, 431-455.

Earthquake Engineering Frontiers in the New Millennium, Spencer & Hu (eds),
© 2001 Swets & Zeitlinger, ISBN 90 2651 852 8

Earthquake damage and countermeasure of industrial lifeline and equipment

Yonglu Li & Xiangdong Xi
Central Research Institute of Building and Construction of Metallurgical Industry

ABSTRACT: This paper describes the aseismic importance of the lifeline and equipment and their applications by means of the earthquake damages to Baotou Iron and Steel Company, etc.

1 GENERAL DESCRIPTION OF BAOTOU EARTHQUAKE DAMAGES

An earthquake with a magnitude of 6.4 occurred on May 3, 1996 in Baotou City of Inner Mongolia Autonomous Region, whose depth of focus and intensity of epicenter were 20 km and eight-degree respectively. Baotou City is located at an area with a seismic intensity of seven-degree. It was an another strong earthquake occurred in a city with a population of more than a million after Tangshan Earthquake in 1976.

The construction of Baotou Iron and Steel Company (BISC) started in 1950s. The aseismic garrison was not taken into account for large amount of the industrial mill buildings and civil constructions. Therefore, from the beginning of 1978, about ¥40,000,000 was used to strengthen the structures that could be reinforced with a strengthened area of 1,720,000 m^2 which played an important role during the earthquake. For instance, No.4 transfer station of BISC having upper devices with a load of 30,000 kN was one of the vital structures for the whole production of the ore dressing system. The original design did not consider earthquake resistance, so the stressed cracks occurred during its use, and it was strengthened after a scientific aseismic appraisal, as a result only slight damages were produced during this earthquake. It was indicated by its state after the earthquake that this transfer station experienced a large seismic horizontal force and an obvious torsional effect. The stressed cracks occurred only on weak points of the structure, which was a normal phenomenon controlled by an aseismic code, and it embodied the principle of "no damages during weak shocks, being repairable after medium earthquakes and no collapses during strong shocks".

Remarkable damages occurred on the unreinforced structures. For example, a brick barrel was used for the floor end of the newly built No. 12 steel frame gallery (west) of No. 5 coke oven system of BISC. It acted in the violation of the national standard "Aseismic Design Code of Structures", as a result, shear failures and dislocation occurred on the brick barrel, which had to be strengthened after just a short period of operation.

For the industrial equipment, one oil pipe of the air blower for No. 3 blast furnace of BISC was broken, which resulted in the stop of the blowing machine and shutdown of the blast furnace. Several ¥10,000,000 was needed to repair the whole system. Damages to the accessories, such as the broken of the stock rod for No. 1 blast furnace, etc can influence the whole.

During the transient of Baotou Earthquake, the defect of the power supply system made BISC have a power cut period of more than 20 min, and the shutdown of BISC was up to 36h due to the stops of water and blowing for the blast furnace system, tuyere slag of No. 1 blast furnace, burning-out of all the 28 tuyeres for No. 4 blast furnace, and the broken of the oil pipe of No. 4 air blower for No. 3 blast furnace. The earthquake damages to the power supply system were not serious and main trouble was the broken of the lightning tower and the damages to the magnetic bottle seals for some oil switches, therefore it was restored very soon.

Among the 133 investigated engineering structures, such as chinmeys, water towers, bridges, substations, gas holders, open-hearth furnaces and reservoirs etc, 50% of them had different degrees of failures. The joint fractures and body break-ups also occurred on the water and heat supplying pipe networks, as well as gas pipelines for the municipal utilities. Damages with different degrees also occurred on the bridges, towers and tanks, etc.

The former "Wusi" steel works (now it is referred to Baotou Rare Earth Iron Alloy Plant) near to BISC was a local enterprise, whose aseismic garrison could not be realized, and the garrison for new projects could not be realized either due to lack of fund, etc.

The earthquake damages to it were completely different, whose blast furnace tilted. The fund required to remove danger, restore production and strengthen structural members after the earthquake was at least 20 times higher than that of the aseismic reinforcement before the earthquake.

A modernized large enterprise is a complicated system in terms of engineering. Having its seismic safety of the house buildings and the structures solved basically, the loss by an earthquake and the effect on the production system depend mainly on the capacity of shockproof and calamity prevention of the equipment systems and lifelines, such as electricity, gas and water etc. It is indicated by the earthquake damages from Baotou Earthquake and recent several earthquakes that the damages to the lifelines from earthquake hazard, which particularly increase the indirect economic loss. One of the outstanding characteristics for a city and an enterprise in modern society is the strong dependence on the lifeline systems.

2 ASEISMIC STRENGTHENING METHOD OF INDUSTRIAL EQUIPMENT

For increasing the aseismic capacity of the equipment, it is first necessary to prevent the equipment that cannot meet the aseismic garrison standard from entering a production line, which must be done at the stages of design, ordering and construction. At the same time, the appraisal and strengthening of the existing devices should be enhanced. "Aseismic Appraisal Standard for Metallurgical Industrial Equipment" (YB/T9260-98) issued on October 1, 1998 provided a basis for appraisal and strengthening of the equipment.

The following procedures are generally used for the earthquake resistance of the industrial enterprise equipment:

1 For very large amount of industrial equipment, first prediction of the earthquake damages and expert riddling shall be carried out on the equipment for a whole factory. This equipment is classified into three categories at a time of the prediction, i.e. process equipment, such as furnaces, kilns, towers, tanks etc; electric equipment, including transformers, accumulator and reactors etc; instruments and computers.

2 According to the prediction of the earthquake damages, aseismic appraisal should be conducted for the important equipment and those may be damaged during an earthquake, whose weak links can be found out to determine which ones need strengthening.

3 The aseismic reinforcement shall be carried out by stages and in-groups and in the light of order of priority, according to the requirements of the process and production. The aseismic measures for some equipment can only be decided properly by combining scientific researches. For partial accurate and precious instruments or special devices, only "earthquake resistance " may not be enough to solve the problem, therefore " Vibration isolation" or "shock absorption" may be used for them.

All sorts of the contact links between the equipment and between the accessories frequently be neglected after solving the aseismic capacity of the devices themselves, whose earthquake damages may influence the whole, i.e. the earthquake resistance of the whole equipment system involving in the production process flow and the dynamic balance in production etc, is the key point and difficult point in the earthquake resistance of the equipment.

3 WAYS FOR ENHANCING ASEISMIC CAPACITY OF LIFELINES

A metallurgical enterprise features many types of pipelines, high density and easy corrosion etc due to its professional characteristics. Different degree of damages to a lifeline system can be produced, such as corrosion, deformations and leakage etc because of external and internal causes with increase in its service period, which resulting in great economic losses. Therefore, modern pipe network technology has a very wide market and an applied prospect in the metallurgical enterprises.

3.1 Detection positioning technology of underground pipe lines

Much attention has been given to the overground pipeline and overlook to the underground pipeline for a long time, and having no scientific and strict management in the construction management of large industrial enterprise pipeline network in China. There are often incomplete file data and drawings for these pipeline network projects due to historic causes. At a time of laying a new pipeline, a blind construction is conducted because of the unclear distribution of the underground pipelines, which brings about the failures of a underground pipeline, stops of water, electricity and gas, as well as interruption of communications, as a result, influencing production and sometimes leading to a shutdown and a hazard accident. Therefore, the state of the underground pipeline for an enterprise should be ascertained as early and soon as possible, which is a problem needs an urgent solution and is also an important way for improving the comprehensive capacity of the calamity prevention of an enterprise and a city.

3.2 Establishment of data base–graph-files information management system for pipeline network

It becomes more important to ensure the completeness, accuracy and dynamic updating of the pipeline network information with the development of the enterprises, more and more types of overground and underground pipelines and also higher and higher density. A data base-graph-files information management system for the pipe network is a computer system that can do acquisition, management, renewal, overall analysis and treatment of pipeline information, using geographic information system (GIS), internet, communication and other special technologies. This management system can realize the resource sharing, dynamic renewal, raise the speed of retrieval and improve the management, so as to make the information of the pipelines reach a visual and controllable state during daily operation, which also provides extreme convenience for rush-repair after an earthquake.

3.3 Enhancement of daily detection and maintenance of pipelines

A paint, replacement and repair shall be done periodically for the pipelines, and the degree of the corrosion on the pipeline, valves and welds shall not exceed that of the design; a large area corrosion shall be replaced timely, while a local corrosion shall be repaired timely in order to avoid an increase in hazard and secondary disaster.

The corrosion, etc are the key factors influencing the reliability and service life of a pipeline system, which can result in not only the leakage losses of the oil, gas and water by holes, wastes of material and manpower due to repairs, as well as losses of shutdown, but also in fire hazard; particularly an as losses of shutdown, but also in fire hazard; particularly an explosion by a gas conduit threatens personal safety, pollutes environment, which leads to very serious consequences.

According to the experiences both at home and abroad, a pipeline will enter into a period of high rate of accidents after being used for 15–20 years. Therefore, detection, appraisal and repair of a pipeline shall be conducted in a planned way after a comprehensive analysis and an overall consideration, which can change a rush-repair into a planned detection and repair, and thus improving the earthquake resistance of the pipeline.

The research on the gas-leaking detection shall be unfolded actively, so the alarm can be done timely, which prevents trouble before it happens.

Urgent measures for shockproof and calamity prevention shall be well prepared from the management, so the valves can be closed timely in case of leakage, which prevents the occurrence of the secondary hazards. The joint positions between the pipeline and indoor or between equipment are the weak points of the earthquake resistance, where the gas can easily be gathered after a leak. Thus, for the pipelines of combustible and frost –prone media, the emptying conditions shall be provided, and a certain area of gas explosion relief shall be taken into account.

3.4 Trenchless technology on in-situ replacement and repair of underground pipeline

An enterprise's underground pipeline network consists of sewage piping, water pipeline, gas pipeline, heat supply pipeline, power cable and communication cables. On the one hand, the service life of all the pipelines is limited, therefore, repair and replacement must be done after a certain time of use; on the other hand, the original underground pipelines that can not meet the needs shall also be replaced with the development of the enterprise. In addition, a non-excavation, in-situ and real-time repair is also required to do to deal with an emergency after an earthquake.

The traditional construction method of a underground pipeline is the "trenching and burying pipe method", whose main defect is to have a great influence on the surface traffic, and at the same time brings about many troubles to the normal operation of an enterprise. Besides, an excavation construction will make the quality of a road go bad and pollute environment. An excavation construction will be the only economic and feasible construction method if a pipeline passes through the bottoms of the structures or equipment.

The trenchless technology consists of in-situ replacement and in-situ repair. The matured method of in-situ repair includes the insertion tube method, soft lining method and spraying method, etc.

In short, the very rapid development of new technology and new material for modern pipeline networks provides a powerful technology support for the normal operation of the lifeline system for an enterprise. In the meantime, a scientific management and a careful maintenance are also the important aspects to ensure the normal operation of a pipeline network. Furthermore, in order to avoid the losses from the damages to a pipeline network, a necessary emergent system shall be set up to improve the capacity of rapid response to burst events, such as earthquakes, accidents, etc.

Urban lifelines effects degree sequence and recovery time sequence

Youpo Su, Yajie Ma, & Ruixing Liu
Earthquake Engineering Research Center, Tangshan, Hebie province, China

ABSTRACT: Urban lifelines benefit, circumscribe, and substitute mutually, and coexist nearby, when earthquake occurs, these lifelines will interact and affect each other on disaster. The research on urban lifelines earthquake disaster during the Tangshan earthquake in China, the Northridge earthquake in USA, and the Hyogoken-Nanbu earthquake in Japan shows that lifelines affect structures and functions mutually, form recovery obstacles for each other and lead to second disaster. According to fact lifelines disaster in many earthquakes, a figure showing how lifelines affect each other is made. Based on this figure, the concept of effects degree, meaning how greatly one lifeline affect others, and calculation formulas, are put forward. According to calculation result, the sequence of effect degree is: $Y_d > Y_j > Y_t > Y_s = Y_p > Y_m$. The sequence of lifeline effects degrees and the time sequence of post-quake recovery are well accordant.

1 INTRODUCTION

Urban lifelines include electricity, communication, transportation, water supply and drainage, gas and so on, and they benefit, circumscribe, substitute mutually and coexist nearby. Although they are different systems and work separately, during an earthquake they affect each other on structures and functions, form recovery obstacles mutually, and lead to secondary disaster. The disaster of all lifelines is enhanced for such complexly interactions. In recent years, some research achievements on urban lifelines mutual effects have been published [1,1], these researches almost focus on qualitative analysis of concrete earthquake disaster. In our research, the effect degree of urban lifelines, their sequence and the relationship between the effect degree sequence and the post-quake recovery time sequence are studied by both qualitative and quantitative analysis methods.

2 EFFECT DEGREE AND THEIR SEQUENCE

2.1 *Effect forms*

Researches on the urban lifelines disaster in the Tangshan earthquake in China, the Northridge earthquake in America, and the Hyogoken-Nanbu earthquake in Japan have been carried out and the results show that urban lifelines affect each other with several different forms, different ranges, and different degrees. The frequent effect forms are showed in Fig. 1.

2.1.1 *Effects in structures*
Mainly because of some different lifelines coexisting nearby, they affect each other in structures. For example: if wires or pipelines of different lifelines are laid together in one channel or trench, when one of them damaged, the others nearby may be affected or damaged. In severe earthquake disaster caused by this form effects is usual and frequent. A representative example of this form effects can be saw when bridges on railways or highways are damaged with lifeline wires and pipelines passing through them damaged too. During the Northridge earthquake, electricity wires through bridges were cut down in 100 locations. During the Hyogoken-Nanbu earthquake, a gray steel pipe through the Kobe bridge, which diameter is 600 mm, and the tie-in of a cast-iron pipe across this bridge, which diameter is 900 mm, were separately bent and damaged; water supply pipes across a bridge were cut off; the retaining walls of a bridge moved and made the moderate pressure gas lines through this bridge damaged.

2.1.2 *Effects in functions*
Mainly because some lifelines benefit and circumscribe mutually, they affect each other in functions. When an earthquake forms functional obstacles of a lifeline, it will followed by functional decrease or loss of other lifelines. Especially when the electrical wires suffered from severe disaster so as to get not available, the functions of other lifelines will be greatly affected. After the Tangshan earthquake, electricity was cut off in the whole city, the equip-

Table 1. The frequent effect forms

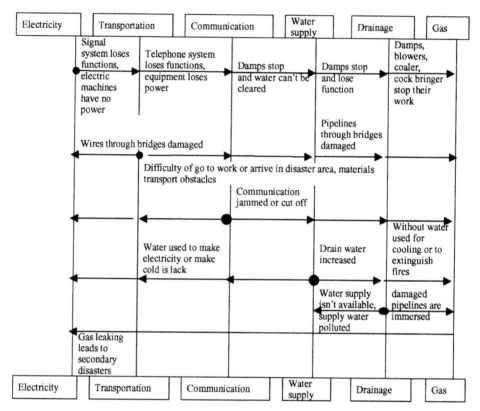

Electricity	Transportation	Communication	Water supply	Drainage	Gas
Signal system loses functions, electric machines have no power	Telephone system loses functions, equipment loses power	Damps stop and water can't be cleared	Damps stop and lose function		Damps, blowers, coaler, cock bringer stop their work
Wires through bridges damaged			Pipelines through bridges damaged		
	Difficulty of go to work or arrive in disaster area, materials transport obstacles	Communication jammed or cut off			
	Water used to make electricity or make cold is lack		Drain water increased		Without water used for cooling or to extinguish fires
			Water supply isn't available, supply water polluted	damaged pipelines are immersed	
Gas leaking leads to secondary disasters					

Electricity	Transportation	Communication	Water supply	Drainage	Gas

ments powered with electricity all lost their functions; The pumps in water supply lines stopped their work so that water was cut off; Communication and transportation are all not completely available, so information could not exchange well, and traffic got jammed. During the Northridge earthquake, 500,000,000 telephone clients were affected for lacking of electricity.

2.1.3 Recovery obstacles

If one lifeline suffers from earthquake disaster, it may affect the recovery process of other lifelines. In another words, the earthquake disaster of one lifeline may come to be the recovery obstacles of other lifelines. If transportation suffers severe earthquake disaster, it will become heavy obstacles for other lifeline recovery especially in the immediate recovery phrase. Because of traffic jammed or cut off, a powerful recovery troop is difficult to be organized for workers not on time or absent; people, equipments and materials that are necessary for recovery can not get in the disaster area or their arriving time was delayed; foreign aid persons and rescue materials cannot quickly get into the disaster area.

2.1.4 Leading to secondary disaster

The post-quake secondary disaster may happen as a result of the complex interaction of some lifelines suffering earthquake disaster. For example: gas leaking out from gas lifeline is a kind of incendive material, when it touches with the high temperature electric machines, short circuit electric wires or breaking high-voltage electric wires in the electricity lifeline, fire or gas blast may take place. Moreover water for the use of fire fighting may not be well supplied because the water supply lifeline having suffered from earthquake disaster. Fire engines may not arrive for traffic jammed or cut off. Fire information may not regularly circulate for the communication losing its functions. All these situations detailed above make it difficult to control or extinguish the fire. During the Kanto earthquake in Japan, 136 fires occurred in Tokyo, which 90 km distant from the earthquake epicenter, they are ignited mostly because the earthquake disaster of electricity and gas lifelines. For water cut off and traffic blocked by the collapsed houses, fires lasted for three days, nearly 2/3 of the houses (about 44,7000) were damaged and 5,6000 people died. These fires having occurred in Tokyo are proven to be the most costly and infrequent post-quake secondary disaster.

2.2 *Effect degree and their sequence*

If the functions of one lifeline decreased with varying degree, it will affect the functional recovery of other relative lifelines with varying degree. How the electricity affects the recovery of water supply is shown with curves in Fig. 2. Curves just like them in Fig. 2, which reflect effects between lifelines, are named lifeline post-quake recovery curves. In this figure, line 1 presents the water supply lifeline recovery curve when electricity suffers no damage; line 2, line 3 and line 4 present the water supply lifeline recovery curve when electricity suffers increasing degree damage. In Fig. 2, it is clear that the more the electricity damaged, the more difficult the water supply lifeline recovery will get, and the lower its recovery efficiency will get. It is not difficult to believe that if the effects from transportation, communication and drainage are considered, the real recovery curve of water supply lifeline will tend to move right, the efficiency of recovery will get lower, and the longer time will be required for recovery.

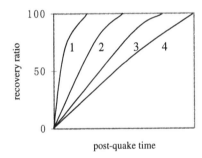

Figure 1. How the earthquake disaster of electricity lifeline affects the functional recovery of water supply lifeline

The concept of lifeline effect degree can quantitatively describe the degree and the range that one lifeline works on another. One lifeline can affect on structures and functions of another lifeline, form recovery obstacles, and lead to secondary disaster. When one lifeline works on another lifeline and causes a form of effects detailed above, we say the effect degree of this lifeline is 1. We can also say that every form of effects has same effect degree. Y_d, Y_m, Y_j, Y_g, Y_p, and Y_t are separately present the effect degree of electricity, gas, transportation, water supply, drainage and communication lifeline. If all the effect degree that electricity lifeline works on gas, transport, water supply, drainage and communication are separately presented by Y_{d-m}, Y_{d-j}, Y_{d-g}, Y_{d-p} and Y_{d-t}, then $Y_d=Y_{d-m}+Y_{d-j}+Y_{d-p}+Y_{d-g}+Y_{d-t}$. According to the relationship shown in Figure 1, we find that $Y_{d-m}=2$, $Y_{d-j}=2$, $Y_{d-g}=1$, $Y_{d-p}=1$, $Y_{d-t}=3$, and then $Y_d=9$.

According to the same theory, we get that $Y_m=1$, $Y_j=7$, $Y_g=3$, $Y_p=3$, and then $Y_t=5$.

Obviously, the sequence of lifeline effect degree is: $Y_d>Y_j>Y_t>Y_g=Y_p>Y_m$.

According to Figure 1, the degree that one lifeline affected by the others, presented by y, can also be calculated, the result is $y_d=4$, $y_j=2$, $y_g=4$, $y_p=4$, $y_m=5$. We can see that the gas lifeline post-quake recovery is affected by the other 5 lifelines. Generally, its recovery efficiency is lowest, and its recovery requires the longest time. Differently, effects that transportation recovery suffered from other lifelines is the smallest.

3 EFFECT SEQUENCE AND POST-QUAKE RECOVERY TIME SEQUENCE

The effect sequence of lifelines is important for the time sequence of lifelines post-quake recovery. If the lifeline, which effect degree is the largest, is rapidly recovered, it will benefit other lifelines recovery. For example: the recovered electricity can power transportation, communication, water supply and gas lifelines, it can make their functions recovered immediately after their structures recovered. The post-quake recovered lifelines will do socialeconomic benefit in helping urban foundational functions recovery and improving the living condition of the people in disaster area. By contraries, if the lifeline with lower effect degree is restored firstly, its functions can't recover after its structural recovery for electricity not available. When gas lifeline structures are restored, it also can't supply gas unless electricity recovered. Although gas lifeline functions get recovered, it can't accelerate other lifeline recovery as electricity can do.

The post-quake recovery curves (transportation not included) of the 1995 Hyogoken-nanbu earthquake in Japan and the San Ferando earthquake in U.S are separate curves (1) and (2) in Figure 2.

From Figure 2, we can see that during many severe earthquakes just like the Tangshan earthquake in China, recovery time sequence of lifelines is almost accordant with their effect degree sequence. the only difference is that gas lifeline get recovered before water supply lifeline. Although this situation is infrequent, it reflects that the recovery time sequence of lifelines is complexly affected by many factors. The research on the Tangshan earthquake in China, the Hyogoken-Nanbu earthquake in Japan, and the Northridge earthquake in the U.S. shows that the time sequence of lifelines recovery is affected by the disaster degree of every lifeline, by the contribution that one lifeline do to the rapid recovery of urban foundational functions and the continuous improvement of victims living condition, and by the labor power and finance contributed to recovery, by the area situation and condition for recovery, by the improvement that one lifeline recovery make in other lifeline recovery, by the characteristic of mate-

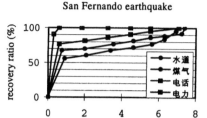

San Fernando earthquake

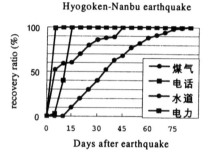

Hyogoken-Nanbu earthquake

Figure 2. The post-quake recovery curves of lifelines

rial exchange between lifelines, and so on. In general situation, electricity, transportation, and communication should be recovered firstly and in the front of the recovery time sequence, and gas should be supply again until all parts of the line suffered from disaster are sure to be well recovered, for gas is the only poisonous, combustible and explosive material flowing in lifelines, and should be cut off in time after earthquake. So gas lifeline should be recovered later, its recovery efficiency is low. Although in the San earthquake gas was recovered before water supply, gas recovery was finished later than other lifelines.

4 CONCLUSION

After an earthquake, one lifeline work on others in structures and functions, form recovery obstacles, and lead to secondary disaster. This not only increases the disaster of all lifelines, but also greatly affects the post-quake recovery of lifelines. The concept of effect degree can be used to describe the effect degree and effect range existing among lifelines. The effect degree sequence of lifelines can reflect the time sequence of their recovery, these two sequences are accordant well. To decide the time sequence of lifeline post-quake recovery, the sequence of effect degree is valuable reference.

REFERENCES

Anshel, J. Schiff 1995. Northridge Earthquake Lifeline Performance and Post--Earthquake Response. New York: American Society of Civil Engineers.
Huixian, Liu 1994. Tangshan Earthquake Disaster (3). Beijing: the earthquake publisher.
Keishi, Shiono 1996. Numerical Evaluation of Inconvenience in Resident's Living in Earthquake. Cultural Province Scientific Research Fund grant-in-aid important research field: Down-Town Earthquake Summary, the Analects of The First Down-Town Earthquake Summary Symposium:307–310.
Nobuoto, Nojima & Hiroyuki Kameda 1996. Relationships between Lifelines. Teizo FUJIWARA. Analysis on Earthquake Disaster Fact Examples Based on Reconnaissance of Hyogoken-Nanbu Earthquake.
The Earthquake Motion Engineering resistance technical center of China Petroleum & Chemistry Group Company. 1998.Petroleum & Chemistry Equipments Anti-earthquake Calculation. Beijing: earthquake publisher.
Urban Disaster Resistance and Environment Symposium. 1998. Earthqake and Lifelines. Kyoto: Kyoto University Publisher.
Youpo, Su & Ruixing, Lui 2000. The Urban Disaster Recovery Time-Sequence in the Emergence Phase. Disaster (2): 33–37.

Earthquake Engineering Frontiers in the New Millennium, Spencer & Hu (eds),
© 2001 Swets & Zeitlinger, ISBN 90 2651 852 8

"Smart" base isolation systems: Theory and experiments

B.F. Spencer, Jr.
University of Notre Dame, Notre Dame, IN 46556, USA

H. Yoshioka
Takenaka Corporation, Chiba, Japan

J.C. Ramallo
Universidad Nacional de Tucumán, Tucumán, Argentina

E.A. Johnson
University of Southern California, Los Angeles, CA 90089, USA

ABSTRACT: Recently proposed "smart" base isolation strategies are shown herein to effectively protect structures against extreme earthquakes, without sacrificing performance during the more frequent, moderate seismic events. This base isolation system is composed of conventional low-damping elastomeric bearings and "smart" controllable (semiactive) dampers, such as magnetorheological fluid dampers. Simulation results comparing the performance of the "smart" isolation system to those for two lead-rubber bearing isolation systems are presented. The effectiveness of these isolation approaches are judged based on computed responses to several historical earthquakes scaled to various magnitudes. The limited performance of passive systems is revealed and the potential advantages of "smart" dampers are highlighted. To demonstrate the efficacy of the "smart" base isolation paradigm, results of an experimental study of a "smart", base isolation system that employs magnetorheological (MR) dampers are presented.

1 INTRODUCTION

One of the most widely implemented and accepted seismic protection systems is base isolation. Seismic base isolation (Skinner *et al.* 1993, Naeim and Kelly 1999) is a technique that mitigates the effects of an earthquake by essentially *detaching* the structure and its contents from potentially dangerous ground motion, especially in the frequency range where the building is most affected. The goal is to simultaneously reduce interstory drifts and floor accelerations to limit or avoid damage, not only to the structure but also to its contents, in a cost-effective manner.

Recent concerns about the effectiveness of base isolation systems under various types of seismic excitations, particularly those of the near-source, highvelocity, long period pulse varieties, has prompted further interest in controllable base isolation (Spencer *et al.* 2000). Indeed, optimal design of base isolation systems depends on the magnitude of the design level earthquake that is considered. For example, the features of an isolation system designed for an El Centro-type earthquake typically will not be optimal for a Kobe-type earthquake and vice versa. To be effective during a wide range of seismic events, an isolation system must be adaptable. Recently proposed "smart" base isolation strategies have been shown to effectively protect structures against extreme earthquakes, without sacrificing performance during the more frequent, moderate seismic events. This base isolation system is composed of conventional lowdamping elastomeric bearings

and "smart" controllable (semiactive) dampers, such as magnetorheological fluid dampers (Spencer and Sain, 1997). For certain applications, this class of dampers have been shown to achieve most of the performance of fully active devices (Dyke *et al.* 1996a-b, Johnson *et al.* 1999), but at a small fraction of the power requirements.

This paper presents both simulation and experimental results to demonstrate the effectiveness of the "smart" isolation systems. Simulation results comparing the performance of the "smart" isolation system to those for two lead-rubber bearing isolation systems are presented. The effectiveness of these isolation approaches are judged based on computed responses to several historical earthquakes scaled to various magnitudes. The limited performance of passive systems is revealed and the potential advantages of "smart" dampers are highlighted. To demonstrate the efficacy of the "smart" base isolation paradigm, results of an experimental study of a "smart", base isolation system that employs magnetorheological (MR) dampers are presented. The experimental structure, constructed and tested at the Structural Dynamics and Control / Earthquake Engineering Laboratory at the University of Notre Dame (*http://www.nd.edu/~quake/*), is a base-isolated two degreeof-freedom building model subjected to artificial ground motion. A sponge-type MR damper is installed between the base and the ground to provide controllable damping for the system. The experimental results for the proposed "smart" base isolation system clearly demonstrates that effective seismic isolation can be achieved for both infrequent

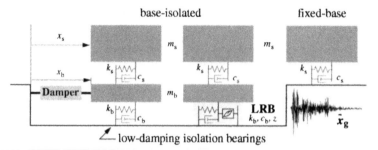

base-isolated   fixed-base

low-damping isolation bearings

Isolation Parameters

m_b=6800 kg
Period 2.5 s $\Rightarrow k_b$=232 kN/m
Damping ratio 2% $\Rightarrow c_b$=3.74 kN·s/m

Structure Parameters

m_s=29485 kg
Period 0.3 s $\Rightarrow k_s$=11912 kN/m
Damping ratio 2% $\Rightarrow c_s$=23.71 kN·s/m

Figure 1. Two degree-of-freedom (2DOF) models.

strong motion earthquakes, as well as for more moderate and frequent events.

2 SIMULATION MODEL FORMULATION

2.1 Structural system model

In this study, the structure is modeled as a single degree-of-freedom (SDOF) system representing the fundamental mode of the five-story building model given in Kelly et al. (1987). Including the isolation layer, the isolated building model has two degrees of freedom (2DOF). The structural parameters of the SDOF fixed-base and 2DOF isolated structure models are given in Fig. 1. The isolation layer gives a fundamental mode with a 2.5 second period and 2% of critical damping. This low-damping, long period, isolation system fits into the *"Class (ii): Lightly damped, linear isolation system"* category of Skinner et al. (1993).

2.2 Damping strategies considered

In this study, two types of isolation damping are considered:

- *Lead-rubber bearing* (LRB):

The horizontal force fLRB in the LRB is modeled as follows (Wen 1976)

$$f_{LRB} = Q_{Pb}z + k_r x_b + c_r \dot{x}_b \qquad (1)$$

where Q_{Pb} is the yield force of the lead plug; k_r and c_r are the horizontal stiffness and viscous damping factor of the rubber composite of the bearing; x_b and $\dot{x}_b$ are the relative displacement and velocity across the bearing. The evolutionary variable z accounts for the hysteretic nature of the force and is governed by

$$\dot{z} = -\gamma \left| \dot{x}_b \right|^{n-1} z - \beta \dot{x}_b \left| z \right|^n + A\dot{x}_b \qquad (2)$$

where γ, β, A, and n are shape parameters of the hysteresis loop which herein are considered time-invariant and adopted as $\gamma = \beta = 0.5, A = 1$, and $n = 1$.

- *"Smart" (semi-active) damper:*

A controllable damper (e.g., variable orifice damper, controllable fluid damper, etc.; see Spencer and Sain 1997, Symans and Constantinou 1999) that may exert only dissipative forces;. *i.e.*, $f_{SA}\dot{x}_b \leq 0$ where f_{SA} is the applied force and x_b is the velocity across the damper. For this portion of the paper, the device is assumed ideal; *i.e.*, it can generate the desired (dissipative) forces with no delay nor actuator dynamics.

3 CONTROL SCHEME FOR SMART DAMPERS

A *clipped-optimal control* strategy, shown to perform well in previous studies involving smart dampers *(e.g.,* Dyke et al. 1996a-b, Johnson et al. 1999, Spencer et al. 1999), is implemented in this study. The strategy is to assume an "ideal" active control device, design an appropriate *primary* controller for this active device, and then, using *a secondary* bangbang-type controller, try to make the smart damper replicate the same forces the active device would have exerted on the structure. Since the force generated by a smart damper is dependent on the structure and its motion, it is not always possible to produce the "desired" force.

Herein, an H₂/LQG control design is adopted as the *primary* controller. To better inform the controller about the frequency content of the ground motions, a Kanai-Tajimi shaping filter is incorporated into the model of the structure. The H₂/LQG *primary* control is then designed using the Control Toolbox in MATLAB®, where the resulting optimal compensator K(s) is dynamic and of order equal to the sum of the orders of the structure and the shaping filter.

For the general smart damping device, the *secondary* control strategy is given by

$$f_{SA} = \begin{cases} f_{opt} & \dot{x}_b < 0 \\ 0, & \text{otherwise} \end{cases} \quad (3)$$

where f_{SA} is the commanded semiactive control force, f_{opt} is the optimal, control force determined by the primary controller, $\dot{x}_b$ is the velocity across the damper.

4 DESIGNING THE "OPTIMAL" ISOLATION SYSTEM

In the selection of a base isolation system employing lead-rubber bearings, the influence of two parameters is considered, namely the total yield force Q_y (expressed as a fraction of the total structural weight) and the stiffness ratio $K_{initial}/K_{yield}$. The post-yield stiffness is fixed at $K_{yield} = k_b = 2.32$ kN/m to obtain a fundamental period of 2.5 seconds once the lead plug has yielded. Q_y is related to Q_{Pb} by

$$Q_{Pb} = \left(1 - \frac{K_{yield}}{K_{initial}}\right) \cdot Q_y \quad (4)$$

Two optimal designs are considered. *LRB1,* with $Q_y = 5$ % of the structural weight and with $K_{initial}/K_{yield} = 6$, is a typical base isolation design and is used herein as a yard-stick for comparison to the other isolation strategies. *LRB2,* with $Q_y = 15\%$ of the structural weight and with $K_{initial}/K_{yield} = 10$, is also studied, as it is a more effective system for nearfield earthquakes such as the Kobe or Northridge earthquakes. This value for Q_y follows the results of Park and Otsuka (1999).

5 SIMULATION RESULTS

The isolation structures considered herein are excited by a suite of ground motions that are intended to encompass moderate (1940 NS El Centro and 1968 NS Hachinohe) and severe (1995 NS Kobe and 1994 NS Northridge (Sylmar County Hospital record) events. Additionally, the earthquakes are scaled to several magnitudes to better understand the effectiveness of the isolation strategies for different earthquake strengths. The moderate records are scaled in the range from 0.50 to 2.00 times the historical record, and the severe ones from 0.50 to 1.50. Although magnifying the severe earthquakes might seem unnecessary, the results using 1.50 times historical are included to show the behavior of different damping devices under truly extreme events.

Table 1 summarizes the percent improvement of LRB2 (the high-yield LRB) and the 'smart' damper over the response of the baseline LRB1 system. As can be seen, the LRB2 system, due to its higher yield level, is capable of substantial reductions in the peak base drift. For the most extreme events, reductions in base drift are on the order of 50%. Note that the improvement of the LRB2 system over the LRB1 system somewhat diminishes in the case of the Northridge earthquake as the scale increases. A similar trend is found for the Kobe earthquake when the scale is increased to 1.50. Not surprisingly, however, this reduction in base drift comes at the price of increased accelerations and interstory drifts. The base acceleration in the LRB2 system was substantially increased for 8 of the 10 ground motions considered. Moreover, for the reduced scale earthquakes (*i.e.,* those that are likely to occur more frequently), the accelerations increased up to 137% (in the case of El Centro) over the LRB1 system. Structural drifts follow a pattern similar to base acceleration.

The 'smart' damping system achieved reductions in peak base drift that were comparable with the LRB2 system. However, in contrast to the LRB2 system, Table 1 shows that the peak base accelerations were reduced (as compared to the LRB1 system) for all except two of the ground motions considered; in these two cases, the increase in the base acceleration was only a few percent. The peak structural drifts were reduced for all cases con-sidered. In particular, note the excellent perfor-man-ce for the reduced scale ground motions where the smart damping system reduces peak base accelration from 24% to 42% and reduces the base drifts by up to 41% in comparison with the LRB1 system. Thus, the decrease in the base drift during the large earthquakes does not necessarily come at the expen-se of larger accelerations and interstory drifts.

With regard to the peak forces exerted, the LRBs attained, in most cases, their yield forces (LRB1: $F_{yield} = 14.83$ kN, LRB2: $F_{yield} = 48.05$ kN). The smart damper employs a peak force of 53.39 kN. Therefore, the performance gains were achieved by the smart damping system using peak damping forces that are only 11% larger than the LRB2 system.

More details of this analytical study are available in Ramallo, *et al.* (2001). The remainder of this paper presents experimental verification of the proposed control strategies.

6 EXPERIMENTAL SETUP

Experimental verification of the proposed smart isolation strategies were conducted on the shaking table at the Structural Dynamics and Control/ Earthquake Engineering Laboratory (SDC/EEL) at the Universi-

Table 1. Percent reduction for different damping mechanisms over LRB1.

Earthquake	Scale	Peak Base Drift		Peak Base (Abs.) Acceleration		Peak Structural Drift		Peak Struct.(Abs.) Acceleration		Peak Applied Force	
		LRB2	Smart Damper	LRB2	Smart Damper	LRB2	Smart Damper	LRB2	Smart Damper	LRB2	Smart Damper
El Centro		4.9	3.2	-137.6	24.2	-125.0	31.3	-117.7	33.1	-193.2	18.6
Hachinohe		15.3	15.8	-91.5	42.2	-61.3	74.2	-89.5	50.2	-181.8	15.9
	0_50										
Kobe		24.3	39.3	-98.0	32.1	-61.3	25.8	-40.4	25.9	-223.6	-98.0
Northridge		50.8	41.6	-12.5	26.8	-8.5	23.4	-0.6	23.6	-223.9	-170.3
El Centro		13.1	4.3	-97.8	4.4	-91.7	8.3	-101.0	8.8	-218.1	-57.8
Hachinohe		31'3	23.8	-70.6	24.6	-13.0	68.5	-74.9	32.4	-219.4	-65.9
	1.00										
Kobe		35.7	39.4	-65.3	6.1	-13.0	14.8	-35.2	13.7	-224.0	-260.0
Northridge		38.0	42.8	4.0	17.1	10.1	25.3	3.1	25.5	-224.0	-260.0
El Centro		35.9	29.0	-79.5	-0.4	-40.5	10.8	-58.7	10.5	-223.0	-136.7
Hachinohe		41.9	37.1	-51.7	17.6	-3.6	69.9	-12.6	31.5	-223.7	-148.9
	1.50										
Kobe		26.8	45.1	-42.6	-4.2	-3.6	22.9	-3.9	22.9	-224.0	-260.0
Northridge		35.5	41.3	17.5	16.2	20.3	31.4	-8.7	31.0	-224.0	-260.0
El Centro		51.7	47.2	-44.4	11.2	-1.7	24.1	9.9	24.3	-223.8	-215.6
	2.00										
Hachinohe		52.8	54.8	-13.8	29.9	-7.1	69.9	20.2	43.6	-224.0	-231.9

ty of Notre Dame (http://www.nd.edu/quake/). The test structure, shown in Fig. 2, is a two-mass model supported by laminated rubber bearings.

This model is a 2 degree-of-freedom representation of a 5-story prototype structure with an isolation period of 2 seconds. The first mass (m_l = 10.5 kg), corresponding to the isolation base of the structure, consists of a 32 cm x 32 cm x 2.5 cm aluminum plate. The second mass (m_2 = 57.5 kg) represents a single degree-of-freedom (one mode) model of the superstructure and consists of one 32 cm x 32 cm x 2.5 cm steel plate and two 30 cm x 32 cm x 2.5 cm steel plates. 20-layer laminated rubber bearings are employed as isolators at each of the four corners of the base. Each layer consists of three neoprene rubber disks with a height of 0.3 cm and a diameter of 1.1 cm attached to a 10.2 cm x 7.7 cm x 0.1 cm steel plate. The top mass, m_2, is mounted on four, 2-layer laminated rubber bearings and attached to the lower mass, m_l. This approach keeps the center of gravity of the structure low, minimizing overturning moments in the model. Additionally, because the vertical stiffness of the isolation bearings is relatively low, a linear guide is installed below the base to restrict vertical and rocking motion. Therefore, only the horizontal motion of the base and the structure are considered in this experiment.

A simple shear-type MR damper is attached between the base and the table for controlling the response of the structure. The characteristics of the damper are: ±3.5 cm maximum stroke, 50 N - maximum force with a current of 0.5 A.

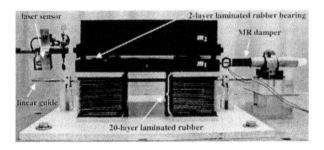

Figure 2. Experimental setup.

7 CONTROLLER IMPLEMENTATION

The experimental implementation of the secondary controller described in Sec. 3 is somewhat different than for the analytical study. To track the optimal force f_{opt}, Dyke et al. (1996a,b) proposed a clippedoptimal switching, defined by

$$v = V_{max} H\left\{\left(f_{opt} - f_{meas}\right) f_{meas}\right\} \qquad (5)$$

where v is the voltage to the current driver associated with saturation of the magnetic field in the MR damper, and $H(\cdot)$ is the Heaviside step function. This control algorithm has the benefit that a model of the damper is not required in the control design. Because the sensor outputs in the experiment include both a small DC offset and noise, the desired force also includes an offset as well as noise. The influence of these errors is significant in the case of small vibration. Thus, the controller may send an incorrect signal to the damper, especially for ambient vibration. In this experiment, an alternative clipped-optimal control with a threshold is employed. The proposed control voltage remains zero below a minimum force, $f_{min} = 4.3$ N.

$$v = \begin{cases} V_{max} H\left\{\left(f_{opt} - f_{meas}\right) f_{meas}\right\} & \left|f_{opt}\right| > f_{min} \\ 0, & \text{otherwise} \end{cases} \qquad (6)$$

8 EXPERIMENTAL RESULTS

The proposed smart base isolation system employing MR dampers was experimentally investigated and compared with the optimal passive base isolation system with the MR damper operated in constant current mode. To demonstrate the effectiveness of this system for different levels and types of seismic events, two earthquake records are considered:
1   Strong El Centro NS: maximum ground motion is scaled to 0.2 g (0.44 g for prototype).
2   Moderate El Centro NS: maximum ground motion is scaled to 0.07 g (0.16 g for prototype).

The passive damper employed in this experiment is experimentally optimized for the strong El Centro NS earthquake record. The smart damping strategies investigated herein were designed to minimize the structural acceleration due to the strong El Centro NS earthquake record.

Figure 3 compare experimental results for optimal passive and smart damping strategies. The passive damper optimized for the strong El Centro NS earthquake record can reduce the acceleration response due to the target earthquake. Compared to the case without the damper, acceleration response and base drift are reduced substantially. The maximum structural acceleration is reduced from 96.2 cm/s² (w/o damper) to 52.5 cm/s² (w/ smart damper). Compared to the input ground accele-

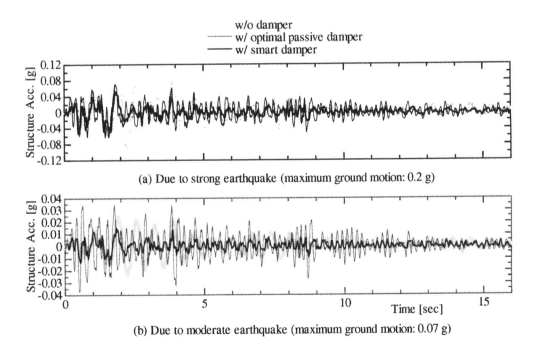

(a) Due to strong earthquake (maximum ground motion: 0.2 g)

(b) Due to moderate earthquake (maximum ground motion: 0.07 g)

Figure 3. Experimental acceleration response of the structure due to the scaled El Centro NS earthquake records.

ration (peak acceleration of 198.5 cm/ s^2), the peak acceleration for the structure employing the smart damper showed an attenuation of 73.6%.

In the case of the moderate El Centro NS earthquake record, the base and structural accele-rations for the optimal passive system are worse than the corresponding responses with no damper installed. On the contrary, the smart isolation sys-tem can reduce responses due to both the strong and moderate El Centro NS earthquake records. Compared to the optimal passive system, acceleration reductions for the smart isolation systems range from 25 to 35% better in the case of the strong earthquake, and 45 to 60% better in the case of the moderate earthquake.

Note that the optimal passive damper generates almost the same maximum force for both earthquake motions. On the other hand, the smart iso-lation system adapts to the situation, generating smaller forces for the case of the moderate earthquake and larger forces for the strong earthquakes. Also note that during the moderate earthquake, although the base drift of the smart isolation sys-tem is larger than for the optimal passive system, it is still smaller than the base drifts of during the se-vere earthquake. Indeed, one would like to have larger drifts during moderate seismic events, so long as the seismic gap of the isolation system is not exceeded.

More details regarding this study can be found in Yoshioka, *et al.* (2001).

9 CONCLUDING REMARKS

A comparison study of three base isolation systems was performed. The responses to a wide range of ground motions were computed using two passive base isolation systems (LRB1, Q_y = 5%, and LRB2, Q_y = 15%, lead-rubber isolators, primarily intended to withstand moderate and severe ground motions, respectively) and a smart (semi-active) isolation system. The selected suite of earthquakes (composed of two 'moderate' and two 'severe' earthquakes) was scaled in magnitude to evaluate the designed supplemental damping systems under ground motions exhibiting different frequency con-tent and/or different magnitudes of applied energy. While for base-drifts, the LRB2 system was capable of important improvements (i. e. reduc-tions) over LRB1 values, the LRB2 system ampli-fies both the base acceleration and interstory drift, as compared to LRB1. This trade-off between the various responses is well-recognized in the lite-rature.

In contrast, the smart damping system, using a controllable passive damper, was able to achieve reductions in the base drift comparable to the LRB2 system over the entire suite of earthquakes, without the large increases in the base acceleration and interstory drift. Moreover, the required dam-ping force levels were only 11% greater than the LRB2 system. The adaptable nature of the smart damper system allows a structure to be protected against extreme earthquakes, without sacrificing performance during the more frequent, moderate seismic events.

The efficacy of this smart base isolation system in reducing the structural responses for a wide range of loading conditions has been demonstrated in a series of experiments conducted at the Structural Dynamics and Control/Earthquake Engi-neering Laboratory at the University of Notre Dame. A modified clippedoptimal control strategy has been proposed and shown to be effective. By applying a threshold to the control voltage for the MR damper, the controller becomes robust to the ambient vibration. The dynamic behavior of this system is also shown to be predictable. Results for the smart isolation system were compared to those where the MR damper was operated in a passive mode (*i.e.,* with a constant current being sent to the MR damper).

In the passive mode, the MR damper behaves as a yielding device and approximates the behavior of lead rubber bearings. An optimization was perfor-med experimentally to obtain the optimal passive damper configuration. As compared to this optimal passive system, the smart isolation system achie-ved significant acceleration reductions over the entire range of earthquake characteristics consi-dered. This study suggests that smart dam-pers, such as magnetorheological fluid dampers, show significant promise for use in a wide range of base isolation applications.

10 ACKNOWLEDGMENTS

The authors gratefully acknowledge the partial support of this research by the National Science Foundation under grant CMS 99-00234 (Dr. S.C. Liu, Program Director), by the LORD Corporation (Dr. J.D. Carlson, Engineering Fellow), by the Takenaka Corporation, and by a fellowship from Consejo Nacional de Investigaciones Científicas y Técnicas (República Argentina).

REFERENCES

Dyke S.J., Spencer jr., B.F., Sam, M.K. and Carlson, J.D. (1996a): Seismic Response Reduction Using Magne-torheological Dampers, *Proc. IFAC World Congress,* San Francisco, CA, L: pp. 145-150.

Dyke, S.J., Spencer Jr., B.F., Sain, M.K. and Carlson J.D. (1996b): Modeling and Control of Magnetorheological Dampers for Seismic Response Reduction, *Smart Materi-als and Struct.,* 5, pp.565-575.

Johnson, E.A., J.C. Ramallo, B.F. Spencer, Jr., and M.K. Sain (1999): Intelligent Base Isolation Systems, *Proc. Second World Conf. Struct. Control,* Kyoto, Japan, 1, pp.367-376.

Kelly, J.M., Leitmann, G. and Soldatos, A.G. (1987): Robust Control of Base-Isolated Structures under Earthquake Excitation, *J. Opt. Theory and Appl.,* 53, pp. 159-180.

Naeim, F. and Kelly, J.M. (1999): *Design of Seismic Isolated Structures: From Theory to Practice,* John Wiley & Sons Ltd, Chichester, England.

Ramallo, J.C., Johnson, E.A. and Spencer Jr., B.F. (2001): 'Smart' Base Isolation Systems, *J. of Engrg. Mech., ASCE,* submitted.

Park, J., and Otsuka, H. (1999): Optimal Yield Level of Bilinear Seismic Isolation Devices, *Earthquake Engrg. Struct. Dyn.,* 28, pp.941-955.

Skinner, R.I., Robinson, W.H. and McVerry, G.H. (1993): *An Introduction to Seismic Isolation,* John Wiley & Sons Ltd, Chichester, England.

Spencer Jr., B.F, and Sain, M.K. (1997): Controlling Buildings: A New Frontier in Feedback, *IEEE Control Systems Mag.,* 17, No.6, pp. 19-35.

Spencer Jr., B.Y, Johnson, E.A. and Ramallo, J.C. (2000): 'Smart' Isolation for Seismic Control. *JSME International Journal, Special issue on Frontiers of Motion and Vibration Control,* Series C, 4(3), pp. 704-711, 2000.

Symans, M.D., and Constantinou, M.C. (1999): Semi-active Control Systems for Seismic Protection of Structures: A State-of-the-Art Review, *Engrg. Struct.,* 21, pp.469-487.

Wen Y.K. (1976): Method of Random Vibration of Hysteretic Systems, *J. Engrg. Mech. Div., Proc., ASCE,* 102(EM2), pp. 249-263.

Yoshioka, H., Ramallo, J.C. and Spencer Jr., B.F (2001): 'Smart' Base Isolation Strategies Employing Magnetorheological Dampers. *J. of Engrg. Mech., ASCE,* submitted.

Earthquake Engineering Frontiers in the New Millennium, Spencer & Hu (eds),
© 2001 Swets & Zeitlinger, ISBN 90 2651 852 8

Magnetorheological fluid and smart damper for structural vibration control

Jinping Ou
Department of Civil Eng., Harbin Inst of Technology, Harbin, P. R. C.

Xinchun Guan
The Research Station on Mater. Science and Eng. for Postdoctoral Fellows, Harbin Inst of Technology, Harbin, P. R. C.

ABSTRACT: In this paper, by using the method of orthogonal experiment, the influence of magnetorheological (MR) fluid's components on its performance and the influence mechanism is studied. The results show that the main function of oleic acid is improving the dissolving capacity of particles and the OP emulsifying agent is improving the disperse capacity of particles. Based on above experimental results and lots of concrete comparative experiments, a kind of high shear yielding stress and low sedimentation MR fluid is got. Based on the performance of MR fluid and characteristics of magnetic circuit, a MR damper that maximum voltage required is 9V, rated current is 2A and peak power required is less than 20W was designed and manufactured. And then, the damping properties and response time of this damper was tested. The results show that the real response time of this damper is about 0.2~0.3 second and the ideal response time is about 10 milliseconds. Finally, the influence factors on the real response time are analyzed.

1 INTRODUCTION

MR fluids normally are comprised of ferromagnetic or paramagnetic particles, typically 1 to about 100 micrometers in diameter, dispersed within a carrier fluid and in the presence of magnetic field, the particles become polarized and are thereby organized into chains of particles within the fluid. The chains of particles act to increase the apparent viscosity or flow resistance of the overall material and in the absence of magnetic field, the particles return to an unorganized or free state and the apparent viscosity or flow resistance of the overall material is correspondingly reduced. Magnetorheological materials are useful in providing variable damping forces within devices, such as dampers, shock absorbers and elastomeric mounts, as well as in controlling torque and pressure levels in various clutch, brake and valve devices (Ashour *et al* 1995, Carlson *et al* 1996, Housner *et al* 1997, Spencer *et al* 1998)

Due to the reasons of high cost, low reliability and great power requirement, the engineering appli-

cations of active control systems in full-scale are still very few. In contrast, the technology of passive control systems are well understood, however they can not change with varying loading conditions. In recent decade, with development of smart materials, a new kind of devices that use those materials, due to possessing the virtue of structure simple, large force, low power requirement, rapid-response, can be used both as passive and semi-active and bring the development of structural control technology to another opportunity. MR damper is one kind of them (Housner *et al* 1997, Spencer *et al* 1997,1998).

In this paper, by using the method of orthogonal experiment, the influence and influence mechanism of the viscosity of carrier fluids, the diameters and volume content of particles, the type and content of surfactant on the performance of MR fluid are studied. A kind of MR fluid whose performance is excellent is got. On the basis of analyzing characteristics of MR fluid and MR devices, a MR damper was designed and manufactured, the damping performance and response time of this damper are tested.

Table 1. Properties of magnetosoft powder carbonyl iron

Type	Average Diameter (μm)	Apparent Density (g/cm^3)	% Iron (Min)	% Oxygen (Max)	% Carbon (Max)	% Nitrogen (Max)
FTF-1	3.67	1.43	96	1.5	1.5	0.3
FTF-4	6.20	2.59	99	0.4	0.15	0.1

2 MAGNETORHEOLOGICAL FLUID DESIGN AND EXPERIMENT

2.1 Materials

The main components of MR fluid are magnetically soft particles, carrier fluid and admixtures (Kordonsky 1993). According to the characteristics of MR fluid, we choose the main components of MR fluid as follows:

1 Carrier fluid: silicon oil that the viscosities are respectively 70, 500 and 2000 centistoke.
2 Admixture: two kind of surfactant, i.e. oleic acid and OP emulsifying agent.
3 Magnetically soft particles: two types of carbonyl iron whose physical and chemical properties are listed in table 1.

2.2 Experiment method

In order to find the best content of each component and to study their influence function, four groups of orthogonal experiment are designed. Table 2 and 3 shows the factors and level of two groups of orthogonal experiment. The factors and level of other two groups are similar to table 2 and 3, only the type of carbonyl iron is FTF-4 instead of FTF-1. During above four groups of experiment, the main index is shear yielding stress, the second index is stability. In order to study the function of surfactant, on the basis of above experiments, lots of concrete comparative experiments are done.

2.3 Experimental results

The yielding stress of the MR fluids manufactured by above four orthogonal tables is tested at the magnetic strength of 0.17, 0.25, 0.31 and 0.35T by a apparatus which is designed and manufactured by ourselves. Figure 1 shows the schematic and photograph of the apparatus. Its characteristics is structure simple and performance reliable.

The experiment results show that the yielding stress of MR fluid is mainly decided by the volume content of carbonyl iron, and the main function of oleic acid is to improve the dissolving capacity of particles and the OP emulsifying agent is to improve the disperse capacity of particles. The results also show that using both oleic acid and OP emulsifying agent as surfactant is better than using only one kind of them.

On the basis of lots of experiments, using silicon oil whose viscosity are 70 and 2000 centistoke, carbonyl iron whose type is FTF-1 and the solution whose component are some proportion of oleic acid and OP emulsifying agent, two kinds of excellent MR fluid are produced. Figures 2 and 3 show the properties of MR fluids. The MR fluid that we manufacture is very stable, it does not have clear setting after stewing for one month.

Table 2. The first group orthogonal experiment

Level	Factor		
	A The volume content of FTF-1 (%)	B Viscosity of carrier fluid (centistoke)	C Ratio of oleic acid and carbonyl iron volume (%)
1	10	70	3
2	20	500	6
3	30	2000	9

Table 3. The second group orthogonal experiment

Level	Factor		
	A The volume content of FTF-1 (%)	B Viscosity of carrier fluid (centistoke)	C Ratio of OP emulsifying agent and carbonyl iron volume (%)
1	10	70	3
2	20	500	6
3	30	2000	9

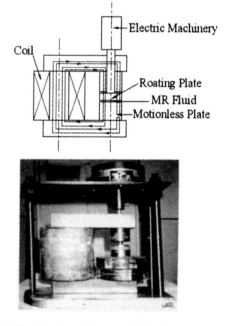

Figure 1. The schematic and photograph of testing apparatus

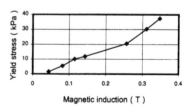

Figure 2. The stress versus induction of MR fluid (the viscosity of carrier fluid is 70 centistoke)

268

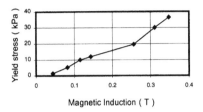

Figure 3. The stress versus induction of MR fluid (the viscosity of carrier fluid is 2000 centistoke)

3 MAGNETORHEOLOGICAL DAMPER CONFIGURATION

According to the strained condition of controllable fluids, the MR fluid devices can be classified as the following four types: flow, shear, squeeze-flow and shear flow modes. Through analyzing the characteristics of those types of devices and the requirement of structural control, we construct a shear flow mode damper. This type of devices possesses the virtue of simple magnetic circuit and large force. Figure 4 shows the schematic and photograph of damper. Table 4 gives its main parameters. The rated current of coils is 2A, the voltage for the electromagnet need supplied by a driver with maximum output voltage is not less than 9V, and the peak power is not less than 20W.

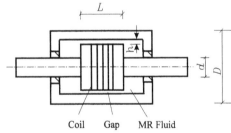

Figure 4. The schematic and photograph of MR damper.

Table 4. Parameters for magnetorheological damper

h(mm)	D(mm)	d(mm)	L(mm)
2	100	40	40

4 EXPERIMENT STUDY ON THE PERFORMANCE OF THE MR DAMPER

In order to realize properties of MR damper and to promote this kind of damper to be used in practical structures, such experiments as magnetic induction, damping fore and response time are done. The experimental results are introduced as follows.

4.1 Magnetic induction of MR damper

The magnetic induction of the damper under the condition of constant electric current direction with maximum current of 2A is tested for 3 times. Figure 5 shows the average tested results.

4.2 Damping force of MR damper

The damping force of MR damper filled with one kind of MR fluid manufactured by us is tested under conditions of that 0.5, 1, 2Hz sinusoid with amplitude of 0.5cm, and 0.25, 0.5, 1Hz sinusoid with amplitude of 1cm, as well as 0.25, 0.5Hz sinusoid with amplitude of 1.5cm. During above experiments, five constant electric current levels, 0, 0.5, 1, 1.5 and 2A is setting to the power amplifier. The damping force versus displacement for the MR damper filled with MR fluid whose properties is shown in Figure 3, due to 0.25Hz and 0.5 Hz sinusoid with amplitude of 1.5cm, is shown in Figure 6. From Figure 6 we can see that:

1 With increasing of electric current, the damping force of MR damper increases too.
2 The ratio of maximum damping force of MR damper under electric current of 2A and 0A is about 2. This means that the controllable multiple is 2.
3 The maximum damping force of MR damper under electric current is about 20kN. The damper can reach higher force if the magnetic induction is improved.

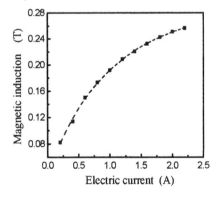

Figure 5. Electric current versus magnetic induction of MR damper

269

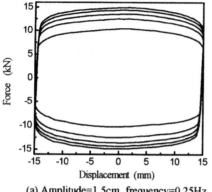

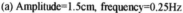

(a) Amplitude=1.5cm, frequency=0.25Hz

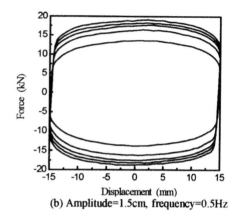

(b) Amplitude=1.5cm, frequency=0.5Hz

Figure 6. Experimentally measured of damping force versus displacement for MR damper (the curves from inner to outer corresponding to electric current of 0, 0.5, 1, 1.5 and 2A)

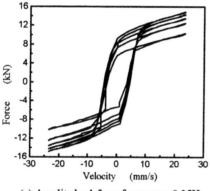

(a) Amplitude=1.5cm, frequency=0.25Hz

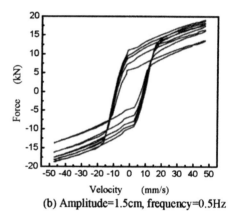

(b) Amplitude=1.5cm, frequency=0.5Hz

Figure 7. Experimentally measured damping force versus velocity for MR damper (the curves of the right half from bottom to top corresponding to electric current of 0, 0.5, 1, 1.5 and 2A)

Figure 7 shows the curves of damping force versus velocity corresponding to the figure 6. From Figure 7, we observe that the damping force produces behavior associated with a plastic material in parallel with a viscous damper, i.e., Bingham plastic behavior.

4.3 Response time of MR damper

As a semi-active control device, besides the magnitude of damping force and controllable range, the response time is another important index for the performance of MR damper. In order to test the response time of MR damper, the following experiments are done:

1 Under the condition of uniform motion of piston shaft, the electric current is sharply changed from one magnitude to another, the force change with displacement is observed.

2 Under the condition of uniform motion of piston shaft, the electric current changes high frequency as shown in Figure 8.

3 Under the condition of piston shaft with sinusoid oscillation, the electric current changes in high frequency as shown in figure 8.

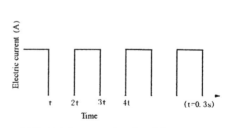

Figure 8. The high frequency changing of the current

270

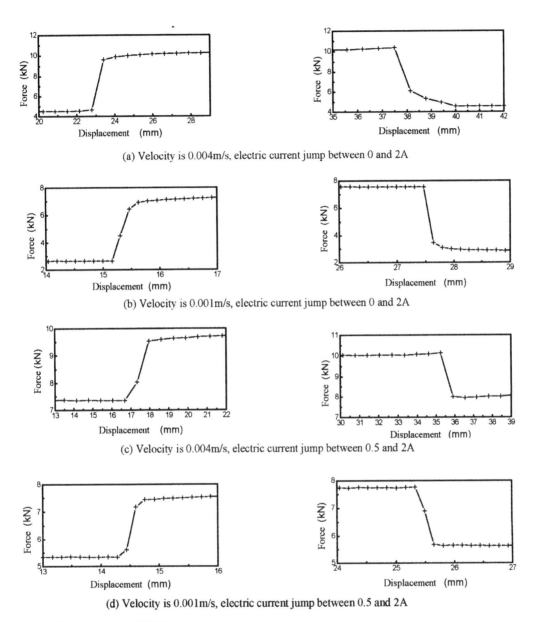

(a) Velocity is 0.004m/s, electric current jump between 0 and 2A

(b) Velocity is 0.001m/s, electric current jump between 0 and 2A

(c) Velocity is 0.004m/s, electric current jump between 0.5 and 2A

(d) Velocity is 0.001m/s, electric current jump between 0.5 and 2A

Figure 9. The response time of MR damper

All of the experiments show almost the same results. Figure 9 shows the force changing with the electric current from 0A to 2A and from 0.5A to 2A, under the condition of that the piston shaft moves with the uniform speed of 0.004m/s and 0.001m/s. In figure 9, the time between each adjacent points is about 0.156 second.

From Figure 9, we can see that:

1 During the changing range of experiment, the velocity almost does not affect response time.

2 When the electric current increases, the force of MR damper can reach stable state in 0.3 second, and can reach 63.2% of the relative force in 0.2 second.

3 When the electric current reduces but does not reach 0A, the response time is almost the same with the condition of electric increase. If electric current reduce to 0A, the time of the force reaches to stable state is longer, about 0.8 second, and reaches to 63.2% of the relative force is about 0.3 second. The above phenomena shows that the

demagnetizing process affects the response time, and it would be better to not reduce the current reduce to 0A.

4 The reason that the response time of MR damper is much longer than those pointed in most references is that some air bubbles are mixed in the damper when MR fluid is filled to the damper (Nakano and Yonekawa 1994). If all of the air bubbles are discharged, the ideal calculation time is about 10 millisecond (except the condition of electric current reduces to 0A).

5 CONCLUSIONS

In this paper, a kind of MR fluid with high shear yielding stress and low setting is compounded, a MR damper is manufactured and its properties are tested. From the experimental results, the following main conclusions are obtained:

1 The shear yielding stress of MR fluid is mainly decided by the volume content of carbonyl iron.
2 The main function of oleic acid is to improve the dissolving capacity of particles, and the OP emulsifying agent is to improve the disperse capacity of particles.
3 For MR fluid, using both oleic acid and OP emulsifying agent as surfactant is better than using only one kind of them.
4 An MR damper with maximum voltage 9V, rated current 2A and peak power less than 20W is designed and manufactured. When filled with MR fluid made by us, the damping force of MR damper can reach about 20 kN.
5 When the current changing, the response time of this damper is about 0.2~0.3 second. The ideal response time of this damper when do not mixed with any air bubbles is about 10 milliseconds.

When the current turn down, in order to reduce the response time and demagnetize effect it would be better for the current not to reach 0A.

6 When the electric current is constant, MR damper can be used as passive control device, however, when the current is adjustable it can also be used as semi-active control device.

6 ACKNOWLEDGMENTS

This research is partially supported by the National Natural Science Foundation under grant No. 50038010 and by the National Outstanding Young Scientist Foundation under grant No. 59625815.

REFERENCES

Ashour O, Rogers C A, Kordonsky W I (1995) Magnetorheological fluids: materials, characterization, and devices *Proc. 6th Int. Conf. on adaptive structures (Key West, Florida)* 23-34.
Carlson J D, Catanzarite D M, St Clair K A (1996) Commercial magneto-rheological fluid devices *Proc. 5th Int. Conf. on ER Fluids, MR fluids and assoc. (Singapore: World Scientific)* 20-28.
Housner G W *et al* (1997) Structural Control: Past, Present, and Future *J. Engineering Mechanics* 123(9) 897-971.
Kordonsky. W I (1993) Magnetorheological Fluids and Their Applications *Materials Technology* 8(11) 240-242.
Nakano M, Yonekawa T (1994) Pressure Response of ER Fluid in a Piston Cylinder—ER Valve Systems. *Proceeding of the 4th International Conference on ER and MR Fluids (ed by Tao R et al)* 477-489.
Spencer B F *et al* (1997) Phenomenological model for magnetorheological dampers *Journal of engineering mechanics* (3) 230-238.
Spencer Jr B F *et al* (1998) Smart dampers for seismic protection of structures: a full-scale study *Proc. 2th Conf. on structural control (Kyoto, Japan)*.

Earthquake Engineering Frontiers in the New Millennium, Spencer & Hu (eds),
© 2001 Swets & Zeitlinger, ISBN 90 2651 852 8

Magnetorheological fluids - ready for real-time motion control

J. David Carlson
Lord Corporation, Materials Division, Cary, North Carolina, USA

ABSTRACT: The past two years have witnessed a blossoming of the commercialization magnetorheological fluid technology. Magnetorheological fluids have now been embraced by a number of manufacturers for inclusion into a diverse spectrum of products. To illustrate the present state of magnetorheological fluid technology, this paper reviews five recent applications wherein the technology has been fully developed into commercial products. As such, these applications represent more than just a laboratory demonstration of basic functionality. Each of these applications fully embodies all of the necessary developments and refinements required to make these magnetorheological fluid devices long-lived in actual service in the field, amenable to mass production techniques, and available at a cost commensurate with the value perceived by the end user.

1 INTRODUCTION

Starting in the mid 1990s, a number of investigators have explored the possibility of using semi-active vibration control devices in civil engineering structures to mitigate the effects of undesirable motions due to hazardous wind or seismic loads. Semi-active devices combine the best features of both passive and active control devices. Semi-active devices offer most of the adaptability of full active systems while retaining much of the simplicity, reliability and cost effectiveness of passive systems. Because of their high force capacity, low power requirement and overall mechanical simplicity, magnetorheological (MR) fluid dampers have been identified as a particularly attractive means of enabling semi-active control in buildings and bridges. Recent laboratory studies have shown that MR fluid dampers can be scaled to sizes appropriate for civil engineering applications.

That MR fluids are suitable for achieving semi-active control beyond the laboratory has been demonstrated (Carlson et al. 1996, Mehri 2000, Conte 2000, Jones 2000). Beginning with the commercialization of MR fluid rotary brakes for use in aerobic exercise equipment in 1995 (Design News 1995), application of MR fluid technology in real-world systems has grown steadily. The past two years have witnessed a rapid expansion of MR fluid technology into commercial applications well beyond the development laboratory. MR fluid technology has been embraced by a number of manufacturers for inclusion into a diverse spectrum of products. To illustrate the present state of MR fluid technology, five recent applications wherein MR flu-

ids have been fully developed into commercial products are reviewed. The MR fluid systems to be described include:

- real-time, semi-active vibration control system for heavy duty truck seating of which there are over 5000 systems presently in use (Design News 1995),
- adjustable MR fluid shock absorber for NASCAR racing automobiles (Carrera 1998),
- rotary MR brake that provides real-time force-feedback in a steer-by-wire system for industrial forklifts (Newton 1999),
- rotary and linear MR dampers that are coupled with industrial pneumatic actuators to enable lowcost position and velocity control (Mehri 2000), and
- MR fluid damper used in an advanced prosthetic knee to enable real-time motion control such that gait automatically adapts to any condition (Jones 2000).

These applications represent more than just a demonstration of basic MR fluid functionality. Each of these applications represents a fully field-proven MR device that embodies all of the necessary refinements required to make it reliable, robust and long-lived in actual service in real-world environments that are quite demanding. Historical concerns about sedimentation, abrasiveness and fluid durability have been solved. These MR devices are also amenable to standard mass production techniques and are commercially available today at prices commensurate with the value perceived by the user. Certainly, much work remains before MR fluid based semi-active control systems could be accepted for protecting civil engineering structural systems against wind and seismic events. However, the proven success of MR fluids in other motion control applications indicates that the possibility of such systems is real.

2 SEMI-ACTIVE MR DAMPERS FOR SEATS

In early 1998 a real-time, semi-active vibration control system became available for use in the seats of Class 8 (eighteen wheeler) trucks (Lord 1998). Manufacturers of premium large trucks such as Western Star and Freightliner offer the MR fluid based Motion Master system as a standard option. A complete semi-active vibration control system including MR fluid damper, sensor, microprocessor, current driver and ancillary cables as shown in Figure 1 is now offered as a retrofit kit to truck operators and is available to anyone over the www for under $300 (Conte 2000). Today, over 5000 MR fluid based, semi-active vibration control systems are in use in heavy-duty, over-the-highway trucks in the United States. These systems receive high praise from the drivers who experience them. They routinely deliver many hundreds of thousands of kilometers of service. Their robustness is illustrated by the fact that there have been no failures in the field.

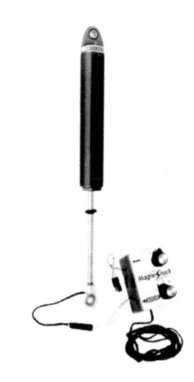

Figure 2. MR fluid based controllable shock absorber for NASCAR racing automobiles.

of maintaining the usual stable of perhaps dozens of different passive hydraulic shock absorbers. Car operators no longer need to physically change shock absorbers to optimize their vehicle for the conditions of the track, weather or tire condition. Rather, they need only perform a few test laps while adjusting the cockpit controls for the MR fluid dampers. A single set of four MR shocks may replace dozens of passive hydraulic dampers. The driver can easily and quickly optimize the left-to-right and fore-to-aft damping in order to optimize car performance on the track. Improved track times by as much as 2 seconds per mile have been realized (Leroy 2000).

Figure 1. MR fluid based semi-active vibration control system for heavy-duty truck seats.

3 MR FLUID SHOCKS FOR RACING CARS

Late in 1998, MR fluid based, adjustable shock absorbers for oval and dirt track racing automobiles were introduced by Carrera (Anderson 1998). A MagneShock by Carrera is shown in Figure 2.

Race-car operators are today able to purchase MagneShock cockpit controlled, primary suspension dampers for about $500. A set of four of these controllable MR fluid dampers eliminate the necessity

4 STEER-BY-WIRE

This is an example of industrial haptics wherein a small MR fluid rotary brake is being used to provide real-time sensory feedback to the operator of an industrial forklift vehicle that is equipped with a steer-by-wire system.

The R14 vehicle, shown in Figure 3, is an all electric forklift introduced by Linde earlier this year (Marjoram 2000). It is intended for close maneuvering and manipulation in confined, clean-spaces such as food handling warehouses with large drivein freezers. There is no mechanical connection between

Figure 3. MR brake provides real-time force-feedback to the operator in conjunction with steer-by-wire operation.

the steering wheel and the ground wheels. Steering is accomplished entirely by electrical control. Rotation of the steering wheel turns an optical encoder which supplies an electrical signal that is transmitted to the drive ground wheel and causes a motor to orient them in the desired direction. The steering wheel and the optical encoder are both mounted the shaft of a MR brake. The brake provides a variable amount of rotational resistance depending on the instantaneous vehicular motion and orientation of the ground wheels. Such tactile feedback to the operator is necessary to insure stable operation. The MR brake and magnetic rotary encoder are packaged into a common package as shown in Figure 4 and mount directly to the dashboard of the forklift.

Figure 4. MR brake and rotary position encoder used in the steer-by-wire application.

5 INDUSTRIAL PNEUMATICS

MR fluid rotary brakes and linear dampers are being used in conjunction with pneumatic actuators to enable open and closed-loop position and velocity control with a precision not normally available with pneumatic systems. MR technology is able to contribute significant added value to pneumatic actuators. The inclusion of a simple MR fluid device is able to add stiffness to an otherwise compliant system overcoming the inherent compressibility of air.

Four different manufacturers have recently introduced pneumatic systems with an MR control option: Parker, phd , Tol-O-Matic and Turn-Act. Figures 5 and 6 show the Turn-Act and Parker embodiments for MR control of rotary position actuators.

Figure 5. Linear pneumatic actuator coupled with MR fluid damper having an external bypass valve to enable precise position and velocity control.

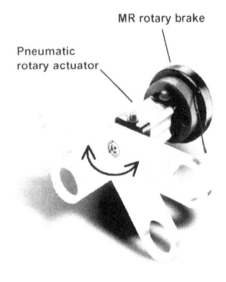

Figure 6. Rotary pneumatic actuator coupled with MR brake to enable precise position and velocity control.

6 ADVANCED PROSTHETICS

Perhaps the most exciting of all of the new MR fluid developments is that of real-time controlled dampers for use in advanced prosthetic devices. The idea here is to use a small MR fluid damper to control, in real-time, the motion of a prosthetic knee. The benefit is a more natural gate that can automatically adapt to virtually any condition.

The HIP or High Intelligence Prosthesis is an above the knee prosthesis that has been introduced by Biedermann Motech this year (Matthis 2000, Carlson 2000, Jones 2000). The basic elements of the system are shown schematically in Figure 7. A group of sensors determines the instantaneous state of the knee: knee angle, swing velocity, axial force and moment. A microprocessor-based controller determines the current needed to be applied to the MR fluid damper to allow for proper motion or locking of the artificial knee based on the instantaneous action being carried out by the user. Once calibrated to a specific user the system automatically adapts in real-time to the users walking speed, stairs or inclination. The HIP system allows the user to move without having to consciously control what is going on with the prosthesis.

As shown in Figure 8, all of the electronics are integrated into the mechanical structure of the prosthetic knee along with the MR damper and battery pack. The entire system operates from a rechargeable battery that provides about two-days of operation between charges. The HIP system with a MR fluid damper provides superior gait control over a wider range of activity than does servo-motor controlled systems costing many times as much.

7 SUMMARY

Five MR fluid-based devices which exemplify the current state of magnetorheological fluid technology have been described. These new applications are fully developed products that have traversed the full course of product development and are today commercially products available to the consumer. All of the cited example applications are based on MR fluid. It is a commentary on the present state of controllable fluid technology that no comparable examples based on electrorheological fluid could be identified.

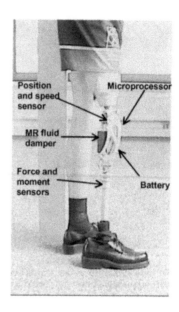

Figure 7. Above-knee prosthesis with real-time gait control provided by MR fluid damper.

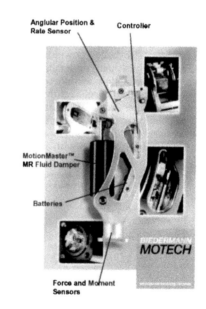

Figure 8. Details of the Prolite Smart Magnetix knee prosthesis.

REFERENCES

Anderson, R. 1998. *Carrera MagneShock^{TM}*, Atlanta: Arre Industries.

Carlson, J.D. 2000. 1st Int. Conf. on Advanced Prosthetics, Newport Beach: Flex Foot.

Carlson, J.D. et al. 1996. Commercial Magneto-Rheological Fluid Devices. In W. A. Bullough (ed), *Proc. 5th Int. Conf. on ER Fluids, MR Fluids and Assoc. Tech.:* 20-28, Singapore: World Scientific.

Conte, F.R. 2000. Lord's Motion Master. *Owner Operator,* Jan/Feb 2000: 39-40.

Design Engineering 1999. A Smoother Ride for Vehicle Seats. *Design Engineering,* Nov. 1999: 54-55.

Design News 1995. Brake Cuts Exercise-Equipment Cost, *Design News,* vol. 50 (25): 28.

Jones, M. 2000. Lord Corp Brings New Life to Knee. *The Cary News,* Oct 28, 2000: 1B-4B.

Leroy, D.A. 2000. Lord Corp., pers. comm.

Lord Corp. 1998. Motion Master Ride Management System. pub no. PB8008a. Cary: Lord Corp.

Matthis, W. 2000. Biedermann Motech Corp.., pers. comm.

Mehri, D. 2000. Economical Actuator Provides Simple Precise Motion Control. *Design News,* vol. 55 (20): 63-65.

Newton, S. 1999. Lansing Linde Corp., pers. comm.

Earthquake Engineering Frontiers in the New Millennium, Spencer & Hu (eds),
© 2001 Swets & Zeitlinger, ISBN 90 2651 852 8

Recent development on seismic isolation, energy dissipation, passive and semi-active control of structures in P. R. China

Fulin Zhou & Qiaoling Xian
Guangzhou University (South China Construction University), Guangzhou, P.R. China

ABSTRACT: This paper describes the most recent state of researches and application development on seismic isolation, energy dissipation, passive tuned mass damper and semi-active control which is based on the energy dissipation for structures. The paper will briefly describe the principle theory, testing and analysis, application on new design or existed buildings, bridges, facilities and other structures using the technique of seismic control of structures in China.

1 INTRODUCTION

This paper considers seismic resistance for structures and seismic control of structures, and suggests a set of new techniques, including isolation, energy dissipation, passive tuned mass damper and semi-active control using in new design or existed buildings, bridges, facilities and other structures in seismic regions in China.

On seismic isolation, many theoretical researches and experiments have been completed in China. A series of tests have been carried out, including cycle loading tests for different kinds of isolators and shaking table tests for structural models with seismic isolation. There are over 400 buildings and 18 bridges with seismic isolation built in China now, and about 80 are being constructed. Some buildings with isolated rubber bearings in Shantou City and Dali City were successfully experienced the earthquake in 1994 and 1995. The results of researches and application of seismic isolation show that the buildings, bridges and industry facilities with seismic isolation are safer also cheaper than traditional seismic resistant systems with base fixed. The struc-tures with base isolation can be ensured to be safe during strong earthquake, also may save the struc-tural cost about 2-10% comparing the structures with base fixed. The seismic isolation will be more wi-dely used in China in coming years. The Chinese design code of seismic isolation has been compiled in 1999.

The energy dissipation has been widely applied in China. Some big projects of high-rise buildings with energy dissipaters (bracing, wall or joints) have been completed. Author and other experts have finished many kinds of full-scale tests of energy dissipaters and model tests on shaking table for structures with energy dissipation systems. The energy dissipation

systems have been used in many kinds of high-rise buildings or other structures to reduce effectively the structural response about 40-60% in earthquake.

Author and other experts in China are doing some researches on other passive control systems such as tuned mass dampers system, which has been used in new design for structures or retrofit for existed structures. This system can be realized by adding one or two stories on the top of existed building to improve the seismic capacity of building structures. One building with this technique has been built in southern China in 1996.

Some tests and researches of active control system, especially semi-active control and hybrid control system have been carried out in China. The results of testing and research show that the semi-control system is more effective, reliable and simple, also save energy input. The semi-active control system may be more suitable to be used in practical structures in China.

2 DEVELOPMENT OF SEISMIC CONTROL OF STRUCTURES IN CHINA

2.1 *Problems existed in traditional anti-seismic system*

1 It is not very safe: It is difficult to control the structural damage level due to the inelastic deformation, and it may be dangerous in severe unpredicted earthquake.

2 It is limited to be used: It is not able to be used in some important structures whose elements are not allowed to work in inelastic range, such as some buildings whose decoration is very expensive, nuclear power plant, museum building, and the buildings with precise instruments in it.

3 It is more expensive: The anti-seismic way of traditional technique is to strengthen the structures: making the columns, beams, shear walls or other structural elements bigger or stronger, increasing the stiffness of the structures. It will cause to increase the seismic load (by increasing the stiffness), then, cause again to strengthen the structures. Finally, it will rise the cost of structure, but the safety level of anti-seismic is still unsure.

The cost of building using traditional anti-seismic technique will rise as bellow comparing with the cost of no-anti-seismic building depending the present design code in China (Zhou, 1995):

For Intensity I-7 (moderate earthquake), increase the cost 3-10%.

For Intensity I-8 (major earthquake), increase the cost 10-20%.

For Intensity I-9 (strong earthquake), increase the cost 15-30%.

For Intensity I>9 (worst earthquake), increase the cost 30-60%, also the design is difficult.

Because the traditional anti-seismic system is not very safe in some cases, also is more expensive, the people do not like to pay more money to "strengthen" the structures for "future unknown" earthquake. It causes the dangerous state of many structures at present in China, including existed buildings, new designing buildings, and the infrastructure structures.

2.2 Application and maturity of seismic of control of structures in China

The urgent task for earthquake engineering is to find out a new system, which needs to satisfy following requirements:

1 Safe: ensure safety for structures in any strong earthquake.
2 Inexpensive: does not increase the cost of structures comparing with the traditional anti-seismic technique.
3 Wide range of application: is able to be used in, both new design structures and existed structures, both important buildings and civil buildings or house buildings, both for protecting the structures and for protecting the facilities inside the buildings.

This new system seems to have been found–the seismic control of structures, which consists of seismic isolation, energy dissipation, passive control, active control and hybrid control.

The range of application and technical maturity of Seismic Control of Structures in China are shown in Table 1 (Zhou, 1997).

3 SEISMIC ISOLATION

3.1 General state of development

The history of application on seismic isolation in China has proved that the isolation system is more safe, economic and reasonable than the traditional structure system, especially suitable for using in the cases where the earthquakes are unpredictable. In mainland China, About 40 existed ancient buildings, towers and temples built thousands years ago with isolation concept have successfully withstood several strong earthquakes.

In recent years, there are 300 buildings and 9 bridges being built with isolation, which isolators mostly were rubber bearings, only a few were sand

Table 1. The range of application and maturity of seismic control of structure

Name of control	Ranges of application	Technical maturity
Seismic isolation	2- 30 stories buildings (New design or existed) bridges, subway Equipment or facilities	mature technique Rich theoretical and testing results Widely application Successfully experienced the strong earthquake in China.
Energy dissipation	20-150 stories high-rise building Long span bridge High tower or skeleton application for seismic or winds	mature technique Rich theoretical and testing results
Passive control	High-rise buildings (TMD,TLD,etc) Long span bridges High tower or skeleton application for seismic or winds	Basically mature technique Some research and testing results
Active control	High rise buildings Long span bridges High tower or skeleton application for seismic or winds	does not become a mature technique some research and testing results

sliding layer or graphite layer. In order to solve some problems for more wide application of isolation, a great numbers of tests for rubber bearings and shaking table tests for building models with rubber bearings were carried out (Zhou, 1994,1997). Based on the testing results, analysis and application, a set of optimal design methods, computer programs were raised in China (Zhou, 1988, 1992, 1997). The "Chinese design Guideline for seismic isolating structures with rubber bearings" will be published (Zhou, 1996).

3.2 Progress in technique and economy using seismic isolation in China

Many significant advantages of structures using rubber bearing isolators in China (Table 2):

Table 2. Isolation structure compares with traditional antiseismic structure.

	Traditional anti-seismic structure	Isolation structure
Acceleration response	1.00	1/4 - 1/12
Working state of structure during earthquake	inelastic	elastic
Building cost	1.00	0.80 - 0.95

1 Safe in strong earthquake: It is very effective to reduce the response of structures in earthquake and is able to prevent the structure from damage or collapse in earthquake. So it can ensure the structures with rubber bearings to be safe in any strong earthquakes. Comparing the seismic isolation structures with the traditional anti-seismic structures, the response of isolating structures can be reduce to 1/4-1/12 of the response of traditional anti-seismic structures, according the testing results and the records in real earthquake also (Zhou, 1997).
2 Save the building cost in many cases: Comparing the seismic isolation structures with the traditional anti-seismic structures, the building cost of isolation structures can be saved 5-20 % of the building cost, because re-designing the supper structure which seismic response is very small, according the final statistics results of 30 buildings with rubber bearings completed in southern, western and northern China.
3 Wide ranges of application: The technique of seismic isolation of structure can be used in, both

new design structures and existed structures, both important buildings and civil buildings or house buildings, both for protecting the structures and for protecting the facilities inside the buildings.
4 Long working life of rubber bearings isolators: According the permanent testing and investigation, the safely working life of rubber bearings are over 70-100 years which is longer than the working life of structure itself (Zhou, 1997).

3.3 Four kinds of location of rubber bearings on structures

There are five kinds of locations putting isolating rubber bearings on structures in China:
1 Put on the base of building if the building is with basement (Fig. 1).
2 Put on the top of the first story of buildings if the building is without basement (Fig. 2).
3 Put on certain stories between different kinds of parts in structure (Fig. 3).
4 Put on the top of piers or columns of bridges or other structures (Fig. 4).
5 Put on the base of tanks or other facilities (Fig.5).

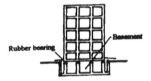

Figure 1. Bearings on the base

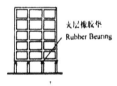

Figure 2. Bearings on the top of the first story

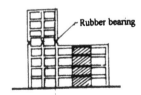

Figure 3. Bearings on certain stories

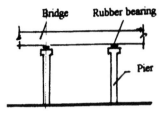

Figure 4. Bearings on top of piers

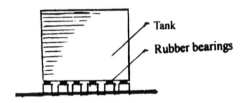

Figure 5. Bearings on base of tank

3.4 *Products of rubber bearings in China*

The rubber bearing is as isolator in most cases in China. A very large factory, "Vibro-Tech" Company, produces high quality also low price rubber bearings, to satisfy the great demands of rubber bearings in China and other countries in the world. The "Vibro-Tech" produces many different types of Rubber Bearings, which main characteristics are following:

Diameter (mm)	200 - 1000
Design compression loads (kn.)	400 - 16000
Design maximum displacement (mm)	130 - 800
Horizontal stiffness (kn./mm)	0.30 - 2.30
Vertical stiffness (kn./mm)	200 - 4300
Equivalent damping ratio (%)	8 – 26.

4 ENERGY DISSIPATION

An energy dissipation system is formed by adding some energy dissipaters into the structure. The energy dissipaters provide the structure with large amounts of damping which will dissipate most vibration energy from vibration sources previous to the structural response reaching the limitation, then ensure the structure to be safe in earthquake or to satisfy the using requirement in wind. The energy dissipaters may be set on the bracing, walls, joints, connection parts, nonstructural elements or any suitable spaces in structures, which may reduce 40-60% of the structural response comparing the traditional structure without energy dissipaters. This technique is very reliable and simple, suitable to be used for general or important new or existed buildings or facilities in seismic regions.

4.1 *Mechanism and testing of energy dissipation system*

For building structure or facilities with energy dissipaters, the energy balance equation in any instant of time during earthquake is:

$$Ein=Ep+Ek+Ep+Eb \qquad (1)$$

where Ein is energy input to the structure, Ep is potential energy in structural vibration, Ek represents kinetic energy in structural vibration, Ed is the energy dissipated by viscous damping of structure or facilities, and Eb is the energy dissipated by energy dissipater.

Testing and research indicates that the energy dissipater can dissipate about 90% of the total energy input at the end of earthquake. The effect of Ed are relatively small which can be neglected in Equation (5), then the energy dissipating design for earthquake resistance need to be satisfied with:

$$Ein<Eb \qquad (2)$$

For calculating energy input Ein, the system can be considered as multi-degree of freedom system. The energy dissipated by energy dissipaters Eb depends on the area enclosed by Load -Displacement loop curve (Zhou, 1982, 1987).

The energy dissipaters provide the structure or facilities with large amounts of damping which will dissipate most vibration energy from earthquake or wind previous to the structural response reaching the limitation (damaging the buildings or loosing the performance function of facilities), then ensure the structure or facilities to be no any damage or to keep the normal performance function during earthquake or strong wind. The shaking table testing of large models with energy dissipaters shown that, the seismic response of structure with energy dissipaters is decreased 20-40% comparing the structure without energy dissipaters.

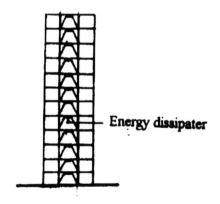

Figure 6. Energy dissipation bracing

4.2 Design and application of energy dissipation system

There are four kinds of energy dissipation system being used now:

1 Energy dissipaters put on the bracing, setting along whole structural height for regular building to rise the capacity of anti-seismic, or only setting in some spaces (or stories) of structure for irregular building to avoid torsion damage during earthquake, shown in Figure 6.

2 Energy dissipaters put on the walls in weak stories of structure to improve the structural working state, avoiding dangerous damage of weak stories during earthquake, shown in Figure 7.

3 Energy dissipaters put on the gaps between adjacent buildings to avoid impact damage during earthquake, shown in Figure 8.

4 Energy dissipaters put on the connection joint between over-bridges and buildings to avoid damage caused by unreasonable connection of different buildings with different response during earthquake, shown in Figure 9.

5 TUNED MASS DAMPER SYSTEM

The system is, on certain position (such as on roof) in new or existed structure (main structure), places an additional Filial Structure (possesses mass, stiffness and damping) which natural period is nearly equal to the natural period of main structure. The dynamic characteristics of structure system are changed. During the earthquake, the Filial Structure will move against the direction of main structures vibration then reduce the response of main structure.

This Filial Structure may be formed by adding one or more stories supported by rubber bearings on the roof of main building structure (Figure 10), or adding a certain mass supported by rubber bearings on the roof or other floors in main building structure (Figure 11).

The shaking table tests show that, the response of main structure adding Filial structure is reduced 30-50% of response of structure without Filial structure (Zhou, 1996,1997).

6 SEMI-ACTIVE CONTROL SYSTEM

Semi-active control system is formed by adding some simple devices on the bracing or walls in structure (Figure 12), which devices provide the structure with variable damping and stiffness to re duce the seismic response of structure in earthquake. The devices will work as energy dissipaters during moderate earthquake or wind, and will work as active controller during unexpected strong earthquake by inputting

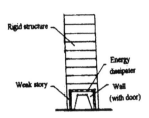

Figure 7. Energy dissipation wall

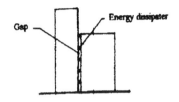

Figure 8. Energy dissipation connection

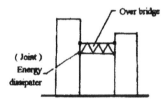

Figure 9. Energy dissipation joints

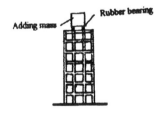

Figure 10. Adding stories on top of building

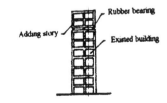

Figure11. Adding mass on stories

283

very small power. This semi-active control system is much more reliable and simple than general active control system, and more effective to reduce the structural response than other passive control system. This technique is suitable to be used in some important buildings or facilities.

Semi-active control system--Active Variable Stiffness/Damper AVSD, corresponding advanced device and algorithm based on the energy dissipation passive control, active control and other semi-control systems (Zhou, 1998). Author and other experts combine Active Variable stiffness (AVS) and Active Variable Damping (AVD) to became one system AVSD, which possesses the all advantages of AVS and AVD. So AVSD is more effective to control the structural response in very wide range of vibration input, more saving energy, more stable and reliable in deferent working state, more simple, economical and practical in application.

The theoretical model of AVSD is shown in Figure 13, which shows that j th AVSD device located on i th story of structure. Where m_i is the mass of i th story of structure, k_i and c_i are the original stiffness and damping of i th story of structure, k_{0j} and c_{0j} are the stiffness and damping of AVSD system. When the system switch on the k_{0j}, the AVSD works as AVS system, while switch on the c_{0j}, the AVSD works as AVD system. How to switch on, is controlled by the control algorithm, according to the sign of value of response of the structure, based on analysis of mechanism of AVSD within every sample period in vibration of structure.

AVSD system could provide two kinds of working state to control the vibration of structure: working as AVS system provided from closing the valve, working as AVD system from opening the valve. Also, this AVSD could amplify the story drift for damper, which make the system to dissipate more energy input from ground motion.

The results of shaking table testing and analysis show that (Zhou, 1998), AVSD is more effective than other control system. Comparing AVS, AVSD is working in that state the original structure is replaced by an energy dissipation system, which is more effect and reliable to control and reduce the structural response (Figure 14). The control algorithm suggested by authors is reasonable. It can reach the optimal state to control the structural response in earthquake or wind load working as AVS system provided from closing the valve, working as AVD system from opening the valve. Also, this AVSD could amplify the story drift for damper, which make the system to dissipate more energy input from ground motiondrift for damper, which make the system to dissipate more energy input from ground motion.

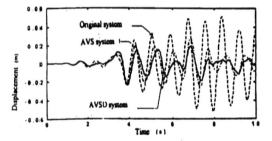

Figure 14. Time history of displacement response for AVSD system

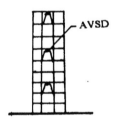

Figure 12. AVSD on certain stories

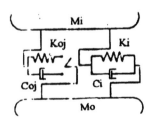

Figure 13. Theoretical model of AVSD system

REFERENCES

Zhou Fu Lin, Yu Gong Hua (1982), "Testing and Design for Energy Dissipation Bracing." *Eport of Earthquake Engineering*, Vol.1. 1982.

Zhou Fu Lin, S.F.Stiemer, S.Cherry (1988), "Design Method of Isolating And Energy Dissipating System for Earthquake Resistant Structures." *Proc. of 9th World Conference on Earthquake Engineering*. Tokyo-Kyoto. Aug.1988.

Zhou Fu Lin, (1989), "Development and Application for Technique of Isolation and Energy Dissipation and Control of Structural Response (1st Part) " *Journal of International Earthquake Engineering*. VoL.4, 1989.

Zhou Fu Lin (1990a), "Development and Application for Technique of Isolation and Energy Dissipation and Control of Structure Response (2nd Part)" *Journal of International Earthquake Engineering*. Vol.1, 1990.

Zhou Fu Lin, Stiemer S.F. and Cherry S. (1990b), "A New Isolation and Energy Dissipating System for Earthquake Resistant Structures". *Proc. of 9th European Conference on Earthquake Engineering*. Moscow, Sept. 1990.

Zhou Fu Lin (1991) " New System of Earthquake Resistant Structures in Seismic Zone." *Recent Development and Future Trends of Computational Mechanics in structural Engineering*. Published by Elsevier Applied Science Publishers LTD, London & New York.

Zhou Fu Lin, et al. (1992), " Recent Research , Application, and development of Seismic Control for Structures ". *Proc. of China National Conference on Seismic Control for Structures*. 1992, Wuhan, China.

Zhou Fu Lin, Kelly J M, Fuller K N G, Pan T C, et al, (1994) " Recent Research Development and Application on Seismic Isolation of Buildings in P R China". *Proc. of International Workshop IWADBI*, Shantou, 1994.

Zhou Fu Lin (1995),"Technical Report on Mission to Santiago, Chile" as Consultant to attend the *International Post-SMiRT conference Seminar on Seismic Isolation Passive Energy Dissipation and Active Control of Vibrations and Structures*, 1995.8.

Zhou Fu Lin (1996), " Discussion on Compiling Chinese Design Code of Seismic Isolation", *Proc.of The PRC-USA Bilateral Workshop on Seismic Codes*, Dec. 1996, Guangzhou, China.

Zhou Fu Lin (1997), "Seismic Control of Structures", *Chinese Seismic Publishing House*.

Zhou Fu Lin, Xie Li Li, Yun Weiming and Tan Ping (1998), "Optimum semi-Active Control System AVSD whis Advanced Device and Algorithm", *Proc. of 2 WCSC,* Kyoto, July 1998.

Earthquake Engineering Frontiers in the New Millennium, Spencer & Hu(eds),
©2001 Swets & Zeitlinger, ISBN 90 265 1852 8

Introducing a benchmark control problem for a cable-stayed bridge subjected to seismic excitation

S.J. Dyke & J.M. Caicedo
Washington University, St. Louis, MO 63130, USA

L.A. Bergman & G. Turan
University of Illinois, Urbana, IL 61801, USA

ABSTRACT: This paper defines the first-generation benchmark structural control problem for cable-stayed bridges. The problem is based on the cable-stayed bridge that will span the Mississippi River and is currently under construction in Cape Girardeau, Missouri, USA. The goal of this study is to provide a test bed for the development of strategies for the control of cable stayed-bridges. A three-dimensional finite element evaluation model has been developed to represent the complex behavior of the full-scale benchmark bridge. Evaluation criteria are presented for the design problem, consistent with the goals of seismic response control of a cablestayed bridge. Control constraints are also provided. Each participant in this benchmark study is given the task of defining, evaluating and reporting on his or her proposed control strategy. Strategies may be passive, active, semi-active, or a combination thereof. The problem has been made available for downloading through the Internet in the form of a set of MATLAB® programs. A sample control design is included to guide participants through the benchmark problem.

1 INTRODUCTION

Benchmark problems have recently been recognized as a means to compare and contrast various structural control strategies (Caughey, 1998). Benchmark structural control problems allow researchers to apply various algorithms, devices, and sensors to a specified problem and make direct comparisons of the system performance on a single structure in terms of a specified set of performance objectives. Additionally, these problems may include control constraints and hardware models to more accurately portray the types of implementation issues and constraints one must consider in reality.

All of the benchmark problems considered so far have focused on the control of buildings (Spencer et al., 1998; Spencer et al., 1999; Yang et al., 1998). During the *Second International Workshop on Structural Control,* held in Hong Kong in December, 1996, working groups were formed to plan the development of a series of benchmark control problems for various classes of civil engineering structures. The working group on bridge control recognized that the control of flexible bridge structures represents a new, difficult, and unique problem, with many complexities in modeling, control design and implementation. Cable-stayed bridges exhibit nonlinear complex behavior in which the translational and torsional motions are often strongly coupled. Clearly, the control of very flexible bridge structures has not been studied to the extent that buildings have. As a result, little exper-

tise has been accumulated. Thus, the control of seismically excited cable-stayed bridges presents a challenging problem to the structural control community. An analytical feasibility study was performed on a well-studied and documented bridge model in order to identify and resolve important issues associated with the control of a flexible bridge structure (Schemmann et al., 1998). Subsequently, a benchmark problem on the control of cable-stayed bridges was initiated. The benchmark problem is based on the cable-stayed bridge that is currently under construction in Cape Girardeau, Missouri, USA. Construction of the bridge is expected to be completed in 2003. Instrumentation will be installed in the Cape Girardeau bridge and surrounding soil during the construction process (Celibi, 1998), allowing for extensive monitoring during its lifetime.

This paper introduces the problem definition for the first generation of benchmark structural control problems for cable-stayed bridges. The problem is available for downloading on the benchmark web site in the form of a set of MATLAB® equations *(http://wusceel.cive.wustl.edu/quake/).* A three-dimensional evaluation model was developed to model the complex behavior of the full scale benchmark bridge. A linear evaluation model, using the equations of motion generated around the deformed equilibrium position, is deemed appro-priate. The effects of soilstructure interaction are neglected. For this initial benchmark problem the ground acceleration is assumed to be applied to the bridge structure longitudinally and acts uniformly at all supports.

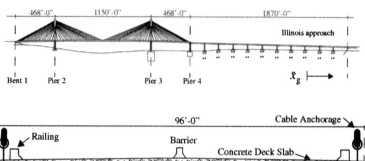

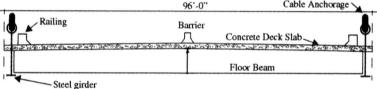

Figure 1. Cape Girardeau Bridge. Cross section of bridge deck.

To evaluate the proposed control strategies in terms that are meaningful for cable-stayed bridges, appropriate evaluation criteria and control design constraints are specified within the problem statement. Designers/researchers participating in this benchmark study will define the devices, sensors and control algorithms used, and evaluate them in the context of their proposed control strategies. These strategies may be passive, active, semi-active or a combination thereof.

2 BENCHMARK BRIDGE

The cable-stayed bridge used for this benchmark study is the Missouri 74-Illinois 146 bridge spanning the Mississippi River near Cape Girardeau, Missouri, designed by HNT13 Corporation[1]. The bridge is currently under construction and is scheduled to be completed in 2003. Earthquake load combinations in accordance with American Association of State Highway and Transportation Officials (AASHTO) division I-A specifications were used in the design. As shown in Figure 1, it is composed of two towers, 128 cables, and 12 additional piers in the approach bridge from the Illinois side.

The bridge has a total length of 1205.8 m (3956 ft). The main span is 350.6 m (1150 ft), the side spans are 142.7 m (468 ft), and the approach on the Illinois side is 570 m (1870 ft). A cross section of the deck is shown in Figure 1. The bridge has four lanes plus two narrower bicycle lanes, for a total width of 29.3 m (96 ft). The deck is composed of steel beams and prestressed concrete slabs. Steel ASTM A709 grade 50W is used, with an f_y of 344 MPa (50 ksi). The concrete slabs are made yof

prestressed concrete with a f'_c of 41.36 MPa (6 ksi). Additionally, a concrete barrier is located at the center of the bridge, and two railings are located along the edges of the deck.

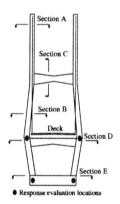

Figure 2. Diagram of the towers.

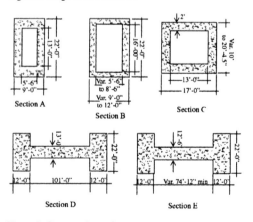

Figure 3. Cross sections of towers.

[1] 1.FTNTB Corporation, 1201 Walnut Suite 700, Kansas City, Missouri, 64106.

288

The 128 cables are made of high-strength, low-relaxation steel (ASTM A882 grade 270). Sixteen 6.67 MN (1,500 kip) shock transmission devices are employed in the connection between the tower and deck. These devices are installed in the longitudinal direction to allow for thermal expansion of the deck. Under dynamic loading these devices are extremely stiff. Additionally, in the transverse direction, earthquake restrainers are employed at the connections between the tower and the deck, and the deck is constrained in the vertical direction at the towers. The bearings at bent 1 and pier 4 are designed to permit longitudinal displacement and rotation about the transverse and vertical axes. Soil-structure interaction is not expected to be significant in this site as the foundations of the cable-stayed portion of the bridge are attached to bedrock.

The H-shaped towers have a height of 100 m (336 ft) at pier 2 and 105 m (356 ft) at pier 3. Each tower supports a total 64 cables. The towers are constructed of reinforced concrete with a resistance, f'_c of 37.92 MPa (5.5 ksi). The cross section of each tower changes five times over its height, as shown in Figures 2 and 3. The deck consists of a rigid diaphragm made of steel, with a slab of concrete at the top.

A three dimensional finite element model of the Cape Girardeau bridge has been developed. The finite element model employs beam elements, cable elements and rigid links. The nonlinear static analysis is performed in ABAQUS® (1998), and the element mass and stiffness matrices at equilibrium are output to MATLAB® for assembly. Subsequently, the constraints are applied, and a dynamic condensation is performed to reduce the size of the model. A linear evaluation model is used in this benchmark study, however, the stiffness matrices used corresponding to the deformed state of the bridge considering dead loads. The first ten natural frequencies of the evaluation model are 0.290, 0.370, 0.468,

0.516, 0.581, 0.649, 0.669, 0.697, 0.710, and 0.720 Hz.

To make it possible for designers/researchers to place devices acting longitudinally between the deck and tower, a modified evaluation model is constructed in which the existing connections between the tower and deck are disconnected. If a designer/researcher specifies devices at these nodes, the modified model will be assembled as the evaluation model, and user-specified control devices must connect the deck to the tower. As expected, the natural frequencies of this modified model are lower than those of the nominal bridge model. The fundamental frequency of this modified model, in the absence of user-specified devices between the tower and deck is 0.162 Hz. Note that the uncontrolled structure used as a basis of comparison for the controlled system corresponds to the original model in which the deck-tower connections are fixed (*i.e.,* the dynamically stiff shock transmission devices are present).

3 BENCHMARK PROBLEM STATEMENT

Designers/researchers are given the task of designing a control system for the benchmark bridge and evaluating its performance based on a specified set of evaluation criteria. The files contained in the MATLAB® problem statement will generate the evaluation models of the bridge. This includes specifying the type and location of each of the control devices and sensors, as well as the control algorithm. The procedure to follow for development and evaluation of a control strategy is described in the flow chart in Figure 4.

Devices may be attached to any active node of the bridge model, including the connection between the deck and the tower. A sample active control design is included to guide benchmark participants through

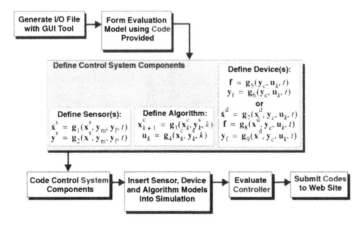

Figure 4. Flow chart of benchmark solution procedure.

289

the required constraints and design criteria. The following sections describe the procedure to be followed in using the benchmark bridge problem statement.

3.1 Interfacing with the benchmark bridge evaluation model

To interface with the benchmark files, the researcher/designer will select the components of each proposed control strategy, including: the control device locations; the measured outputs sent to the sensors, y_m ; the evaluation outputs y_e; and the connection outputs y_c. These inputs and outputs are specified within an input/output file provided with the benchmark problem statement. A MATLAB® graphical user interface is provided to simplify this procedure (see Figure 5). However, this information can be inserted directly into the input/output file in terms of the node numbers, if preferred.

The evaluation model and earthquake inputs are fixed for this benchmark problem. Models of the sensors and control devices are to be defined by the par ticipants. Dynamics of control devices may be neglected. However, dynamic device models may be employed by participants if they so choose.

3.2 Benchmark problem simulation

Each proposed control strategy is evaluated through simulation with the evaluation model. The SIMULINK® program shown in Figure 6 has been developed for evaluation of the control strategies. Participants should insert models of their sensors, devices, and algorithms into the simulation to evaluate their proposed control strategies. The simulation uses the analysis tool provided by Ohtori and Spencer (1999). This tool allows the user to implement the compiled C code from within the MATLAB® environment through a SIMULINK® block to simulate the responses of a seismically-excited structural system. This tool solves the incremental equations of motion using the Newmark-β method in combination with the pseudo-force method.

3.3 Evaluation criteria

The seismic response of cable-stayed bridges is highly coupled. Lateral excitations can generate vertical and longitudinal motions. For cable-stayed bridges subjected to earthquake loading, the relevant responses are related to the structural integrity of the bridge rather than to serviceability issues. Thus, in

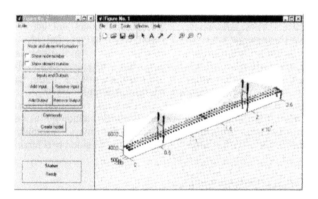

Figure 5. GUI interface for actuator/sensor placement.

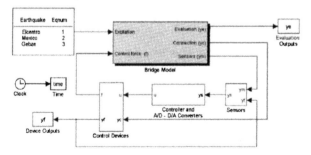

Figure 6. SIMULINK model for benchmark cable-stayed bridge problem.

Table 1: Evaluation Criteria.

	Base Shear	Shear at Deck Level	Overturning Moment	Moment at Deck Level
PEAK RESPONSE	$J_1 = \max\limits_{\substack{El\ Centro \\ Mexico\ City \\ Gebze}} \dfrac{\max\limits_{i,t}\|F_{bi}(t)\|}{F_{0b}^{max}}$	$J_2 = \max\limits_{\substack{El\ Centro \\ Mexico\ City \\ Gebze}} \dfrac{\max\limits_{i,t}\|F_{di}(t)\|}{F_{0d}^{max}}$	$J_3 = \max\limits_{\substack{El\ Centro \\ Mexico\ City \\ Gebze}} \dfrac{\max\limits_{i,t}\|M_{bi}(t)\|}{M_{0b}^{max}}$	$J_4 = \max\limits_{\substack{El\ Centro \\ Mexico\ City \\ Gebze}} \dfrac{\max\limits_{i,t}\|M_{di}(t)\|}{M_{0d}^{max}}$
	Stay Cable Tension	**Displacement at Abutments**	**Base Shear**	**Shear at Deck Level**
	$J_5 = \max\limits_{\substack{El\ Centro \\ Mexico\ City \\ Gebze}} \max\limits_{i,t} \left\|\dfrac{T_{ai}(t)-T_{0i}}{T_{0i}}\right\|$	$J_6 = \max\limits_{\substack{El\ Centro \\ Mexico\ City \\ Gebze}} \max\limits_{i,t} \left\|\dfrac{x_{bi}(t)}{x_{0b}}\right\|$	$J_7 = \max\limits_{\substack{El\ Centro \\ Mexico\ City \\ Gebze}} \dfrac{\max\limits_{i}\|F_{bi}(t)\|}{\|F_{0b}(t)\|}$	$J_8 = \max\limits_{\substack{El\ Centro \\ Mexico\ City \\ Gebze}} \dfrac{\max\limits_{i}\|F_{di}(t)\|}{\|F_{0d}(t)\|}$
	Overturning Moment	**Moment at Deck Level**	**Stay Cable Tension**	**NORMED RESPONSE**
	$J_9 = \max\limits_{\substack{El\ Centro \\ Mexico\ City \\ Gebze}} \dfrac{\max\limits_{i}\|M_{bi}(t)\|}{\|M_{0b}(t)\|}$	$J_{10} = \max\limits_{\substack{El\ Centro \\ Mexico\ City \\ Gebze}} \dfrac{\max\limits_{i}\|M_{di}(t)\|}{\|M_{0d}(t)\|}$	$J_{11} = \max\limits_{\substack{El\ Centro \\ Mexico\ City \\ Gebze}} \max\limits_{i} \dfrac{\|T_{ai}(t)-T_{0i}\|}{T_{0i}}$	
	Control Force	**Device Stroke**	**Device Peak Power**	**Device Total Power**
CONTROLLER	$J_{12} = \max\limits_{\substack{El\ Centro \\ Mexico\ City \\ Gebze}} \max\limits_{i,t} \dfrac{f_i(t)}{W}$	$J_{13} = \max\limits_{\substack{El\ Centro \\ Mexico\ City \\ Gebze}} \max\limits_{i,t} \dfrac{\|y_i^d(t)\|}{x_0^{max}}$	$J_{14} = \max\limits_{\substack{El\ Centro \\ Mexico\ City \\ Gebze}} \dfrac{\max\limits_{t}\left[\sum\limits_i P_i(t)\right]}{\dot{x}_0^{max}\,W}$	$J_{15} = \max\limits_{\substack{El\ Centro \\ Mexico\ City \\ Gebze}} \dfrac{\int_0^{t_f}\sum\limits_i P_i(t)\,dt}{x_0^{max}\,W}$
	$J_{16} = $ #of control devices	$J_{17} = $ #of sensors		$J_{18} = \dim(\mathbf{x}_k^c)$

evaluating the performance of each control algorithm, the shear forces and moments in the towers at key locations (see Figure 2) must be considered. Additionally, the tension in the cables should never approach zero and should remain close to the nominal pretension.

A set of eighteen criteria have been developed to evaluate the capability of each control strategy. The first five evaluation criteria consider the ability of the controller to reduce peak responses; the second five consider normed responses over the entire time record; and the last seven consider the requirements of the control system itself.

A one dimensional ground acceleration is applied in the longitudinal direction. For each control design, the evaluation criteria should be deteremined for each of three earthquake records provided in the benchmark package: *i) EL Centro*. The North-South component recorded at the Imperial Valley Irrigation District substation in El Centro, California, during the Imperial Valley, California earthquake of May, 18, 1940; ii) *Mexico City*. Recorded at the Galeta de Campos station with site Geology of Meta-Andesite Breccia in Sep. 19, 1985; iii) *Gebze*. The NorthSouth component recorded at the Gebze Tubitak Marmara Arastirma Merkezi on Aug. 17, 1999. These three earthquakes are each at or below the design peak ground acceleration level for the bridge. The evaluation criteria are summarized in Table 1.

The values of the uncontrolled responses for the three earthquakes required to calculate the evaluation criteria are provided in the MATLAB® files and in Dyke et al. (2000). All eighteen criteria should be reported for each proposed controller. The Mexico City, El Centro, and Gebze earthquakes should all be considered in determining the evaluation criteria. However, designers/researchers are encouraged to include additional criteria in their results if, through these additional criteria, their results demonstrate an overall desirable quality.

3.4 *Control constraints and procedures*

To allow researchers/designers to compare and contrast various control strategies, each of the controllers must be subjected to a uniform set of constraints and procedures. A set of constraints has been formulated for this benchmark problem to ensure that the proposed control strategies can realistically be implemented on the real structure. For instance, the precision and span of the A/D and D/A converters used for implementation of control algorithms are fixed, and the amount of noise in the sensors is specitied. When active systems are implemented, participants should discuss the robust stability of their control systems. Additionally, participants must provide justification for sensors and devices that are employed in their proposed control strategies. Force and stroke limitations of the devices used should be provided, and the controller should meet these limitations upon implementation in the system. Furthermore, tension in the stay cables must remain within a specified range of values to reduce the possibility of unseating or failure of the cables. These constraints and procedures are described in detail in Dyke et al. (2000).

4 SUMMARY

A summary of the benchmark problem for control of a seismically-excited cable-stayed bridge has been presented. Evaluation criteria, consistent with the goals of controlling cable-stayed bridges subjected to earthquake loading, have been presented. The evaluation model of the Cape Girardeau cable-stayed bridge, the MATLAB$^{®}$ files used for the sample control design, and the simulation model are available at:
http://wusceel.cive.wustl.edu/quake/
If you cannot access the World Wide Web or have questions regarding the benchmark problem, please contact Dr. Shirley Dyke via e-mail at:
sdyke@seas.wustl.edu.
Participants in the benchmark study will submit their control designs and supporting MATLAB$^{®}$ files electronically for inclusion on the benchmark web page.

5 ACKNOWLEDGMENTS

This research is supported in part by National Science Foundation Grant No. CMS 97-33272 (Dr. S.C. Liu, Program Director). The authors thank Shyam Gupta and Bill Strossener from the MODOT for information on the Cape Girardeau Bridge. The helpful advice of Mr. Steven Hague (HNTB Corp.), Prof. Yozo Fujino (Univ. of Tokyo), Prof. Masato Abe (Univ. of Tokyo), Prof. Hirokazu Iemura (Kyoto Univ.), Prof. Joel Conte (UCLA), and Prof. Fabio Biondini (Politecnico di Milano) is gratefully acknowledged. Additional input provided by members of the ASCE Task Group on Benchmark Problems and member-at-large of the structural control community is also recognized.

REFERENCES

ABAQUS$^{®}$ (1996). Hibbitt, Karlsson & Sorensen Inc. Pawtucket, RI.

Caicedo, J.M., Dyke, S.J., Turan, G and Bergman, L.A., (2000). "Comparison of Modeling Techniques for Dynamic Analysis of a Cable-Stayed Bridge." *Proceedings of the Engineering Mechanics Conference,* ASCE, Austin, Texas, May 21-23.

Caughey, T. K., (1998). "The Benchmark Problem." *Earthquake Engineering and Structural Dynamics,* Vol. 27, pp. 1125.

Celebi, M., (1998), *Final Proposal for Seismic Instrumentation of the Cable-Stayed Girardeau (MO) Bridge,* U.S. Geological Survey.

Dyke, S.J., Turan, G, Caicedo, J.M., and Bergman, L.A.(2000). "Benchmark Control Problem for Seismic Response of Cable-Stayed Bridges," *<http://wusceel.cive.wustl.edu/quake/*

Hague, S. (1997). "Composite Design for Long Span Bridges." *Proceedings of the XV ASCE Structures Congress,* Portland, Oregon.

MATLAB$^{®}$ (1997). The Math Works, Inc. Natick, Massachusetts.

Ohtori, Y. and Spencer, B.F., Jr. (1999). "A MATLAB$^{®}$-Based Tool for Nonlinear Structural Analysis," *Proceedings of the 13th Engineering Mechanics Conference,* Baltimore, Maryland, June 13-16.

Schemmann, A.G., Smith, H.A., Bergman, L.A., and Dyke, S.J. (1998), "Feasibility Study: Control of a Cable-Stayed Bridge Model, Part 1: Problem Definition" *Proceedings of the Second International Conference on Structural Control,* Vol. 2, John Wiley & Sons, England, pp. 975-979, Kyoto, JAPAN, June 30-July 2.

Spencer Jr., B.F., Dyke, S.J. and Deoskar, H.S. (1998). "Benchmark Problems in Structural Control - Part 1: Active Mass Driver System," and "Part 11: Active Tendon System," *Earthquake Engineering and Structural Dynamics,* Vol. 27: 1127-1147.

Spencer Jr., B.F., Christenson, R. and Dyke, S.J. (1999). "Next Generation Benchmark Control Problem for Seismically Excited Buildings," *<http://www.nd.edu/~quake/>.*

Yang, J.N., Wu, J.C., Samali, B. and Agrawal, A.K. (1998). "A Benchmark Problem For Response Control of Wind-Excited Tall Buildings," *Proc. of the 2nd World Couf. on Structural Control* <also see *http://www.eng.uci.edu/ anil/benchmark.html>.*

Earthquake Engineering Frontiers in the New Millennium, Spencer & Hu(eds),
©2001 Swets & Zeitlinger, ISBN 90 265 1852 8

Ultimate vs. serviceability limit state in designing bridge energy dissipation devices

F. Casciati & L. Faravelli
Dept. of Structural Mechanics, University of Pavia, Pavia, Italy

M. Battaini
ALGA spa,via Olona 12, Milan, Italy

ABSTRACT: Energy dissipation devices are used in bridge design against ultimate limit states, i.e. against the strongest design response spectrum the code introduces. They are conceived to serve as high stiffness components below a force threshold and to undergo large hysteretic cycles when the threshold is crossed. This paper discusses the selection of this threshold as the result of a cost-benefit analysis: lower the threshold is, higher the bridge reliability becomes. On the other side, lower the threshold is, higher the probability that the bridge devices undergo permanent deformation, under the serviceability excitation, becomes. The recent earthquake in Turkey produced a significant case study, which will be illustrated and discussed.

1 INTRODUCTION

When designing an energy dissipation device (Casciati et al., 1999; Housner et al., 1997; Soong & Dargush, 1997) for a bridge in a seismic region, one meets two basic problems:

1. the structural codes impose material partial safety factors which alter the actual dynamic behaviour of the structural system;
2. one must design the device accounting for both serviceability and ultimate limit states.

Such a situation suggests to investigate the feasibility of semi-active devices which can switch from one behaviour to another according to the severity of the ground motion.

The goal of the present study can be summarised as follows:

- To emphasise the need for the availability in a damper of different force-displacement curves, each of them suitable for different levels of seismic intensity;
- To specify an optimal value for the per cent variation of the yielding level;
- To investigate the temporal transient in moving from one situation to another;
- To clarify the optimal sequence for the transient: either down from a higher value of the yielding or up from a lower value.

In this paper only the first point will be reported while the others are currently under investigation.

The use of dissipation devices introduces nonlinearities in the structural system and requires a time step dynamic analysis, the seismic excitation being assigned as an acceleration time history.

In order to evaluate the structural system seismic response, the dynamic analyses should be performed by several accelerograms compatible with the elastic spectrum of Eurocode; the results shown in this paper have been obtained by using a single accelerogram compatible with Eurocode 8.

2 THE SEISMIC EXCITATION

The selected accelerogram spectrum is shown in Figure 1. The peak ground acceleration is 0.3093 g at 3.24 seconds.

The response spectrum for damping values of 2% and 5% have been calculated and shown in Figure 1

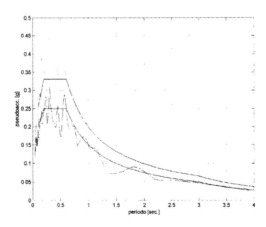

Figure 1: Accelerogram response spectrum for damping 2% and 5%; they are compared with the corresponding Eurocode 8 design spectrum for the ultimate state limit check

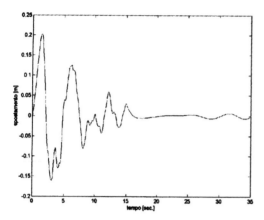

Figure 2: Base displacement time history

where also the Eurocode 8 spectrum for ultimate state load condition (ULS) is shown.

For the stress checks at the elastic limit the accelerogram has to be scaled by a factor of 3.5.

The dynamic analyses have been performed by the finite element code MARC. The assigned base accelerogram has been processed in order to obtain a zero mean displacement signal by filtering the frequencies between 0.1 and 0.15 Hertz and between 25 and 27 Hertz. The filtered displacement signal is shown in Figure 2.

3 THE STRUCTURAL SYSTEM

The simplified structural system used in the dynamic analyses is shown in Figures 3 and 4. It represents a simply supported bridge deck including the pier model. The dissipation device is located between the pier top and the deck while the other deck support is simply a vertical support representing a sliding bearing free to move along the longitudinal direction.

This simplified model represents the static scheme of an highway viaduct in Turkey recently subject to an exceptional earthquake with a peak ground acceleration more that twice the expected one. The viaduct is seismically isolated by elasto-plastic energy dissipation devices manufactured by ALGA S.p.A.

The seismic isolation system showed a good behaviour during the earthquake. Despite the very highexcitation level, the structure did not collapse and the main damages were located in the energy dissipation devices as expected.

The study presented in this paper uses the same isolation concept of the highway viaduct applied in the model shown in Figure 3.

The seismic analysis is performed supposing the acceleration direction parallel to the structure longi-tudinal axis even if the devices can react in every direction of the horizontal plane.

In the finite element model the dissipation device has been modelled as an elasto-plastic truss located between the deck and the pier cap. The force-displacement response law of the elasto-plastic device is shown in Figure 5. Three levels of yielding force have been considered; the actual value and two further ones, decreased to 75% and increased to 125%, respectively as shown in Figure 5.

For each of the three devices, two dynamic analyses have been performed considering two levels of external excitation corresponding to service load condition (SLS) and ultimate load condition (ULS).

The device response diagrams are shown in Figures 6, 7and 8. For each yielding levels the devices react as an elastic spring for SLS excitation while they dissipate energy for ULS excitation (see for instance Figure 9).

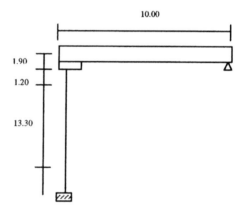

Figure 3: Structural system Scheme used in the dynamic analyses

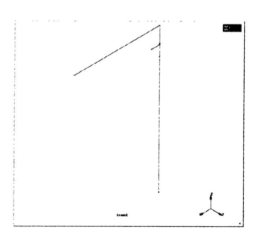

Figure 4: Structural model implemented in the finite element code MARC

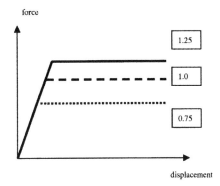

Figure 5: Force-displacement diagram for different yielding F_y values

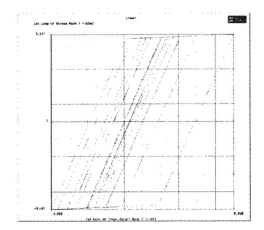

Figure 6: Response diagram of the device with 75% of the actual yielding force for the ULS load condition. Abscissa range: -6.726, +7.919; Ordinate range:-3.897, +3.932

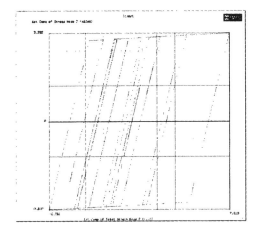

Figure 7: Response diagram of the actual device for the ULS load condition. Abscissa range:-4.533, +5.985; Ordinate range:-5.091, +5.147

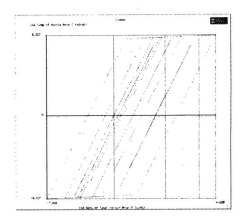

Figure 8: Response diagram of the actual device with 125% yielding force for the ULS load condition. Abscissa range: -7.402, +4.105; Ordinate range:-6.476, +6.317

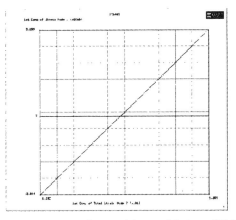

Figure 9: Response diagram of the actual device at SLS load condition

At ULS, in particular, the problem of the designer is to select a suitable force-displacement response curve: increasing the yielding it is possible to reduce the device displacement (i.e. the relative displacement between deck and piles) but higher shear force are transmitted to the substructures. Decreasing the yielding force it is possible to reduce the shear forces, but the displacement is increased.

Figures 10, 11 and 12 show the deck acceleration using the three yielding levels at ULS. The lower acceleration is obtained by increasing the damping of the structural system i.e. by adopting the lower elasto-plastic curve (see Figure 10).

The optimal design must reach the best compromise between the two previous aspects. Big displacements certainly increase the device cost but they also decrease the cost of the substructure. By contrast, accepting a higher value of the shear force, it is possible to reduce the device cost but the substructure will cost more.

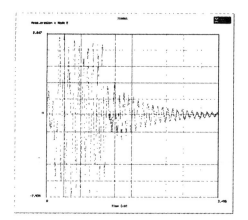

Figure 10: Deck acceleration using the device with 75% yielding force for the ULS load condition. The absolute peak values is 2.847 ms⁻²

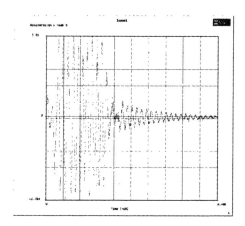

Figure 11: Deck acceleration using the actual device for the ULS load condition. The absolute peak value is 3.61 ms⁻²

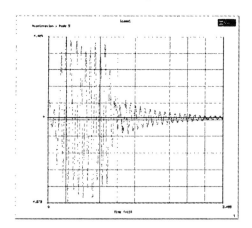

Figure 12: Deck acceleration using the device with 125% yielding force for the ULS load condition. The absolute peak value is 4.466 ms⁻²

The previous considerations show that the availability of devices able to adapt (Soong, 1991) their properties to the structure requirement could optimise the structural design in order to minimise both dynamic responses and costs.

Indeed in designing bridge energy dissipation devices, it turns out that:

- the energy dissipation device must be ineffective for the SLS load condition, so that no permanent damage (i.e. plastic deformations) is due to excitation;
- the energy dissipation device must reduce the structural response for the ULS load condition;
- its yielding level should be selected to obtain the maximum dissipation of energy.

It is worth nothing that the code requirement of introducing suitable material partial safety factors involves the risk of reducing the efficacy of the device for the ULS.

If one requires a safety coefficient of 1.5 for the yielding level at SLS, it can happen that the yielding force is too big at ULS and the device is only able to dissipate a small amount of energy. On the other side a material safety factor for a design at ULS involves damages for the SLS load conditions.

This can be achieved by adopting semi-active devices as for instance the ones conceived on the basis of electro-inductive principles (Marioni et al.,1999).

The open problems of the semi-active solution are:

- the definition of the optimal yielding values;
- the device response time in varying the device properties;
- adapting the plastic response in term of force-displacement diagram after the yielding point has been achieved.

The previous points can be investigated introducing in general-purpose finite-element codes suitable elements in order to model devices whose properties can be changed during the analysis by suitable algorithms.

4 CONCLUSION

The design of bridge energy dissipation devices is achieved as the result of two conflicting requirements no damage under serviceability limit state load condition and maximum dissipation under ultimate limit state load condition.

A suitable compromise is offered by the adoption of semiactive devices.

5 ACKNOWLEDGEMENT

This research is supported by the Italian Ministry of the University and of the Scientific and Technological Research (MURST).

REFERENCES

Casciati F., Maceri F., Singh M.P. and Spanos P. (eds.),1999, *Civil Infrastructure Systems, Intelligent Renewal,* World Scientific, Singapore.

Housner G.W., Bergman L.A., Caughey T.K., Chassiakos A.G., Claus R.O., Masri S.F., Skelton R.E., Soong T.T., Spencer B.F. Jr. and Yao J.T.P., 1997, Structural Control: Past, Present and Future, *Journal of Engineering Mechanics, ASCE;* 123 (9), 897-971

Marc, 1993, Reference *Manuals*

Marioni A., Battaini M., Del Carlo G., 1999, Development and Application of Electro-inductive Dissipators, International Post-SmiRT Conference Seminar on Seismic Isolation, Passive Energy Dissipation and Active Control of Vibrations of Structures, Cheju, Korea ,347-354

Soong T.T., 1991, *Active Structural Control: Theory and Practice,* Longman Scientific and Technical, Essex, UK.

Soong T.T. and Dargush G.F., 1997, Passive Energy Dissipation Systems in Structural Engineering, J. Wiley, London.

Earthquake Engineering Frontiers in the New Millennium, Spencer & Hu (eds),
© 2001 Swets & Zeitlinger, ISBN 90 2651 852 8

Active/semiactive seismic response control of cable-supported bridges: Current research status and key issues

Y.Q. Ni
University of Notre Dame (USA) & The Hong Kong Polytechnic University (Hong Kong)

B.F. Spencer, Jr.
University of Notre Dame, Notre Dame, IN 46556, USA

J.M. Ko
The Hong Kong Polytechnic University, Hung Hom, Kowloon, Hong Kong

ABSTRACT: This paper reviews the current research status of active/semiactive vibration control of cable-supported bridges and explores some implementation issues. Although active control of cable-stayed bridges has been proposed at the end of the seventies, control of long-span bridge structures has not been studied to the extent that tall buildings have. Studies of applying active control to cable-supported bridges have been revived in the last decade, mainly focusing on active tendon control and decentralized control techniques in terms of simple structural models. Research of active/semiactive seismic response control in terms of detailed bridge models and using various control devices has only recently been considered by the ASCE Task Committee on Structural Control Benchmarks. Some control-design issues such as control oriented modeling and control device adoption for cable-supported bridges are also addressed in this paper.

1 INTRODUCTION

With the Tatara Bridge of 890 m main span being a landmark, the construction of cable-stayed bridges is now entering a new era, with main spans reaching 1000 m (Russell 1999). Increasing bridge span and flexibility renders cable-stayed bridges vulnerable to dynamic loading such as earthquakes and strong wind. Structural control provides an efficient means for protection of cable-stayed bridges against earthquakes, wind and other hazards. It has been accepted that active control is a viable technology for enhancing structural functionality and safety against natural hazards. However, it has also been recognized that the active/semiactive control of long-span bridge structures represents a new, difficult, and unique problem, with many complexities in modeling, control design, and implementation (Shoureshi et al. 1995, Schemmann et al. 1998, Dyke et al. 2000b). To date, active control of bridge structures has not been studied as extensively as buildings.

Although the concept of active control of cable-stayed bridges has been proposed as early as the end of the seventies (Yang and Giannopolous 1979a,b), substantial research on this subject did not progress in the eighties. Studies of applying active control to cable-supported bridges have revived in the last decade. While active control of wind-induced flutter vibration has been extensively studied using both mechanical and aerodynamic measures, active seismic control of cable-supported bridges has essentially focused on exploring active cable tendon control and decentralized control techniques by referring to simple structural models. Investigation of active/semiactive seismic response control in terms of detailed bridge models and allowing for diverse control devices has only recently been considered by the ASCE Task Committee on Structural Control Benchmarks (Dyke et al. 2000a,b). Although active systems have been used temporarily in construction of long-span bridges (free-standing bridge towers or cantilever deck segments) against wind loading (Spencer and Soong 1999), the full-scale implementation of active/semiactive devices for seismic response control of cable-supported bridges was not reported.

This paper reviews the current research status of active/semiactive vibration control of cable-supported bridges and explores some issues related to active control implementation. After an overview of the literature, two important control-design issues are addressed. The issue of developing reduced-order models for active control design is first explored after identifying several challenging problems specific for cable-supported bridges. A mode-based model reduction approach based on both modal response analysis and evaluation of observability and controllability grammians is proposed. Another issue explored is control devices applicable to cable-supported bridges. Both active and semiactive control devices are discussed. Active mass drivers and active tendons are addressed for active control, and magnetorheological (MR) dampers and variable stiffness devices are considered for semiactive control.

2 ACTIVE TENDON CONTROL

Yang and Giannopolous (1979a,b) first proposed active tendon control of cable-stayed bridges subject to strong wind gusts. With respect to a simple model of a continuous beam suspended by four stay cables, they performed the feedback control by using the four cables as active tendons to which servo-hydraulic actuators were connected. Sensors were installed at the anchorage of each cable to detect the motion of bridge deck. The sensed motions were passed through a linear feedback operator and fed to the actuators. The actuators actively changed the cable tension, applying time-varying forces to the deck. Very promising results were obtained from their theoretical study.

Warnitchai et al. (1993) carried out an analytical and experimental study of active tendon control of cable-stayed bridges employing a cable-supported cantilever beam model. Sinusoidal vertical excitation and a velocity feedback control were adopted. An electromagnetic exciter was used to feed dynamic forces to the cable-supported cantilever girder, and the signal from an analog controller was converted to an axial displacement of the cable anchorage using a piezoelectric actuator. The active tendon control was found to be very effective in suppressing girder vertical vibration when the cable local vibration was small.

Hutton (1993) studied the active cable-tension control of a planar double-stayed guide-way subject to moving vehicles. Tensions in the lower cables are controlled in response to changes in upper stay-cable tensions caused by the external loading. Numerical simulation results showed that the guideway deflection could be effectively reduced through tension control in the lower stay cables.

Kobayashi et al. (1994) studied the tendon control of cable-stayed bridges by setting control cables parallel to stay cables and connecting the control cables from tower top to actuators at the anchorage or girder. Experiments on a 1/100 scale half-span model of a 410 m center span cable-stayed bridge were conducted to demonstrate damping control with the tendon control force proportional to the velocity of the girder.

Achkire and Preumont (1996) considered the active vibration control of cable-stayed bridges with an active tendon controlling the axial displacement of the cable anchor point. By using a piezoelectric actuator collocated with a force sensor measuring tension in the cable, a force feedback control law, called *integral force feedback*, is applied to offer an active damping (energy absorbing) control. An experimental setup consisting of a cable in connection with a spring-mass system was tested to evaluate control efficiency.

Recently, Preumont and Bossens (2000) reported an experimental study of a large-scale model using active tendon control. The experimental setup was a 30 m-long cable stayed cantilever structure, equipped with hydraulic actuators on the two longest stay-cables. The control system used collocated force sensors and tendon displacement actuators (hydraulic actuators). High-pass filters were included after the force sensors to eliminate the static load in the active cables, while low-pass filters were introduced to eliminate the internal resonance of the hydraulic actuators. The experimental results showed that the active tendon control brought a substantial reduction in the deck vibration amplitude, especially for the first global bending mode.

Active tendon control of cable-stayed bridge models using decentralized control schemes has been studied by a number of researchers. In this approach, the structure to be controlled is divided into a set of subsystems, and interactions between the subsystems are viewed as unknown external disturbances. Each subsystem is then controlled independently using only local information. Volz et al. (1994) presented a decentralized active controller for a two-cable, cable-stayed beam model subjected to vertical seismic excitation. Magana et al. (1997) developed a nonlinear decentralized controller for the same structural model. The proposed controller consists of a linear part that introduces damping into the structure and a nonlinear part that further reduces the vibration caused by the external disturbance. Magana and Rodellar (1998) extended this study by taking into account nonlinearities of the structure due to the geometry of the stay cables and their sag effect.

Based on the sliding mode control principle, Luo et al. (1998) designed an active decentralized stabilizing controller for a multi-cable, cable-stayed beam structure subject to parametric uncertainties and earthquake excitation. Magana and Rodellar (1999) proposed a fuzzy logic controller for decentralized active tendon control of a simple cable-stayed bridge model. By using a two-subsystem scheme, Cao et al. (2000) developed an active decentralized controller for a cable-stayed beam model that minimizes the performance index of each subsystem and guarantees the stability of the overall structure. It should be noted that in all aforecited decentralized active tendon control studies, only simply idealized structural models were used for the cable-stayed bridge deck and only vertical seismic excitation was considered.

3 ACTIVE FLUTTER CONTROL

Flutter instability is the most important concern in the design of a long-span cable-supported bridge because it may lead to total collapse of the bridge. Active flutter control has been widely investigated by means of both mechanical and aerodynamic measures. Active mechanical control devices actively modify the dynamic properties of a bridge structure itself to enhance the flutter suppression capacity. Miyata et al. (1994) proposed the application of an H∞ robust control strategy to increase the bridge flutter resistance based on the active twisting moment. The active twisting moment was generated by driving an eccentric weight or by changing the speed of a gyroscope installed on the bridge deck. Since the flutter was essentially related to the system instability, the control performance index was formulated in terms of the H∞ norm of the transfer function, instead of specific response quantities. A numerical example on a 2D sectional model showed that the proposed control scheme using the active moment provided a good solution for the flutter problem. Dung et al. (1996) extended this study to active flutter control of 3D full models by using the method of model reduction based on flutter modes.

Wilde et al. (1996) conducted an analytical study on active flutter control using a variable eccentricity method. A movable auxiliary mass was placed inside the bridge deck, and its position was actively controlled so that the induced eccentricity and stabilizing moment were used to prevent the onset of flutter. Rational function approximation of the unsteady aerodynamic forces was introduced to derive a time domain equation for the aeroelastic system. A linear-saturated control law was applied, and stability of the system was determined by describing function and phase portrait methods.

Achkire et al. (1998) considered the use of active cables for flutter control of cable-supported bridges. The control system used a decentralized integral force feedback which consisted of an active tendon (displacement actuator) collocated with a force sensor. Experiments on a simple model were conducted to obtain the flutter gain with and without the active tendon control.

Other kinds of devices for active flutter suppression include aerodynamic measures that actively modify the flow around the bridge deck or generate active stabilizing aerodynamic forces from the flow. Kobayashi and Nagaoka (1992) presented an active aerodynamic method for flutter suppression by using control wings attached to the bridge deck. The aerodynamic forces acting on the control wings provided additional aerodynamic force to the bridge deck. By appropriately controlling the pitching motion of the control wings, the additional aerodynamic force produced by the control wings became positive aerodynamic damping, thus suppressing the flutter. Wind tunnel test on a section model showed that the active control wings based on a simple direct control law could increase the flutter speed by a factor of two. Ostenfeld and Larsen (1992) proposed the application of an active control surface system for flutter suppression. This system was based on on-line monitoring of the deck movements and use of control surface movements to generate stabilizing aerodynamic forces through a proper feedback control

law. The aerodynamic forces acting on the control surfaces were proportional to the wind speed squared, and thus proportional to the forces acting on the deck. As the stabilizing forces were driven by the airflow, and the actuators were only used to control rotation of the control surfaces, the energy required for control was small.

Wilde et al. (1994) applied optimal control theory for active flutter suppression by means of control surfaces. By representing the unsteady aerodynamic forces with rational function approximations, the aeroelastic equation of motion was formulated in state space so that modern control theory could be applied. A variable-gain output feedback control law was implemented for active flutter control. Kobayashi and Nitta (1996) conducted wind tunnel tests on a 2D-sectional model to verify the efficiency of active flutter control using control surfaces. The control surfaces adopted two winglets mounted above both edges of the deck and two flaps installed at the deck edges, which were actively driven by DC servo motors inside the deck. Preidikman and Mook (1997) proposed the use of a single wing mounted below the deck for active control of the flutter instability. A linear control law was implemented to actively regulate the angle of attack of the wing to enhance the critical flutter speed.

Hansen and Thoft-Christensen (1998) studied active flutter control using flaps. The flap control system consisted of flaps integrated in the bridge girder so that each flap was the streamlined part of the edge of the girder. The flags were actively regulated according to the measurements from sensors placed inside the deck. A numerical study of a bridge section model using different flap configurations showed that the trailing flap was more efficient than the leading flap. Kobayashi et al. (1998) investigated active flutter control of a bridge deck employing ailerons. The ailerons were attached to the deck girder at both leading and trailing edges and were actively driven with suitable phase differences to exert stabilizing aerodynamic forces. Both analytical and experimental investigations were performed to verify the control efficiency. Wilde and Fujino (1998) studied in detail active aerodynamic control methodology for flutter suppression using control surfaces, including time-domain modeling of aerodynamic forces and design of control law accommodating the dependence of the flutter equation on wind speed.

4 ACTIVE SEISMIC RESPONSE CONTROL

Studies on active seismic response control of cable-supported bridges referring to detailed bridge models have evolved only in recent years. Miyata et al. (1996) conducted a study of applying active tuned mass dampers (ATMDs) to the Akashi-Kaikyo suspension bridge for seismic response mitigation. A finite element model of the bridge was formulated with 1444 DOFs. To control the bridge subject to the transverse El Centro earthquake, 3 ATMDs with mass 5×10^4 kg each were installed (one on top of each tower and one at the mid-span point of the central span), and an acceleration feedback control law was implemented. The main difficulty encounte-red was computational intractability due to the large bridge model with numerous DOFs. Shoureshi and Bell (1996) performed an analytical and experimental study on active seismic control of a 1:150 scale cable-stayed bridge model. Based on a 328-DOF finite element model of this test structure, a reduced-order analytical model was further derived for control implementation. A feedforward/feedback control strategy was used where the feedforward control signal was acquired from a dynamic predictor (seismometer) reading at a distant location while the feedback control signals were obtained from accelerometer readings on the structure. The experimental results showed substantial response attenuation.

Paulet-Crainiceanu (1998) studied full state optimal linear control of the Tatara cable-stayed bridge subject to the transverse and longitudinal El-Centro earthquakes, respectively. A 3D finite element model of the bridge with 1370 DOFs was developed for the control design. Under transverse seismic excitation, a total of seven ATMDs were used: three 8×10^4 kg ATMDs at the mid-span point of the central span and at 71.2 m from the mid-span point; four 4×10^4 kg ATMDs in pairs on top of the towers. Under longitudinal seismic excitation, a total of eleven ATMDs were used, with additional four 4×10^4 kg ATMDs in pairs at the half heights of the towers. The simulation results indicated that significant reduction in the structural response could be achieved by the proposed control strategy. However, the amount of energy required to produce the control forces was large, and the active mass displacements were prohibitive.

Based on an analytical feasibility study of a well-studied and documented cable-stayed bridge model, Schemmann et al. (1998) explored some important issues and problems associated with attempting to actively control long-span flexible bridge structures. This study was initiated with the ultimate goal of developing a bridge control benchmark problem.

Schemmann and Smith (1998a,b) performed a detailed analytical study on active seismic response control of the Jindo cable-stayed bridge in Korea using a linear quadratic regulator control strategy. A finite element model of the bridge was established with 316 DOFs. Based on an analysis of the structure in response to uniform-support Northridge earthquake motion and asynchronous multiple-support El Centro earthquake motion, a reduced-order model including 21 modes was obtained for control design through modal reduction. Two control schemes, full state feedback control and output

feedback control with a Kalman-Bucy filter estimator, were implemented to evaluate the control effectiveness. Simulation results showed that significant reduction in the internal force and displacement response could be achieved through both full state and output feedback control, and an investigation of various control configurations concluded that forces were most effective when applied to the center of the bridge span.

Under the coordination of the ASCE Task Committee on Structural Control Benchmarks (http://www.nd.edu/~quake/bench.html), Dyke et al. (2000a,b) developed the first generation of benchmark structural control problems for cable-stayed bridges. The benchmark bridge is the Cape Girardeau cable-stayed bridge over the Mississippi river in the USA. Development of this benchmark problem aims to provide a testbed for comparing different control strategies for cable-stayed bridges. A 3D finite element model with 909 DOFs has been developed in MATLAB for the full-scale bridge. A reduced-order evaluation model with 419 DOFs was also derived from the full model. A total of eighteen criteria were developed to evaluate the performance of various control strategies, and a set of control design constraints were defined. For each control design, the evaluation criteria are required to be evaluated for each of three earthquake records provided in the benchmark problem. A sample active control design using a linear quadratic Gaussian control algorithm has been provided to serve as a guide to participants in the study of this benchmark bridge.

Recently Ni et al. (2001) studied the active seismic response control of the Ting Kau cable-stayed bridge in Hong Kong. The Ting Kau Bridge is one of the limited instances of the use of multi-span cable-stayed bridges in practice. Due to adopting a separated deck system and monoleg bridge towers, the bridge is one of the most flexible cable-stayed bridges in the world. A three-level modeling scheme has been proposed for developing control-oriented analytical models of this bridge. Then active seismic response control was investigated by using stochastic optimal control algorithms (*Report No. TKBD-4* 2001). The analytical results obtained under different active mass driver (AMD) configurations and under different seismic excitation directions indicated good response control effectiveness for both the bridge deck and towers.

5 CONTROL-DESIGN ISSUES

5.1 *Control oriented modeling*

Because a 3D full model for cable-supported bridges usually involves more than one thousand DOFs, model reduction is necessary to achieve a reduced-order analytical model suitable for control design. Among the existing model reduction methods, the mode-based reduction technique (Schemmann and Smith 1998a) seems most attractive for the control analysis of large-scale cable-supported bridges because it can significantly reduce the size of the model while retaining the important dynamic properties. However, some challenging problems still remain when applying this technique to cable-supported bridges. First, a cable-supported bridge usually has complicated modal properties. It can exhibit not only coupled modes among lateral, vertical and torsional components, but also modes with strong coupling (interaction) among the deck, cables and towers (Ni et al. 2000). These coupled modes should be appropriately accommodated in the reduced-order model. Second, deflection is not a design criterion in cable-supported bridges (Tang 1995). Instead, structural internal forces govern the control design of cable-supported bridges, which involve significant participation of higher-order vibration modes.

For seismic control, the modes involved in the reduced-order model are usually determined through a modal analysis in response to historical earthquakes. An array of earthquake records that are representative of ground motion features at the bridge site should be used. An analysis of both deterministic and random seismic response is preferable for this purpose. However, such a way to select the modes is based solely on their participation in the seismic response, without any consideration from the control viewpoint. A model reduction technique based on both modal analysis and as observability/controlla-bility grammian evaluation is suggested here. Following this approach, the modes that contribute much to overall response are first selected from modal response analysis. Then the H_∞ norms for each modal transfer function corresponding to optimal actuator positioning (Henriot et al. 2000) and/or the H_∞ norms for global transfer functions corresponding to prescribed actuator positions (Ni et al. 2001) are evaluated. The former is beneficial for determining adequate locations of actuators and sensors, while the latter is useful for evaluating observability/con-trollability grammians of the selected modes. The modes included in the reduced-order model are determined through a comprehensive evaluation of both participation factors and grammians.

5.2 Control devices

Adequate active/semiactive control devices are important for practical implementation of this technique to cable-supported bridges. Although nearly forty buildings throughout the world have been installed with various types of active control systems, most of these systems serve to improve occupant comfort and serviceability. Instead, active control of cable-supported bridges is developed for mitigating any potential damage. The reliability of control devices for such structures is therefore particularly important. For bridge structures, control systems are preferred that can withstand earthquakes and enable switching to other strategies in case of an imminent typhoon. Apart from fully active control devices such as AMDs and active tendons, recently developed semiactive control devices such as magneto-rheological dampers (Spencer et al. 1998) and variable stiffness devices (Kobori et al. 1998) should be appealing for such applications in view that: (i) they require low-power; (ii) they are fail-safe devices that serve as passive dampers even in the event that control system fails or is not activated; and (iii) they can be easily integrated with or designed as structural components to form intelligent bridge systems.

6 CLOSURE

The idea of applying active/semiactive control as a means for hazard mitigation of civil engineering structures has grown increasingly popular. However, only limited research has been directed towards cable-supported bridges. In this paper, a state-of-the-art review on active/semiactive control of cable-supported bridges is presented. Based on a discussion on some control-design issues, it is recognized that a model reduction technique based on modal analysis as well as observability/controllability evaluation is effective for control design, and the intelligent bridge system integrating active/smart devices with structural components is promising for practical implementation.

REFERENCES

Achkire, Y., Bossens, F. & Preumont, A. 1998. Active damping and flutter control of cable-stayed bridges. *Journal of Wind Engineering and Industrial Aerodynamics* 74-76: 913-921.

Achkire, Y. & Preumont, A. 1996. Active tendon control of cable-stayed bridges. *Earthquake Engineering and Structural Dynamics* 25: 585-597.

Cao, D.Q., Ko, J.M., Ni, Y.Q. & Liu, H.J. 2000. Decentralized active tendon control and stability of cable-stayed bridges. In J.M. Ko & Y.L. Xu (eds.), *Advances in Structural Dynamics*, 2: 1257-1264. Oxford: Elsevier.

Dung, N.N., Miyata, T. & Yamada, H. 1996. A method for flutter control in the long span bridge. In *Proceedings of the 3rd International Conference on Motion and Vibration Control*, Chiba, Japan, 2: 47-52.

Dyke, S.J., Caicedo, J.M., Bergman, L.A. & Turan, G. 2000a. Introducing a benchmark control problem for a cable-stayed bridge subjected to seismic excitation. In *Proceedings of the China-US Millennium Symposium on Earthquake Engineering*, Beijing, China.

Dyke, S.J., Turan, G., Caicedo, J.M., Bergman, L.A. & Hague, S. 2000b. Benchmark control problem for seismic response of cable-stayed bridges. *Research Report*, September 15, 2000, 42p.

Hansen, H.I. & Thoft-Christensen, P. 1998. Active control of long bridges using flaps. In *Proceedings of the 2nd World Conference on Structural Control*, Kyoto, Japan, 2: 851-858.

Henriot, P., Verge, M. & Coffignal, G. 2000. Model reduction for active vibration control. In N.M. Wereley (ed.), *Smart Structures and Materials 2000: Smart Structures and Integrated Systems*, SPIE Vol. 3985, 780-789.

Hutton, D.V. 1993. Dynamic analysis of active-control, cable-stayed guideway. *ASCE Journal of Structural Engineering* 119: 2403-2420.

Kobayashi, H. & Nagaoka, H. 1992. Active control of flutter of a suspension bridge. *Journal of Wind Engineering and Industrial Aerodynamics* 41-44: 143-151.

Kobayashi, H., Hosomi, M. & Koba, K. 1994. Tendon control for cable stayed bridge vibration. In *Proceedings of the 1st World Conference on Structural Control*, Los Angeles, California, 3: (FA1)23-29.

Kobayashi, H. & Nitta, Y. 1996. Active flutter control of suspension bridge by control surfaces. In *Proceedings of the 3rd International Conference on Motion and Vibration Control*, Chiba, Japan, 2: 42-46.

Kobayashi, H., Ogawa, R. & Taniguchi, S. 1998. Active flutter control of a bridge deck by ailerons. In *Proceedings of the 2nd World Conference on Structural Control*, Kyoto, Japan, 3: 1841-1848.

Kobori, T. 1998. Mission and perspective towards future structural control research. In *Proceedings of the 2nd World Conference on Structural Control*, Kyoto, 1: 25-34.

Luo, N., Rodellar, J., de la Sen, M. & Magana, M.E. 1998. Decentralized model reference variable structure control of multi-cable-stayed bridges. In *Proceedings of the 2nd World Conference on Structural Control*, Kyoto, Japan, 2: 1011-1020.

Magana, M.E. & Rodellar, J.J. 1998. Active nonlinear robust control of cable-stayed bridges in the presence of strong vertical ground motion due to earthquakes. In *Proceedings of the 2nd World Conference on Structural Control*, Kyoto, Japan, 3: 1947-1955.

Magana, M.E. & Rodellar, J.J. 1999. Fuzzy logic control of cable-stayed bridges in the presence of seismic excitation. In *Proceedings of the 17th International Modal Analysis Conference*, Kissimmee, USA, 1: 738-744.

Magana, M.E., Volz, P. & Miller, T. 1997. Nonlinear decentralized control of a flexible cable-stayed beam structure. *ASME Journal of Vibration and Acoustics* 119: 523-526.

Miyata, T., Yamada, H., Dung, N.N. & Kazama, K. 1994. On active control and structural response control of the coupled flutter problem for long span bridges. In *Proceedings of the 1st World Conference on Structural Control*, Los Angeles, California, 1: (WA4)40-49.

Miyata, T., Yamada, H. & Paulet-Crainiceanu, F. 1996. Active structural control for cable bridges under earthquake loads. In *Proceedings of the 3rd International Conference on Motion and Vibration Control*, Chiba, Japan, 2: 53-58.

Ni, Y.Q., Spencer Jr., B.F. & Ko, J.M. 2001. Feasibility of active control of cable-stayed bridges: an insight into Ting Kau Bridge. To appear in S.C. Liu (ed.), *Smart Structures and Materials 2001: Smart Systems for Bridges, Structures, and Highways*, SPIE Vol. 4330.

Ni, Y.Q., Wang, J.Y. & Ko, J.M. 2000. Modal interaction in cable-stayed Ting Kau Bridge. In J.M. Ko & Y.L. Xu (eds.), *Advances in Structural Dynamics*, 1: 537-544. Oxford: Elsevier.

Ostenfeld, K. & Larsen, A. 1992. Bridge engineering and aerodynamics. In A. Larsen (ed.), *Aerodynamics of Large Bridges*, 3-22. Rotterdam: A.A. Balkema.

Paulet-Crainiceanu, F. 1998. Seismic response control of cable-stayed bridges. In *Proceedings of the 2nd World Conference on Structural Control*, Kyoto, Japan, 2: 959-964.

Preidikman, S. & Mook, D.T. 1997. A new method for actively suppressing flutter of suspension bridges. *Journal of Wind Engineering and Industrial Aerodynamics* 69-71: 955-974.

Preumont, A. & Bossens, F. 2000. Active tendon control of cable-stayed bridges. In S.C. Liu (ed.), *Smart Structures and Materials 2000: Smart Systems for Bridges, Structures, and Highways*, SPIE Vol. 3988, 188-198.

Report No. TKBD-4. 2001. Stochastic optimal control of Ting Kau Bridge under seismic excitations. *Department of Civil and Structural Engineering, The Hong Kong Polytechnic University* (http://www.cse.polyu.edu.hk/~dynamics).

Russell, H. 1999. Hong Kong bids for cable-stayed bridge record. *Bridge Design and Engineering* (issue No. 15): 7-7.

Schemmann, A.G. & Smith, H.A. 1998a. Vibration control of cable-stayed bridges-part 1: modeling issues. *Earthquake Engineering and Structural Dynamics* 27: 811-824.

Schemmann, A.G. & Smith, H.A. 1998b. Vibration control of cable-stayed bridges-part 2: control analyses. *Earthquake Engineering and Structural Dynamics* 27: 825-843.

Schemmann, A.G., Smith, H.A., Bergman, L.A. & Dyke, S.F. 1998. Feasibility study: control of a cable-stayed bridge model, Part I: Problem definition. In *Proceedings of the 2nd World Conference on Structural Control*, Kyoto, Japan, 2: 975-979.

Shoureshi, R.A. & Bell, M.J. 1996. Vibration control of cable-stayed bridges: control system development and experimental results. In Y.K. Lin and T.C. Su (eds.), *Engineering Mechanics: Proceedings of the 11th Conference*, ASCE, New York, 2: 902-905.

Shoureshi, R., Wheeler, M., Bell, M., Alves, G. & Maguire, D. 1995. On implementation of active control systems. In *Proceedings of the 1995 American Control Conference*, Seattle, Washington, 3: 2364-2368.

Spencer Jr., B.F., Yang, G., Carlson, J.D. & Sain, M.K. 1998. "Smart" dampers for seismic protection of structures: a full-scale study. In *Proceedings of the 2nd World Conference on Structural Control*, Kyoto, Japan, 1: 417-426.

Spencer Jr., B.F. & Soong, T.T. 1999. New applications and development of active, semi-active and hybrid control techniques for seismic and non-seismic vibration in the USA. In *Proceedings of the International Post-SMiRT Conference Seminar on Seismic Isolation, Passive Energy Dissipation and Active Control of Vibrations of Structures*, Cheju, Korea, 1: 467-488.

Tang, M.-C. 1995. Multispan cable-stayed bridges. In M. Sanayei (ed.), *Restructuring: America and Beyond*, ASCE, New York, 1: 455-458.

Volz, P., Magana, M.E., Hernied, A.G. & Miller, T.H. 1994. A decentralized active controller for cable-stayed bridges. In *Proceedings of the 1st World Conference on Structural Control*, Los Angeles, California, 3: (FA1)13-22.

Warnitchai, P., Fujino, Y., Pacheco, B.M. & Agret, R. 1993. An experimental study on active tendon control of cable-stayed bridges. *Earthquake Engineering and Structural Dynamics* 22: 93-111.

Wilde, K. & Fujino, Y. 1998. Aerodynamic control of bridge deck flutter by active surfaces. *ASCE Journal of Engineering Mechanics* 124: 718-727.

Wilde, K., Fujino, Y. & Prabis, V. 1996. Effects of eccentric mass on flutter of long span bridges. In *Proceedings of the 2nd International Workshop on Structural Control*, Hong Kong, 564-574.

Wilde, K., Masukawa, J., Fujino, Y. & Bhartia, B. 1994. Active control of flutter instability of bridge deck. In *Proceedings of the 1st World Conference on Structural Control*, Los Angeles, California, 1: (WA4)50-59.

Yang, J.N. & Giannopolous, F. 1979a. Active control and stability of cable-stayed bridge. *ASCE Journal of the Engineering Mechanics Division* 105: 677-694.

Yang, J.N. & Giannopolous, F. 1979b. Active control of two-cable-stayed bridge. *ASCE Journal of the Engineering Mechanics Division* 105: 795-810.

Earthquake Engineering Frontiers in the New Millennium, Spencer & Hu (eds),
© 2001 Swets & Zeitlinger, ISBN 90 2651 852 8

Study on seismic response reduction for floor increased structural system with friction bedding and dampers

Hongnan Li, Guanglei Liu, & Yongwei Yin
Shenyang Architectural and Civil Engineering Institute

Liyan Lin
Liaoning Architectural Design and Research Institute

ABSTRACT: This paper presents a new type of energy dissipation floor increased structural system, which uses the friction bedding set between the lowest increased floor of outer frame structure and the top floor of original building. It also applies energy dissipation dampers between columns of outer frame structure and each lower floor at their connecting points. The structural model of this system and a new Coulomb friction representation are introduced to the nonlinear dynamic analysis for the first time in this paper. Special attention is also paid to the study of several parameters that affect the seismic response reduction rate. At last, an existing engineering project is presented to illustrate the application and delineate the advantages of the method.

1 INTRODUCTION

Recently, with the growth of economy as well as the urban construction and real estate market, problems like how to make a good use of land attract more and more attention. Facing the old buildings that have stood for years and were designed according to dated building codes which seldom took seismic protection into real considerations, both the land developers and engineers have to decide whether to demolish them first and then set up larger and better designed ones or reuse the original structures and make some changes in order to meet the present need. During renovation of existing buildings, many engineering projects and papers have adopted floor increased structural system, which is an investment saving, demolition and relocation free as well as urban environment protecting method. And successful examples can be found in many countries, though their styles are different from one to another. Because every such kind of system has its own complex characteristics, numerous structural styles during the design and construction of flooring increasing projects have been emerged in China. However, most of current practicing methods fall under the following categories—direct floor increased system, outer-frame-structure floor increased system, interior floor increased system and underground floor increased system.

As we know, in non-seismic region it is easy to apply floor-increasing projects, but in seismic region many floor increased structural styles are restricted on a certain scale. On the other hand, China hasn't published any general seismic accreditations or de-

sign standards for such structural system. Most of the engineering projects rely entirely on the strength of the structure and its ability to dissipate energy to withstand seismic load so that they can satisfy the seismic protection standards. With good design and under normal loading conditions, the response of these structures will remain in the safe domain. However, since we cannot estimate accurately the intensity and characteristics of an earthquake until now, there is no guarantee that in-service loads experienced by the building system will be in the expected range. At this point, the whole floor increased structural system cannot change in accordance with environmental loadings. Under unexpected severe earthquake, it may become impossible to satisfy the safety demand and result in the weakening of the system integrity as well as other damages, even collapse.

The concept of structural control, first presented to the structural engineering profession by Yao (1972), has been widely used to mitigate the effects of structural response under dynamic loadings. Also, it's a good use in finding appropriate control strategies for the building of floor increased structural systems. This paper discusses the application of a new control technique using friction bedding between the lowest increased floor of outer frame structure and the top floor of existing building. It also applies energy dissipation dampers between columns of outer frame structure and each lower floor at their connecting points. And through the computing analysis it also provides several rational control parameters and methods to direct the engineering design and construction work.

2 PROBLEM FORMULATION

2.1 Basic assumptions

A simplified diagram of floor increased structural system is showed in Fig. 1. Loading the lowest floor of outer frame structure on the friction bedding between it and the top floor of existing building, in which way the compression force produces the frictional forces. At the same time, energy dissipation dampers are installed between columns of outer frame structure and each lower floor at their connecting points. But these dampers don't have to be put at each of the floor line. The actual layout scheme depends on different cases.

Considering the simplest situation, the computation in this paper is based on the following assumptions: (1) the stiffness of floor in the horizontal plane is infinite; (2) both the deformation between floors and those of the whole system can be regarded as a shear pattern; (3) the mass of each floor concentrate at the floor elevation; (4) any torsional or vertical displacement are not taken into considerations in this very case; (5) force in the friction bedding is sliding friction, and it obeys the Coulomb friction law.

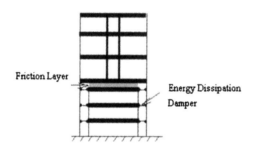

Figure 1. Structure system model

2.2 Analysis model and equation of motion

Consider an N degrees of freedom floor increased structural system divided into two structures which are identified as the I (existing structure) and II (outer frame structure). The model of this system is illustrated in Fig. 2. The number of degrees of freedom of the I and II structures are n and m, respectively, and N=n+m. The equation of motion of the two structures in free vibration can be written as,

$$\begin{cases} [M]_I\{\ddot{x}\}_I + [C]_I\{\dot{x}\}_I + [K]_I\{x\}_I + \{F\}_I = -[M]_I\{I\}\ddot{x}_g \\ [M]_{II}\{\ddot{x}\}_{II} + [C]_{II}\{\dot{x}\}_{II} + [K]_{II}\{x\}_{II} + \{F\}_{II} = -[M]_{II}\{I\}\ddot{x}_g \end{cases}$$

(1)

Eq.1 can be expressed in one equation as the following form:

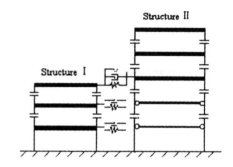

Structure II

Structure I

Figure 2. Structural system computing model

$$[M]\{\ddot{x}\}+[C]\{\dot{x}\}+[K]\{x\}+\{F\}=-[M]\{I\}\ddot{x}_g \qquad (2)$$

Where [M], [C] and [K] are, respectively, the mass, damping (Rayleigh), and stiffness matrices of the system; $\ddot{x}_g$ =ground acceleration; {x}=relative displacement response vector and dot over a quantity represents its time derivative as:

$$\{\ddot{x}\}=\begin{Bmatrix} \{\ddot{x}\}_I \\ \{\ddot{x}\}_{II} \end{Bmatrix}, \quad \{\dot{x}\}=\begin{Bmatrix} \{\dot{x}\}_I \\ \{\dot{x}\}_{II} \end{Bmatrix}, \quad \{x\}=\begin{Bmatrix} \{x\}_I \\ \{x\}_{II} \end{Bmatrix}$$

and:

$$\{I\}=\underbrace{\{1 \quad \cdots \quad 1 \quad \cdots \quad 1\}^T}_{n+m}, \{F\}=\begin{Bmatrix} \{F\}_I \\ \{F\}_{II} \end{Bmatrix}$$

in which {I} is a column vector of ones; {F}, {F}$_I$ and {F}$_{II}$ are the vectors of control force within the system, structure I and structure II respectively. They can be written in details as:

$$\begin{cases} \{F\}_I = \{f\}_I + \{R\} \\ \{F\}_{II} = \{f\}_{II} + \{R'\} \end{cases} \qquad (3)$$

$$\{f\}_I = \{f_{1,1} \quad \cdots \quad f_{1,i} \quad \cdots \quad f_{1,n}\}^T,$$
$$\{f\}_{II} = \{f_{2,1} \quad \cdots \quad f_{2,i} \quad \cdots \quad f_{2,n} \, 0 \quad \cdots \quad 0\}^T$$

$$\{R\}=\mu \, m_0 g \, \text{sgn}(\dot{x}_{1,n} - \dot{x}_{2,n}) \qquad (4)$$

$$\{R\}=\mu \, m_0 g \, \text{sgn}(\dot{x}_{2,n} - \dot{x}_{1,n}) \qquad (5)$$

where {R} and {R'} are the column vectors of control force in the friction bedding between the outer frame structure and the existing building; μ is the friction coefficient, g is the acceleration of gravity, and m_0 is the floor mass set on the friction bedding in order to provide control force; $\dot{x}_{1,n}$ and $\dot{x}_{2,n}$ are the relative velocities of the nth degree of structure I and structure II. And sgn(s) is a function that is equal to +1 when s is positive, and −1 when s is negative. The value of sgn(0) is zero. {f}$_I$ and {f}$_{II}$

are the vectors of control force provided by the energy dissipation dampers between the two structures. And the representations of the ith vector of $\{f\}_I$ and $\{f\}_{II}$ are shown as the following:

$$\begin{cases} f_{1,i} = k_{0i}(x_{1,i} - x_{2,i}) + c_{0i}(\dot{x}_{1,i} - \dot{x}_{2,i}) \\ f_{2,i} = k_{0i}(x_{2,i} - x_{1,i}) + c_{0i}(\dot{x}_{2,i} - \dot{x}_{1,i}) \end{cases} \quad (6)$$

where k_{0i} = the stiffness of the ith energy dissipation damper; c_{0i}=the damping of the ith energy dissipation damper; $x_{1,i}$ and $x_{2,i}$ are the ith relative displacement response vector of structure I and structure II respectively. The dots over a vector represent its time derivative.

2.3 Stick-slip criteria and introduction of a friction representation

For the floor increased structural system discussed in this paper, friction is the main part of the control force. Before finding the sliding transition criteria, let

$$I1 = \left| \sum_{i=1}^{n} m_{1,i}(\ddot{x}_{1,i} + \ddot{x}_g) + k_{1,1}x_{1,1} + c_{1,1}\dot{x}_{1,1} + \sum_{i=1}^{n} k_{0,i}(x_{1,i} - x_{2,i}) + \sum_{i=1}^{n} c_{0,i}(\dot{x}_{1,i} - \dot{x}_{2,i}) \right|$$

$$I2 = \left| \sum_{i=1}^{n+m} m_{2,i}(\ddot{x}_{2,i} + \ddot{x}_g) + k_{2,1}x_{2,1} + c_{2,1}\dot{x}_{2,1} + \sum_{i=1}^{n} k_{0,i}(x_{2,i} - x_{1,i}) + \sum_{i=1}^{n} c_{0,i}(\dot{x}_{2,i} - \dot{x}_{1,i}) \right|$$

(7)

where $I1$ and $I2$ are the sum of total horizontal inertia forces, elastic and damping forces of the first floor as well as the passive control forces produced by energy dissipation dampers of structures I and II, respectively.

Under the assumption of the Coulomb friction law, non-sliding phase continues as long as

$$I \le F_s = \mu_s m_0 g \text{ and } x_{1,n} - x_{2,n} = 0 \quad (8)$$

where I is the larger value between $I1$ and $I2$; F_s is the maximum of static friction force; μ_s is the static friction coefficient. During a stick period, the friction equals to the initial force of the outer frame structure.

As soon as the condition

$$I > F_s = \mu_s m_0 g \quad (9)$$

is met, the sliding phase starts and the friction force equals to slip friction

$$F_d = \mu_d m_0 g \text{ sgn}(\dot{x}_{1,n} - \dot{x}_{2,n}) \quad (10)$$

where μ_s is the slip friction coefficient.

Due to the strong nonlinear property of this structural system, the time step of integral has to be very small in order to approach each of the transitions, or else the inaccuracy can be accumulated to a very large value. However, no matter how tiny the time step will be, it is impossible to identify the exact time at which sliding velocity passes the zero point. This paper introduces a continuous function to take the place of the discontinuous friction and sliding

velocity relationship, in which way it may avoid the necessity to trace carefully each stick-slip phase and its transitions boundary. Therefore, through a thorough and reasonable selection, a continuous function which comes close arbitrarily to the exact discontinuous frictional is given below,

$$f_1(\alpha_1, v) = \text{Erf}(\alpha_1 v) \quad (11)$$

$$\text{sgn}(v) = f_1(\alpha_1, v), \quad F = \mu m_0 g f_1(\alpha_1, v) \quad (12)$$

where v is the sliding velocity, α is a non-unit parameter, Erf is the error function. In order to illustrate the relationship between the changing of α and the function exactitude, f_1 and v is shown in Fig. 3.

Based on the fourth-order Runge-Kutta scheme, a MATLAB program for numerical evaluation of the equations of motion has been developed in this paper.

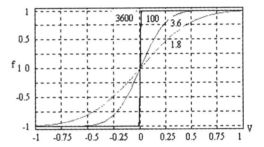

Figure 3. f(α=1.8,3.6,100,3600)

3 PARAMETER ANALYSIS

To make a deep and thorough study on searching the characteristics which affect the controlled system vibration reduction rate, first this paper takes a relatively simple case which regards both the original and newly built outer-floor increasing frame structure as a single degree of freedom system. Here, the weight of the original structure and the outer frame structure is 316 and 111.7 tons respectively, and the system damping ratio is 5%.

Let's define reduction rate as

$$\delta = \frac{D - D_1}{D} \times 100\% \quad (13)$$

Where δ is the reduction rate of seismic responses for the model structure system, D and D_1 are the response peaks without control and with control.

Then, parameters including the ratio of periods (same as that of stiffness) of two structures, the friction coefficient as well as the stiffness and damping of the energy dissipation devices are discussed below.

3.1 Ratio of periods

Average floors of the present dated buildings are no more than three, so generally the periods are short

307

too. Take this into consideration, six different original structure periods (T1) are selected here, that is T1=0.1s, 0.25s, 0.3s, 0.4s, 0.5s, 1.0s and 1.5s. Because the outer frame structure is more flexible than the original structures, their periods (T2) are longer too. So, nine ratios of periods are studied here, that is T2/T1=0.2, 0.5, 1, 2, 3, 4, 5, 10 and 15. Some basic characteristics are shown in Table 1.

Table 1. Structural parameters

Earthquake Records	El-Centro
Duration	6.0s
Friction Coefficient	0.25
Compression Force	40t
Stiffness and Damping of the Energy Dissipation Damper	$k_0=0$, $c_0=0$

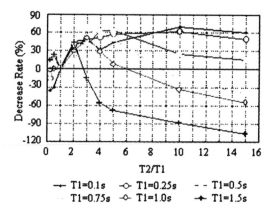

Figure 4. Ratio of period and seismic reduction rate of outer frame structure

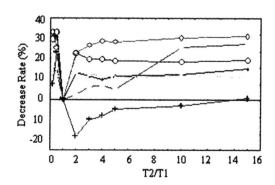

Figure 5. Ratio of period and seismic reduction rate of original structure

Through calculation, the results of reduction rate of seismic responses are shown in Fig. 4 and Fig. 5. It can be seen from the two figures that when the ratio is less than 1, δ of the original structures is decreasing as the ratio is increasing, and δ of the outer frame structure is not very obvious while it increases at certain time. When the ratio equals to 1, the two structures respond simultaneously, and the friction layer doesn't work during the process, so δ is zero. When the ratio of periods is bigger than 1, especially when it ranges through 2 to 5, δ of the original structures is decreasing as the ratio increases, and the reduction rate of seismic responses can be raised to a satisfactory value. After the ratio of periods continues to increase, δ begins to decrease. Therefore, by adjusting the ratio of period for the outer frame structure and the original structure, a good result can be achieved to the seismic reduction.

3.2 Friction coefficient

The value of friction coefficient, μ, is a key to determine the frictional force produced by the friction layer. The larger the friction coefficient, the greater the frictional resisting force will be when sliding begins. Here let the ratio of periods equal to four and use the characteristics studied in section 3.1. According to the practical characteristics of the commonly used engineering materials, a serial of friction coefficients have taken into consideration during the computation. Let μ=0, 0.1, 0.2, 0.25, 0.3, 0.4, the results can be seen from Fig. 6.

The reduction rate increases with the friction coefficient, especially for the outer frame structure the decrease rate seems more noticeable. However, the reduction rate of the original structure is increasing smoothly.

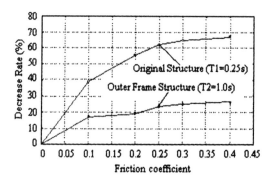

Figure 6. Friction coefficient and decrease rate

3.3 Damping and stiffness of the energy dissipating damper

The energy dissipating dampers at the connecting points between the original structure and the outer frame structure also play an important role for the seismic reduction. Same characteristics are chosen as section 3.1 and ratio of period is 4 in this case. Figs. 7 and 8 illustrate the relative results after computing. It can be seen from these figures that when the damping is the main factor during the reduction of seismic response, as the changing of damping value, the effectiveness of stiffness is not quite obvious. And this result sounds reasonable, because it is the damping that absorbs and dissipates the dynamic energy during an earthquake.

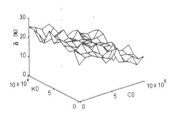

Figure 7. Reduction rate of the original structure

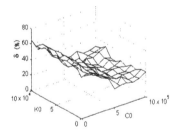

Figure 8. Reduction rate of the outer frame structure

3.4 Earthquake intensity

As we know, different site characteristics could affect the effectiveness of the structural control applications too. In this section, three different earthquake records, Qian An, El-Centro, Ning He, are input to prove the productiveness of this passive control method. All the structural characteristics are the same as those of section 3.1, and the ratio of period is four. Results are shown in Table 2.

From Table 2, it can be seen that by using friction layer between the outer frame structure and the original structure, they all achieve a satisfactory seismic reduction rate. With the increasing of the earthquake intensity, the reduction rate decrease in both structures, and it often follows a visible law. Under the action of El-Centro, the reduction rate stands stably above 60%, which is advantageous for the outer frame structure which usually has a flexible first floor.

Table 2 Seismic reduction rate of friction layer

Earthquake Records	Intensity	Original Structure	Outer Frame Structure
Qian An, t=10s	VII	17.23%	42.43%
Δt=0.01s,	VIII	15.91%	35.32%
period 0.1s	IX	9.56%	22.67%
El-Centro, t=6s	VII	30.33%	88.56%
Δt=0.02s,	VIII	28.59%	76.56%
period 0.5s	IX	20.52%	60.93%
Ning He, t=10s	VII	19.79%	84.58%
Δt=0.01s,	VIII	15.34%	53.82%
period, 0.9s	IX	5.90%	27.98%

4 AN ENGINEERING PROJECT

In this section an office building (sketch plan in Fig. 9), set up in 1950s' and added one floor in 1980s', has been studied using the passive control method of this paper. The building is a brick-and-concrete composite construction, with 4 floors in total. It undergoes a floor increasing of four floors all together. The original structure didn't apply the seismic protection. The seismic protection intensity of this building site is VII, soil type of the site is II.

The structural characteristics of the outer frame structure and the original structure are listed in Table 3. The parameters of the dampers are shown in Table 4. The parameter information is shown in Table 5. Two different earthquake inputs have been implemented and the relative results have been illustrated in Table 6.

Figure 9. Building layout

Table 3 Structural parameters

Floor No.	Outer Frame Structure		Original Structure	
	Floor Weight (t)	Floor Stiffness (10^3 KN/m)	Floor Weight (t)	Floor Stiffness (10^3 KN/m)
1	336.6	1420	1285.54	3620
2	321.8	1420	1170.93	3620
3	321.8	1420	1170.93	3580
4	427.0	1856	987.68	3580
5	1071.2	2252		
6	1106.2	2252		
7	1106.2	2252		
8	1158.0	2252		
Period	T1=0.8502, T2=0.2027 T1=0.3153, T2=0.1119 T3=0.1331, T4=0.0989			

309

Table 4. Damper parameters

Floor No.	Stiffness (KN/m)	Damping (KN·s/m)
1	1.0×10^7	1.0×10^5
2	1.0×10^7	1.0×10^5
3	0.0	0.0
4	0.0	0.0

Table 5. Friction layer parameter

Material	μ	Above Weight
sand	0.25	300t

Table 6. Seismic reduction rate under different earthquake records earthquake records (Qian An)

Original Structure	Floor No.	Uncontrolled	Controlled	Reduction Rate
	1	-3.03	-1.81	40.38
	2	-5.57	-3.18	42.96
	3	10.24	-4.21	58.85
	4	12.76	-4.57	64.15
Outer Frame Structure				
	1	-2.93	-1.79	38.78
	2	-4.57	-3.16	30.87
	3	6.60	-3.89	41.13
	4	13.53	5.15	61.89
	5	17.37	5.16	70.27
	6	14.51	-3.86	73.42
	7	7.01	2.44	65.24
	8	11.26	3.67	67.38

Earthquake records (EL-Centro)

Original Structure	Floor No.	Uncontrolled	Controlled	Reduction Rate
	1	3.4	3.63	-6.31
	2	6.14	6.41	-4.34
	3	10.24	8.64	15.60
	4	12.76	9.63	24.52
Outer Frame Structure				
	1	3.37	3.59	-6.43
	2	5.82	6.28	-7.94
	3	6.60	7.82	-18.41
	4	13.53	8.80	34.99
	5	17.37	8.40	51.64
	6	14.5	5.95	59.01
	7	7.01	-3.02	56.97
	8	11.26	-4.70	58.22

From Table 6, it can be seen that under seismic action the reduction rate of each floor on the original structure increases from the lowest floor to the top floor. This means that the friction layer cuts down

the seismic response of the original building. Because the horizontal stiffness of the outer frame structure is not uniformly distributed, the reduction rate of each floor varies from floor to floor without obvious ranging tendency, but generally speaking, the closer to the friction layer the greater the response reduction, take 4th, 5th and 6th floor for instance. On the other hand, it is also found that the greater the basic frequency of the earthquake records, the larger the response reduction of the structural system.

5 CONCLUSIONS AND DEVELOP DIRECTIONS

The new method of setting friction layer and energy dissipation dampers in the floor increased structural system to reduce the structural responses to seismic loading is presented in this paper. Some conclusions with the practical significance are given by
(1) The ratio of period of the outer frame structure and the original structure is the main factor, which can greatly affect the effectiveness of seismic reduction
(2) The greater the friction coefficient of the friction layer, the better the seismic reduction rate to both the two structures
(3) The damping of the energy dissipation dampers is the main factor during the reduction of seismic response
(4) Under different site conditions, the controlled floor increasing system will all perform well and produce a good seismic reduction fruit.

With the development of the science and technology as well as the accumulation of the practical engineering experiences, more and more characteristics will be emerged in the future. The following could be even more promising.
(1) There will be more and more available schemes of the floor increasing structural type.
(2) Structural control strategies, passive and active control, will be used widely.

REFERENCE

Hengzai Wang. 1993.The Structural Planning Selection Methods of the Floor Increasing Projects,1,2,3. Industry Building(4),(7),(9).
Hengzai Wang. 1994.The Result Analysis and Protection Methods in the Floor Increased Cases Home and Abroad. Building Structure (10) .
Yao, J.T.P. 1972. Concept of Structural Control. ASCE J. Struct. Div. 98:1567-1574.
N. Mostaghel & T. Davis. 1997.Representations of Coulomb Friction for Dynamic Analysis. Earthquake Engineering and Structural Dynamics 26(5).

Earthquake Engineering Frontiers in the New Millennium, Spencer & Hu (eds),
© *2001 Swets & Zeitlinger, ISBN 90 2651 852 8*

Variable-damping-based semiactive control: Analogy between structural and suspension systems

Akira Nishitani
Department of Architecture, Waseda University, Shinjuku, Tokyo, 169-8555 Japan

Yoshihiro Nittta
JSPS Post-Doctoral Research Associate,RISE, Waseda University, Shinjuku, Tokyo, 169-8555 Japan

ABSTRACT: Since the birth of the first actively controlled building, structural control technology has been changing the conventional image of buildings. Now more than thirty buildings in Japan and a couple of buildings and towers in China and Europe have active-control systems implemented. Such a computer-control-based strategy for structural response reduction is stepping forward to more and more advanced levels. This paper discusses the basic philosophy and current view of semiactive control scheme regarded as one of the most promising strategies for protecting structures against large earthquake, demonstrating some analogies with car suspension control.

1 INTRODUCTION

A large number of events have marked remarkable milestones in the history of the development of earthquake engineering. One of such most significant events in the 20th century is perhaps the world's first practical application of automatic response control to a real building structure in 1989 (Kobori et al 1991). This building has dramatically altered the traditional concept of civil structures. The great future possibility of scheme of intelligently ensuring the structural safety in case of severe seismic event has been recognized.

It has passed more than ten years since the first full-scale implementation of active control in civil engineering field was achieved. At the initial stage of active structural-response control implementation, most of the active-controlled buildings were mainly aimed at the response reduction to small/moderate seismic excitation or strong wind excitation. Structural engineers community had appeared quite satisfied with such a gradual or unharried progress of structural control technology until the Kobe earthquake occurred in 1995. Since then, however, the structural engineers have been requested to search a solution to the problem of how to establish the active control schemes for civil structures against severe seismic excitation. In this regard, among several possible strategies accounting for severe seismic loading, semiactive control strategy seems to be promising for the ultimate goal of active structural control. With such a current state of active structural control, this paper discusses a semiactive structural control scheme based upon variable damping system demonstrating some analogies between structural seismic response control and car suspension system control.

2 CAR SUSPENSION SYSTEMS

It may look like that there are differences between structural seismic response control and car suspension control. In terms of a concept of active vibration control, however, these two systems, structural and car suspension systems, are quite similar to each other. This means that the control of building responses to seismic excitation has some similarity with the control of motor vehicle suspension systems. Active control characteristically needs three items - first, control computer; second, sensors; and third, external power supply. The difference is that structural control mainly deals with the horizontal movement and on the other hand car suspension control in principle deals with vertical movement.

In looking back the history of development of structural control, active structural response control is not totally a new idea as a conceptual philosophy. It was more than forty years ago that Kobori (1956) proposed the significance and necessity of integrating automatic control concept into the seismic design of building structures. At that time, of course, control engineering was not matured enough to put into practice the response control of civil structures. In this respect, it took us more than forty years to have the world's first actively-controlled building, the Kyobashi Seiwa, in Tokyo, Japan (Kobori et al 1991).

When it comes to motor vehicle suspension, the practical application of active control was initiated in early 1980s. In view of this, car suspension control has only ten years ahead of structural control in civil engineering field. At the current stage of car suspension control, after attempting a variety of passive and active control schemes, one of the most advanced strategies for car suspension control is seemingly based upon semiactive control scheme. According to the most commonly or normally accepted definition, semiactive control is the one that controls only a part of characteristics of a control actuator. In such semiactive control scheme, therefore, only a limited amount of power is expected to be supplied to the actuator, and only at appropriate points of time to regulate the characteristics of the actuator.

This section presents car suspension systems in a very simplified manner. Figure 1 is a schematic illustration of car suspension system. This suspension system consists of four subsystems. Each subsystem is composed of one spring and one dashpot. A simple and plain way of modeling the suspension system is to deal with each subsystem on the individual basis by considering that any of the subsystems can represent the movement of car body. Based upon such a simplified manner, car suspension is represented by a two-degree-of-freedom (2DOF) model as shown in Figure 2, in which the upper mass represents the car body and the lower mass represents the tire (Kayaba Industry 1991). Further simplicity can be introduced into the modeling of each subsystem by assuming that the vertical movement of the tire is almost identical with the ground irregularities or roughness. Having such simplicity introduced, the above 2DOF system turns out to be a single-degree-of-freedom (SDOF) system like Figure 3 (Kayaba Industry 1991), which of course represents the vertical movement of the car body.

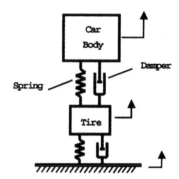

Figure 2. 2DOF model representing car suspension system

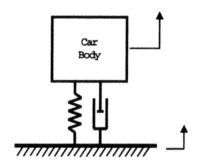

Figure 3. SDOF model representing car suspension system

It is known that the ultimate ride comfort would be provided with zero absolute acceleration or zero absolute displacement of the car body. In the case of steady-state sinusoidal input excitation, smaller absolute acceleration leads to smaller absolute displacement. The transfer function from the ground irregularities to the absolute acceleration of the car body is shown in Figure 4, in which ω_0 represents the natural frequency of the system and co the frequency of input excitation. On the other hand, the transfer function from the ground irregularities to the absolute displacement is given in Figure 5. This figure also shows the transfer function from the seismic ground acceleration to the absolute response acceleration in regard to a SDOF structural system representing the horizontal movement. As can be recognized in Figure 5, for the range in which the ratio of the input excitation frequency ω to the natural frequency ω_0 is smaller than $\sqrt{2}$,larger damping leads to smaller response. On the contrary, for the range in which the ratio is larger than $\sqrt{2}$,smaller damping provides larger response. This result indicates that larger damping does not necessarily provide smaller responses. In view of this, it would be desirable for good ride comfort with variable damping. In passing, better driving stability would be obtained if the movement of the car body is closer to the ground irregularity. This means having good driving stability scarifies ride comfort.

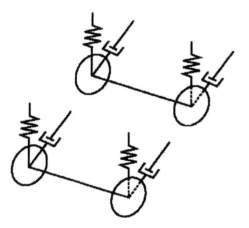

Figure 1. Schematic figure of car suspension system

312

A skilful way of employing variable damping is the scheme based upon so-called sky-hook damper. Its fundamental concept is schematically illustrated in Figure 6. This damper provides a damping force proportional to the absolute velocity of the mass. In employing sky-hook damper, larger damping coefficient always leads to smaller response. The skyhook damper, of course, cannot be produced in direct manner, but can be brought about by changing the damping coefficient of the dashpot in the system shown in Figure 4 so as to provide the same effect as absolute velocity-proportional damping. With z and x_g denoted as the absolute vertical displacement and ground vertical irregularity effect respectively, the equation of motion for the system having an idealistic sky-hook damper is written as:

$$\ddot{z} + H\dot{z} + \omega_0^2 z = \omega_0^2 x_g \quad (1)$$

in which H indicates the coefficient of the sky-hook damper. As already mentioned, sky-hook damper provides smaller response with larger damping coefficient. If the damping coefficient, c, of the real-installed dashpot providing the damping force proportional to the relative velocity to the ground irregularity, $\dot{z} - \dot{x}_g$, can be controlled in such a way that

$$c = \frac{H\dot{z}}{\dot{z} - \dot{x}_g} \quad (2)$$

then the sky-hook damper effect is obtained. In practice, however, when the sign of the absolute velocity, $\dot{z}$, is not the same as that of the relative velocity, $\dot{z} - \dot{x}_g$, the sky-hook damper effect, Hz, is not possible because no negative damping coefficient is available in providing the damping force proportional to the relative velocity, $\dot{z} - \dot{x}_g$.

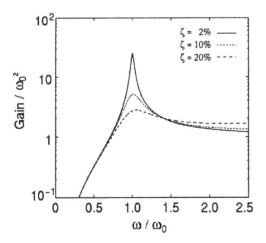

Figure 4. Transfer function of SDOF model from ground irregularity to absolute acceleration

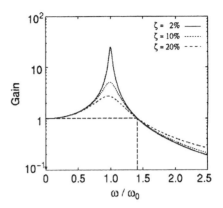

Figure 5. Transfer function of SDOF model from ground irregularity to absolute displacement

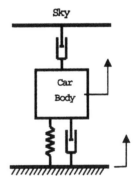

Figure 6. Sky-hook damper

3 CONTROL OF BUILDINGS

3.1 Current state of active structural control

The first full-size implementation of active structural control has in fact opened the door to innovative structural engineering incorporating with advanced modern technologies. Inspired by this first practical application of active control in civil engineering field, more than 30 buildings in Japan have now installed a variety of active, semiactive and hybrid control systems. At present, most of the full-scale implementations of such automatic structural control are found in Japan. The information on these structural-controlled buildings in Japan has been provided in several papers and articles (for instance, Nishitani 1998, Spencer and Soong 1999, Nishitani 1999, and Soong and Spencer 2000).

The control actuators most widely employed in the controlled buildings in Japan are added-mass type of systems such as active mass damper (AMD) or active tuned mass damper (ATMD). The principal philosophy of these systems is to drive them in such a way as to transfer more vibration energy of a main

313

structure to AMD or ATMD. However, when it comes to the discussion about active response control for buildings against severe seismic excitation, the use of such dynamic vibration absorbers of added mass type does not seem to be fit to the purpose of intelligently ensuring the seismic safety. In case of strong seismic event, a large amount of power and a large stroke would be needed for AMD or ATMD. From this reason, most of the employed AMDs have been designed in such a way as to intentionally stop the operation during severe earthquake.

3.2 Semiactive structural control with variable damping systems

In considering the future development toward establishing the control strategy against severe seismic excitation, one of the key points is how to make a system with less reliance on external power. In this regard, as mentioned in INTRODUCTION, semiactive control is one of the most promising control strategies in civil engineering field as well. As already mentioned, semiactive control regulates only partial characteristics of a control actuator with a limited amount of power. At the current stage in Japan, three buildings are semiactively controlled. They are: Kajima Research Institute (KRI) Building No.21 (laboratory building constructed in 1990); the Kajima Shizuoka Building (office building constructed in 1998); and a new office and laboratory building in Keio University School of Science and Technology (constructed in 2000).

Among these three buildings, while only the KRI building employs a variable stiffness system to avoid resonance vibration with ground shaking, the Kajima Shizuoka building and the new building in Keio University have installed semiactively-controlled variable damping systems utilizing oil-hydraulic dampers. However, the new building in Keio Engineering School is fundamentally a base-isolated building, in which semiactively-controlled hydraulic dampers are integrated into the base isolation system. The Kajima Shizuoka building (Kurata et al 1999), on the other hand, has installed only a semiactive-control-based variable damping system, which accounts for the final goal of intelligently ensuring the structural safety against severe seismic loading. The central computer in this five-story building regulates the damping coefficients of eight dampers, two dampers installed on each floor from the first to fourth floor, so as to provide the building with the optimal damping force.

In addition to the Kajima Shizuoka building, the Gifu Regional Office of Chubu Electric Power, an eleven-story office building to be completed in early 2001, will install a semiactive-controlled variable damping system in which each damper is in principle regulated by each micro-computer.

3.3 Semiactive control with variable friction dampers

In the final part of this section one of the authors' recent attempts for semiactive structural control is presented very briefly. It is a semiactive-controlled variable friction damper (Nishitani et al 1999, Nishitani et al 2000) of which only the slip-force level is controlled so as to exhibit a hysteresis with a constant ductility factor no matter how small or how moderate or how severe an earthquake excitation is. Advantageously, a constant ductility factor would make an equivalent linear structural model regardless of the magnitude of earthquake excitation and thus the same equivalent damping ratio would expectedly be obtained. The following results are all from the paper presented by Nishitani et al (2000) at the 2000 American Control Conference held in Chicago.

The basic concept and effectiveness of this damper is demonstrated utilizing a SDOF model. The damper is controlled in such a way as to start to slip at the time of observing the peak response velocity so as to have the ductility factor of two. Figure 7 presents the simulated hysteresis of the damper in case of sinusoidal excitation with the amplitude changing from cycle to cycle. Figure 8 demonstrates the simulated hysteresis of the damper installed in a seismically-excited SDOF model with the natural frequency of vibration equal to 1 Hz.

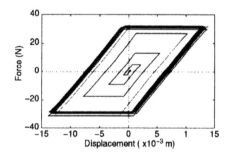

Figure 7. Hysteresis of damper subject to sinusoidal excitation

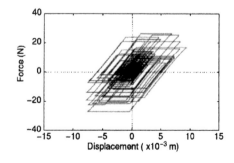

Figure 8. Hysteresis of damper subject to the Taft earthquake

Then, a sample simulation analysis is presented for an equivalent shear-structure model representing a 20-story building. This 20-story building has the first, second and third modal periods of vibration equal to 1.9, 0.63, and 0.36 sec, respectively, without considering any effect of dampers. The simulation is conducted under the assumption that the structural damping is proportional to the stiffness matrix with 1% damping factor of the first mode, the variable friction dampers are installed on every floor, and the building is excited by the NS component of the El Centro earthquake with Peak Ground Acceleration (PGA) of 200 cm/sec^2. Each damper is individually controlled so that the slip should start at the time of peak interstory velocity response. The initial stiffness of the damper in each story is set equal to the stiffness of the story. Figures 9 and 10 compare the controlled and uncontrolled response displacements and absolute accelerations, respectively. The response reduction can be recognized in Figures 9 and 10. By conducting ARX model-based identification based on the input and output data, the equivalent-linear viscous damping ratio has been found 10% (Nishitani et al 2000). Further detailed discussion about the semiactive control strategy utilizing variable friction dampers has been made in the paper presented at the 2000 ACC (Nishitani et al 2000).

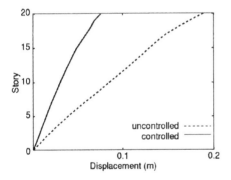

Figure 9. Maximum displacements relative to the base

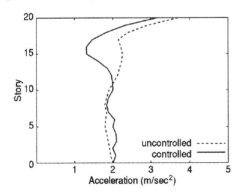

Figure 10. Maximum absolute accelerations

4 CONCLUDING REMARKS

A lot of efforts have been devoted to conducting the research and development of active, semiactive and hybrid structural control. In considering a strategy for intelligently ensuring the safety of buildings against severe seismic loading, at the present stage, semiactive control scheme appears promising. There seem to be a lot of things that we should learn from car suspension control in conducting structural response control against seismic loading. Along with the rapid progress of computer technology and information science, structural engineering field will expectedly incorporate with more and more sensing and control technologies in many ways. In a decade or so, active and semiactive control schemes might entirely change the significance of civil structures having a deeper relationship with advanced technologies.

The author acknowledges the partial support of this work by JSPS Research for the Future Program (96R15701).

REFERENCES

Kayaba Industry. 1991. *Automotive Suspension,* Sankaido
Kobori, T. 1956. Quake resistant and nonlinear problems of the structural vibrations to violent earthquake, *Journal of Kyoto University Disaster Prevention Laboratory,* 5th Anniversary edition: 116-124 (in Japanese)
Koori. et al. 1991. Seismic-response-controlled structure with active mass driver system: Part 1-Design, Part 2Verification, *Earthquake Engineering and Structural Dynamics,* and 20: 133-161
Kurata, N. et al 1999. Actual seismic response controlled building with semi-active damper system, *Earthquake Engineering and Structural Dynamics, 28:* 1427-1447
Nishitani, A. 1998. Application of active structural control in Japan, *Progress in Structural Engineering and Materials,* 1(3): 301-307
Nishitani, A. 1999. Application of active, hybrid and semiactive structural control in Japan: Recent state, *Proceeding of the International Conference Seminar on Seismic Isolation, Passive Energy Dissipation and Active Control of Vibrations of Structures,* Cheju, Korea, 1: 461-466
Nishitani, A., Nitta, Y., Ishibashi, Y. & Itoh, A. 1999. Semiactive structural control with variable friction dampers, *Proceedings of the American Control Conference,* San Diego, 2: 1017-1021
Nishitani, A., Nitta, Y., Itoh, A. & Ikeda, Y. 2000. Semi-active variable-friction damper control with simple algorithm, *Proceedings of the American Control Conference,* Chicago, 1: 503-507
Soong, T.T. & Spencer, B.F., Jr. 2000. Active, semiactive and hybrid control of structures, *Proceedings of the 12th World Conference on Earthquake Engineering* (CD-ROM), Auckland, New Zealand, No.2834
Spencer, B.F., Jr. & Soong, T.T. 1999. New application and development of active, semi-active and hybrid control techniques for seismic and non-seismic vibration in the USA, *Proceeding of the International Conference Seminar on Seismic Isolation, Passive Energy Dissipation and Active Control of Vibrations of Structures,* Cheju, Korea, 1: 467-488.

Earthquake Engineering Frontiers in the New Millennium, Spencer & Hu (eds),
© 2001 Swets & Zeitlinger, ISBN 90 2651 852 8

Optimum and simulation of semi-active control system based on MRFD[*]

L.H. Xu & Z.X. Li
School of Civil Engineering, Tianjin University, Tianjin, 300072, PR China

Y. Zhou
Department of Civil Engineering, Guangzhou University, Guangzhou, 510405, PR China

ABSTRACT: In this paper, the optimum and simulation for a semi-active control system based on the mag-neto-rheological fluid damper (MRFD) are performed. The optimal distribution of the control devices on the structure is systematically investigated, and the simulated analysis for a tall building excited by El-Centro earthquake is studied when a passive control system based on MRFD or a semi-active control system based on MRFD is installed on the building respectively. The research results provide a powerful reference for the design and application of the semi-active control system based on MRFD to control the seismic responses of tall buildings.

1 INTRODUCTION

Magneto-rheological (MR) fluid is a kind of smart material that responds to an applied magnetic field with a dramatic change in rheological behavior. The essential characteristics of this kind of fluid is its ability to reversibly change from free-flowing, linear viscous liquid to semi-solid having controllable yield strength in milliseconds when a magnetic field is exposed. Because of their mechanical simplicity, high dynamic range, low power requirement, large force capacity and robustness, MR fluid damper (MRFD) has shown to be a semi-active control device that meshes well with application demands of protecting civil engineering structures against severe earthquake and wind loading (Spencer et. al. 1998).

According to the feedback information of struc-tural response and/or external excitation, the semi-active control system can real-time adjust structural parameter to make the structural responses to reach the optimal state, which has the merit of passive control system and active control system and can achieve the majority of the performance of full ac-tive system using certain control laws. In this paper, the optimum design of a semi-active control system based on MRFD is studied and the simulated analy-sis is made for an 11-story frame structure under earthquake excitations.

2 STATE EQUATIONS OF SYSTEM

The equations of motion of a MDOF linear structure installed by r control devices based on MRFD, ex-cited by one-dimension earthquake, can be written in the standard form as:

$$M\ddot{Y}(t) + C\dot{Y}(t) + KY(t) = -MI_1\ddot{u}_g(t) + EU(t) \qquad (1)$$

where M, C and K denote the mass, damping and stiffness matrices of the structure respectively. $Y(t)$ denotes the displacement vector of the structure rela-tive to the ground. $\ddot{u}_g(t)$ is the seismic accelerations excitation. I_1 is an unity vector. $U(t)$ denotes the r-dimension vector of control forces. E is the $n \times r$ ma-trix representing the placements of control devices on the structure.

Based on Equation 1, the state equations of sys-tem can be written as follows:

$$\dot{X}(t) = AX(t) + BU(t) + W\ddot{u}_g(t) \qquad (2)$$

in which,

$$X(t) = \begin{bmatrix} Y(t) \\ \dot{Y}(t) \end{bmatrix}, A = \begin{bmatrix} 0 & I_n \\ -M^{-1}K & -M^{-1}C \end{bmatrix}, B = \begin{bmatrix} 0 \\ M^{-1}E \end{bmatrix}, W = \begin{bmatrix} 0 \\ -I_1 \end{bmatrix} \qquad (3)$$

where I_n denotes the n-dimension unity matrix.

By given the sampling interval Δt and dividing the duration of earthquake input, t_f, into N intervals,

_____
*Supported by NNSF of China

the discrete state equations of system may be written as:

$$X(k+1) = GX(k) + HU(k) + W_1(k) \qquad (4)$$

in which,

$$G = e^{A\Delta t}, \quad H = (e^{A\Delta t} - I)A^{-1}B$$
$$W_1(k) = (e^{A\Delta t} - I)A^{-1}W\ddot{u}_g(k) \qquad (5)$$

where H, G and $W_1(k)$ are the constant matrices only related to the sampling period Δt, and $k=1, 2, \cdots, N$.

3 OPTIMAL CONTROL

To obtain satisfied control effects with minimum control devices, the optimal distribution of the control devices on the structure should be determined. The performance index of system is the evaluation function to reach the optimal control. To minimize the performance index can realize the objective of optimal control.

The performance index of system can be defined as:

$$J = \sum_{k=0}^{\infty} [X^T(k)Q_1X(k) + U^T(k)RU(k)] \qquad (6)$$

where Q_1 and R denote the weighting matrices of the state vector of system and the vector of control forces respectively.

Applying the feedback control law (Soong 1990), the control force in Equation 4 can be expressed as:

$$U(k) = -LX(k) \qquad (7)$$

in which $L = R^{-1}B^T P$ is the feedback gain matrix, where P is a matrix to satisfy the following Riccati equation:

$$PA + A^T P - PBR^{-1}B^T P + Q_1 = 0 \qquad (8)$$

Substitutes Equation 7 into Equation 6, get:

$$J = \sum_{k=0}^{\infty} X^T(k)QX(k) \qquad (9)$$

in which,

$$Q = Q_1 + L^T RL \qquad (10)$$

Substitutes Equation 7 into Equation 4, get:

$$X(k+1) = (G - HL)X(k) + W_1(k) \qquad (11)$$

Define

$$\Psi(k) = (G - HL)^k \qquad (12)$$

where $\Psi(k)$ denotes the state transfer matrix.

Thus, Equation 11 can be rewritten as:

$$X(k+1) = \Psi(1)X(k) + W_1(k) \qquad (13)$$

Given the initial state vector $X(0)$ and the external excitation $\ddot{u}_g(k)$, and applying the iterative method, the solution of Equation 13 can be solved as follows:

$$X(k) = \Psi(k)X(0) + \sum_{i=0}^{k-1}\Psi(k-i-1)W_1(i) \qquad (14)$$

Substituting Equation 14 into Equation 9 and ignoring the non-diagonal items, get:

$$J = \sum_{k=0}^{\infty}[X^T(0)\Psi^T(k)Q\Psi(k)X(0)] +$$
$$\sum_{k=0}^{\infty}[\sum_{i=0}^{k-1}W_1^T(i)\Psi^T(k-i-1)Q\Psi(k-i-1)W_1(i)]$$
$$= X^T(0)[\sum_{k=0}^{\infty}\Psi^T(k)Q\Psi(k)]X(0) + \qquad (15)$$
$$\sum_{i=0}^{N}W_1^T(i)[\sum_{k=i+1}^{\infty}\Psi^T(k-i-1)Q\Psi(k-i-1)]W_1(i)$$

Define:

$$S = \sum_{k=0}^{\infty}\Psi^T(k)Q\Psi(k) \qquad (16)$$

where S satisfies Lyapunov equation. Let $j=k-1$, get:

$$S = \sum_{j=0}^{\infty}\Psi^T(j+1)Q\Psi(j+1) + Q$$
$$= \Psi^T(1)S\Psi(1) + Q \qquad (17)$$

Thus, Equation 15 can be turned into:

$$J = X^T(0)SX(0) + \sum_{i=0}^{N}W_1^T(i)SW_1(i) \qquad (18)$$

When the position indication matrix of control devices, H, changes ΔH, the performance index of system, J, has an increment (Soong 1990):

$$\Delta J = J(H + \Delta H) - J(H) = \text{tr}(R_0 + P_0)\Delta S \qquad (19)$$

where tr denotes the sign of matrix trace, and

$$R_0 = X(0)X^T(0)$$
$$P_0 = \sum_{i=0}^{N}W_1(i)W_1^T(i) \qquad (20)$$

From Equation 12, the increment of $\Psi(1)$ can be written as:

$$\Delta\Psi(1) = -\Delta HL \qquad (21)$$

From Equation 17, the increment of S can be written as:

$$\Delta S = \Psi^T(1)\Delta S\Psi(1) - [L^T\Delta H^T S\Psi(1) + \Psi^T(1)S\Delta HL] \qquad (22)$$

Comparing Equation 22 with Equation 17, get:

$$Q = -[L^T\Delta H^T S\Psi(1) + \Psi^T(1)S\Delta HL] \qquad (23)$$

Substituting Equation 23 into Equation 16, get:

$$\Delta S = -\sum_{k=0}^{\infty} \Psi^T(k)[L^T \Delta H^T S \Psi(1) + \Psi^T(1)S\Delta HL]\Psi(k) \quad (24)$$

Substituting Equation 24 into Equation 19, obtain:

$$\Delta J = -2\mathrm{tr}\{[\sum_{k=0}^{\infty} \Psi^T(k)R_0 \Psi(k) \\ + \sum_{k=0}^{\infty} \Psi^T(k)P_0\Psi(k)]\Psi^T(1)S\Delta HL\} \quad (25)$$

Suppose:

$$R_1 = \sum_{k=0}^{\infty} \Psi^T(k)R_0 \Psi(k) \\ P_1 = \sum_{k=0}^{\infty} \Psi^T(k)P_0\Psi(k) \quad (26)$$

and their solutions satisfy the following Lyapunov equations:

$$R_1 = \Psi^T(1)R_1\Psi(1) + R_0 \\ P_1 = \Psi^T(1)P_1\Psi(1) + P_0 \quad (27)$$

Then, Equation 25 can be turned into:

$$\Delta J = -2\mathrm{tr}[(R_1 + P_1)\Psi^T(1)S\Delta HL] \quad (28)$$

According to the matrix theory (Franklin 1968), exist:

$$\Delta J = \mathrm{tr}[(\frac{\partial J}{\partial H})^T \Delta H] \quad (29)$$

Comparing Equation 29 with Equation 28, get:

$$\frac{\partial J}{\partial H} = -2[L(R_1 + P_1)\Psi^T(1)S]^T \quad (30)$$

In Equation 30, $\partial J/\partial H$ represents the sensitivity of the performance index of system to the placements of control devices on the structure, which is an important aspect to judge the performance of control system. To obtain the optimal control effects, the minimum sensitivity of the performance index of system should be reached, i.e. optimal distribution of control devices on the structure should be reached. Thus, in the design of control system, the control device whose control action has little influence on the sensitivity of the performance index of system can be removed to obtain more economic distribution of control devices.

Exactly, at the beginning of the design, suppose the structure are fully installed with the control devices, and then remove the control devices which have smaller influence on the sensitivity of the performance index of system. When the i-th control device is removed, the position indication matrix, H, will occur a change, ΔH_i. Thus, from Equation 29, a decrement of the performance index, ΔJ, will be gained, which means the loss of the performance index of optimal control because of removing the ith control action.

Define the contribution index of the i-th control action as:

$$\xi_i = \frac{\Delta J_i}{\sum_{i=1}^{n} \Delta J_i} \times 100\% \quad (31)$$

which represents the contribution of the i-th control action on the performance index of optimal control. Thus, in the process of design of control system, the control devices should be installed on the positions where the contribution indexes, ξ_i, $i=1, 2, \cdots, r$, are larger.

4 NUMERICAL SIMULATION

In order to investigate the feasibility and effectiveness of the semi-active control system based on MRFD to tall buildings under earthquake excitations, an 11-story frame structure with 8 MRFDs, excited by El-Centro (1940 NS) earthquake, will be simulated, and the performance of the passive control system based on MRFD (with the maximum magnetic field) and the semi-active control system based on MRFD will be compared. The stiffness and mass parameters of the structure are illustrated in Table 1, and the damping ratio of the structure is determined as 5%. The maximum value of acceleration of earthquake excitation is scaled to 200 cm/s^2.

Table 1. Structural parameters

Floor No.	High (m)	Stiffness (10^7 N/m)	Mass (10^4kg)
1	3.6	9.16	4.42
2	3.6	9.16	4.42
3	3.3	8.83	4.42
4	3.3	8.92	4.42
5	3.3	7.91	4.42
6	3.3	7.31	4.42
7	3.3	6.61	4.42
8	3.3	5.80	4.42
9	3.3	4.88	4.42
10	3.3	3.81	4.42
11	3.1	2.55	5.42

The selection of weighting matrices of Q_1 and R is more difficult for MDOF structures. In general, R is symmetrical positive definite matrix, and Q_1 is a symmetrical matrix. Here, Q_1 and R are selected as:

$$Q_1 = \begin{bmatrix} K & C/2 \\ C/2 & M \end{bmatrix}, \quad R = 10^{-9}I_2 \quad (32)$$

where I_2 is $r \times r$ dimension unit matrix.

Figure 1 shows the contributions of the control devices on the performance index of optimal control. Thus, the 8 control devices based on MRFD should be installed on the 1st, 3rd, 4th, 5th, 6th, 9th, 10th and 11th stories. To simulate the effectiveness of

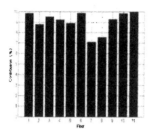

Figure 1. Contributions of control devices

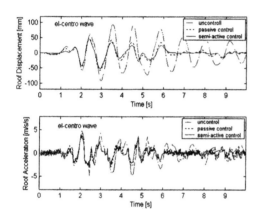

Figure 2. Time-history at the top floor

Table 2. Displacement amplitudes (mm)

Floor No	Without control	Passive control		Semi-active control	
		Disp.	Reduction	Disp.	Reduction
1	9.1	5.3	41.8%	4.7	48.4%
2	17.5	10.5	40%	9.1	48%
3	26.2	15.7	40.1%	13.6	48.1%
4	34.5	20.8	39.7%	17.8	48.4%
5	43.4	26.4	39.2%	22.6	47.9%
6	52.3	31.9	39%	27.5	47.4%
7	60.9	37.5	38.4%	32.9	46%
8	69.5	43.4	37.6%	38.2	45%
9	78.0	48.7	37.6%	43.1	44.7%
10	86.4	53.6	38%	47.7	44.8%
11	94.5	57.7	38.9%	51.7	45.3%

control based on MRFD, three control cases are considered as: (1) without control, (2) passive control based on MRFD, and (3) semi-active control based on MRFD. Table 2 shows the displacement amplitudes of the structure in the different control cases, and Figure 2 shoes the curves of time history of the displacement and acceleration responses at the top floor of the structure.

From Table 2 and Figure 2, it is seen that the control effectiveness is changed for passive control system from 37.6% to 41.8% and for semi-active control system from 44.7% to 48.4%. The results show that the variation range of the control effectiveness is narrower, and the upper and lower limited values for the case of semi-active control based on MRFD are larger than for the case of passive control based on MRFD. And, it is seen that either the passive control based on MRFD or the semi-active control based on MRFD can reduce the structural responses effectively, but the total control effectiveness of the semi-active control is more significant than that of the passive control, which is approved that MRFD can display greater superiority in the semi-active control system than in the passive control system.

5 CONCLUSIONS

Through the study on the optimum design and simulation of the semi-active control system based on MRFD in this paper, some conclusions can be given that either the passive control based on MRFD or the semi-active control based on MRFD can reduce the structural responses remarkably, but the total control effectiveness of the semi-active control is more significant than that of the passive control, which shows greater superiority of the semi-active control system than the passive control system. This research shows the prospect of engineering application of the semi-active control system based on MRFD to control the seismic responses of tall buildings.

REFERENCES

Franklin J.N. 1968. *Matrix theory*. New York: Prentice-Hall.
Leipholz H.H.E. & Abdel-Rohman M. 1988. *Control of structures*. Boston: Martinus Nijhoff Publishers.
Soong T.T. 1990. *Active structural control: theory and practice*. London: Longman Scientific & Technical.
Spencer B.F., Yang G., Calson J.D. & Sain M.K. 1998. 'Smart' dampers for seismic protection of structures: a full-scale study. Proc. 2nd Intern. Conf. on Struct. Control, Kyoto.
Yang J.N. & M.J. Lin (1982). Optimal critical-mode control of building under seismic load. *J. of Eng. Mech.*, ASCE.

Earthquake Engineering Frontiers in the New Millennium, Spencer & Hu (eds),
© *2001 Swets & Zeitlinger, ISBN 90 2651 852 8*

A new type of base-isolator: A vertical shock absorber

Masao Muraji
Department of Architectural Engineering, Osaka University, Japan

Toru Shimizu
Japan Research Institute, Japan

Philip Esper
Ove Arup & Partners, United Kingdom

Eizaburo Tachibana
Department of Architectural Engineering, Osaka University, Japan

ABSTRACT: The most significant recent advances in seismic engineering have been the development of a range of variable isolator devices. These include various combinations of simple rubber sheets, steel plates, lead extrusion elements and steel torsional and flexural curved beams. However all these devices cater for reducing the effect of horizontal ground motion on the structure. In the latest Northridge and Kobe earthquakes, it was evident the vertical component of a strong ground motion can have very damaging effects on reinforced and steel columns. This paper is proposing for the first time a vertical shock absorber that is backed up by experimental as well as analytical results which illustrate its functionality and performance in reducing the effect of vertical vibration on a structure. A review of vertical ground motion effects on building columns is given first, then testing of plain concrete samples under cyclic dynamic loading is presented. A two dimensional analysis of a frame structure with vertical elasto-plastic bearing is discussed and then a prototype of the new vertical shock absorber is introduced together with the testing and analysis work which its satisfactory performance.

1 INTRODUCTION

When the collapse patterns of buildings were examined following the Hyogo-ken Nambu earthquake (1995), it was clear that (see Plate 1) the columns standing on the same floor did not always collapse at the same time or in the same direction. An example is shown in Plate 2. In this case, a partial collapse of the same level was observed. Many structures collapsed similar to this building, before developing the predicted unstable mechanism by hinge theory. One explanation may be that high frequency vertical vibrations caused the destruction of the columns. However, vertical vibrations are not regarded as being very onerous in Japanese Building Seismic Design Codes. Seismic records in the past have showed that both values of acceleration and displacement in the vertical direction are less than those in the horizontal direction, but it was observed in the Hyogo-ken Nambu earthquake that the maximum vertical acceleration exceeded horizontal accelerations in some sights (see Table 1). Moreover Japanese seismic records cut off components of frequency above 50Hz-100Hz. Vertical natural frequencies of buildings, in general, are higher than the horizontal ones. By investigating the reasons behind the partial collapse of buildings observed in the Hyogo-ken Nambu earthquake and a more detailed examination of the strength of the structural materials under dynamic loading, it was possible to highlight some important points which is discussed in this paper.

Plate 1 Example of story collapse

Plate 2 Example of vertical collapse

321

Table 1 Maximum accelerations recorded in Hyogoken Nanbu earthquake

	Direction	Max of Acceration (gal)
Kobe university	NS	270
	EW	305
	UD	446
Kobe Port Island	NS	341
	EW	284
	UD	556
JMA Kobe	NS	818
	EW	617
	UD	332

2 EXPERIMENT OF PLAIN CONCRETE UNDER DYNAMIC CYCLIC LOADING

2.1 General

There is a considerable amount of research work on relationships between strain rate and axial compressive strength of structural materials [see Refs. 1-3]. The research work discussed in this paper showed that the strength of structural materials is sensitive to the rate of loading. However, there are not many studies on material strength under dynamic cyclic loading, especially on the strength of concrete material.

In this section, experimental investigations of the uniaxial compressive strength and the fracture behavior of plain concrete under dynamic cyclic loading are discussed.

2.2 Loading methods

Dynamic cyclic loading of 40cycles in 4Hz with adjusted loading time was applied to test samples of plain concrete as shown in Fig. 1. Because the frequency of about 4Hz represents the natural frequency of low rise residential buildings, and because approximately the first 10 seconds might be very important for the effects of vertical ground motions, load curves can be divided into two parts. One is adjusting loading time and another is steady loading time, in which the upper and the lower bound values of cyclic loads are coincident with the assigned values. The maximum static load of test samples was 22.7tonf (maximum static compression strength; Fc=289.0kgf/cmz). The upper-bound forces of loading were 25.0tonf (about 1.1Fc), 27.5tonf (about 1.2Fc) and 30.0tonf (about 1.3Fc). The lower-bound force was 5tonf in all the cases. The plain concrete test pieces were 100 mm in diameter, 200 mm in height.

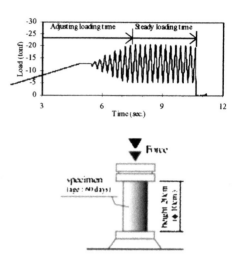

Fig. 1 A chopping waves of 40 cycles in 4Hz and configuration of test samples

2.3 Experimental results

Experimental results are shown in Table 2. A load-displacement curve of series F2 is shown in Fig. 2. The values of compression are under zero. Displacement of this test sample did not proceed under cyclic loading, but the sample was destroyed suddenly.

In summary, these axial compression experimental results of plain concrete under cyclic loading of 4Hz, 40cycles show that;

1) 4 of 5 test samples were not destroyed under cyclic loading when upper-bound force was 1.1Fc.
2) All 5 test samples were destroyed during steady loading time under cyclic loading when upper force was 1.2Fc.
3) All 5 test samples were destroyed during adjusting loading time under cyclic loading when upper force was 1.3Fc.
4) The process of destruction of test samples under cyclic loading is completely different from that under static loading through video (Plate 3).

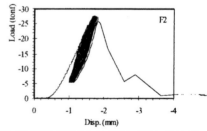

Fig. 2 Load-displacement curve (Series F-2 in Table 2)

2.4 Observation

The test samples were destroyed within 5 seconds under cyclic loading when upper force was 1.1Fc to 1.2Fc. The theory that strength grows up as the veracity of loading grows up should not be overestimated from this result.

3 DYNAMIC ANALYSIS OF TWO-DIMENSIONAL MODEL

3.1 Outline

When a dynamic load is applied to a building, it is possible that repetitive loading, such as, cyclic loading will be applied to columns. Because stress

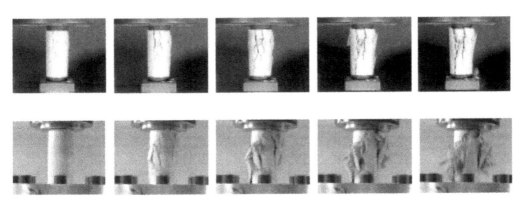

Plate 3. Consecutive shot by video of fracture by static loading (uper) and dynamic cyclic loading in 4 Hz (lower)

Table 2 Results of dynamic cyclic loading experiment

Series	Test type Upper force (day)	Lower force (tonf)	Upper force (tonf)	Limit cycle	Cycle	Fractured load (tonf)	Fractured period (in Fig. 3.1)	Comment
C-1		-2	-5	20	20	-	-	σ_t=24.1kgf/cm^2
C-2	Dynamic	-2	-5	20	20	-	-	(7.58tonf)
C-3	splitting	-2	-5	20	20	-	-	
C-4	0.66σ_t	-2	-5	20	20	-	-	Loading statically
C-5	(97.9.26)	-2	-5	20	20	-	-	after this test
D-1		-2	-10	20	0	-10.00	Adjusting loading time	
D-2	Dynamic	-2	-10	20	0	-9.80	Adjusting loading time	
D-3	splitting	-2	-10	20	0	-9.98	Adjusting loading time	
D-4	1.32σ_t	-2	-10	20	0	-9.98	Adjusting loading time	
D-5	(97.9.26)	-2	-10	20	19	-10.00	Adjusting loading time	
E-1		-5	-25	40	0	-12.60	Steady loading time	σ_c=289.3kgf/cm^2
E-2	Dynamic	-5	-25	40	40	-	-	(22.72tonf)
E-3	compression	-5	-25	40	40	-	-	
E-4	1.1σ_c	-5	-25	40	40	-	-	Loading statically
E-5	(97.9.27)	-5	-25	40	40	-	-	after this test
F-1		-5	-28	40	14	-27.25	Steady loading time	
F-2	Dynamic	-5	-28	40	7	-27.55	Steady loading time	
F-3	compression	-5	-28	40	4	-27.35	Steady loading time	
F-4	1.21σ_c	-5	-28	40	1	-27.55	Steady loading time	
F-5	(97.9.27)	-5	-28	40	8	-27.60	Steady loading time	
G-1		-5	-30	40	0	-28.85	Adjusting loading time	
G-2	Dynamic	-5	-30	40	0	-28.85	Adjusting loading time	
G-3	compression	-5	-30	40	0	-28.40	Adjusting loading time	
G-4	1.32σ_c	-5	-30	40	0	-29.10	Adjusting loading time	
G-5	(97.9.27)	-5	-30	40	0	-29.05	Adjusting loading time	

wave of the dynamic loading repeatedly reflect in the frame of the building.

In this section, the effect of impulsive loading that is applied to a building is studied by using finite element code LS-DYNA.

3.2 *Analysis model*

The analysis model is a two-dimensional model shown in Fig. 3. Material property assumed to be those of a typical reinforced concrete structure. All sections of columns are 600mm*600mm. All sections of beams are 400 mm (width)*600mm (height).

The left-hand side and center columns are divided to 20 elements. All beams and right column are divided to 40 elements. All elements are elastic.

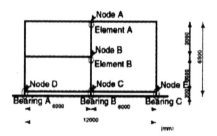

Fig. 3. Frame model

3.3 *Input wave*

The input wave is a Ricker wavelet shown in Fig. 4, where Tp=0.1 sec A=5G. This wave is applied to the base of the FE model in the vertical direction. The value of 5G is much large. Seismic recorder, however, cut off components of over 100Hz in seis-

mic wave. The maximum acceleration of this input wave through Low Pass Filter is 630gal as shown in Fig. 5.

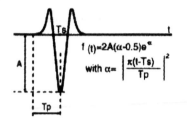

Fig. 4 Ricker wavelet

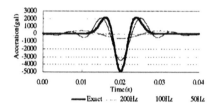

Fig. 5 Ricker wavelet through Low Pass Filter

3.4 *Analysis results*

Diagrams of axial force from 0.0214sec after acceleration of input wave attains maximum by 0.0004sec are shown in Fig. 6. Process that stress waves pass through columns is observed.

3.5 *Summary*

The axial force at the top of a column may exceed that at the bottom of the column.

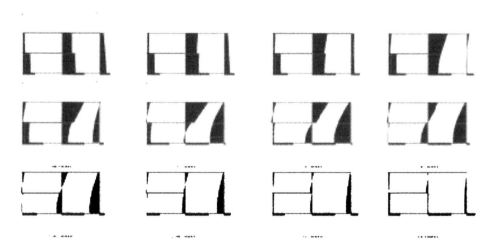

Fig. 6 Axial force of columns from 0.0214 seconds after Ts (in Fig.4.2) to 0.0258 seconds

4 DYNAMIC ANALYSIS OF TWO-DIMENSIONAL MODEL WITH ELASTO-PLASTIC BEARING

4.1 Outline

In this section the effect of elasto-plastic bearing to reduce impulsive vertical loading is studied. Analysis method, model and conditions are the same as the one discussed in section 4 (except for the bearings).

4.2 Property of elasto plastic bearing

Yield stress is different each other. Each bearing yields at a stress of 1.2 times of that of a static axial stress. Hardening moduli are all 1% of Young's modulus.

The total rigidity of three bearings is nearly equal to that of three columns on the first floor. Vertical each rigidity of three bearings is decided, as all static displacements are same. All bearings are assumed that they are fixed horizontally.

4.3 Analysis result and observation

History of axial force of Element A of the FE model illustrated in Fig.4 is shown in Fig 7. The amplitude of axial force decreases by elasto-plastic bearing.

History of relative vertical displacement of Node D and E to Node C is shown in Fig.8. The maximum relative vertical displacement is about 0.2mm. Angle of beam by relative vertical displacement is 0.00003. Consequently additional stresses of beam by installing elasto-plastic bearings seem to be negotiable. Relative displacement is small, because shear force of beam carries axial force to other bearings.

5 VERTICAL SHOCK ABSORBER

5.1 Outline

In Section 4, it was mentioned that installing elasto-plastic bearing to structural model reduces the amplitude of axial force. In this section, a new type of vertical shock absorber is introduced, experimental and analytical results of this prototype are described.

5.2 Configuration of vertical shock absorber and loading experiment

A prototype of vertical shock absorber has 4 steel pipes and a rubber mat packed with 2 steel disks as shown in Fig.9 and 10. Steel pipes and rubber mat carry axial force and a center projection carry shear force. When excessive axial force occurs by vertical resonance or impulsive external vertical force, steel pipes buckle and axial force is distributed to rubber mat in order that the acceleration response and the axial force on the building reduce.

The height of steel pipes is 100mm, with a thick top and bottom and a middle section as shown in Fig. 11. The shape of rubber mat is like a disk which central part and the positions of 4 steel pipes are cut off.

Static loading experiment was carried out first. Vertical load-displacement curve of this prototype is shown in Fig. 15. The stiffness of this model does not decrease by increase of stress of rubber mat after steel pipes buckle at around 5mm of vertical displacement.

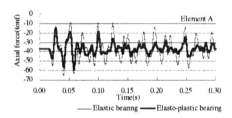

Fig. 7 History of axial force (Element A in Fig 4)

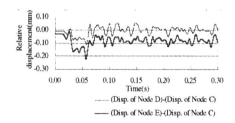

Fig. 8 Relative vertical displacement-time histories

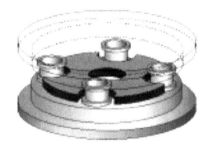

Fig. 9 Configuration of vertical shock absorber

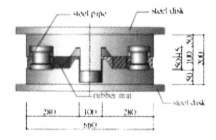

Fig. 10 Section of vertical shock absorber (mm)

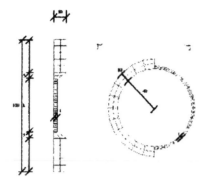

Fig. 11 FE model of a steel pipe

5.3 *Analysis of steel pipe and rubber mat*

The FE model of steel pipe is shown in Fig. 11. Parameter (t) that presents central thickness of steel pipe is varied as follows 2.7mm, 3.0 mm and 3.5 mm. The method of loading used in analysis is that the rigid loading board is moved vertically by Omm-30mm in order to achieve the same condition of experiment Contact elements modeled the condition of the roller support.

    Material properties used are;
Mass density: 8.0 kg fs^2/cm^2
Young's modulus: $2.1*10^6$ kgf/cm^2
Poisson's ration: 0.3
Hardening modulus: 0.75%, 1.00 % and 1.25% of Young's modulus

    These material properties were verified by experimental results of steel pipe. Mass density, however, is larger than proper value, in order to shorten the calculation time of LS-DYNA.

    In the setting of material property of rubber mat, Mooney-Rivlin equation is used to find strain energy density function was shown in equation (1)

$$w=C_1(I_1-3)+C_2(I_2-3) \qquad (1)$$

w: Strain energy density function
C_i: Constant
I_i: Invariant of Cauchy-Green deformation tensor
where and are expressed in terms of axial strech ration (I=1,2 and 3), as follows:

$$I_1=\lambda_1^2+\lambda_2^2+\lambda_3^2 \qquad (2)$$

$$I_2=\lambda_2^2\lambda_3^2+\lambda_3^2\lambda_1^2+\lambda_1^2\lambda_2^2 \qquad (3)$$

$$\lambda_i=(1+\varepsilon_i) \qquad (4)$$

ε_i: Principal strain

    Material constants C_1 and C_2 of rubber are set 1.545(kgf/cm^2) and 0.053(kgf/cm^2) in this analysis.

These values are calculated in two directional stretching experiment of mbber [E. Tachibana et al]. Analysis that C_1 and C_2 are set to 5.0 (kgf/cm^2) and 0.5 (kgf/cm^2) also were performed in order to examine sensitivity of these constants. Mass density is set to $1.0*10^{-6}$ (kgfs^2/cm^4). Analysis method is same as steel pipe.

5.4 *Analysis result*

Both experimental and analytical results are shown in Fig. 12. This graph shows results of steel pipe with thickness of 2.7mm (t=2.7mm). Vertical axis shows the value that compression axial force is divided by yield stress and section area of middle thickness of steel pipe. Horizontal axis shows vertical displacement. Analysis results show good agree with experimental result. Compression axial force, however, varies from 10% to 15% by hardening modulus. Self-contact start, from about vertical displacement of 26mm in experiment, but in analysis it start, from less displacement by about 4m m. This seems due to be rough division of FE model of steel pipe (see Fig.13).
Vertical load-displacement curves of rubber mat obtained by both experiment and analysis are shown in Fig. 14. This analysis using MooneyRivlin equation agrees with rubber's property that a load-displacement

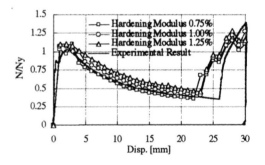

Fig. 12 Load-displacement curves of steel pipes(t-2.7mm)

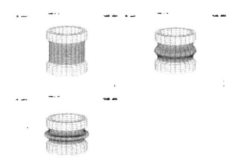

Fig. 13 Buckling process of a steel pipe (t=3.Omm)

curve of rubber shows a second curved line. This curve is more sensitive about constant C_1 than C_2.

Fig. 15 shows a load-displacement curve super-imposed analysis results of steel pipe (t=3.0mm) and rubber mat in order to compare with the experimental result of the shock absorber. Analysis results in this case show that hardening modulus is 1.25% almost agrees with experimental result.

5.5 *Summary*

The static analysis of vertical shock absorber was performed as a first step to examine the effect of introducing the vertical shock absorber on the dynamic response of the building studied. Consequently the historical property including the behavior after buckling was almost realized in this analysis. Dynamic analysis of vertical shock absorber is planed to execute in future.

6 CONCLUSIONS

1 Experiment of plain concrete under cyclic loading shows that plain concrete has tendency to fracture in a brittle manner when it is under cyclic loading.

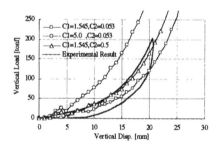

Fig. 14 Load-displacement curves of rubber

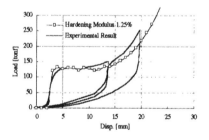

Fig. 15 Load-displacement curve of vertical shock absorber

2 Finite element analysis of a twodimensional frame model showed axial force of columns in two-dimensional framed model is not uniform, axial force at the top of the column may exceeds that at the bottom of the same column and pulling force may develop at the bottom of the column.

3 Finite element analysis of a twodimensional frame model with elasto-plastic bearing shows that installing elasto-plastic bearing will decrease the amplitude of the axial force of columns.

4 In developing vertical shock absorber, the analysis results of the prototype by finite element method agree well with the experimental result.

Vertical displacement of seismic vibration is less than horizontal one. High frequency components characteristic of vertical seismic vibration, however, effects the behavior of buildings. The research in this area is still at its early stage.

7 ACKNOWLEDGMENT

The authors gratefully acknowledge the support of Takehito Kitano and Akio Nakayama of Kansai Electric Power Corporation, Takashi Kashiwade of Ataka Industries, Ltd., Shizuo Tujioka of Fukui Institute Unversity, Takao Nakagawa of Osaka University and Tatsuji Matsumoto of Sumitomo Rubber Industries, Ltd.

REFERENCE

P. H. Bischoff and S. H. Perry, "Compressive behaviour of concrete at high strain rates.", Mat. and Struct., 24(144).

W. Suaris and S. P. Shah, "Constitutive model for dynamic loading of concrete.", J. Struct. Engrg., ASCE, 3(111).

Shirai K., Ito C. and Funahashi M. (1992), "Evaluation of the impact behavior of the contents of reprocessing radioactive waste shipping cask subjected to drop impact" PATRAM ' 92.

Toru Shimizu and Eizaburo Tachibana (1998), "The study of the strength of a plane concrete subjected to dynamic cyclic loading."

Toru Shimizu and Eizaburo Tachibana (1999), "Finite element method of vertical shock absorber installing to the base of a building."

Akio Nakayama (1998), "The effect of the property of vertical restoring force of bearings to structural responses."

The committee of seismic engineering of civil engineering (1997), "The problems of earthquake of level 2 and seismic load for design."

Japan Research Institute (1997), "LS-DYNA ver.940 USER'S MANUAL", Japan Research Institute.

Earthquake Engineering Frontiers in the New Millennium, Spencer & Hu (eds),
© 2001 Swets & Zeitlinger, ISBN 90 2651 852 8

Dynamic analysis for structures supported on slide-limited friction base isolation system

Jian Fan & Jiaxiang Tang
Dept. of Civil Engineering, Huazhong Univ. of Sci. & Tech., Wuhan, China

ABSTRACT: This paper presents a new type base isolation system, i.e., Slide-limited friction base isolation system (S-LF). Based on this system, harmonic and subharmonic periodic response of S-LF subjected to harmonic motions is investigated by using Fourier-Galerkin-Newton method with Floquet theory. Dynamic response of S-LF subjected to earthquake ground motions is calculated with a high order precise direct integration method, and the numerical results are presented in maximum acceleration response spectra of superstructure and maximum sliding displacement response spectrum form. By comparison of isolating effects of S-LF, Pure-friction base isolation system (P-F) and Resilient-friction base isolation system (R-FBI), it shows that the isolating property of S-LF is superior to P-F and R-FBI. Finally through analyzing an engineering example, it is observed that the distribution of the maximum shear between floors and absolute acceleration of S-LF to earthquake ground motion are very different from that of traditional structures.

1 INTRODUCTION

Base isolation is regarded as a practical and economical way to protect structures from damages subject to earthquake motion. Isolators implemented between the base raft of the structure and foundation can dissipate and absorb earthquake energy, and reduce the transmission of ground motion to structure obviously. In a short, the principle of base isolation includes two factors. i) The natural period of structures with fixed base is usually 0.2 ~ 1.2 s, close to the natural period of foundation which is 0.2 ~ 0.8 s. By using base isolator, the natural period of structures increases and is apart from the resonance area of ground excitations. ii) Normal structures absorb the earthquake energy at the expense of large plastic deformation of structural and non-structural elements, but base isolation structures dissipate earthquake energy through continuous motions of isolator with high damping to do work. In the past, numerous schemes for base isolators have been suggested. Among these schemes, frictional base isolators are the most popular. This kind of base isolation system includes Pure-friction base isolation system (P-F), Resilient-friction base isolation system (R-FBI), Electricite de France base isolation system (FDF), Friction pendulum system (FPS), and so on. In recently years, the rolling frictional base isolators have been also developed such as Free circular rolling rods for base isolation, Elliptical rods for base isolation, and Ball system with restoring property. Among these frictional base isolation systems, P-F

and R-FBI have been developed widely both in theory and practice.

P-F is the simplest base isolation system, the relative slippage between base raft and supporting interface can dissipate the earthquake energy, and resonance almost does not occur. This system is effective for a wide frequency range of input ground motion. Now, there has been a large body of theoretical and test work on its performance. However, such a system suffers from re-centering capability, resulting in large peak and residual base displacement. To overcome this, Mostaghel and Khodaverdian proposed the Resilient-friction base isolator(R-FBI). R-FBI consists of concentric layers of teflon coated plates which can slide on each other and a central core of rubber which provide elastic resilient force for re-centering. But the isolating effect of R-FBI is poorer than that of P-F due to the slide of steel plates is restrained by the stiffness of the rubber cores, and the large distortion of isolators may make isolators destroy to strong earthquake ground motion.

This paper integrates the merit of P-F and R-FBI, and presents a new type base isolation system, i.e., Slide-limited friction base isolation system (S-LF), as shown in figure 1. The slide-limited device includes symmetrical rigid stops and limitrs that consist of springs and dampers. The spring stiffness of limitrs is less than that of the superstructure. When the limitr collides with the stop, the structure becomes R-FBI and its natural period is 2 ~ 4s, as same as that of the normal base isolation system. The initial distance d between limitr and stop is

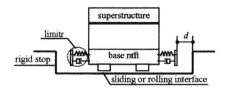

Figure 1. Slide-limited friction base isolation system

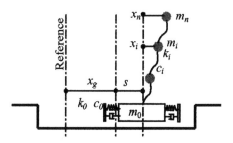

Figure 2. Computational model of S-LF

equal to the maximum sliding displacement of base raft to medium earthquake. Therefore, when the ground motion is weak or medium, this base isolation system acts as P-F. The slide-limited device will take effect to strong earthquake ground motion or repeated ground motion. Based on this kind of base isolation model, the Fourier-Galerkin-Newton (FGN) method with Floquet theory is used to study the dynamical characteristic of S-LF to harmonic excitation. Dynamic response of S-LF subjected to earthquake ground motions is calculated with a high order precise direct integration method, and the numerical results are presented in maximum acceleration response spectra of superstructure and maximum sliding displacement response spectrum form. The result of the study shows that the isolating property of S-LF is superior to P-F and R-FBI. Finally through analyzing an engineering example, it is observed that the distribution of the maximum shear between floors and absolute acceleration of S-LF to earthquake ground motions are very different from that of traditional structures.

2 GOVERNING MOTION EQUATIONS

Consider the structural model of an idealized N-story linear shear type structure supported on slide-limited friction base isolation system, as shown in figure 2. Some assumptions made in this study include: (1) The frictional resistance between the base raft and supporting interface is assumed to be continuous hysteretic type. (2) The frictional coefficient remains constant throughout the motion of base raft, the dynamic and static friction are taken to be the

same. (3) Only horizontal ground motion is considered. The governing equation of motion of base-isolated N-story structure subjected to ground excitation is written by

$$[M] \cdot \{\ddot{x}\} + [C] \cdot \{\dot{x}\} + [K] \cdot \{x\} = -(\ddot{s} + \ddot{x}_g) \cdot [M]\{I\} \quad (1)$$

$$\ddot{S} = -\ddot{x}_g - \mu_s \cdot g \cdot Z - \sum_{i=1}^{n} \gamma_i \ddot{x}_i -$$

$$[\omega_0^2 \cdot (S - sign(S) \cdot d) + 2\xi_0 \omega_0 \dot{S}] \cdot H(|S| - d) \quad (2)$$

$$A\dot{Z} = \dot{S} - 0.9Z \cdot |\dot{S}| - 0.1\dot{S} \cdot |Z| = \dot{S} - F1(\dot{S}, Z) \quad (3)$$

where $[M]$, $[K]$ and $[C]$ represent mass, stiffness and damping matrices of the superstructure, respectively, of the order $n \times n$, k_0 and c_0 represent stiffness and damping of the limitr, μ_s is coefficient of sliding (or rolling) friction, $\gamma_i = m_i \Big/ \sum_{j=0}^{n} m_j$, $\omega_0^2 = k_0 \Big/ \sum_{j=0}^{n} m_j$, $2\xi_0 \omega_0 = c_0 \Big/ \sum_{j=0}^{n} m_j$, A is a constant determined by experiment, Z is a factor describing the condition for the separation and reattachment of the sliding base raft. Sign($\cdot$) and H($\cdot$) are the signum function and the Heaviside unit step function respectively.

Since the response of superstructure is linear, it can be represented by

$$\{x\} = [\phi]\{y\} \quad (4)$$

where $[\phi]$ represents the fixed base modal shapes matrix of superstructure, of the order $n \times m$, $\{y\} = \{y_1, y_2, \ldots, y_m\}^T$ is the modal amplitudes vector, m is the number of modes of vibration considered ($m < n$).

Substitution of the above equation into equations (1) and (2) yields:

$$\ddot{y}_i + 2\xi_i \omega_i \dot{y}_i + \omega_i^2 y_i = -\lambda_i (\ddot{x}_g + \ddot{S}) \quad (i=1,2,\ldots,m) \quad (5)$$

$$\ddot{S} = -\ddot{x}_g - \mu_s \cdot g \cdot Z - \sum_{i=1}^{m} \beta_i \ddot{y}_i -$$

$$[\omega_0^2 \cdot (S - sign(S) \cdot d) + 2\xi_0 \omega_0 \dot{S}] \cdot H(|S| - d)$$

$$= -\ddot{x}_g - \mu_s \cdot g \cdot Z - \sum_{i=1}^{m} \beta_i \ddot{y}_i - G1(S, \dot{S}) \quad (6)$$

where, ω_i and ξ_i are the frequency and damping for the ith mode, and

$$\beta_i = \sum_{j=1}^{n} \gamma_j \cdot \phi_{ji}, \lambda_i = \sum_{j=1}^{n} (m_j \cdot \phi_{ji}) \Big/ \sum_{j=1}^{n} (m_j \cdot \phi_{ji}^2).$$

Equations (3), (5) and (6) are the governing motion equation of S-LF.

3 PERIODIC RESPONSE AND STABILITY ANALYSIS USING FGN METHOD WITH FLOQUET THEORY

The motion equation (3)(5)(6) is a strongly nonlinear differential equation. This paper uses FGN method to obtain the periodic response of S-LF to a harmonic base motion $\ddot{x}_g = a\cos\omega t$, let $\omega t = k\tau$ ($k = 1, 2, \ldots$, resprent kth order subharmonic),

$$(\cdot)' = \frac{d(\cdot)}{d\tau}, \quad (\cdot)'' = \frac{d^2(\cdot)}{d\tau^2},$$

equation (3)(5)(6) can be rewritten as

$$\omega^2[M_p]\{V''\} + \omega k[C_p]\{V'\} + k^2[K_p]\{V\} = -(\{P\} + \{E\}) \quad (7)$$

here

$$[M_p] = \begin{bmatrix} 0 & \{0\}^T \\ \{0\} & [M_q] \end{bmatrix}_{(m+2)\times(m+2)}; \quad [C_p] = \begin{bmatrix} A & \{0,\cdots,0,-1\} \\ \{0\} & [C_q] \end{bmatrix}_{(m+2)\times(m+2)};$$

$$[K_p] = \begin{bmatrix} 0 & 0 \\ -\mu_s g\begin{Bmatrix}\{\lambda\}\\-1\end{Bmatrix} & [K_q] \end{bmatrix}_{(m+2)\times(m+2)}; \quad \{P\} = \left\{\{0\}^T_{(m+1)\times1}, a\cdot k^2 \cdot \cos(k\tau)\right\}^T;$$

$$\{\lambda\} = \{\lambda_1, \lambda_2, \cdots, \lambda_m\}^T; \quad \{\beta\} = \{\beta_1, \beta_2, \cdots, \beta_m\}^T;$$

$\{V\} = \{Z, y_1, y_2, \ldots, y_m, S\}^T$; $[M_q]$, $[C_q]$, $[K_q]$, $\{E\}$ are given in the appendix.

Upon assuming a solution of the form:

$$\{V\} = \{V_0\} + \sum_{j=1}^{N}[\{V_{2j-1}\} \cdot \cos(j\tau) + \{V_{2j}\} \cdot \sin(j\tau)] \quad (8)$$

The spectral components of the non-linear function $G(S, S')$, $F(S', Z)$ of $\{E\}$ are given by the Fast Fourier transform (FFT):

$$FFT(G(S, S')) = a_0 + \sum_{j=1}^{N}[a_{2j-1}\cos(j\tau) + a_{2j}\sin(j\tau)] \quad (9)$$

$$FFT(F(S, Z)) = b_0 + \sum_{j=1}^{N}[b_{2j-1}\cos(j\tau) + b_{2j}\sin(j\tau)] \quad (10)$$

Substituting equation (8)(9)(10) into equation (7), by applying harmonic balance, one obtains (2N+1) (m+2) nonlinear algebraic equations about $\{U\} = \{\{V_0\}^T, \{V_1\}^T, \ldots, \{V_{2N}\}^T\}^T$

$$\{g(\{U\})\} = \{\{g_0\}^T, \{g_1\}^T, \cdots \{g_{2N}\}^T\}^T = \{0\} \quad (11)$$

This equation can be solved by the Newton-Raphson procedure

$$([JL] + [JN]) \cdot \{\Delta U\} + \{g(\{U^k\})\} = 0 \quad (12)$$

$$\{U^{k+1}\} = \{U^k\} + \{\Delta U\} \quad (13)$$

where, $\{g\}$, $[JL]$, $[JN]$ are given in the Appendix, and the convergence is checked through the error estimates ε, given by

$$\sqrt{\|\{g_0\}\| + \sum_{j=1}^{N}(\|\{g_{2j-1}\}\| + \|\{g_{2j}\}\|)} \le \varepsilon,$$

the iterations are repeated until a solution with an acceptable error estimate.

Stability of the periodic solution is investigated by using the Floquet theory facilitating bifurcation analysis. When equation (7) is perturbed and transformed into a state variable form it can be written as

$$\{\Delta\dot{W}\} = [B] \cdot \{\Delta W\} \quad (14)$$

where,

$$[B] = \begin{bmatrix} -\frac{\partial F}{\partial Z} \cdot \frac{1}{A} & \{0\}^T & \{0,\cdots,0,1-\frac{\partial F}{\partial Z}\} \cdot \frac{1}{A} \\ \{0\} & [0] & [I] \\ \mu_s \cdot g[M_q]^{-1}\begin{Bmatrix}\{\lambda\}\\-1\end{Bmatrix} & -\frac{1}{\omega^2}[M_q]^{-1}(k^2 \cdot [K_q] + [\frac{\partial\{E1\}}{\partial\{V1\}}]) & -\frac{1}{\omega^2}[M_q]^{-1}(\omega \cdot k \cdot [C_q] + [\frac{\partial\{E1\}}{\partial\{\dot{V}1\}}]) \end{bmatrix};$$

$\{\Delta W\} = \{\Delta Z, \{\Delta V1\}^T, \{\Delta\dot{V}1\}^T\}^T$;

$\{\Delta V1\} = \{\Delta y_1, \cdots, \Delta y_m, \Delta S\}^T$; $\left[\frac{\partial\{E1\}}{\partial\{V1\}}\right], \left[\frac{\partial\{E1\}}{\partial\{\dot{V}1\}}\right]$ are given in the Appendix.

Let $[\Phi(t)]$ be the matrix solution satisfying

$$[\dot{\Phi}(t)] = [B] \cdot [\Phi(t)] \quad (15)$$

$$[\Phi(0)] = [I] \quad (16)$$

where, $[\Phi]$ is the monodromy matrix and $[I]$ is a unit matrix. For stable solutions the eigenvalues of $[\Phi(2\pi)]$ matrix lie within the unit circle, and one of the eigenvalues moves across the boundary of the unit circle when a bifurcation occurs.

4 DYNAMIC ANALYSIS USING A PRECISE DIRECT INTEGRATION METHOD

The governing motion equation (3)(5)(6) can be rewritten in a state space form

$$\{\dot{W}\} = [B] \cdot \{W\} + \{Q\} \quad (17)$$

where

$$[B] = \begin{bmatrix} 0 & \{0\}^T & \{0,\cdots,0,1\}\cdot\dfrac{1}{A} \\ 0 & [0] & [I] \\ \mu_s\cdot g[M_q]^{-1}\begin{Bmatrix}\{\lambda\}\\-1\end{Bmatrix} & -[M_q]^{-1}[K_q] & -[M_q]^{-1}[C_q]\end{bmatrix},$$

$$\{Q\} = \begin{Bmatrix} F1(\dot{S},Z) \\ \{0\} \\ -[M_q]^{-1}\cdot\left(\begin{Bmatrix}\{0\}\\\ddot{x}_g(t)\end{Bmatrix}+\begin{Bmatrix}-G1(S,\dot{S})\cdot\{\lambda\}\\G1(S,\dot{S})\end{Bmatrix}\right)\end{Bmatrix}$$

$\{W\} = \{Z, \{V1\}^T, \{\dot{V}1\}^T\}^T$, $\{V1\} = \{y_1,\cdots,y_m,S\}^T$.

$\{Q\}$ includes the discontinuous function Sign($\cdot$) and H($\cdot$). In order to reduce the accumulated errors of numerical integration method, Sign($\cdot$) and H($\cdot$) can be approached by continuous function:

$$sign(\cdot) \approx \tanh(\alpha\cdot), \quad H(\cdot) \approx \frac{\tanh(\alpha\cdot)+1}{2} \quad (\alpha > 100) \quad (18)$$

Substitution of the above equation into $\{Q\}$, it becomes a continuous function about $Z, S, \dot{S}, t$. In each digitized internal $t\in[t_k, t_k+\Delta t]$ (Δt is time step), $\{Q\}$ can be expanded in a Taylor series about t, and take the first two terms:

$$\{Q(Z,S,\dot{S},t)\} \approx \{Q(Z_k,S_k,\dot{S}_k,t_k)\}+$$

$$\{Q'(Z_k,S_k,\dot{S}_k,t_k)\}(t-t_k) \quad (19)$$

where, ' $'$ ' is the global derivative of function about t. $\{Q'\}$ can be expressed in the form of difference:

$$\{Q'(Z_k,S_k,\dot{S}_k,t_k)\} = \{\{Q(Z_{k+1},S_{k+1},\dot{S}_{k+1},t_{k+1})\}-$$

$$\{Q(Z_k,S_k,\dot{S}_k,t_k)\}\}/\Delta t \quad (20)$$

Substituting equation (19) (20) into (17), non-linear differential equations are transformed to linear differential equation. The solution is shown in the following form:

$$\{W(t_{k+1})\} = e^{[B]\cdot\Delta t}\cdot[\{W(t_k)\}+[B]^{-1}(\{Q(Z_k,S_k,\dot{S}_k,t_k)\}+$$

$$[B]^{-1}\{Q'(Z_k,S_k,\dot{S}_k,t_k)\})]-[B]^{-1}(\{Q(Z_k,S_k,\dot{S}_k,t_k)\}+$$

$$[B]^{-1}\{Q'(Z_k,S_k,\dot{S}_k,t_k)\}+Q'(Z_k,S_k,\dot{S}_k,t_k)\Delta t) \quad (21)$$

Equation (21) is the non-linear algebraic equations about $\{W(t_{k+1})\}$, using the iteration method to solve. In order to increase the converging velocity, the following equation will be taken to estimate the initial value of $\{W(t_{k+1})\}$

$$\{W(t_{k+1})\} = e^{[B]\cdot\Delta t}\{W(t_k)\}+(e^{[B]\cdot\Delta t}-[I])\cdot[B]^{-1}\cdot$$

$$\{Q(Z_k,\dot{Z}_k,\dot{S}_k,t_k)\} \quad (22)$$

$e^{[B]\cdot\Delta t}$ in equation (21)(22) can be calculated with 2^W arithmetic, i.e., precise direct integration method that was presented in reference[5].

5 NUMERICAL RESULTS AND ANALYSIS

Take a two degree of freedom(TDOF) model for example, the effects of some parameters of the model and base isolation device on the dynamic characteristic of S-LF are investigated.Due to the MDOF model can be changed into TDOF model through mode transform, the study of TDOF model is representative.The slide-limited device of S-LF just takes effect at strong earthquake, so the peak ground acceleration will be taken to equal 1g in the following.

5.1 Harmonic responses

Parameters of equations (3)(5)(6) are listed as following, $A = 0.0001$m, $\omega_1 = 5\pi$ rad/s, $\xi_0 = \xi_1 = 0.05$, $\gamma_1 = 0.2, 0.5, 0.8$ (the ratio of the mass of the superstructure to the total mass), $\mu_s = 0.1$, $\ddot{x}_g = 1$g $\cos(6t)$. Harmonic responses of structure and stability of the periodic solution are obtained by using the FGN method and Floquet theory. In Figure 3 and 4,the maximum acceleration a_{max} response spectra of superstructure and the maximum sliding displacement S_{max} response spectra of the base raft of R-FBI($d = 0$), S-LF($d = 0.20$m) and P-F($d = \infty$) are plotted respectively. The horizontal coordinate T_0 is

given by $T_0 = 2\pi\sqrt{\sum_{i=0}^{n} m_i / k_0}$, Σm_i is the total mass,

k_0 is the stiffness of limitr of S-LF or the stiffness of isolator of R-FBI in previous figures. As seen from this figure, the level of a_{max} of R-FBI and S-LF depends on the value of γ_1 and T_0, larger γ_1 and T_0 yield lower a_{max}, but the level of a_{max} of P-F is independant of T_0. The level of a_{max} of S-LF is lower than that of R-FBI,and slightly larger than P-F. It indicates that the isolating effect of S-LF is better than that of R-FBI, and almost as good as P-F. But the maximum sliding displacement of P-F exceeds 5m (S_{max} of P-F is too large, so it is not shown in Figure 4), this can not content with the engineering need, moreover it has no capability to limit the sliding displacement of base raft.The value of S_{max} of S-LF is the lowest,the distortion of limitr S_{max}-d is only $0.07 \sim 0.10$ m,and it can limit the base raft sliding within a definite scope. The distortion of isolator of R-FBI is $0.3 \sim 0.5$ m,the large distortion may make the isolator destory.As known from above analysis, the isolating property of S-LF is superior to P-F and R-FBI subjected to harmonic motion.

5.2 *Subharmonic responses*

Parameters of equation (3) (5) (6) are as seem as the above except $\gamma_1 = 0.5$ and d = 0.20m.The maximum acceleration a_{max} response spectra of superstructure is shown in figure 5. For $T_0 \in (1.52,1.61)$ and $(2.08, 2.48)$,S-LF may exhibits harmonic response or 1/2 and 1/3 subharmonic response from different initial conditions, as shown in figure 6.

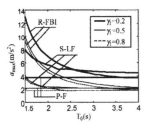

Figure 3. Comparison of spectra of maximum absolute acceleration of superstructure of S-LF, R-FBI and P-F to harmonic excitation

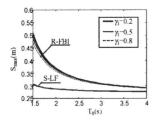

Figure 4. Comparison of spectra of maximum sliding displacement of base raft of S-LF and R-FBI to harmonic excitation

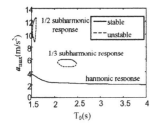

Figure 5. Spectra of maximum absolute acceleration of harmonic and subharmonic response of superstructure of S-LF

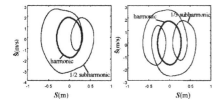

Figure 6. The phase of harmonic and subharmonic response of the fase raft

5.3 *Earthquake responses*

EL-Centro earthquake record (1940,NS) normalized to a peak ground acceleration of 1g is chosen as the input motion in this paper.Using the precise direct integration method, the effects of some structural parameters on the earthquake responses of S-LF is investigated.

For TDOF model, parameters of equation (3) (5) (6) are, $A = 0.0001\text{m},\omega_1 = 3.33\pi\text{rad/s}, \xi_0 = \xi_1 = 0.05$, $\mu_s = 0.08$, $d = 0.18$m. The effects of γ_1 and T_0 on the maximum acceleration a_{max} of superstructure and the maximum displacement S_{max} of base raft are obtained, as shown in figure 7 (a)(b). The value of γ_1 affects the level of a_{max} and S_{max} significantly, the level of a_{max} is smaller for larger γ_1. It indicates that the larger the mass of the superstructure as compared to the mass of the base raft, the better the isolating effect of S-LF, but the larger the value of S_{max} . The effects of T_0 on the level of a_{max} and S_{max} are less significant than that of γ_1. For the structure with small γ_1, variations of T_0 hardly affect the level of a_{max} and S_{max}. For the large γ_1, the level of a_{max} decreases with the increase of T_0, but the level of S_{max} is reverse. So it can be concluded that, for $\gamma_1 > 0.5$, the limitr of S-LF with small T_0 should be chosen in case of the value of S_{max} is too large, for $\gamma_1 < 0.5$, the value of T_0 chosen in the range of [1.5, 4] is all available.

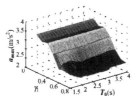

(a) Spectra of maximum absolute acceleration of superstructure

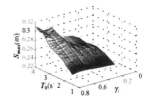

(b) Spectra of maximum sliding displacement of base raft

Figure 7. The effect of γ_1 on the aseismic behavior of base isolation system for EL-Centro 1940 earthquake with adjusted peak acceleration to 1g

For $A = 10^{-4}$m,$\omega_1 = 3.33\pi$rad/s, $\xi_0 = \xi_1 = 0.05$, $\mu_s = 0.08$, $\gamma_1 = 0.8$., The effects of d on a_{max} and S_{max} are shown in figure 8(a)(b). For $d = 0$, $0 < d < 0.3$m, $d \geq 0.3$, the structure is R-FBI, S-LF, P-F respectively. With the increase of d , the level of a_{max} reduces but the S_{max} increases. For $d \geq 0.2$m, the curve of $d \sim$

a_{max} and $T_0 \sim a_{max}$ become smooth, which indicates that the effect of d on isolation is not remarkable when $d \geq 0.2$m, and T_0 chosen in the range of [1.5, 4] is available. As known from the figure 8, the isolating effect of S-LF is better than that of R-FBI, for $d \geq 0.2$m, is as good as P-F. But P-F has no capability to limit the sliding displacement of base raft, resulting in large peak and residual base displacement, the base raft of structure may collide with the nearby object subjected to aftershock or next earthquake ground motions. Although R-FBI has the function of replacement, the maximum distortion of its isolator exceeds 0.3m. The maximum distortion of limitr of S-LF (S_{max}-d) is less than 0.1m and S-LF has capability to limit the sliding displacement of base raft. So it can be conclude that the isolating property of S-LF is also superior to P-F and R-FBI to earthquake ground motions

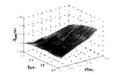

(b) Spectra of maximum sliding displacement of base raft

Figure 9. The effect of C on the aseismic behavior of base isolation system for EL-Centro 1940 earthquake with adjusted peak acceleration to 1g

For A=0.0001m, ξ_0=ξ_1=0.05, μ_s=0.08, γ_1=0.8, d=0.18m, the effects of the natural period T_1 (T_1=$2\pi/\omega_1$) of superstructure on a_{max} and S_{max} of S-LF are shown in figure 10(a)(b). The effects of T_1 on a_{max} and S_{max} are not remarkable, but with the increase of T_1, the isolating effect of S-LF reduces. So this base isolation system is fit for structures whose T_1 is small.

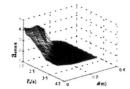

(a) Spectra of maximum absolute acceleration of superstructure

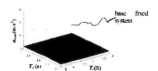

(a) Spectra of maximum absolute acceleration of superstructure

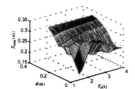

(b) Spectra of maximum sliding displacement of base raft

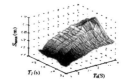

(b) Spectra of maximum sliding displacement of base raft

Figure 8. The effect of d on the aseismic behavior of base isolation system for EL-Centro 1940 earthquake with adjusted peak acceleration to 1g

Figure 10. The effect of $\mathbf{T_1}$ on the aseismic behavior of base isolation system for EL-Centro 1940 earthquake with adjusted peak acceleration to 1g

For $A = 10^{-4}$m,$\omega_1 = 3.33\pi$rad/s, $\xi_0 = \xi_1 = 0.05$, $\mu_s = 0.08$, $\gamma_1 = 0.8$ $d = 0.25$m. The base raft of S-LF will produce residual base displacement $C \in [0, 0.25]$ under earthquake ground motions. The effects of C on a_{max} and S_{max} are shown in figure 9 (a) (b). The levels of a_{max} and S_{max} don't significantly vary with C. For example, for $T_0 = 3$s and $C = 0$, a_{max} = 1.9m/s^2, S_{max} = 0.32m, for $T_0 = 3$s and $C = 0.25$m, a_{max} = 2.1m/s^2, S_{max} = 0.37m.

6 ENGINEERING EXAMPLE

In figure 2, take a four-degree model for example, $m_0 = m_1 = m_2 = m_3 = 3\times10^5$kg; $k_0 = 5.3\times10^6$N/m; $k_1 = k_2 = k_3 = 1\times10^9$N/m; the damping matrix is Rayler form; $\xi_1 = \xi_2 = 0.05$; $d = 0$ (R-FBI), 0.2m (S-LF), ∞(P-F) respectively; μ_s=0.08; EL-Centro earthquake record(1940, NS) normalized to a peak ground acceleration of 1g is chosen as the input motion. The maximum floor shear Q_{max} and the maximum absolute acceleration a_{max} on condition of base isolated and fixed base systems are shown in figure 11. The value of the first floor Q_{max} of R-FBI, P-F and S-LF is reduced to the order of 12%, 8% and 8% of that of fixed base system respectively. So the isolating effects of S-LF and P-F are superior to R-FBI to

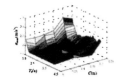

(a) Spectra of maximum absolute acceleration of superstructure strong earthquake ground motions. Comparing figure 11, it may be observed that the value of the second floor Q_{max} of P-F and S-LF is larger than that of top and first floor. This is different from the fixed base system and R-FBI system whose first floor Q_{max} is larger than top floor and second floor, distributed as ladder. The value of a_{max} of base isolated systems are all distributed as "K" form, i.e., the middle floor a_{max} is smaller than first and top floor, this is different from fixed base system whose a_{max} is distributed as triangle. Figure 12 shows the loop of base friction of S-LF under earthquake ground motion. The shape of the loop is rectangle, which has strong capability of absorbing energy, so it can be concluded that S-LF can isolate the earthquake wave to transmit into the superstructure.

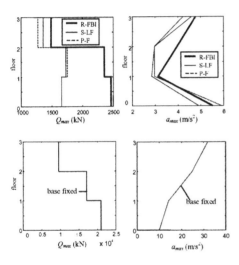

Figure 11. Comparison of the maximum floor shear and the maximum absolute acceleration of base isolation system and fixed base system for EL-Centro 1940 earthquake with adjusted peak acceleration to 1g

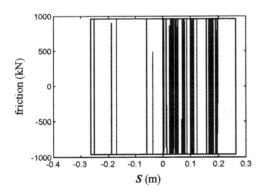

Figure 12. The loop of friction with respect to the base raft displacement for EL-Centro 1940 earthquake with adjusted peak acceleration to 1g

7 CONCLUSIONS

This paper presents a new type base isolation system, i.e., Slide-limited friction base isolation system based on synthesizing the merit and demerit of P-F and R-FBI. The motion equations of S-LF are derived. Because the equations are strong nonlinear differential equations, the FGN method is used to calculate harmonic and subharmonic responses to harmonic excitation. Stability of periodic solution is investigated by using the Floquet theory. Earthquake responses of S-LF are calculated with a high precise direct integration method. The dynamical characteristics of S-LF are the following.

1. Under harmonic support motion, the larger the ratio of the mass of the superstructure to the total mass and the natural period of the limitr, the better the isolating effects of S-LF, but the larger the maximum displacement of base raft. For special parameters, S-LF may exhibits harmonic response or 1/2 and 1/3 subharmonic responses under different initial conditions.

2. The maximum acceleration response spectra of superstructure and the maximum sliding displacement response spectra to earthquake ground motion show that parameters γ_l, d, C, T_1 and T_0 all affect the isolating effects of S-LF. For the structure with large γ_l, small T_0 of limitr should be chosen in case of the value of S_{max} is too large. The isolating effect is better for large value of d, but d is restricted by exterior condition, moreover, the large d is uneconomical. The analysis of this work shows that when parameter d exceeds a special value, the increase of d can not bring remarkable benefit on isolating effect, thereby d must be adopted properly. The isolating effect of S-LF reduces with the decreasing of stiffness of structures. So S-LF is not suitable for structures whose stiffness is small.

3. Engineering example shows that the maximum floor shear Q_{max} and the maximum absolute acceleration a_{max} of structures base-isolated by S-LF can be reduced significantly. The maximum middle floor shear of S-LF is larger than that of top floor and base floor, this is different from the fixed base system whose Q_{max} distributed as ladder. The value of a_{max} of S-LF is distributed as "K" form, i.e., the middle floor a_{max} is smaller than base and top floor, this is also different from fixed base system whose a_{max} is distributed as triangle. The isolating effect of S-LF is better than that of R-FBI, and almost as good as P-F. But P-F has no capability to limit the sliding displacement of base raft. R-FBI has the function of replacement, but the maximum distortion of its isolator is too large under strong earthquake. It can be conclude that the isolating property of S-LF is superior to P-F and R-FBI. S-LF is not only suitble for the sliding friction base isolating systems, but also is suitble for rolling friction base isolating systems.

8 APPENDIX

$[M_q]$, $[C_q]$, $[K_q]$, $\{E\}$ are ginven by

$$[M_q] = \begin{bmatrix} [I] - \{\lambda\}\{\beta\}^T & \{0\} \\ \{\beta\}^T & 1 \end{bmatrix}_{(m+1)\times(m+1)} \quad ,$$

$$[C_q] = \begin{bmatrix} diag(2\xi_1\omega_1, 2\xi_2\omega_2, \cdots, 2\xi_m\omega_m) & \{0\} \\ \{0\}^T & 0 \end{bmatrix}_{(m+1)\times(m+1)} \quad ,$$

$$[K_q] = \begin{bmatrix} diag(\omega_1^2, \omega_2^2, \cdots, \omega_m^2) & \{0\} \\ \{0\}^T & 0 \end{bmatrix}_{(m+1)\times(m+1)} \quad ,$$

$$\{E\} = \begin{Bmatrix} F(S',Z) \\ -G(S,S')\cdot\{\lambda\} \\ G(S,S') \end{Bmatrix} = \begin{Bmatrix} F(S',Z) \\ \{E1\} \end{Bmatrix} \cdot$$

$$F(S',Z) = \omega \cdot k(0.9Z \cdot |S'| + 0.1S' \cdot |Z|),$$

$$G(S,S') = k^2[\omega_0^2(S - sign(S)\cdot d) + 2\xi_0\omega_0\omega S / k]H(|S| - d)$$

$\{g\}$, $[JL]$, $[JN]$ are ginven by

$$\{g_0\} = k^2[K_p]\{V_0\} + \begin{Bmatrix} b_0 \\ -a_0\cdot\{\lambda\} \\ a_0 \end{Bmatrix} = 0$$

$$\{g_{2j-1}\} = (-j^2\cdot\omega^2\cdot[M_p] + k^2\cdot[K_p])\cdot\{V_{2j-1}\} +$$

$$j\omega k[C_p]\{V_{2j}\} + \begin{Bmatrix} b_{2j-1} \\ -a_{2j-1}\cdot\{\lambda\} \\ a_{2j-1} + a\cdot\delta_j^k \end{Bmatrix} = 0$$

$$(j = 1,2,\cdots,N)$$

$$\{g_{2j}\} = -j\cdot\omega\cdot[C_p]\{V_{2j-1}\} + (k^2\cdot[K_p] -$$

$$j\omega^2[M_p])\{V_{2j}\} + \begin{Bmatrix} b_{2j} \\ -a_{2j}\cdot\{\lambda\} \\ a_{2j} \end{Bmatrix} = 0 \quad (j=1,2,\dots,N)$$

$$[JL] = \begin{bmatrix} k^2\cdot[K_p] & & & & \\ & [JL_1] & & & \mathbf{0} \\ & & \ddots & & \\ & & & [JL_j] & \\ & \mathbf{0} & & & \ddots \\ & & & & [JL_N] \end{bmatrix},$$

$$[JL_j] = \begin{bmatrix} -j^2\omega^2\cdot[M_p] + k^2\cdot[K_p] & j\omega\cdot k\cdot[C_p] \\ -j\omega\cdot k\cdot[C_p] & -j^2\omega^2\cdot[M_p] + k^2\cdot[K_p] \end{bmatrix}$$

$$(j=1,2,\cdots,N)$$

$$[JN] = \begin{bmatrix} [J_{0,0}] & [J_{0,1}] & \cdots & [J_{0,2N}] \\ [J_{1,0}] & [J_{1,1}] & \cdots & [J_{1,2N}] \\ \vdots & \vdots & \ddots & \vdots \\ [J_{2N,0}] & [J_{sN,1}] & \cdots & [J_{2N,2N}] \end{bmatrix}$$

$$[J_{0,0}] + \sum_{J=1}^{n}([J_{2j-1,0}]\cos j\tau + [J_{2j,0}]\sin j\tau) = FFT\left[\frac{\partial\{E\}}{\partial\{V\}}\right]$$

$$[J_{0,2k-1}] + \sum_{j=1}^{N}([J_{2j-1,2k-1}]\cos j\tau + [J_{2j,2k-1}]\sin j\tau) =$$

$$FFT\left[\frac{\partial\{E\}}{\partial\{V\}}\cos k\theta_i + \frac{\partial\{E\}}{\partial\{\dot{V}\}}(-k\sin k\theta_i)\right]$$

$$[J_{0,2k}] + \sum_{j=1}^{N}([J_{2j-1,2k}]\cos j\tau + [J_{2j,2k}]\sin j\tau) =$$

$$FFT\left[\frac{\partial\{E\}}{\partial\{V\}}\sin k\theta_i + \frac{\partial\{E\}}{\partial\{\dot{V}\}}(+k\cos k\theta_i)\right]$$

$k=1,2,\dots N$, $i=1,2,\dots,M$, $\theta_i = 2\pi(i-1)/M$, $M \ge 4N$, N is harmonic number, M is collocation nodes in the FFT, in this paper, $N=11$, $M=256$.

$$\frac{\partial\{E\}}{\partial\{V\}} = \begin{bmatrix} \dfrac{\partial F}{\partial Z} & \{0\}^T \\ \{0\} & \dfrac{\partial\{E1\}}{\partial\{V1\}} \end{bmatrix}$$

$$\frac{\partial\{E\}}{\partial\{\dot{V}\}} = \begin{bmatrix} 0 & \{0,\cdots,0,-\dfrac{\partial F}{\partial\dot{Z}}\} \\ \{0\} & \dfrac{\partial\{E1\}}{\partial\{\dot{V}1\}} \end{bmatrix}$$

$\left[\dfrac{\partial\{E1\}}{\partial\{V1\}}\right]$, $\left[\dfrac{\partial\{E1\}}{\partial\{\dot{V}1\}}\right]$ are ginven by

$$\frac{\partial\{E1\}}{\partial\{V1\}} = \begin{bmatrix} [0] & -\dfrac{\partial G}{\partial S}\cdot\{\lambda\} \\ \{0\}^T & \dfrac{\partial G}{\partial S} \end{bmatrix}$$

$$\frac{\partial\{E1\}}{\partial\{\dot{V}1\}} = \begin{bmatrix} [0] & -\dfrac{\partial G}{\partial\dot{S}}\cdot\{\lambda\} \\ \{0\}^T & \dfrac{\partial G}{\partial\dot{S}} \end{bmatrix}$$

$$\frac{\partial G}{\partial S} = \begin{cases} \omega_0^2 k^2 & |S| \geq d \\ 0 & |S| < d \end{cases}$$

$$\frac{\partial G}{\partial \dot{S}} = \begin{cases} 2\xi_0 \omega_0 \omega k & |S| \geq d \\ 0 & |S| < d \end{cases}$$

$$\frac{\partial F}{\partial Z} = (0.9|\dot{S}| + 0.1\dot{S} \cdot sign(Z)) \cdot \omega k$$

$$\frac{\partial F}{\partial \dot{S}} = 0.9Z \cdot sign(\dot{S}) + 0.1|Z|$$

REFERENCES

B.Westermo & F.Udwadia, 1983. Periodic response of a sliding oscillator system to harmonic excitation. *Earthquake Engng Struct. Dyn.* 11(1): 135-146.

C. S. Hsu, 1974. On approximating a general linear periodic system. *J. Mathematical analysis and applications* 45(2): 234-251.

J. M. Kelly & K. E. Beucke, 1983. A friction damped base isolation system with fail-safe characteristics. *Earthquake Engng Struct. Dyn.* 11(1): 33-56.

L.Su et al, 1989. A comparative study of performances of various base isolation system part I: shear beam structures. *Earthquake Engng Struct. Dyn.* 18(1): 11-32.

Lin J H et al, 1995. A high precision direct integration scheme for structures subject to transient dynamic loading. Computers and structures.56(1): 113-120.

Lin, T. W. & Hone, C. C., 1993. Base isolation by free rolling rods under basement. *Earthquake Engng Struct. Dyn.* 22(2): 261-273.

Lin, T. W.et al, C. C., 1995. Experimental study of base isolation by free rolling. *Earthquake Engng Struct. Dyn.*24(12): 1645-1650 .

Masashi Iura et al, 1992. Analytical expressions for three different modes in harmonic motion of sliding structures. *Earthquake Engng Struct. Dyn.*.21(9): 757-769.

N.Mostaghel & J.Tanbakuchi, 1983. Response of sliding structures to earthquake support motion. *Earthquake Engng Struct. Dyn.* 11(6): 729-748.

N.Mostaghel & M.Khodaverdian, 1987. Dynamics of Resilient-Friction Base Isolator (R-FBI). *Earthquake Engng Struct. Dyn.* 15(3): 379-390.

N.Mostaghel & M.Khodaverdian. 1988. Seismic Response of Structures Supported on R-FBI System. *Earthquake Engng Struct. Dyn.*16(7): 839-854.

N.Mostaghel et al, 1983. Response of sliding structures to harmonic support motion. *Earthquake Engng Struct. Dyn.*11(3): 355-366.

Qiang Zhou et al, 1998 . Dynamic analysis on structures base-isolated by a ball system with restoring property. *Earthquake Engng Struct. Dyn.*27(7): 773-791.

R. S. Jangid, & Y. B. Londhe, 1998. Effectiveness of elliptical rolling rods for base isolation. *J.struct.Eng.*124(3): 469-472.

S.Narayanan et al, 1998. A frequency domain based numeric-analytical method for non-linear dynamical systems. *J. Sound Vib.*211(3): 409-424 .

V. Zayas & S. Low, 1990. A simple pendulum technique for achieving seismic isolation. *Earthquake Spectra* 6(3): 317-333.

Xiong Zhongming et al, 1995. Dynamic analysis on sliding structures. *.J. Civil Eng.* 28(5): 21-30 (in chinese).

Yu MaoHong et al, 1996. Analysis of Base Isolation of Brick Masonry Building Structure. *J.struct.Eng.*17(4): 52-59 (in chinese).

Earthquake Engineering Frontiers in the New Millennium, Spencer & Hu (eds),
© 2001 Swets & Zeitlinger, ISBN 90 2651 852 8

The development of super high strength bolts

Kozo Wakiyama
Osaka Sangyo University, Osaka, Japan

Keiji Hirai
Nishi Nippon Inst. of Tech, Fukuoka, Japan

Nobuyoshi Uno
Nippon Steel Corp., Tokyo, Japan

ABSTRACT:There was much damage in the welding joints of the steel frame building, but there was a little damage of the high strength bolted joints as a result of the earthquake damage investigations. We think with the thing that the adoption of the high strength bolt joints contribute to the improvement of the safety of the steel frame building from these results. If a building becomes larger or taller, the number of the bolts for the joint increases and the bolted joint will be not suitable for building construction. High strength-ization of the bolt solves this problem. If a bolt is made more high strength, the bolt number of the joint decreases as shown in the following figure. But the risk of the delayed fracture is entailed for higher strength of the high strength bolts. The development of the super high strength bolt which delay fracture doesn't occur in was begun by my proposal in Nippon Steel Corp. in 1989. We succeeded in the development of the super high strength bolts about five years ago, and started to use them carefully this year. This paper introduces the summary of the result of the examination about delayed fracture of these bolts.

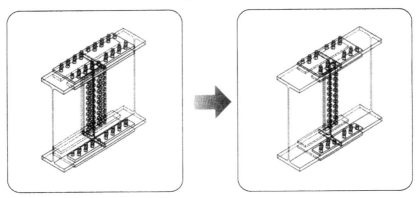

1 INTRODUCTION

This paper describes the results of exposure tests on newly developed super high strength bolts having a tensile strength of about 150 kgf/mm^2 (1470 N/mm^2). We named this bolt the super-high tension bolt (SHTB). The SHTB have an improved thread shape, and the elastic stress concentration factor at the bottom of thread is reduced to 60% of that of conventional high strength bolts. Two kinds of exposure tests were conducted: accelerated exposure test by Eto-type ferris wheel and actual exposure test at four climatically different locations in Japan. Both conventional F10T bolts and the SHTB were put to these exposure tests.

The steels used for the exposure tests are two kinds of steels for the F10T high strength bolts

specified by JIS B1186 (steel W and steel X) and two kinds of steels experimentally manufactured for the SHTB (steel Y and steel Z). The strength level of the specimen bolts was varied in several steps from 980 N/mm^2 to 1470 N/mm^2.

2 SPECIMEN BOLTS

2.1 *F10T high strength bolts specified by JIS B1186*

The F10T high strength bolts specified by JIS B1186 are considered to suffer no delayed fracture in the ordinary environment, and they are at present the most commonly used high strength bolts in Japan. In the tests, the F10T high strength bolts were used for making a comparative study with the SHTB and also for evaluating their own delayed fracture characteristics. The bolts used as specimens in the tests are commercially available Torque-Shear- type F10T high strength bolts (Steel X). Some JIS-type hexagon-head high strength bolts whose tensile strength was heightened by changing the tempering temperature from that for ordinary F10T high strength bolts were also prepared as specimen bolts (Steel W). The nominal size of all the bolts is M22.

2.2 *Super high strength bolts (SHTB)*

The steels for the SHTB are of two kinds, steel Y and steel Z, whose critical diffusible hydrogen contents by the previously reported method are evaluated to be above that of ordinary F10T high strength bolts. And these steels are evaluated to be hard to suffer delayed fracture. Although JIS-type hexagon-head high strength bolts were used at the outset of the tests, Torque-Shear- type bolts were finally adopted for the tests. Fig.1 compares the shape and dimension of the Torque-Shear- type M22 SHTB and those of ordinary F10T high strength bolts. The R under the neck of the SHTB is somewhat larger, that is 2.5R, and the incomplete thread of the bolts is provided with a part shifting to the shank.

Figure 2 shows the elastic stress concentration factor at the bottom of thread analyzed by the finite element method. While the factor at the bottom of thread of the JIS B1186 high strength bolts is 2.54, that of the SHTB is 1.52 or about 40% lower. The washer for the SHTB is the same as that for ordinarybolts, but the nut thickness is made 4.4 mm larger considering the bite of the bolt and the nut.

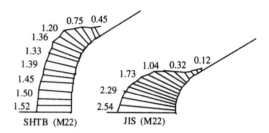

Figure 2. Stress concentration of thread

3 EXPOSURE TESTS

3.1 *Accelerated exposure test using Eto-type ferris wheels*

Putting bolts to actual exposure tests in actual exposure environments is the surest way to confirm the presence or absence of delayed fracture, but it takes many days and great cost to conduct such tests. Therefore, it is rational to make an accelerated exposure test first and then to make an actual exposure test only on high strength bolts which are found to be hard to suffer delayed fracture in the accelerated exposure test. The accelerated exposure test was conducted using Eto-type ferris wheels.

Figure 3 shows the Eto-type ferris wheel.

This device is designed to cause an exposure-accelerating effect by repeating the cycle of immersing the specimens in a 3.5% salt solution for 10 minutes and drying them for 50 minutes.

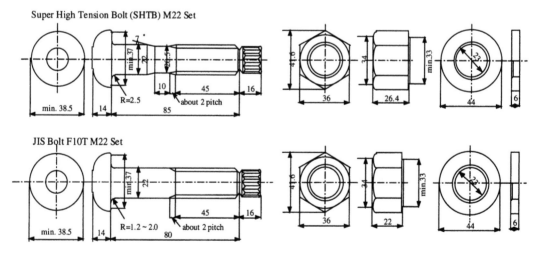

Figure 1. Dimensions of high tension bolt set and JIS bolt set

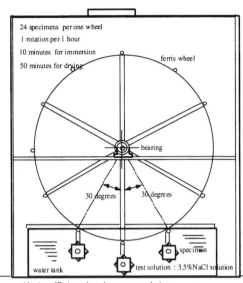
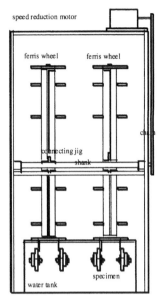

24 specimens per one wheel
1 rotation per 1 hour
10 minutes for immersion
50 minutes for drying

ferris wheel

bearing

30 degrees   30 degrees

specimen

water tank

test solution : 3.5%NaCl solution

side view of Eto's accelerated exposure test device
(Eto's ferris wheel)

speed reduction motor

ferris wheel   ferris wheel

chain

connecting jig

shank

specimen

water tank

front view of the device

Figure 3. Eto-type ferris wheel testing machine

Figure 4 shows the shape and dimension of the specimens. The specimen is provided with a water groove to allow entry of water into it.

3.2 *Actual exposure tests*

The facilities for actual exposure test were installed at four different locations in Japan with the aim of testing the specimens in different environments. They are Okinawa Prefecture (Gushikami, Shimajiri

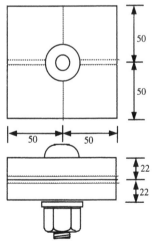

Figure 4. Demensions of specimens for ferris wheel

District), Fukuoka Prefecture (Minoshima, Yuku-hashi City), Tokyo (Yoyogi, Shibuya Ward) and Hokkaido (Muroran City). Gushikami in Okinawa, located in the south of Japan, is very hot and humid in the summer period. In addition, the exposure test site is on the coast exposed to the salt-bearing sea breezes. At times of typhoons, etc., the spray will drench the specimens. The site at Minoshima in Fukuoka is in a severe environment, and the specimens are immersed in the sea water at high tide twice every day. In the case of Yoyogi in Tokyo, the site is on the roof of a three-story building in the neighborhood of an urban expressway assuming the midtown environment. Muroran in Hokkaido is located in the north of Japan. It is very cold and the temperature is below zero in winter. The site is on the roof of a two-story building.

Figure 5 shows an example of actual exposure test.

4 RESULTS AND CONCLUSIONS

Table 1 shows a list of specimens and the results of the accelerated exposure test.

Table 2 shows a corresponding list for the actual exposure test. The results of the accelerated exposure test show that in period 1, the occurrence of delayed fracture becomes more frequent with the increasing nut rotation angle at the time of introducing

Table 1. Specimens for accelerated exposure tests and interim test results

Specimen Series	Material	Forging	Tensile Strength (N/mm2)	Blast Treatment of Bolt	Radius of Neck	Tread Type	Nut Turn or Clamping Force (kN)	Test Period (Day)	Number of Specimen	Number of Fracture	Ratio of Fracture (%)	Neck	Shank	Incompleat Thread	Thread	Un-known	Remarks
AY47C1	Y	Cold	1560	Not Exist	2.5R	SHTB	120 deg.	203	6	0	0.0	0	0	0	0	0	The First Period (HEX Type)
AY47C2							180 deg.	203	6	1	16.7	0	1	0	0	0	
AY47C3							374	131	8	8	100.0	6	1	0	1	0	
AY51C1			1530				120 deg.	203	6	4	66.7	2	1	0	0	1	
AY51C2							180 deg.	203	6	6	100.0	2	2	2	0	0	
AY51C3							366	131	8	7	87.5	7	0	0	0	0	
AY55C1			1500				120 deg.	203	6	6	100.0	2	1	2	1	0	
AY55C2							180 deg.	203	6	6	100.0	2	1	2	1	0	
AY55C3							359	131	8	6	75.0	5	0	0	1	0	
AY59C1			1450				120 deg.	203	6	3	50.0	0	1	1	0	1	
AY59C2							180 deg.	203	6	5	83.3	0	0	2	3	0	
AY59C3							347	131	8	4	50.0	0	0	1	3	0	
AY47CB1			1560	Exist			120 deg.	203	6	1	16.7	0	0	1	0	0	
AY47CB2							180 deg.	203	6	6	100.0	0	0	2	4	0	
AY47CB3							374	131	8	6	75.0	0	1	0	5	0	
AY51CB1			1530				120 deg.	203	6	6	100.0	3	0	2	1	0	
AY51CB2							180 deg.	203	6	6	100.0	1	0	2	3	0	
AY51CB3							366	131	8	8	100.0	2	0	0	6	0	
AW5C1	W		1300	Not Exist	JIS	JIS	120 deg.	274	12	0	0.0	0	0	0	0	0	
AY51C4	Y	Cold	1490	Not Exist	2.5R	SHTB	364	263	8	7	87.5	3	0	1	3	0	The Second Period (HEX Type)
AY56C			1460				339	263	8	8	100.0	4	1	2	1	0	
AY59C4			1400					263	8	8	100.0	3	0	4	1	0	
AY61C1			1360					263	8	5	62.5	1	2	1	1	0	
AY51CB4			1510	Exist			364	263	16	14	87.5	2	3	1	7	1	
AY56CB1			1460					263	16	14	87.5	2	1	6	5	0	
AY59CB1			1400				339	263	16	7	43.8	2	3	1	1	0	
AY61CB1			1360					263	8	4	50.0	1	0	0	2	1	
AY51HB1		Hot	1510				364	263	16	11	68.8	1	2	6	1	1	
AY56HB1			1460					263	16	13	81.3	6	1	5	1	0	
AY59HB1			1400				339	263	16	9	56.3	2	2	4	1	0	
AY61HB1			1360					263	8	7	87.5	1	0	4	2	0	
AY59CB2	Y	Cold	1430	Exist	2.5R	SHTB	335	194	43	18	41.9	5	3	6	4	0	The Third Period
AY61CB2			1330				342	194	12	3	25.0	3	0	0	0	0	
AZ58CB1	Z		1460				366	>950	13	2	15.4	1	1	0	0	0	(Torque Shear Type)
AZ60CB1			1440				369	>950	92	0	0.0	0	0	0	0	0	
AX42C1	X		1040	Not Exist			222	>950	18	0	0.0	0	0	0	0	0	
AW35C1	W		1300		JIS	JIS	120 Rot. of Nut	>950	6	0	0.0	0	0	0	0	0	(HEX Type)

342

Table 2. Specimens for actual exposure tests and interim results

Specimen Series	Place of Tests	Material	Forging	Tensile Strength (N/mm2)	Blast Treatment of Bolt	Radius of Neck	Tread Type	Nut Turn or Clamping Force (kN)	Test Period (Month)	Number of Specimen	Number of Fracture	Ratio of Fracture (%)	Position of Fracture — Neck	Shank	Incomplete Thread	Thread	Unknown	Remarks
EW35C	Okinawa (Gushikami)	W	Cold	1300	Not Exist	JIS	JIS	180deg.	13	60	0	0.0						(HEX Type)
EY55C3.5R		Y		1500		3.5R	SHTB	374	17	30	2	6.7					2	
EY59C3.5R				1450					>43	60	18	30.0			1	13	4	
EY63C3.5R				1340				312	17	30	0	0.0						
EY55C4.5R				1500		4.5R		334		30	10	33.3					10	
EY59C4.5R				1450					>43	60	29	48.3	4			19	6	
EY63C4.5R				1340				312	17	30	0	0.0						
EY59CB3.5R				1450	Exist	3.5R		334	>43	14	9	64.3					9	
EY63CB3.5R				1340				312	17	15	0	0.0						
EY59CB4.5R				1450		4.5R		334	>43	15	11	73.3	1			5	5	
EY63CB4.5R				1340				312	17	15	0	0.0						
EY45HB			Hot	1560		2.5R		364	13	10	3	30.0					3	
EY51HB				-						60	4	6.7					4	
EY51CB			Cold	1510	Not Exist					60	1	1.7					1	
EY51C				1490						60	8	13.3					8	
EY56CB1				1450	Exist			339	>39	60	3	5.0				2	1	
EY56CB2				1400						40	1	2.5				1		
EY59CB				1360						60	1	1.7				1		
EY61CB				1430						60	0	0.0						
EZ60CB		Z		1430		2.5R		329	>24	800	0	0.0						(Torque Shear Type)
										36	0	0.0						(For Hydrogen)
										14	0	0.0						(For Hydrogen)
EX42C	Tokyo (Shibuya)	X		1070	Not Exist	JIS	JIS	222	13	14	0	0.0						
EW35C		W	Cold	1300	Not Exist	JIS	JIS	180deg.		60	0	0.0						(HEX Type)
EY51HB		Y	Hot	-	Exist	2.5R	SHTB	364	>36	60	0	0.0						
EY51CB			Cold	1510	Not Exist					60	0	0.0						
EY51C				1490	Exist					60	0	0.0						
EY56CB1				1450				339		60	0	0.0						
EY56CB2				1400						30	0	0.0						
EY59CB				1360						60	0	0.0						
EY61CB				1430		2.5R		329	>19	60	0	0.0						
EZ60CB		Z								800	0	0.0						(Torque Shear Type)
EX42C	Fukuoka (Minoshima)	X		1070	Not Exist	JIS	JIS	222	>15	36	0	0.0						(For Hydrogen)
EZ60CB		Z	Cold	1430	Exist	2.5R	SHTB	333		14	0	0.0						(Torque Shear Type)
										800	0	0.0						(For Hydrogen)
X42C	Hokkaido (Muroran)	X		1070	Not Exist	JIS	JIS	222	>9	36	0	0.0						(For Hydrogen)
EZ60CB		Z	Cold	1430	Exist	2.5R	SHTB	331		14	0	0.0						(Torque Shear Type)
EZ60CB(M20)				1410				331		800	0	0.0						(M20)
								263		33	0	0.0						
EX42C		X		1070	Not Exist	JIS	JIS	222		100	0	0.0						(For Hydrogen)
										14	0	0.0						

Figure 5. Actual exposure test condition example in Okinawa

the bolt tension. Whether bolts were shot-blasted or not was parameterized in order to check the effect of bolt phosphoration by bonderizing. Although shot-blasting to some extent tends to decrease the delayed fracture occurrence ratio in period 2, almost no significant difference is noted in period 1. Regarding the difference between cold forging and hot forging, no significant difference in the delayed fracture occurrence ratio is noticed, either. The tensile strength is considered to have the greatest effect on the delayed fracture occurrence ratio, but the ratio not always increases with the increasing strength because various other factors also influence the ratio. The kind of steel is also a great influencing factor. Delayed fracture occurred in steel Y specimens which has tensile strength,1400 N/mm^2, but it has not occurred up to the present in steel Z specimens whose tensile strength,1430N/mm^2. The total average ratios of delayed fracture occurrence during the whole period (three periods) of accelerated exposure test as classified by part of the bolt are 29.7% for the part under the neck of bolt, 13.7% for the shank (including the part shifting from the incomplete thread to the shank), 28.3% for the incomplete thread, 26.9% for the thread (including the free thread) and 1.4% for an unknown part. These figures show that delayed fracture, like fatigue fracture, occurs frequently at parts where stress concentration occurs. Now to look at the actual exposure test results, even the specimens prepared under the same conditions suffer delayed fracture in some cases and do not in other cases due to the difference in site environment and test period. Delayed fracture has not occurred up to the present in specimens tested at any of the locations except Okinawa. In Okinawa, delayed fracture does not occur in the steel Z specimens or the specimens of the steels for F10T bolts (EZ60CB, EW35C and EX42C), but it occurred in all steel Y specimens except those strength are below 1360 N/mm^2. As it is clear from these results, no significant difference is noted in delayed fracture occurrence due to the difference in shape of the R part under the neck, whether shot-blasting was performed or not, forging condition, etc. And delayed fracture occurrence is greatly influenced by the tensile strength and the kind of steel as in the case of the accelerated exposure test.

REFERENCES

Hirai,K. & Uno,N. 1995. Experimental Study on Delayed Fracture of High Strength Bolts. Nordic Steel Construction Conference '95, Proceedings Volume 1, 177-184.

Earthquake Engineering Frontiers in the New Millennium, Spencer & Hu (eds),
© 2001 Swets & Zeitlinger, ISBN 90 2651 852 8

Optimal seismic design of building structures with friction dampers

M.P. Singh
Department of Engineering Science and Mechanics, Virginia Tech, Blacksburg, USA

L.M. Moreschi
Bechtel Power Corporation, Frederick, Maryland, USA

ABSTRACT: The effectiveness of passive energy dissipation systems to improve seismic performance of structural systems is now well established through extensive analytical and experimental investigations. The focus of this paper is on the optimal design of friction dampers for the minimization of a performance function of a structural system. The design parameters that control the effectiveness of the devices are the damper locations in the building, slip loads of different dampers, and the stiffness of the braces in which the dampers are installed. Since the structural systems with friction dampers behave in a highly nonlinear manner, a genetic algorithm is used to calculate the global optimal values for the design parameters of the devices. Numerical results are presented to demonstrate the effectiveness of the proposed optimal design procedure for different forms of performance functions.

1 INTRODUCTION

A large amount of energy is imparted into a structure during earthquake ground motions. Conventional design philosophy seeks to prevent collapse by allowing structural members to absorb and dissipate the transmitted earthquake energy by inelastic cyclic deformations in specially detailed regions. This strategy implies that some damage may occur, possibly to the extent that the structure is not longer repairable.

In the last two decades, several protective systems have been developed to enhance safety and to reduce damage of structures against earthquakes. The use of passive devices to control the structural seismic response and energy dissipation demand on the structural members is now well accepted. In particular, friction devices relying on the resistance developed between two solid interfaces sliding relative to one another have been shown to be reliable and sturdy for structural applications. During severe seismic excitations, the device slips at a predetermined load, providing the desired energy dissipation by friction while at the same time shifting the structural fundamental mode away from the earthquake resonant frequency. Friction dampers are not susceptible to thermal effects, have a reliable performance and possess a stable hysteretic behavior for a large number of cycles under a wide range of excitation conditions. The latter characteristic is a desired feature for a device aimed to protect a structural system during long duration earthquake loadings.

The maximum force developed in the friction devices is controlled by the design slip-load. Virtually any desired combination of limiting loads is feasible. However, by considering high limiting loads the energy dissipated (area under the force-deformation curve) will be minimal since there will be no incursion of the devices into their slippage or inelastic ranges. In this case, the structure will behave as a braced frame. If the limiting loads are low, large incursion in the inelastic and slippage phases will be expected but again the amount of energy will be negligible.

The installation of friction dampers in a structure would render it to behave nonlinearly even if all other members were designed to remain linear. Therefore, one must resort to parametric nonlinear dynamic analysis to determine the slip-load distribution that gives the best structural response reduction. The problem of determining the optimal slip-load distribution in a friction-damped structure can also be solved using gradient-based nonlinear optimization procedures. However, these techniques may lead to a locally optimal solution near the starting design guess. Moreover, the additional information required by the search procedure, such as the calculations of the gradients of the objective functions and constraints, can be cumbersome. In addition, the force-deformation relationships of these devices may introduce discontinuities in the gradient functions depending on the model used to characterize their hysteretic cyclic behavior. Therefore, the implementation of these approaches for optimal design with highly nonlinear energy dissipation devices can be a difficult task.

In this paper, the use of genetic algorithms is proposed to obtain the optimal slip-load distribution in a friction-damped structure subjected to seismic disturbances. The genetic approach is quite flexible to accommodate any desired form of performance function. Nonlinear step-by-step nonlinear time history analyses are carried out for performance evaluations of the system. In the following sections, the details of implementing this approach are presented.

2 METHODOLOGY

2.1 Problem formulation

The main purpose of installing friction devices in structures is to reduce their dynamic response. It is desired to determine the optimum slip-load distribution throughout a building structure such that the energy dissipation capability of each device is fully exploited while providing at the same time the best response reduction.

The effectiveness of a given slip-load distribution can be measured in terms of an optimality criterion or performance index. Thus, the optimal design problem can be stated as:

$$\underset{\mathbf{d}}{\text{minimize}} \quad f\left[\mathbf{R}(\mathbf{d},t)\right]; \qquad t \in [0,t_f] \qquad (1)$$

subject to

$$g_j(\mathbf{d},\mathbf{n},t) \leq 0 \qquad j = 1,\ldots,m; \qquad t \in [0,t_f] \qquad (2)$$

where $\mathbf{R}(\mathbf{d},t)$ is the desired structural response vector in terms of which the performance function $f(\ )$ is defined, $\mathbf{d}$ is the vector of design variables representing the parameters of the added friction dampers, and m is the number of inequality constraints g_j which may include upper and lower bounds on the design variables.

A number of alternate performance indices can be used to evaluate the improvement in the seismic performance of a building structure. Depending upon the chosen criteria, different design solutions can be obtained for the same problem. Moreover, a solution obtained by reducing some measure of the structural response may increase some other response quantities. It is clear that there is no unique way of defining an optimal problem. In this paper two different forms of performance indices are considered to determine the designs that produce the best overall behavior.

2.2 Optimization procedure

Genetic algorithms are robust search and optimization techniques that are based on the principles of natural biological evolution where stronger individuals are likely to be the winners in a competing environment (Goldberg (1989), Holland (1975). Genetic algorithms explore the design space by operating on a population of potential solutions (designs) simulating evolution by means of random genetic changes that produce successively better approximations to a design solution. Since many design points are considered simultaneously in the search space, genetic algorithms have a reduced chance of converging to local optima. Moreover, they do not require any computations of gradients of complex functions to guide their search; the only information needed is the response of the system to calculate the objective or fitness function. These attributes make genetic algorithms well suited for solving optimization problems such as the one stated by Eq. 1. This approach has been effectively used by the authors earlier (Singh and Moreschi, 1999).

2.3 Analytical modeling of friction devices

The cyclic force-deformation response of friction devices is characterized by rectangular hysteresis loops. This behavior has been represented in practice by rigid-perfectly-plastic models, as shown in Figure 1. The threshold force at which the device starts to deform continuously is called the slip-load. The value of this parameter, denoted here as P_s, provides a complete definition of the idealized model of the device.

The above description is sufficient to portray the behavior of a friction damper in which the elements used to support and connect the device to the main structural members are considered as rigid. The flexibility of the bracings can also be introduced in the analysis by considering the ratio SR between the stiffness k_b of the bracings and the structural stiffness k_s of the story in which the damper is located. That is,

$$SR = \frac{k_b}{k_s} \qquad (3)$$

For design purposes, the slip-load can then be related to the deformation Δ_y experienced by the device-brace assembly in terms of the stiffness parameter SR as

$$P_s = SR\,k_s\,\Delta_y \qquad (4)$$

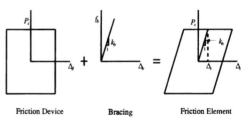

Friction Device   Bracing   Friction Element

Figure 1. Idealized hysteretic behavior of friction dampers

Eq. 4 is the basic expression relating the mechanical parameters of a friction element. From this equation, it can be observed that the behavior of a friction element is governed by the slip load P_s, the stiffness ratio SR, and the displacement of the bracing Δ_y at which the device starts to slip. However, only two of these variables are independent since the third one can be determined from Eq. 4.

The hysteretic behavior of the friction element can be characterized using a continuous Bouc-Wen's model (Wen 1980). Recognizing the absence of any post-yielding or strain-hardening effect, the force $P(t)$ developed in a friction element can be obtained as

$$P(t) = P_s h(t) \tag{5}$$

where the time dependent auxiliary variable $h(t)$ of the model is defined by the following first order differential equation

$$\Delta_y \dot{h}(t) - H \dot{\Delta}(t) + \gamma \left| \dot{\Delta}(t) \right| h(t) \left| h(t) \right|^{\eta-1}$$
$$+ \beta \dot{\Delta}(t) \left| h(t) \right|^{\eta} = 0 \tag{6}$$

where $\Delta(t)$ is the story drift or the deformation of the brace and the device. The model parameters H, γ, β and η are adjusted to approximate the shape of the hysteresis loops. The remaining model parameter, Δ_y, can be related to the mechanical properties of the friction element. This can be done by considering that at the slipping condition, the hysteretic variable $h(t)$ takes values of ± 1, and the friction element force $P(t)$ is equal to the slip-load P_s. Thus, it can be shown that

$$\Delta_y = \frac{P_s}{SR\, k_s} \tag{7}$$

2.4 Response calculations

The equations of motion for a plane shear building with a single friction device per story, can be written as,

$$\mathbf{M}\ddot{\mathbf{x}}(t) + \mathbf{C}_s \dot{\mathbf{x}}(t) + \mathbf{K}_s \mathbf{x}(t) + \sum_{d=1}^{n_l} \mathbf{r}_d P_d(t) = -\mathbf{M}\,\mathbf{E}\,\ddot{X}_g(t) \tag{8}$$

where $\mathbf{M}$, $\mathbf{C}$ and $\mathbf{K}$ represent, respectively, the mass, stiffness and inherent structural damping matrices of the structure; $\mathbf{E}$ is the vector of ground motion influence coefficients; and $\ddot{X}_g(t)$ is the ground acceleration at the base of the structure. The force $P_d(t)$ due to the friction damper installed at the dth location is considered through the influence vector $\mathbf{r}_d$, with n_l being the number of possible locations for a device in the structure. The force in the dth damper, represented by the continuous hysteretic Bouc-Wen model of Eq. 5, can be written as:

$$P_d(t) = P_s^d h_d(t) \tag{9}$$

$$P_s^d \dot{h}_d(t) - SR_d\, k_s^d (\dot{\Delta}_d(t) - \gamma \left| \dot{\Delta}_d \right| h_d(t) \left| h_d(t) \right|^{\eta-1}$$
$$- \beta \dot{\Delta}_d \left| h_d(t) \right|^{\eta}) = 0 \tag{10}$$

The expressions (8), combined with the equations (9) and (10) for the element forces, can be rewritten as a set of first-order differential equations of the form:

$$\begin{Bmatrix} \ddot{\mathbf{x}}(t) \\ \dot{\mathbf{x}}(t) \\ \dot{\mathbf{h}}(t) \end{Bmatrix} = g\left[\mathbf{x}(t), \dot{\mathbf{x}}(t), \mathbf{h}(t), \ddot{X}_g(t), t \right] \tag{11}$$

The differential Eq. 11 constitutes a set of coupled nonlinear differential equations and can be conveniently integrated using several accurate and efficient solvers.

3 NUMERICAL RESULTS

This section illustrates the application of the genetic algorithm approach to the design of the friction dampers for seismic protection of building structures. A ten-story building is considered in this section for retrofitting purposes. For the numerical calculations, it is assumed that a single device is installed at each story, with mechanical properties to be determined by the optimization procedure. For optimization purposes, the slip-loads P_s^d and stiffness ratios SR_d are considered here.

For numerical calculations, a set of four artificially generated accelerograms is used. These synthetic earthquakes were generated for a Kanai-Tajimi (Kanai 1961; Tajimi 1960). Power spectral density function the parameters of which were determined to represent the characteristics of the 1971 San Fernando ground motion with peak ground acceleration of 0.315g.

As the first example, the design objective is to reduce a relative performance index, RPI, as defined by Filiatrault and Cherry (1990):

$$RPI = \frac{1}{2} \left(\frac{SEA}{SEA_{(o)}} + \frac{U_{max}}{U_{max(o)}} \right) \tag{12}$$

where SEA and U_{max} are, respectively, the area under the elastic strain-energy time history and the maximum strain energy for a friction-damped structure; $SEA_{(o)}$ and $U_{max(o)}$ are the respective quantities of the original uncontrolled structure. The selection of this performance index is motivated by the direct relation that exists between the amount of elastic strain energy imparted into a building and the resulting structural response.

First, the friction devices are designed under the assumption of uniform distribution of slip-loads and bracing element stiffness over the height of the structure. The stiffnesses of the supporting braces are assumed proportional to those of the main structural frame with the stiffness ration $SR_d = 2$. The friction-damped structure is then subjected to the set of four artificially generated earthquakes, and the genetic algorithm is used to find the optimal design solution. For a proper implementation of this search procedure, the slip-load design space has to be discretized. In this regard, it is considered that the ratio between the slip-load and the total weight of the building structure, W, can take on any multiple value of 0.005 between 0.0 and $0.125W$, with zero corresponding to the situation of no device. Nonlinear time history response analyses are performed, and for every potential design solution the RPI index of Eq. 12 is evaluated. The value of slip-load P_s that minimizes this RPI index, averaged over the four earthquakes, is presented in Column (2) of Table 1. These results are expressed as percentages of the total building weight.

Next, the same design problem is solved considering the slip-load at each dth location as an independent variable, and not the same at all locations as assumed before. Therefore, the assumption of uniform slip-load distribution is removed, and the genetic algorithm is used to find the optimal design solution. The same load interval and load increment used in the previous design example are adopted for the numerical calculations. Therefore, the discretization scheme leads to twenty-six possible values of slip-load for each floor, and a total of 26^{10} possible combinations for the device loads P_s^d. To compare the results for this case with that of the previous case, a constraint is added to the optimization problem such that the total friction load in this case is the same as the one previously obtained for the uniform slip-load distribution case.

Column (3) of Table 1 presents the slip-load distribution obtained using the genetic algorithm approach averaged over the four earthquakes. These results have been obtained for a population of 20 individuals after 500 generations. The additional input parameter for the optimization runs are as follows: probability of crossover $p_c = 0.9$, and probability of slip-load mutation = 0.03. It is noted that performance, as measured by the value of the performance index in the last row of the table, is improved in this case.

The design of the previous friction-damped structure is repeated here adding a second variable per device. In this case, the parameter SR_d is set free and able to take on any integer value ranging from 1 to 10. The genetic algorithm is then used to search for the best design solution. Since the number of possible combinations has increased, a larger population of 30 individuals has been considered for the numerical calculations. The parameters of the optimization algorithm have been taken as follows: prob-

Table 1. Optimal design of friction devices according to the RPI performance index of Eq. 12

Story (1)	Uniform Slip-Load P_s [%W] (2)	Variable Slip-Load P_s^d [%W] (3)	Slip-Load P_s^d and SR_d ratios	
			P_s^d [%W] (4)	SR_d (5)
1	3.25	5.30	5.9	9.75
2	3.25	4.50	4.9	9.25
3	3.25	3.80	4.6	8.75
4	3.25	3.60	3.8	7.25
5	3.25	3.00	3.4	9.50
6	3.25	2.40	2.9	9.25
7	3.25	2.40	2.5	9.50
8	3.25	2.10	2.1	9.00
9	3.25	2.10	1.5	9.00
10	3.25	3.40	1.0	8.00
Total load	32.5	32.5	32.5	
RPI	0.2681	0.2347	0.1060	

Note: the results of Columns (2) and (3) have been obtained for a stiffness ratio $SR = 2$.

ability of crossover $p_c = 0.9$, probability of slip-load mutation $p_{ms} = 0.15$, and probability of SR ratio mutation $p_{mr} = 0.15$. Columns (4) and (5) of Table 1 show, respectively, the values of the slip-loads P_s^d and stiffness ratios SR_d for each story. These results have been obtained after 700 generations. Although the same total friction load is used, the RPI index is further reduced by a 60%. Such dramatic improvement in the seismic performance of the structural system can be attributed to the additional stiffness contributed by the friction elements, and to a better utilization of their energy dissipation capabilities.

Figure 2 investigates the reduction achieved in the maximum inter-story drifts, displacements and absolute accelerations for the original building and

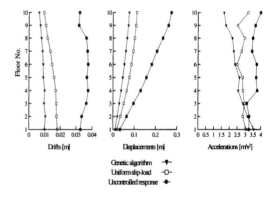

Figure 2. Comparison of maximum response quantities along the building height (RPI performance index)

the friction-damped structure designed using the uniform slip-load distribution and the design solution obtained by considering two design variables per friction device. These quantities have been obtained by averaging the responses obtained for the simulated acceleration records.

As evidenced from the design solutions presented above, the friction dampers reduce the structural response through a combination of improved energy dissipation capabilities and increased lateral stiffness of the building. However, the presence of additional stiffness may also induce larger floor accelerations and structural members stresses. Therefore, a performance index that considers simultaneously the reductions in the maximum inter-story drifts and maximum story acceleration can be expressed as:

$$ f\left[\mathbf{R}(\mathbf{d},t)\right] = \frac{1}{2}\left\{ \frac{\max_i \Delta_i(t)}{\max_i \Delta_{i(o)}(t)} + \frac{\max_i \left[\ddot{x}_i(t) + \ddot{X}_g(t)\right]}{\max_i \left[\ddot{x}_{i(o)}(t) + \ddot{X}_g(t)\right]} \right\} \quad (13) $$

where the deformation experienced at the ith story, denoted as Δ_i, and absolute floor acceleration are normalized with respect to the corresponding values of the original building. It is noted that this index gives equal weights to the deformation and acceleration related responses. If desired, different weights can also be assigned.

The design of the same ten-story friction-damped building is now repeated. The goal is to determine the slip-load and stiffness ratio distribution required to minimize the index of Eq. 13. As before, the slip-load P_s^d is considered first as the only design parameter per device. The bracings are designed proportional to the stiffness of the building stories in which the device is placed, and a value of $SR = 2$ is adopted for the numerical calculations. Columns (2) and (3) of Table 2 present the results obtained under these conditions by the genetic algorithm optimization procedure. Columns (4) and (5) of the same table show the design solution obtained when the stiffness ratio of each device, SR_d, is also considered as a design variable. For this design solution, Figure 3 compares the maximum inter-story drifts, maximum displacements and maximum absolute accelerations obtained at different stories of the original and friction-damped structures. These responses have been averaged over the four simulated earthquakes. This figure also shows the corresponding maximum responses obtained for the structure designed using the uniform slip-load distribution that minimized the RPI index. These responses quantities were previously presented in Figure 2. It can be observed from Figures 2 and 3 that the design solution obtained by minimizing the performance index of Eq. 13 provides comparable reductions in the maximum inter-story drifts and displacements, while reducing substantially the maximum accelerations at all building levels. Figure 4 shows the evolution of

the best design in successive generations and the convergence characteristics of the genetic algorithm used in this study.

Table 2. Optimal design of friction devices according to the performance index of Eq. 13

Story	Slip-Load P_s^d		Slip-Load P_s^d and SR_d	
	SR_d	P_s^d [%W]	SR_d	P_s^d [%W]
(1)	(2)	(3)	(4)	(5)
1	2.0	1.6	5.50	3.5
2	2.0	3.6	3.00	3.8
3	2.0	3.1	7.25	4.3
4	2.0	2.6	7.25	4.5
5	2.0	2.1	7.50	2.5
6	2.0	2.0	6.25	2.5
7	2.0	2.4	6.00	3.5
8	2.0	2.1	4.75	3.8
9	2.0	1.1	7.75	2.8
10	2.0	0.8	5.50	2.5
$f\mathbf{R}(\mathbf{d}^*,t)]$	0.5226		0.4617	

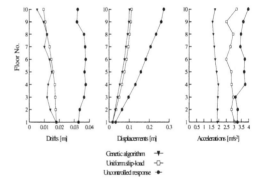

Figure 3. Comparison of maximum response quantities along the building height (performance index of Eq. 11)

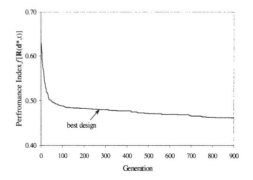

Figure 4. Optimization history for maximum response reduction using genetic algorithm

4 CONCLUSIONS

The paper presents a methodology for optimal design of friction damper for multi-story buildings exposed to seismic motions. The procedure defines the optimal locations and slip loads for the dampers and the stiffness of the bracings that must be used. Although a continuous analytical model is used to characterize the hysteretic behavior of the friction dampers, any other form or model can also be used. Since a structural system installed with friction dampers behaves in a strongly nonlinear manner, a genetic algorithm based search has been implemented to calculate the optimal values of the design parameters for a given performance function. Numerical results demonstrate the effectiveness of such an optimal design procedure for performance improvement of structures.

5 ACKNOWLEDGEMENTS

This study was sponsored by National Science Foundation through grant numbers CMS-9626850 and CMS-9987469. This financial support is gratefully acknowledged.

REFERENCES

Filiatrault, A. and Cherry, S., (1990). "Seismic Design Spectra for Friction-Damped Structures," *Journal of Structural Engineering*, 116, 1334–1355.

Goldberg, D. E., (1989). *Genetic Algorithms in Search, Optimization and Machine Learning*, Addison-Wesley, Reading, MA.

Holland, J. H., (1975). *Adaptation in Natural and Artificial Systems*, University of Michigan Press, Ann Arbor, MI.

Kanai, K., (1961). "An Empirical Formula for the Spectrum of Strong Earthquake Motions," *Bulletin Earthquake Research Institute, University of Tokyo*, 39, 85–95.

Singh, M. P. & Moreschi, L.M. (1999). "Genetic Algorithm-based Control of Structures for Dynamic Loads", *Proceedings of IA-99*, November 15-16, Osaka University, Japan.

Wen, Y. K., (1980). "Equivalent Linearization of Hysteretic Systems Under Random Excitation," *Journal of Applied Mechanics, ASME*, 47(EM2), 150–154.

Tajimi, H., (1960). "A Statistical Method of Determining the Maximum Response of a Building Structure During an Earthquake", *Proceeding of II World Conference in Earthquake Engineering*, Tokyo, 781–797.

Earthquake Engineering Frontiers in the New Millennium, Spencer & Hu (eds),
© 2001 Swets & Zeitlinger, ISBN 90 2651 852 8

Study on the control algorithm based on maximum energy dissipation criteria for semi-active fluid dampers

Hui Li & Bo Wu
Professor, School of Civil Engineering, Harbin Institute of Technology, Harbin, 150090, China

Lambros Katafyoist
Assistant Professor, The Hong Kong University of Science and Technology, Kowloon, Hong Kong, China

Yoshiyuki Suzuki
Professor, Disaster Prevention Research Institute, Kyoto University, Gokasho, Uji, Kyoto 611-0011, Japan

Xuesong Yuan
Ph.D student, School of Building Engineering, Harbin University of Civil Engineering and Architecture, Harbin, 150090, China

ABSTRACT: The semi-active fluid dampers can dissipate the energy and tune the frequency of a structure, so they can suppress the structural vibration. A control algorithm based on maximum energy dissipation for the semi-active fluid dampers is presented in this study. The algorithm indicates that the voltage is only put on maximal value or minimal value according to the switching line. Further study indicates that there is not exist switching line and the voltage is only put on maximum value, which state is called as "passive on state". In this way, the voltage does not need to be instantaneously changed during vibration. The effects of the semi-active fluid dampers on tuned frequency are investigated by using LQR in the semi-active control. The result that the semi-active dampers can tune the frequency to a very small extent is found, so the energy dissipation of the semi-active fluid dampers affects the reduction of vibration of structures significantly, however, the tuned frequency does not influence on the control performance significantly and the performance of the control algorithm based on maximum energy dissipation criteria presented in this study is nearly same as that of LQR.

1 INTRODUCTION

The semi-active control has caused a lot of interests in civil engineering. Many researchers have studied the performance of the semi-active fluid dampers such as variable orifice viscous dampers, ER dampers and MR dampers and developed some control algorithms for them. However, in actual, most of the algorithms for the semi-active fluid dampers are that for the active control, such as LQG/LQR or slide mode control approach. The active control algorithms cannot describe the vibration reduction of semi-active dampers in physics. In another aspect, the research results indicated that the semi-active dampers reduce responses of a structure, significantly, and the performance of the semi-active damper is nearly same as that of active control and better than passive control under the same control force. However, some experimental results indicate that passive dampers can achieve the nearly same effectiveness as semi-active dampers.

In physics, semi-active fluid dampers can dissipate energy and tune frequencies of a structure, so they can suppress the structural vibration. The difference between semi-active control and "passive on" control that the value of the voltage is fixed at maximum value is that semi-active fluid dampers can tune frequencies of a structure besides energy dissipation. This can be proved through the active

control algorithms such as LQR and free vibration of a structure with semi-active fluid dampers. If the semi-active control can tune frequencies over a narrow range, the reason that it can reduce the vibration of a structure, significantly, is that it can dissipate energy. At this viewpoint, the performance of "passive on" dampers is nearly same as that of semi-active fluid dampers and semi-active fluid dampers are no evidently advantage to reduce responses comparing with "passive on" dampers. The conclusion above has been pointed out by Symans (1997).

A control algorithm based on maximum energy dissipation criteria for semi-active fluid dampers is presented in this study. The algorithm indicates that the voltage is only fixed at maximum value or minimum value according to the switching line. Further study indicates that switching line does not exist and the voltage is only fixed at maximum value, which state is called as "passive on state". From the viewpoint, the voltage does not need to be instantaneously adjusted during vibration.

The effects of semi-active fluid dampers on tuned frequency are investigated by using LQR. The result that semi-active fluid dampers can tune frequencies of a structure over a narrow range is found, so, in fact, semi-active fluid dampers act as passive energy dissipation dampers. The performance of the control algorithm based on maximum energy dissipation criteria presented in this study is nearly same as that of LQR.

Finally, the conclusion above is verified by using the numerical examples, the numerical results indicate that passive on dampers can achieve the nearly same control effectiveness as semi-active dampers.

2 A CONTROL ALGORITHM FOR SEMI-ACTIVE FLUID DAMPER BASED ON MAXIMUM ENERGY DISSIPATION

The state equation of a structure with a semi-active fluid damper can be written:

$$\dot{X} = AX + BU + GW \tag{1}$$

in which X is the state variable, U is the control force vector and can be expressed by the voltage and the state variable, $U = a(X) + b(X)v$, v is the voltage vector and $0 \le v \le v_{max}$, and $a(X)$ and $b(X)$ are functions of the state variable, respectively. W is the disturbance, A is the system matrix, B is the location matrix and its elements are zeros or ones, and G is a vector.

The following performance index (PI) is employed to get the optimal control force that maximize the PI:

$$J = \int_0^{t_f} U^T \dot{x} dt = \int_0^{t_f} U^T N X dt \tag{2}$$

in which $\dot{x}$ is the velocity vector. $N = [0_{n \times n} \quad I_{n \times n}]$, I is the unit matrix, t_f is the finial time of interest. The performance index denotes energy dissipation by a damper.

The optimal control problem is to determine a control force to maximize the PI and also satisfies the constraint of Eq. 1.

According to maximum theorem, defining the Hamilton function:

$$H = U^T N X + \lambda^T (AX + BU) \tag{3}$$

in which λ is Lagrange multipliers.

Assuming that the optimal control force U^* makes the optimal state of a structure be X^*, the following inequality can be found:

$$H^* = U^{*T} N X^* + \lambda^{*T} (AX^* + BU^*)$$
$$> H = U^T N X^* + \lambda^{*T} (AX^* + BU) \tag{4}$$

The following equation can be obtained from both of the Eq. 4 and the expression of U

$$(a(X^*) + b(X^*)v^*)^T (NX^* + B^T \lambda^*)$$
$$> (a(X^*) + b(X^*)v)^T (NX^* + B^T \lambda^*) \tag{5}$$

Because the voltage can be varied from 0 to v_{max}, in order to satisfy Eq. 5, the voltage must satisfy:

$$\begin{aligned} &if \quad b(X^*)(NX^* + B^T \lambda^*) > 0 \\ &then \qquad v^* = v_{max} \\ &if \quad b(X^*)(NX^* + B^T \lambda^*) < 0 \\ &then \qquad v^* = 0 \\ &if \quad b(X^*)(NX^* + B^T \lambda^*) = 0 \\ &then \qquad v^* = [0 \quad \sim \quad v_{max}] \end{aligned} \tag{6}$$

The Eq. 6 indicates that the voltage does not need to be instantaneously adjusted during vibration. It only needs to be changed on switching line, which is $b(X^*)(NX^* + B^T \lambda^*)$.

For linear variable orifice damper, the control force can be written as follows by Symans (1996) and Li (2000):

$$U = c(v)\dot{x} = (c_0 + c_1 v)\dot{x} \tag{7}$$

in which c_0 and c_1 are damping coefficient matrices, respectively. In Symans' test, c_1 is negative, the passive on state corresponds to the minimum voltage. However, in Li's test, c_1 is positive, the passive on state corresponds to the maximum voltage. In this study, the case that c_1 is positive will be discussed. The same conclusion can be obtained for c_1 to be negative. The item of $(b(X^*))^T NX^*$ in Eq. 6 is always positive. If the item of $(b(X^*))^T B^T \lambda^*$ in Eq. 6 is also always positive, the voltage can only be fixed at the maximum value.

The co-state λ can be obtained from the following equation:

$$-\dot{\lambda} = \frac{\partial H}{\partial X} = N^T U + A^T \lambda \tag{8}$$

The necessary and sufficient condition that guarantees the solution stability of Eq. 8 should be as follows:

$$U = d\lambda \tag{9}$$

in which $d = [0_{n \times n} \quad A_1]$, A_1 is a diagonal matrix and its elements are positive. The relation between the co-state and the control force can be rewritten:

$$\lambda = d^{-1} U \tag{10}$$

Considering the Eqs. 10 and 7, the item of $(b(X^*))^T B^T \lambda^* = \dot{x}^{*T} c_1^T B^T d^{-1} (c_0 + c_1 v)\dot{x}^*$ to be found.

Because $c_1^T B^T d^{-1} (c_0 + c_1 v)$ is positive definition, the second item in Eq. 6 is positive. As a result, the

352

voltage only needs to be fixed at the maximum value. From the viewpoint, the semi-active control strategy is the same as the "passive on state" and the voltage does not need to be instantaneously adjusted during vibration.

For a MR damper, the relation between the control force and the voltage can be written by Marathe (1998):

$$U = (\tau_{y0} + \tau_{y1}v)A_d \tanh(A\dot{x}) \tag{11}$$

in which A_d is the area of a MR damper, τ_{y0} and τ_{y1} are the coefficients of the yield force, respectively, and both of them are positive, A is a constant. The Eq. 4 becomes following expression:

$$(\tau_{y1}v^*A_d \tanh(A\dot{x}^*))^T (NX^* + B^T\lambda^*)$$
$$> (\tau_{y1}vA_d \tanh(A\dot{x}^*))^T (NX^* + B^T\lambda^*) \tag{12}$$

Because the voltage can be varied from 0 to v_{max}, in order to satisfy Eq. 12, the voltage must satisfy:

$$if \quad (\tanh(A\dot{x}^*))^T (NX^* + B^T\lambda^*) > 0$$
$$then \qquad v^* = v_{max}$$
$$if \quad (\tanh(A\dot{x}^*))^T (NX^* + B^T\lambda^*) < 0$$
$$then \qquad v^* = 0$$
$$if \quad (\tanh(A\dot{x}^*))^T (NX^* + B^T\lambda^*) = 0$$
$$then \qquad v^* = \begin{bmatrix} 0 & \sim & v_{max} \end{bmatrix} \tag{13}$$

Because the first item is always positive and the control force has similar expression with that in Eq. 9, only the first case in Eq. 13 can be found and the voltage should be fixed at the maximum value. As a result, the performance of a semi-active MR damper is the same as that of a "passive on state" MR damper.

The same conclusion can be obtained for an ER damper.

3 COMPARISON BETWEEN SEMI-ACTIVE CONTROL AND ACTIVE CONTROL

The fact that semi-active fluid dampers can change damping and frequencies of a structure can be proved through the algorithm of LQR that has widely been used in semi-active control. The performance index of LQR is as follows:

$$J = \int_0^{t_f} (X^TQX + U^TRU)dt \tag{14}$$

Let $Q = \begin{bmatrix} K & 0 \\ 0 & 0 \end{bmatrix}$. When the weighing matrix R is fixed, to make the performance index in Eq. 10

minimum means to make the displacements of a structure minimum.

The optimal control force can be obtained as follows:

$$U = -R^{-1}B^TPX = K_Lx + C_L\dot{x} \tag{15}$$

in which P can be obtained from the following Ricatti equation:

$$PA + A^TP - PBR^{-1}B^TP + Q = 0 \tag{16}$$

For a SDOF structure, K_L and C_L can be determined by the following expression:

$$K_L = -k_s + \sqrt{k_s^2 + k_sR^{-1}} \tag{17}$$

$$C_L = -c_s$$
$$+ \sqrt{c_s^2 + 2R^{-1}(-k_s + \sqrt{k_s^2 + k_sR^{-1}})} \tag{18}$$

in which k_s and c_s are the stiffness and damping coefficient of a SDOF structure, respectively.

The force of a semi-active fluid damper can be assumed to be the same as that in Eq. 15, so the variation of frequencies of the controlled structure by a semi-active damper is:

$$\frac{\omega}{\omega_s} = \sqrt{\frac{K_L + k_s}{k_s}} = \sqrt{\frac{-k_s + \sqrt{k_s^2 + k_sR^{-1}} + k_s}{k_s}}$$
$$= \sqrt{\sqrt{1 + \frac{R^{-1}}{k_s}}} \tag{19}$$

The variation of the frequencies of a structure by a semi-active fluid damper is related with the ratio of the inverse of the weighing coefficient to the stiffness of a structure. However, because of the value of the weighing coefficient limited by the control force, the ratio of the frequency of a structure with a semi-active damper to the structure without damper is very small in general.

The same conclusion can be obtained for a MDF structure through numerical examples.

4 NUMERICAL STUDIES

The feasibility of the algorithm and the variation of the frequency of a structure with a semi-active damper will be proved through the numerical examples in this section.

4.1 The variation of the frequency of a structure with a semi-active fluid damper

Consider a SDOF structure, its frequency $f = 1Hz$, the damping ratio $\zeta = 0.02$, the mass of the structure is 1 000 kg and the maximum control force is limited to 500 N. The disturbance is harmonic excitation with frequency varying from 0.5Hz to 2Hz. The control force of a semi-active damper is determined by LQR. The weighing coefficient can be obtained through numerical examples and is tabulated in Table 1. The variation of the frequency and the damping ratio of the controlled structure are also listed in Table 1.

Table 1. The variation of frequency for a SDOF structure

f_d / f	0.5	0.8	1.0	1.2	1.5	2.0
ω / ω_s	1.17	1.02	1.00	1.01	.103	1.1
ζ / ζ_s	178.9	7.12	1.97	5.16	9.19	14.3
R^{-1} / k_s	0.905	0.08	0.004	0.04	0.14	0.36

It can be shown from Table 1 that a semi-active fluid damper can tune frequencies over a very narrow range. However, it can change the damping ratio, significantly. From this point of view, a semi-active damper suppresses vibration of a structure mainly because it can dissipate energy.

Consider a three-story structure as the second example. The mass of each floor is 1 000kg, respectively. The stiffness of the first story, the second story and the third story is assumed to be 4×10^5 N/m, 2×10^5 N/m and 4×10^5 N/m, respectively. The maximum control force is limited to 1 000N. The disturbance is harmonic excitation with frequency varying from 0.5Hz to 2Hz. The control force of a semi-active damper is determined by LQR. The semi-active damper is installed on second story. The weighing coefficient can be obtained through numerical examples and is tabulated in Table 2. The variation of the frequency of the controlled structure is also listed in Table 2.

The results in Table 2 indicate that a semi-active fluid damper can tune frequencies over a smaller range for a MDOF structure than that for a SDOF.

4.2 *The comparison of the control effectiveness of a semi-active fluid damper and a passive on damper*

A three-story structure is employed to calculate the responses of the structure with a semi-active damper or a passive on damper installed at the second story under earthquake. The mass of each floor is 1 920kg. The stiffness of the first story, the second story and the third story is 2.5×10^5 N/m, 2.5×10^5 N/m and 1.75×10^5 N/m, respectively. The damping coefficient provided by the semi-active damper varies from 100Ns/m to 10 000Ns/m. The damping coefficient of the passive on damper is 10000Ns/m. El Centro is used as the input. The response of the structure is listed in Table 3.

The results in Table 3 indicate that the responses of the structure with a semi-active damper are nearly same as that with a passive on damper under the same control force.

5 CONCLUSIONS

A control algorithm based on maximum energy dissipation criteria is presented and the algorithm indicates that the semi-active damper has no obviously advantage in reduction of vibration to compare with the passive on state damper. The reason is that the semi-active fluid damper can tune the frequency, insignificantly, however, it can dissipate energy significantly. From this viewpoint, the performance of a semi-active fluid damper is the same as that of a passive fluid damper.

Table 2. The variation of frequencies for MOF structure

f_d / f	0.5	0.8	1.0	1.2	1.5	1.8	2.0
ω_1 / ω_{s1}	0.999	1.000	1.000	1.00	1.000	1.000	1.000
ω_2 / ω_{s2}	0.991	0.999	1.000	0.999	0.997	0.994	0.994
R^{-1} / k_{s2}	0.227	0.047	0.006	0.0357	0.0833	0.135	0.156

Table 3. The responses of a structure with a semi-active damper or a passive on damper

	Max. Dis. (cm)			Max. Vel. (cm/s)			Max. U (N)
Without dampers	2.46	4.32	5.61	15.57	24.32	30.86	
With a semi-active damper	2.01	3.67	5.20	11.95	21.06	25.20	556
With a passive on damper	2.04	3.68	5.16	11.19	19.73	25.49	552
	Max. Dis. (cm)			Max. Vel. (cm/s)			Max. U (N)
Without dampers	2.46	4.32	5.61	15.57	24.32	30.86	
With a semi-active damper	2.01	3.67	5.20	11.95	21.06	25.20	556
With a passive on damper	2.04	3.68	5.16	11.19	19.73	25.49	552

REFERENCES

M.D. Symans, M.C. Constantinou. Experimental testing and Analytical Modeling of Semi-Active Fluid Dampers for Seismic Protection. Journal of Intelligent Material Systems and Structures, Vol.8, 1997(8), 644-657

B.F. Spencer Jr., S.J.Dyke, M.K.Sain and J.D. Carlson. Phenomenological Model for Magnetorheological Dampers. Journal of Engineering Mechanics, Vol.123, No.3, 1997, 230-238.

S.J.Dyke, B.F. Spencer Jr, M.K. Sain and J.D. Carlson. Modeling and Control of Magnetorheological Dampers for Seismic Response Reduction. Smart Material Structure, No.5, 1996, 565-575.

S.S. Marathe, K.W.Wang and F. Gandhi. The effect model uncertainty on magnetorheological fluid damper based systems under feedback linearization control. Adaptive Structures and Materials Systems, AD-Vol.57/MD-Vol.83, 1998, 129-140.

H. Li, X.S. Yuan and B. Wu. The vibration control of structures with semi-active fluid dampers. Proceedings of International workshop on structural control and health monitoring (CD-ROM), 2000, China (in Chinese).

Earthquake Engineering Frontiers in the New Millennium, Spencer & Hu (eds),
© 2001 Swets & Zeitlinger, ISBN 90 2651 852 8

Seismic assessment and retrofit of Beijing engineering in the capital area of Beijing

Yayong Wang
Institute of Earthquake Engineering, China Academy of Building Research, Beijing 100013

ABSTRACT: The paper presents the new technology of the seismic assessment, evaluation, retrofitting, strengthening and the practices in the capital area of Beijing for the land-mark public buildings, such as The Beijing Railway Station, The Historic Museum of China, The Beijing Hotel, The Agriculture Exhibition Hall of China, The Beijing Exhibition Hall and The Beijing Library which were designed and constructed in 1950's or even earlier without earthquake resistance measures. These buildings are still being used over and above their service-life. Same problems remain in schools and hospitals, which hold crowds and function a key role for the emergency rescue work of post-earthquakes. The Chinese government allocated since 1997 special funds for the overall and comprehensive assessment, evaluation, retrofit and strengthening on public buildings in Beijing University, Beijing Aero-space University, Beijing Pedagogical University, University of Chemical Industry, as well as the hospitals affiliated to China Academy of Medical Science, University of Medical Science of Beijing, etc. Some recently developed technology and instruments have been employed to ensure capacity of earthquake resistance to the buildings. For example, the supersonic wave surveying, the electronic scanning, core-sampling approaches for the quality investigation of R.C. components; the seismic assessment and the strengthening design according to the national standards; the small pile named as 'tree-root pile', the R.C. attached wall named as 'sandwich wall', the steel jacket and band, the R.C. shear wall, the vacuum-high pressure grouting, and the energy dissipation damper, etc. have been used for the retrofitting and strengthening of the existing buildings not only in the capital area of Beijing but also nationwide.

1 STANDARD AND REGULATION FOR SEISMIC STRENGTHENING AND RETROFITING

- Seismic Evaluation Standard for Existing Buildings GB50023-95, national code
- Technical Code for Seismic Strengthening of buildings, ministerial code
- Classification Standard for Earthquake Damaged Buildings, ministerial code
- Standard for Classification of Seismic Protection of Buildings GB 50223-95, national code
- Seismic Technical Specification for Multistory Masonry Buildings with Reinforced Concrete Tie Column JGJ/T 13-94, ministerial code
- Seismic Evaluation Standard for Industrial Facilities, national code
- Seismic Evaluation Standard for Industry Special Structures, national code
- Seismic Evaluation Standard for Water Supply, Water Drain Away Facilities Outdoor, national code

- Seismic Evaluation Standard for Gas and Heating Engineering Facilities, national code
- Seismic Evaluation Standard for Water Transport and Hydraulic Engineering, ministerial code
- Regulation of Seismic Strengthening and Acceptance for Communication Facilities, ministerial code
- Seismic Evaluation Standard for Petroleum Refinery Facilities, ministerial code.

2 THREE PROGRESSIVE STAGES OF STRENGTHENING AND RETROFITTING IN CHINA

- Primary Stage: Emergent-Temporary Measures and Simply Strengthening Technology
- Emergent repair and temporary support after earthquake
  (1) 'Sandwich' concrete wall
  (2) Attached tie-column
  (3) Attached tie beam and steel tensile bar

- Advanced Stage: Pre-event Strengthening and Post-event Retrofitting
  (1) Vacuum-high pressure grouting technology

(2)'Tree-root piles' for strengthening of base and foundation

(3) R.C. enveloping and steel binding (jacket and band) technology for R.C. frame structure

(4) Attached wind wall or shear wall to R.C. frame structure

(5) Attached steel brace to spacious R.C. workshop

- Comprehensive Stage: in addition to seismic strengthening and retrofitting
  (1) Extension of the existing building (office building of 5-story to 8-story)

(The old building of 5-story)

(Retrofit and extension work)

(The renewed building of 8-story)

(2) Decoration and facilities (The Conference Hall)

(3) High-tech usage (base isolator, energy dissipation facility, TMD, etc.)

The spring-viscous dampers are installed inside the Central Hall of Beijing Exhibition Center

359

The Taylor Devices' fluid viscous dampers are installed in The Beijing Railway Station

EXAMPLES

3 EFFICIENCY ANALYSIS AND INVESTMENT OPTIMIZATION

- Direct loss and Induced Loss
- Cost for and Benefit from Strengthening
- Efficiency Ratio J

$$J = \frac{2Z'}{Z}$$

where Z' is the benefit (reduction of loss), Z is the cost for strengthening, it is assumed that the cost of reconstruction equals to the earthquake loss. It is obvious that the efficiency ratio J depends on the regional seismic risk (zoned intensity), the original value of the building, the acceptable damage or failure to the objective building and the cost of construction. The ratio J is generally ranged in 20-100.

The Beijing Hotel equipped with the French Jarret's dampers

The Capital Library of Beijing

The Beijing Exhibition Hall

Earthquake Engineering Frontiers in the New Millennium, Spencer & Hu (eds),
© 2001 Swets & Zeitlinger, ISBN 90 2651 852 8

Space nonlinear time-history analyzing for base-isolated Suqian city gymnasium

Weiqing Liu, Jun Dong, Shuguang Wang, & Zongjian Lan
Nanjing Architectural and Civil Engineering Institute, Nanjing 210009, China

Shigong Yin & Yougen Zhou
Jiangsu Architectural Design and Research Institute, Nanjing 210029, China

ABSTRACT: Suqian City Gymnasium (SCG) is situated in a region of 9 degree of fortification intensity, 4500 seats, about 13000m^2. The principal structure includes base-isolation layer (BIL), reinforced-concrete space frame, and steel spatial grid roof. BIL located between pile caps and the super-structure, mainly consisting of laminated rubber bearings and viscous dampers. By SAP2000N software and a nonlinear space time-history analyzing method, thorough investigations for the base-isolated system have been implemented. Important results are as follows: (1) The horizontal earthquake action upon the super-structure in base-isolated system can be greatly reduced, and the discount rate is small than 0.25; (2) Viscous dampers effectively restrict the BIL displacement meanwhile do not increase the earthquake action; (3) Technical and economic comparison demonstrates that base-isolated SCG possess evident social and economic benefit.

1 INTRODUCTION

Suqian City Gymnasium (SCG) locates at Suqian City, Jiangsu province, China, where is a region of 9 degree of fortification intensity (R9DFI). The building includes 4500 seats, about 13000 m^2. The principal structure is a reinforced concrete space frame, which exhibits an elliptic plan form, 80 m length, 60 m width, maximum 23.6 m height, and the roof is a steel spatial grid. The main plan is shown in Figure 1.

The principal structure is symmetric in the structural stiffness and mass distribution express obvious non-homogenization. The building site attributes to R9DFI and the predominant period is 0.34s. Owing to lack of engineering experience of constructing large gymnasium in R9DFI by traditional structural techniques but lot of difficulty (Housner et al. 1997, Tang & Liu 1997, Zhou 1997), we consider using base isolation techniques in SCG.

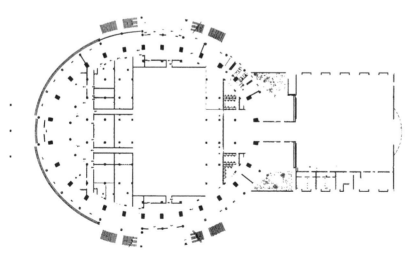

Figure1. Plan of Suqian city gymnasium

2 BASE ISOLATION DESIGN GOAL, SCHEME AND COMPONENT PARAMETERS

According to the principle of leading technique incorporating with economy, the base isolation design goal is set up to reduce the earthquake action upon the super-structure in base-isolated system to lower than 25% of the value in the counterpart base fixed system.

Table 1. Optimal parameters of bearings

Type	LS5	S5	S6
D mm	510	510	620
R_v kN	2700	2500	4000
K_v kN/mm	1800	1500	2000
d_k mm	100	100	130
d_{max} mm	275	275	299
F_y kN	40	/	/
Number	36	76	10
50% shear strain			
K_h kN/mm	1.7	0.75	1.0
$K_h^{'}$ kN/mm	0.85	/	/
ς %	25	4	4
100% shear strain			
K_h kN/mm			1.0
$K_h^{'}$ kN/mm			/
ς %			4
250% shear strain			
K_h kN/mm	0.9	0.7	
$K_h^{'}$ kN/mm	0.75	/	
ς %	12	3	

The isolation system in SCG is constituted of laminated rubber bearings and viscous dampers. Laminated rubber bearings are installed at the bottom of every grounded column. The elevations of top of all bearings are the same. In order to ensure the isolation effect and maintainability, the equipment channels are expanded to form an integral isolation layer. There are 28 large columns in the principal frame, in which 18 columns are installed two 500 mm diameter bearings with lead core (LS5) at each of the bottoms, and the other 10 columns are installed one 600 mm diameter bearings without lead core (S6) at each bottoms. One 500 mm diameter bearing without lead core (S5) is installed at each bottom of other 76 grounded small columns. 12 high-property viscous dampers are arranged at each bottom of 12 of the 28 large columns. The layout of bearings and dampers is shown in Figure 2.

Table 2. Sections of main components

Super structure				
Height/width or diameter /mm				
C1	C2	C3	C4	C5
1500/ 1000	600/4 00	450/4 50	Φ800	Φ600
B1	B2	B3	B4	B5
800/ 600	600/ 400	600/ 300	500/ 300	450/ 250
Ground floor				
Height/width /mm				
GB1	GB2	GB3	GB4	GB5
800/ 600	800/ 450	700/ 500	700/ 400	600/ 400

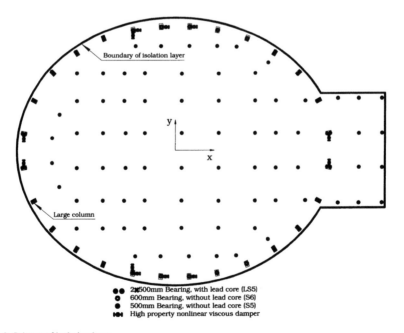

Figure 2. Scheme of isolation layer

Optimal parameters for laminated bearings are shown in Table 1. Optimal parameters for the viscous dampers are as follows: damping coefficient c = 55 kNs/mm, damping exponent ξ = 0.35, stroke ±300 mm, design maximum damping force 600 kN. Parameters of beams and columns are shown in Table 2.

3 ANALYSES OF BASE-ISOLATED SYSTEM

3.1 Analyzing method and software

Since SCG includes a complex space structure and the viscous dampers possess high degree of non-linearity, a nonlinear spatial finite element time-history analyzing method should be used to get reasonable results. Based on strict comparison we choose the SAP2000N (V7.10) as main analyzing tool. SAP2000N can be used conveniently to establish complex spatial model including nonlinear isolation bearings and viscous dampers, meanwhile the local nonlinear analysis strategy in SAP2000N ensures calculation efficiency, thus the integral nonlinear time-history analyzing becomes practical.

3.2 Nonlinear isolator element and viscous damper element

SAP2000 includes a biaxial hysteretic isolator element (Nagarajaiah et al. 1991, Wen 1976, Park et al. 1986) that has couple plasticity for the two shear deformations. The coupled force-deformation relationship is given by:

$$f_{u2} = \eta_2 \cdot k_2 \cdot d_{u2} + (1 - \eta_2) F_{y2} \cdot z_2$$
$$f_{u3} = \eta_3 \cdot k_3 \cdot d_{u3} + (1 - \eta_3) F_{y3} \cdot z_3$$

where k_2, k_3 are the elastic spring constant, F_{y2}, F_{y3} are the yield force, η_2, η_3 are the ratio of post-yield stiffnesses to elastic stiffnesses (k_2, k_3), z_2, z_3 are internal hysteretic variables. These variables have a range of $\sqrt{z_2^2 + z_3^2} \leq 1$, with the yield surface represented by $\sqrt{z_2^2 + z_3^2} = 1$. The initial values of z_2, z_3 are zero, and they evolve according to the differential equations:

$$\begin{Bmatrix} \dot{z}_2 \\ \dot{z}_3 \end{Bmatrix} = \begin{bmatrix} 1 - a_2 z_2^2, & -a_3 z_2 z_3 \\ -a_2 z_2 z_3, & 1 - a_3 z_3^2 \end{bmatrix} \begin{Bmatrix} \frac{k_2}{\eta_2} \dot{d}_{u2} \\ \frac{k_3}{\eta_3} \dot{d}_{u3} \end{Bmatrix}$$

where $a_2 = \begin{cases} 1 & if \ \dot{d}_{u2} \cdot z_2 > 0 \\ 0 & otherwise \end{cases}$,

$a_3 = \begin{cases} 1 & if \ \dot{d}_{u3} \cdot z_3 > 0 \\ 0 & otherwise \end{cases}$

SAP2000N provides a nonlinear damper element based on the Maxwell model of viscoelasticity (Malvern 1969) having a nonlinear damper in series with a spring. The nonlinear force-deformation relationship is given by:

$$f = K d_K = c \cdot \dot{d}_c^{\xi}$$

where K is the spring constant, c is the damping coefficient, $\dot{d}_c$ is the deformation rate across the damper, ξ is the damping exponent (ξ must be positive, the practical range is between 0.2 to 2.0), d_k is the deformation across the spring. The spring and damping deformations sum to the total internal deformation: $d = d_k + d_e$

If pure damping behavior is desired, the effect of the spring can be negligible by making it sufficiently stiff.

3.3 Earthquake waves for time-history analyzing

According to <code for seismic design of building> (GB50011- xx, for trial design), two real site records and one artificial ground motion must be considered (GB50011-xx.for trial design, in Chinese) Since the site soil attribute to class II, the most common used earthquake records, EL-Centro NS and Taft NE are adopted. Artificial wave is synthesized from the local earthquake parameters. All waves are adjusted to satisfy the code requests. Waves and their spectra for intensity of frequently occurred earthquakes (IFOE) are in Figures 3 & 4.

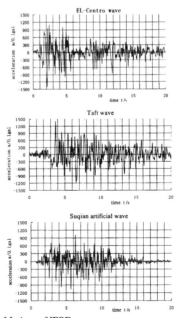

Figure 3. Motions of IFOE

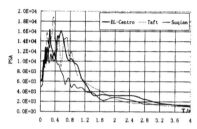

Figure 4. Spectra of motions of IFOE

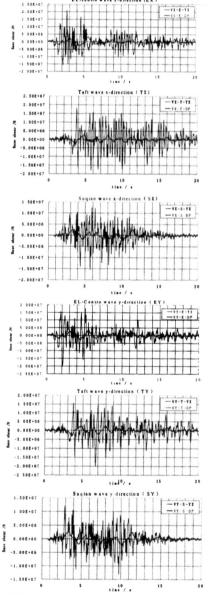

Figure 5. Layer shear forces of ground floor under motions of IFOE (YX: base-fixed/DP: base-isolated)

3.4 *Analyzing results*

A number of analyses have been carried out. The most important results of base-fixed system and optimal base-isolated system are shown in Table 3, 4; Layer shear forces of the ground floor under excitations of IFOE are contrasted in Figure 5; Displacements of the isolation layer under excitations of intensity of seldom occurred earthquakes (ISOE) See Figure 6.

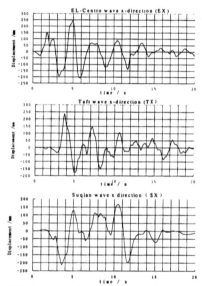

Figure 6. Displacement of isolation layer under ISOE

Table 3. Important results: base fixed system

Periods	T1	T2	T3	T4		
	0.5562	0.4650	0.3425	0.3344		
Maximum responses under motions of IFOE						
Waves	V_x^i	V_y^i	a_x^i	a_y^i	d_x^i	d_y^i
	kN	kN	0.1gal	0.1gal	mm	mm
EL-Centro	21140	20080	4280	5605	18	32
Taft	18200	19940	5237	6263	19	35
Suqian	15640	13300	4250	4625	16	29
Average max	V	a	d	V_{base}	α_i	
	kN	0.1gal	mm	kN		
	18063	5044	25	18063	0.160	
Maximum responses under motions of ISOE						
Waves	V_x^i	V_y^i	a_x^i	a_y^i	d_x^i	d_y^i
	kN	kN	0.1gal	0.1gal	mm	mm
EL-Centro	88217	90874	18954	24822	80	142
Taft	84763	94417	23192	27745	84	188
Suqian	64701	56996	18821	20482	71	128
Average max	V	a	d	V_{base}	α_i	
	kN	0.1gal	mm	kN		
	79993	22338	111	79993	0.705	

Table 4. Important results: base isolated system

Periods	T1	T2	T3	T4
	0.5562	0.4650	0.3425	0.3344

Maximum responses under motions of IFOE						
Waves	V_x^i kN	V_y^i kN	a_x^i 0.1gal	a_y^i 0.1gal	d_x^i mm	d_y^i mm
EL-Centro	3391	3387	601	423	51	51
Taft	2900	3035	511	418	43	44
Suqian	2609	2560	527	358	36	37

Average maximum						
V kN	a 0.1gal	d mm	V_{base} kN	α_i	σ^i_{Bearng} MP	F_{damper} kN
2980	473	44	4434	0.026	<14.4	301

Maximum responses under motions of ISOE						
Waves	V_x^i kN	V_y^i kN	a_x^i 0.1gal	a_y^i 0.1gal	d_x^i mm	d_y^i mm
EL-Centro	13540	13320	1414	1418	230	244
Taft	13230	13140	1524	1352	222	226
Suqian	12240	11390	1475	1310	197	204

Average maximum						
V kN	a 0.1gal	d mm	V_{base} kN	α_i	σ^i_{Bearng} MP	F_{damper} kN
12810	1415	220	18853	0.112	<14.4	568

Summing up results can infer conclusions as follows: (1) Base-isolated structural system greatly reduce the horizontal seismic action upon super structure, the related discount rate small than 0.25; (2) Large-capacity viscous dampers behavior very excellent and can be effectively used to resolve the contradiction between reducing earthquake actions and restricting isolation layer displacements; (3) Integral spatial models are necessary and practical; (3) Nonlinear biaxial hysteretic isolator elements and nonlinear damper elements can reasonably express characteristic of a isolation layer and more reliable results than from the simplified methods can be obtained; (4) Technical and economic comparison demonstrates that base-isolated SCG possess evident social and economic benefit.

4 CONCLUSIONS

Careful studies prove that it is suitable adopting base-isolation techniques in SCG. High property viscous dampers are very effective to enhance the capability of isolated system. The nonlinear spatial finite element time-history analyzing method is not only necessary but also practical. In the whole, a united of leading techniques and economy is implemented in SCG.

REFERENCES

G. W Housner, L A Bergman, T K Caughey, et al. *Structural control: Past, present, and future* [J]. *J. Engng. Mech*, 1997, 123(9), 897–971.

J. X. Tang, Z. H. Liu, *Base isolations of building structures* (M). Taibei, Shuxin Press, 1997 (in Chinese).

F. L. Zhou, *Vibration-reduced controls of engineering structures.* (M). Beijing: Seismic Press, 1997 (in Chinese).

S. Nagarajaiah, A. M. Reinhorn, M. C. Constantinou. *3D-Basis: Nonlinear Dynamic Analysis of Three Dimensional Base Isolated Structures: Part II* [R], Technical Report NCEER-91-0005, National Center for Earthquake Engineering Research, State University of New York at Buffalo, Buffalo, N. Y., 1991.

Y. K. Wen. *Method for Random Vibration of Hysteretic Systems* [J], *Journal of the Engineering, Mechanics Division*, ASCE, 1976, 102(EM2).

Y. J. Park, Y. K. Wen, A. H-S Ang. *Random Vibration of Hysteretic Systems under Bi-Directional Ground Motions* [J], *Earthquake Engineering and Structural Dynamics*, 1986, 14.

Malvern L E. *Introduction to the Mechanics of a Continuous Medium* [M], Engle-wood Cliffs: N.J. Prentice-Hall, 1969.

GB50011-xx. *Code for seismic design of buildings.* (S) (for trial design, in Chinese).

Earthquake Engineering Frontiers in the New Millennium, Spencer & Hu (eds),
© 2001 Swets & Zeitlinger, ISBN 90 2651 852 8

High performance, cost effective structural systems for seismic-resistant buildings

L.W. Lu, R. Sause, J.M. Ricles, & S.P. Pessiki
Lehigh University, Bethlehem, Pennsylvania, USA

ABSTRACT: Presented in this paper are brief descriptions of three structural systems which were developed recently for construction of buildings in seismic regions. The first is a steel frame system, the second a precast concrete frame system and the third a precast concrete wall system. These systems utilize the concept of post tensioning and have the unique ability to control damage and to reduce permanent structural distortion after a severe earthquake. Initial experiences of field applications and limited cost studies have shown that the new systems may even have cost advantages.

1 INTRODUCTION

The observations and analyses made after the major recent earthquakes (e.g. the 1994 Northridge earthquake, 1995 Kobe earthquake, and 1999 Chi-Chi earthquake) have shown that serious deficiencies exist in certain structural systems commonly used in construction of buildings in seismic regions. These studies also indicate to the engineering profession, as well as the general public, that the level of performance and quality of construction of some of the so-called seismic-resistant buildings need to be substantially improved. Extensive work has been carried out to develop high-performance, cost effective structural systems, utilizing the concept of post-tensioning. These systems have the ability to control damage and to reduce (or even eliminate) permanent structural distortion, after the building is subjected to a severe earthquake. Unlike the presently used systems, they do not depend solely on inelastic deformation of the individual members and connections for hysteretic energy absorption.

The post-tensioned systems to be described are: (1) a steel (or composite) frame system with post-tensioned moment connections, (2) a precast concrete frame system with unbonded post-tensioned moment connections, and (3) a precast concrete wall system with unbonded post-tensioned tendons.

2 POST-TENSIONED STEEL FRAME SYSTEM

This system utilizes a connection consisting of bolted top and seat angles and post-tensioning strands. The strands run through the column and are anchored outside the connection region. The details of the connection are shown in Figure 1. The con-nection avoids the use of field welding, reduces beam damage, and results in a relative small amount of permanent distortion (or drift) following an earthquake.

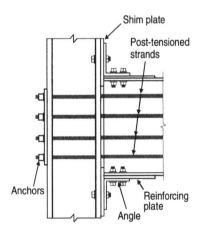

Figure 1. Post-tensioned steel connection

The flexure behavior of the connection is characterized by a gap opening and closing at the beam-column interface under cyclic loading. An idealized moment-rotation (M-θ_r) relationship for a PT steel connection is shown in Figure 2 where θ_r is the relative rotation that develops between the beam and column when gap opening occurs. Under the action of the initial post-tensioning force, contact stresses develop at the interface of the beam and column. When a moment is applied to the beam, the initial stiffness of the connection is the same as that of a welded moment connection (i.e., fully-restrained). That is, θ_r is zero until the gap opens as the beam tension flange separates from the column face. The

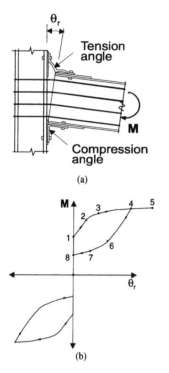

(a)

(b)

Figure 2. Cyclic moment-rotation behavior of a post-tensioned steel connection

moment at gap opening is the decompression moment. Decompression is identified as event 1. The stiffness of the connection immediately after decompression occurs is associated with the elastic stiffness of the angles and the post-tensioning strands. With continued loading, the angles will yield (event 2). Full plastic yielding of the angles occurs at event 3.

The post-tensioning strands will eventually yield at event 5 if loading is continued. The M-θ_r relationship is linear between events 3 and 5; where the stiffness is associated with the elastic stiffness of the strands and the post-elastic stiffness of the angles. If unloading occurs at event 4, the angles will dissipate energy (between events 4 and 8) until the gap between the beam flange and the column face is closed at event 8 (i.e., when θ_r is equal to zero). A reversal in yielding of the angles (beginning at event 6) is needed to close the gap. A complete reversal in moment will result in similar connection behavior in the opposite direction of loading.

An extensive program of investigations has been conducted to study experimentally the cyclic behavior of this connection and to develop analytical models that can be used to predict the performance of the connection and seismic response of overall structural frames. This work has been summarized by Garlock et al (2000) and by Ricles et al (2000).

3 POST-TENSIONED PRECAST CONCRETE FRAME SYSTEM

In this system, the precast beams and columns are joined together by post-tensioning the embedded steel, which is left unbonded through the column and through portions of the beams. Figure 3 shows a beam-and-column subassemblage with a post-tensioned connection. The flexural behavior of connection is characterized by gap opening/closing at the beam-column interface upon loading/unload-ing. Unlike a cast-in-place connection, the inelastic deformations are concentrated in the connection region where a "crack" already exists between the beam and column. Furthermore, because the post-tensioning steel is unbonded, no additional flexural cracks will form in the beams in the connection region. The unbonded length can be selected to allow the lateral displacement demand of the design level ground motion to be reached without yielding of the post-tensioning steel. Consequently, the prestressing force can be maintained through the loading/unloading cycles. A wide gap is expected at the beam-column interface, and the associated concrete compression strains near the gap are likely to be large. Therefore, spiral reinforcement is necessary to confine the concrete.

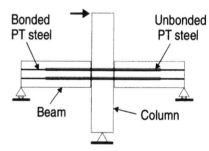

Figure 3. Unbond post-tensioned precast concrete subassemblage

The cyclic lateral load vs. deflection response of a post-tensioned subassemblage is shown schematically in Figure 4. The response provides very limited hysteretic energy dissipation; but the structure has the unique ability to self center. The behavior can therefore be considered as essentially non-linear elastic. Models which can provide accurate prediction of the behavior of the connection have been developed and incorporated into a non-linear structural analysis program. For design, a trilinear relationship, defined by three limit states, the linear state, the yield limit state, and the ultimate limit state, can be used to define the moment-rotation characteristics of the connection. Detailed analyses and design procedures for this system have been presented in papers by El-Sheikh et al (1999, 2000).

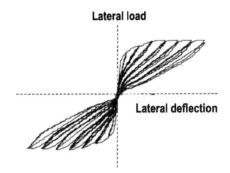

Lateral load

Lateral deflection

Figure 4. Lateral load-deflection relationship of subassemblage

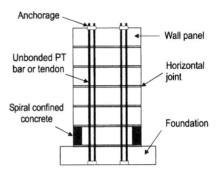

Figure 5. Unbonded post-tensioned walls

This system has been adopted recently in two construction projects: a 4-story building in Los Angeles and a 39-story building in San Francisco (Shuster, 2000). The 4-story building is essentially complete. The connections used in the building are the hybrid type, which include both bonded mild steel reinforcement and unbonded post-tensioning cables. The post-tensioning of all the connections at each floor level was carried out in one operation, resulting in substantial savings in construction cost. It has been reported that for the 39-story building a total of $4 million to $5 million could be saved by using the hybrid post-tensioned system.

4 POST-TENSIONED PRECAST CONCRETE WALL SYSTEM

An unbonded post-tensioned wall is constructed by post-tensioning wall panels across horizontal joints at the floor levels using post-tensioning steel, which is not bonded to the concrete (Figure 4). The behavior of the wall is very different from that of a cast-in-place wall. The lateral load resistance is provided by the post-tensioning steel (bars or tendons), located inside ducts which are not grouted. Spiral reinforcing steel is used to confine the concrete in the wall panel near the base of the wall. Wire mesh is used as bonded reinforcement in the panels.

The behavior of an unbonded post-tensioned wall under lateral load is governed by the behavior along the horizontal joints. Figure 5 shows the two types of behavior that can occur along the joints, namely, gap opening and shear slip. In the case of gap opening, the post-tensioning force and axial force due to gravity load provide a restoring force that tends to close the gaps upon unloading. In the case of shear slip, however, there is no restoring force to reverse the slip. Shear slip should therefore, be prevented by proper design and detailing of the wall.

The base shear (equal to the sum of all the lateral loads) vs. roof drift,), relationship of the wall is shown in Figure 7. This relationship can be established by

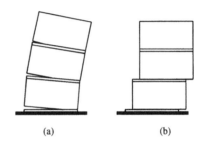

(a)   (b)

Figure 6. Behavior along horizontal joints: (a) Gap Opening and (b) Shear Slip

considering the axial-flexural behavior (i.e., behavior under combined axial force and flexure) and gap opening. As the wall displaces, it goes through four limit states. The first is the decompression state, which represents the initiation of gap opening along the horizontal joint between the wall and the foundation represents the initiation of gap opening along the horizontal joint between the wall and the foundation and is the beginning of non-linear behavior. However, the effect of this non-linear behavior on the lateral stiffness of the wall is small until the gap opening extends over a significant portion of the joint. The second is the softening state, which signifies the beginning of an appreciable reduction in the lateral stiffness of the wall due to gap opening and non-linear behavior of the concrete in compression. The reduction in the lateral stiffness of the wall occurs in a smooth and continuous manner. Therefore, an effective linear limit is used to identify this state. The third limit state is related to the beginning of yielding of the post-tensioning steel. A properly designed wall does not reach the yielding state until a large non-linear drift has occurred. The final state is the failure or ultimate state when axial-flexural failure of the wall occurs as a result of crushing of the spiral confined concrete. Sufficient spiral reinforcement is provided in the wall panels such that the failure state is reached at a drift significantly larger than the drift at the yielding state.

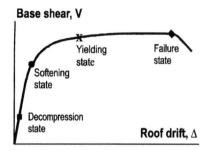

Figure 7. Base shear vs. roof drift relationship of unbonded post-tensioned wall

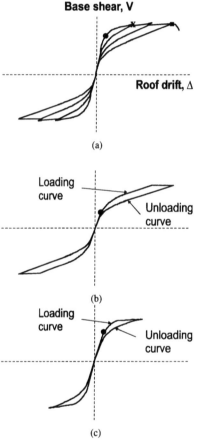

(a)

(b)

(c)

Figure 8. Hysteretic behavior of wall under lateral load: (a) entire behavior, (b) loading cycle just reaching yielding state, and (c) loading cycle beyond yielding state

The behavior of the wall under cyclic lateral load is illustrated in Figure 8. Figure 8(b) shows the loading and unloading behavior during a load cycle with a drift equal to yielding state drift. Figure 8(c) shows the behavior during a subsequent cycle with a

maximum drift between the yielding state drift and the failure state drift. The hysteresis loops indicate that the behavior of the wall is nearly non-linear elastic, characterized by loading and unloading curves that are very close to each other. This behavior results in a self-centering capability.

Procedures have been developed for the analysis and design of this system and are described by Kurama et al (1999a, 1999b).

5 CONCLUSIONS

Three seismic-resistant structural systems, one for steel construction, two for precast concrete construction, have been described. These systems utilize the post-tensioning concept and their performance under lateral load (seismic) has been found to be superior to the conventional systems. When properly designed and detailed the systems have well defined limit states. The level of structural damage corresponding to each limit state can be controlled in design. The overall structure will have a much reduced permanent lateral displacement (sway) after a severe earthquake because of the self-centering capability provided by the post-tensioning force. Construction experience, limited at the present time, and prototype structural cost studies seem to indicate that the costs of these systems are competitive, when compared with the costs of some of the conventional systems.

REFERENCES

El-Sheikh, M.T., Sause, R., Pessiki, S., and Lu, L.W., 1999. Seismic behavior and design of unbonded post-tensioned precast concrete frames. *PCI Journal*. Vol. 44, No. 3, May-June, pp. 54–71.

El-Sheikh, M.T., Pessiki, S., Sause, R., and Lu, L.W., 2000. Moment-rotation behavior on unbonded post-tensioned precast concrete beam-column connections. *ACI Structural Journal*. Vol. 97, No. 1, Jan-Feb., pp. 122–131.

Garlock, M.M., Ricles, J.M., Sause, R., Zhao, C., and Lu, L.W., 2000. Seismic behavior of post-tensioned steel frames. *Proc. Conference on Behavior of Steel Structures in Seismic Areas*. Montreal, pp. 593–599.

Kurama, Y., Pessiki, S., Sause, R., and Lu, L.W., 1999a, Seismic behavior and design of unbonded post-tensioned precast concrete walls. *PCI Journal*. Vol. 44, No. 3, May-June, pp. 72–89.

Kurama, Y., Sause, R., Pessiki, S., and Lu, L.W., 1999b, Lateral load behavior and seismic design of unbonded post-tensioned precast concrete walls. *ACI Structural Journal*. Vol. 96, No. 4, July-August, pp. 622–632.

Ricles, J.M., Sause, R., Garlock, M.M., Peng, S.W., and Lu, L.W., 2000. Experimental studies on post-tensioned seismic resistant connections for steel frames. *Proc. Conference on Behavior of Steel Structures in Seismic Areas*, Montreal, pp. 231–238.

Shuster, L.A., 2000. Keeping it together, Civil Engineering, ASCE, March, pp. 44–47.

Earthquake Engineering Frontiers in the New Millennium, Spencer & Hu (eds),
© 2001 Swets & Zeitlinger, ISBN 90 2651 852 8

The optimal fortification load and reliability of aseismic structures

Guangyuan Wang & Dagang Lu
School of Civil Engineering, Harbin Institute of Technology, Harbin, 150090, China

ABSTRACT: A two-stage optimum design method for aseismic structures is presented, in which the decision of the optimal fortification load is made in the first stage, and the minimum-cost design under the optimal fortification load is made in the second stage. A more scientific objective function of decision-making for the optimal fortification load of aseismic structures is presented by taking the fortification intensity as the typical representative of the seismic fortification load. The objective can consider both the short-term investments, that is, structural cost, and the long-term benefits, that is, the loss expectation under earthquakes. Two methods for solving the minimum-cost function are presented, which include the precise method and the simplified one. A macroscopic analysis method for calculating the fuzzy probabilities of seismic damaged structures is developed according to the provisions of the current design codes and the three-level fortification principal of aseismic structures. On the basis of these, the decision-making method and the corresponding procedure for the optimal fortification load of aseismic structures are put forward. A unified relationship between the optimal fortification load and the optimal fortification reliability is deduced, and then, the method of decision-making of the optimal fortification reliability based on the optimal fortification load is advanced.

1 INTRODUCTION

The seismic fortification level is directly related to the seismic performance of structures, and so, how to determine the optimal fortification level for aseismic structures plays an important role in the theory of performance-based seismic design (Bertero V.V., 1996). There are two kinds of the optimal fortification level of aseismic structures, which include the optimal fortification load and the optimal fortification reliability (WANG Guangyuan, 1992, 1999). At present, the determination of the optimal fortification load is basically made based on the subjective experience of the leaders and the code-makers, and this method is still in the stage of low-grade decision-making owing to lack of quantitative analysis. Meanwhile, it is also difficult to decide the optimal fortification reliability through the theory of structural optimization. A rational fortification level should be determined by rational optimization procedure based on the comprehensive consideration of both the current investments and the loss expectation damaged by the earthquakes during the design reference period. In this paper, a scientific method and procedure of decision-making of the optimal fortification load for aseismic structures will be presented. And then, a unified relationship between the optimal fortification load and the optimal fortification reliability will be given, therefore, the optimal fortifica-

tion reliability can be determined through the decision-making of the optimal fortification load.

2 TWO-STAGE OPTIMUM DESIGN METHOD OF ASEISMIC STRUCTURES

In this paper, the fortification intensity I_d is taken as the fortification load. In the variable design phase of structures, the structural design scheme, that is, the design vector $\bar{x}$, can be denoted as the function $\bar{x}(I_d)$ of the fortification intensity I_d. The optimum design process of aseismic structures can be divided into the following two stages:

2.1 *Stage 1: Decision-making for the optimal fortification load of aseismic structures*

Determine the optimal fortification load of aseismic structures according to the following optimization model:

$$\begin{cases} Find & I_d \\ \min & W[\bar{x}(I_d)] = C_{\min}[\bar{x}(I_d)] + \theta L[\bar{x}(I_d)] \end{cases} \quad (1)$$

where, $\bar{x}(I_d)$ represents the minimum-cost design scheme under the fortification intensity I_d, $C_{min}[\bar{x}(I_d)]$ denotes its minimum-cost, and $L(\bar{x})$ denotes the loss expectation caused by the structural failure under the strong earthquake action; the adjusting parameter θ is a coefficient defined when considering the different importance of the cost C and the loss expectation L. In general, we can let $\theta=1$ if there are no specific requirements.

In the objective function of model (1), the minimum-cost $C_{min}[\bar{x}(I_d)]$ is an increasing function of the fortification intensity I_d, while the loss expectation $L[\bar{x}(I_d)]$ is a decreasing one of I_d, so the curve $W[\bar{x}(I_d)]$ generated by the sum of these two functions has a lowest point. The fortification intensity I_d corresponding to this point is the optimal fortification intensity I_d that we want to obtain (shown as in Fig.1).

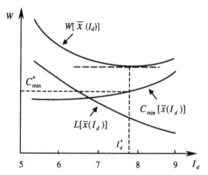

Figure1. Decision-making for the optimal fortification load

The decision-making of the optimal fortification load of aseismic structures includes the following two main aspects:
(1) Find the function relationship $C_{min}[\bar{x}(I_d)]$ of the structural minimum-cost C_{min} with the fortification load I_d.
(2) Find the function relationship $L[\bar{x}(I_d)]$ of the loss expectation L with the fortification load I_d.

2.2 Stage2: Optimum design under the optimal fortification load

After the optimal fortification load I_d^* has been obtained, the minimum-cost design can be made under I_d^*, that is, find the design scheme $\bar{x}(I_d^*)$, so as to make the structural cost

$$C[\bar{x}(I_d^*)] \rightarrow min \qquad (2)$$

subjected to all constraints and requirements of codes.

The final result is the optimal design scheme in consideration of the loss expectation $L(\bar{x})$. The loss expectation has been taken into consideration when deciding I_d^*, so in this stage it is only necessary to counteract the resistance I_d^* by the minimum-cost design scheme.

3 THE OPTIMAL FORTIFICATION LOAD OF ASEISMIC STRUCTURES

3.1 Two methods for finding the curves of the minimum-cost

There are two methods for finding the curves $C_{min}[\bar{x}(I_d)]$ of the structural minimum-cost, i.e. the function $C_{min}[\bar{x}(I_d)]$ as follows:

3.1.1 The precise method

Firstly, under the given fortification load I_d, find the design scheme $\bar{x}(I_d)$, so as to make the structural cost

$$C[\bar{x}(I_d)] \rightarrow min \qquad (3)$$

subjected to all constraints and requirements of codes.

Secondly, give some different fortification load I_d, and then find the corresponding minimum-cost C_{min}, thus, some points of the function $C_{min}[\bar{x}(I_d)]$ or the relationship $C_{min}[\bar{x}(I_d)]$–I_d can be obtained, and consequently, the minimum-cost curve C_{min}–I_d shown as in Fig.1 can be drawn by using the methods of statistical regression. The minimum-cost $C_{min}[\bar{x}(I_d)]$ increase with the increase of the fortification load I_d, that is to say, $C_{min}[\bar{x}(I_d)]$ is a monotonously increasing function.

3.1.2 The simplified method

During the optimum design process of aseismic structures, although the calculation of $C_{min}[\bar{x}(I_d)]$ does not bring about any more difficulties, it still needs a lot of workload. For saving workload, some empirical relationships of C–I_d can substitute the minimum-cost curve C_{min}–I_d. These empirical relationships can be statistically found based on the experience of design and calculation for some often used structural forms. Although the simplification method is approximate, it has the advantage of convenience. The reason is that the iteration calculation will be made until the structural cost is formulized as the analytical function $C(\bar{x})$ of the design vector $\bar{x}$ in the objective function (11) to find the minimum-cost $C_{min}[\bar{x}(I_d)]$, while it is difficult for the analytical function $C(\bar{x})$ to represent the real cost of structures. On the contrary, the simplified method can calculate the real cost $C_{min}(I_d)$ of structures through the routine methods such as budgetary estimate and budget, after the design scheme $\bar{x}$ is found according to the fortification load I_d. Therefore, this method is more suitable to the practical engineering problem.

3.2 A macroscopic analysis method for calculating the fuzzy probabilities of seismic damaged structures

In the current seismic code of China (GBJ11-89), a definite three-level seismic fortification requirement is proposed, that is, "Do not be damaged under minor earthquake, Be repaired under moderate earthquake, Do not collapse under major earthquake". The relative relations of the three-level intensity with the basic intensity are summed up as shown in Table1.

According to the requirements of the current seismic code, the grades of earthquake damaged engineering structures include five ones including intact, slight damage, moderate damage, severe damage and collapse. In fact, the seismic damage grade is a fuzzy concept, because it is difficult to completely separate the different damage grades by some specific value of structural responses. In this paper, we denote $\tilde{B}_i$ as the ith fuzzy seismic damage grade, that is to say, the universe set of the seismic damage grade is:

$$[\tilde{B}_1, \tilde{B}_2, \tilde{B}_3, \tilde{B}_4, \tilde{B}_5] = \qquad (4)$$

[intact, slight damage, moderate damage, severe damage, collapse]

If the symbol $\tilde{B}_i^*$ represents the fuzzy seismic damage grade greater than the grade of $\tilde{B}_i$, then $\tilde{B}_i^*$ is the fuzzy failure domain of structures. Under the multi-level failure criteria of "Do not be damaged under minor earthquake, Be repaired under moderate earthquake, Do not collapse under major earthquake", the fuzzy failure domain $\tilde{B}_i^*$ is related to the failure criteria including "Do not be damaged", "Be repaired" and "Do not collapse". For aseismic structures, it can be approximately thought that:

"Do not be damaged" means that structures do not have a seismic damage grade greater than $\tilde{B}_1$, that is to say, the fuzzy failure domain is $\tilde{B}_1^*$;

"Be repaired" means that structures do not have a seismic damage grade greater than $\tilde{B}_3$, that is to say, the fuzzy failure domain is $\tilde{B}_3^*$;

"Do not collapse" means that structures do not have a seismic damage grade greater than $\tilde{B}_4$, that is to say, the fuzzy failure domain is $\tilde{B}_4^*$.

Table 1. Relationship of three-level seismic levels with the basic intensity (the design reference period is 50 years)

seismic levels	minor earthquake	moderate earthquake	major earthquake
exceedance probability in 50 years	0.632	0.10	0.02-0.03
relationship with the basic intensity	1.55 degree lower than the basic intensity	equals the basic intensity	about 1 degree higher than the basic intensity
design requirement	do not be damaged	be repaired	do not collapse

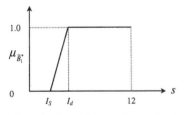

(a) Do not be damaged under minor earthquake

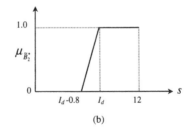

(b)

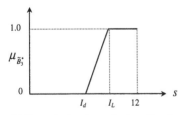

(c) Be repaired under moderate earthquake

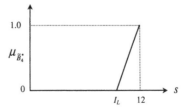

(d) Do not collapse under major earthquake

Figure 2. Membership function curve of fuzzy failure domain

According to the regulations of the multi-level fortification criteria as shown in Table 1, the membership function curves of all kinds of fuzzy failure domains of structures can be given as shown in Fig.2. Fig.2 (a) represents "Do not be damaged under minor earthquake", Fig.2 (c) represents "Be repaired under moderate earthquake", and Fig.2 (d) represents "Do not collapse under major earthquake". The membership function curve shown in Fig.2 (b) is approximately drawn out between Fig.2 (a) and Fig.2 (c).

If the probability density function $f_I(s)$ of the maximum seismic intensity during the 50-year reference period has been obtained through the seismic risk analysis, then the fuzzy probability of different failure level of the design scheme $\bar{x}(I_d)$ can be calculated by the following equation:

$$P_f[\tilde{B}_i^*, \bar{x}(I_d)] = \int_0^{12} \mu_{\tilde{B}_i^*} \cdot f_I(s)\, ds \qquad (5)$$

The fuzzy probability of seismic damage grade $\tilde{B}_i$ of the design scheme $\bar{x}(I_d)$ can be solved by the following formula:

$$\left.\begin{aligned}
P_f[\tilde{B}_1, \bar{x}(I_d)] &= 1 - P_f[\tilde{B}_1^*, \bar{x}(I_d)] \\
P_f[\tilde{B}_i, \bar{x}(I_d)] &= P_f[\tilde{B}_{i-1}^*, \bar{x}(I_d)] - P_f[\tilde{B}_i^*, \bar{x}(I_d)] \quad (i=2,3,4) \\
P_f[\tilde{B}_5, \bar{x}(I_d)] &= P_f[\tilde{B}_4^*, \bar{x}(I_d)]
\end{aligned}\right\} \quad (6)$$

This kind of probability is only related to two parameters, that is, the fortification intensity I_d and the seismic damage grade $\tilde{B}_i$. This is completely reasonable, because once the site has been selected, the future environment disturbance or the seismic action has been determined, so the failure probability of different grades will be dependent on the $\tilde{B}_i$ and the overall resistance of structural design scheme $\bar{x}(I_d)$, that is, the fortification intensity I_d. Therefore, this is a kind of macroscopic analysis method. According to the fuzzy probability curve shown as in Fig.2 and formula (5) and (6), the fuzzy probability $P_f[\tilde{B}_i, \bar{x}(I_d)]$ $(i=1,2,\cdots,5)$ can be finally solved.

The loss of structures with seismic damage grade $\tilde{B}_i$ can be expressed as follows:

$$D_i = D_i^{(1)} + D_i^{(2)} \quad (i=1,\cdots,5) \qquad (7)$$

where, $D_i^{(1)}$ represents the direct loss of structures, while $D_i^{(2)}$ is the indirect loss caused by the structural seismic damage.

The loss D_i must be evaluated according to the specific situation of structures and the seismic damage grades. The total loss expectation value under multi-level failure criteria can be obtained by the following formula:

$$L[\bar{x}(I_d)] = \sum_{i=1}^{5} P_f[\tilde{B}_i, \bar{x}(I_d)] \cdot D_i \qquad (8)$$

Therefore, given some different fortification intensity I_d, the loss expectation curve L-I_d shown as in Fig.1 can be obtained with respect to the specific problem. The loss expectation curve L-I_d decreases as the fortification intensity I_d increases, that is to say, $L[\bar{x}(I_d)]$ is a monotonously decreasing function of I_d.

3.3 Decision-making of the optimal fortification load

Since we have the two curves C_{min}-I_d and L-I_d shown in Fig.1, the optimal fortification intensity I_d^* of aseismic structures can be obtained with the schematic method shown in Fig.1, or with the unconstraint minimum optimization method according to the following objective function:

$$W(I_d) = C_{min}(I_d) + L(I_d) \to \min \qquad (9)$$

The solution to formula (9) is the optimal fortification load I_d^*, it is corresponding to the minimum point of the curve $W[\bar{x}(I_d)]$. Then, the minimum-cost design can be made under this intensity I_d^*.

4 THE OPTIMAL FORTIFICATION RELIABILITY OF ASEISMIC STRUCTURES

4.1 The unified relationship between the optimal fortification load and the optimal fortification reliability

Load Roughness Index (LRI) is a non-dimension parameter to calculate the degree of relative dispersion between structural resistance and load effect. It is defined as follows:

$$LRI = \frac{\sigma_S}{\sqrt{\sigma_S^2 + \sigma_R^2}} \qquad (10)$$

where, σ_S and σ_R are the standard deviations of load effect and structural resistance.

When $\sigma_S \ll \sigma_R$, the load is called Smooth Load because the value of LRI is very small; on the contrary, when $\sigma_S \gg \sigma_R$, the load is called Rough Load because the value of LRI is very large. When the values of σ_S and σ_R are at the same scale, the load is called General Rough Load. As two especial examples, when $LRI=0$, the load is called Infinite Smooth Load; on the contrary, when $LRI=1$, the load is called Infinite Rough Load. The research results show that the non-disaster load is General Rough Load, while the disaster load can be approximately regarded as Infinite Rough Load because the Load Roughness Index of the disaster load is close to one.

374

Suppose that both load effect S and structural resistance R all satisfy the norm distribution, and the limit state equation is

$$g(R, S) = R - S \qquad (11)$$

then the reliability index β of structural members is

$$\beta = \frac{\mu_R - \mu_S}{\sqrt{\sigma_R^2 + \sigma_S^2}} = \frac{\sigma_S}{\sqrt{\sigma_R^2 + \sigma_S^2}} \frac{\mu_R - \mu_S}{\sigma_S} = LRI \frac{\mu_R - \mu_S}{\sigma_S} \qquad (12)$$

For the Infinite Rough Load that satisfies $LRI = 1$, the resistance of structural members is a determined variable R, that is to say, $\sigma_R = 0$, therefore, the reliability index β takes the form as follows:

$$\beta = LRI \frac{\mu_R - \mu_S}{\sigma_S} = \frac{R - \mu_S}{\sigma_S} \qquad (13)$$

The above formula shows that the reliability index of structural members under the infinite rough load is mainly dependent on the statistical property of the random load effect.

Since the disaster load can be approximately dealt with as an infinite rough load, the reliability of structures under disaster loads is mainly dependent on the property of disaster loads, i.e., the failure of structures is mostly due to the excessive load.

On the condition of single failure criteria, the fortification load I_d is the overall resistance R of the structure during the stage of conceptual design, while the random seismic intensity R is the overall load effect of structures. According to the results of formula (13), the relationship between the reliability index β_d of structural system during the stage of conceptual design and the fortification load I_d can be obtained through the following equation:

$$\beta_d = \frac{I_d - \mu_I}{\sigma_I} \qquad (14)$$

From the formula (14) we can conclude that the reliability index β_d is corresponding to the fortification load I_d one by one. Therefore, β_d can be regarded as the fortification reliability index, or the target reliability index of structural system.

Because structural reliability Ψ and the reliability index β have the following relationship:

$$\Psi = \Phi(\beta) \qquad (15)$$

Therefore, there exists a unified relationship between the reliability index β_d and the fortification load I_d of structural system as follows:

$$\Psi_d = f(I_d) = \Phi(\frac{I_d - \mu_I}{\sigma_I}) \qquad (16)$$

4.2 Determine the optimal fortification reliability through the optimal fortification load

According to Eq. (16), the optimal fortification reliability Ψ_d^* can be obtained through the optimal fortification load I_d^*:

$$\Psi_d^* = f(I_d^*) = \Phi(\frac{I_d^* - \mu_I}{\sigma_I}) \qquad (17)$$

On the condition of the multi-level fortification criteria "Do not be damaged under minor earthquake, Be repaired under moderate earthquake, Do not collapse under major earthquake", the overall resistance of conceptual structures is relative to the three-level seismic fortification load including "minor earthquake I_S", "moderate earthquake I_M" and "major earthquake I_L". Correspondingly, aseismic structures have the following three-level fortification reliability:

$$\Psi_{Sd} = f(I_S) = \Phi(\frac{I_S - \mu_I}{\sigma_I}) = \Phi\left(\frac{(I_d - 1.55) - \mu_I}{\sigma_I}\right) \qquad (18)$$

$$\Psi_{Md} = f(I_M) = \Phi(\frac{I_M - \mu_I}{\sigma_I}) = \Phi\left(\frac{I_d - \mu_I}{\sigma_I}\right) \qquad (19)$$

$$\Psi_{Ld} = f(I_L) = \Phi(\frac{I_L - \mu_I}{\sigma_I}) = \Phi\left(\frac{(I_d + 1) - \mu_I}{\sigma_I}\right) \qquad (20)$$

After the optimal fortification load I_d^* is determined, the three-level optimal fortification reliability $\Psi_{Sd}^*, \Psi_{Md}^*, \Psi_{Ld}^*$ can be obtained through the formula (18)–(20).

From the above analysis, it can be concluded that the two kinds of the optimal fortification levels, i.e. the optimal fortification load I_d^* and the optimal fortification reliability Ψ_d^* are identical to each other. However, it is very difficult to directly find the optimal fortification reliability, on the contrary, it is relatively easy to obtain the optimal fortification reliability by using the above-mentioned method.

5 CONCLUSION

The method of decision-making of the optimal fortification load for aseismic structures has the following characteristics:

(1) It is practical to use and consistent with the national design codes.

(2) Both the cost and the loss expectation of aseismic structures are considered in the decision-making for the optimal fortification intensity.

(3) The multi-level failure criterion for aseismic structures is considered in the determination of the loss expectation.

The optimal fortification load and the optimal fortification reliability are identical to each other, and they have a unified function relationship. It is easy to determine the optimal fortification reliability by finding the optimal fortification load, while it is difficult to directly decide the optimal fortification reliability by the optimization method.

REFERENCES

Bertero V.V.: Overview of Seismic Risk Reduction in Urban Areas: Role, Importance, and Reliability of Current U.S. Seismic Codes —— Performance Based Seismic Engineering. *Proc. of China–United States Bilateral Workshop on Seismic Codes*, Guangzhou, China, 1996.

GBJ11-89: *Aseismic Design Code of Buildings*, Beijing, Building Engineering Press of China, 1990.

Wang Guangyuan: *Theory of Soft Design in Engineering*, Beijing, Science Press, 1992.

Wang Guangyuan, etc.: *The Optimal Fortification Intensity and Reliability of Aseismic Structures*, Beijing, Science Press, 1999.

Earthquake Engineering Frontiers in the New Millennium, Spencer & Hu (eds),
© 2001 Swets & Zeitlinger, ISBN 90 2651 852 8

Shaking table test study on a model of steel-concrete hybrid structure for tall buildings

G.Q. Li & X. Ding
Department of Structural Engineering, Tongji University, Shanghai, P. R. China

X.M. Zhou
Department of Civil Engineering, Hong Kong University of Science and Technology

ABSTRACT: Shaking table test on a 1:20 model of a typical steel-concrete hybrid structure of tall building was conducted successfully at Tongji University. The dynamic properties, seismic responses and damage features of the model were recorded and analyzed. The behavior of the model subjected to 3-D earthquakes is discussed in detail in this paper, which establishes the base for the earthquake-resistant design and analysis of hybrid structures, and for compiling the design code.

1 INTRODUCTION

With the advantages such as high speed of construction, strong stiffness and low cost, hybrid structures, composed of steel frames (SF) and concrete core-tube (CT), have been widely used for the (super) high-rise in China in recent years.

However, hybrid structures are rarely used in earthquake zone abroad. Because SF share only 2~5% of the total horizontal loads of the whole structure according to elastic analysis, it is generally considered that the aseismic behavior of hybrid structures depends mainly on concrete core-tubes, and is not so advantageous as that of steel structures. So it is necessary to conduct research on aseismic capability of hybrid structures.

The Structural Institute of China Academy of Building Science has accomplished a series of experimental studies on a 1:20 model of 23-story steel-concrete hybrid structure, including vertical load test, mechanic behavior test, dynamic property test[4]. Pulsation method, hammer driving method and initial displacement release method (free vibration method) are employed to attain its dynamic properties such as flexibility, frequency and damping. However, no shaking table test has ever been done on hybrid structures.

In this paper, through shaking table test on a reduced-scale (1:20) model of a typical high-rise hybrid structure, the influence of seismic wave frequency spectrums and earthquake intensity on it is explored; its aseismic behavior, seismic response characteristic and damage feature are obtained. Accordingly the experimental data and theoretical criteria for appraising the modeling and aseismic design of hybrid structure are provided.

2 DESIGN OF THE MODEL

Based on the structure of Technology Center Mansion in East Shanghai Shipyard, the test model adopted length similitude coefficient $S_l = 1/20$, density similitude coefficient $S_\rho = 4$. Because CT contributes mainly to the lateral stiffness of high-rise hybrid structure, the similitude coefficients of the whole structure are deduced from CT as follows (Table 1).

Table 1. Similitude coefficients of the hybrid structure model

Variable	Symbol	Value
Length	S_l	1/20
Time	S_t	1/8
Frequency	S_f	8
Velocity	S_v	0.4
Acceleration	S_a	3.2
Displacement	S_u	1/20
Density	S_ρ	4
Strain	S_ε	1
Stress	S_σ	0.64
Elastic modulus	S_E	0.64
Concentrated force	S_F	1/625
Face-distributed Mass	S_w	1/5

The 25-story test model is 5.0 m in height, with the plan dimension 1.3×2.0 m. The rectangular concrete core-tube is located in the center of the model with the plan dimension 0.885×0.465 m. The core-tube of the model is poured with fine graded aggregate concrete reinforced with galvanized iron wire. In the frame columns, three kinds of box-shaped sections are used, which are 30×30×2.2 mm, 30×30×1.75 mm and 30×30×1.5 mm. Two channel steels are welded back to back to form I-shaped frame beams. The plan and section of the model are demonstrated in figure 1 and figure 2.

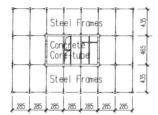

Figure 1. Plan of the model

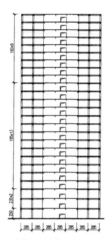

Figure 2. Elevation of the model

Figure 3. The model on the table

The steel frames adopt Q235 structural steel made in China as the same as the prototype. Twelve standard specimens of steel frames are divided equally into four groups for material test. As the results of the material test, the average yield strength f_y is 236.1 Mpa; the average ultimate strength f_u is 350.5 Mpa; the average percentage of elongation is 25.4%. Concrete grade C40 is applied in the core-tube of the

prototype. The actual strength for the model core-tube is 26.5 Mpa, obtained by tests.

To satisfy the requirement of density similitude, additional iron blocks are attached to all the floors, including dead loads and 50% live loads. The additional mass of each floor is 220 kg and the total additional mass of the model is 5.50 ton. The total mass of the test model is 13.839 ton.

3 PROCESS OF SHAKING TABLE TEST

3.1 Sensor arrangement and load plan

The test is conducted on the shaking table in the State Key Disaster-Prevention Lab of Tongji University. The test measures displacements, acceleration responses of the whole model and vertical dynamic strain at the bottom of core-tube and the frame columns. During the test, the cracking process and distribution of concrete core-tube, and the damage process of the whole model are observed and recorded; the data of the interaction between SF and CT are acquired.

There are 3 displacement transducers and 20 accelerometers and 8 strain sensors set on the model. At the sixth floor of the model, two displacement transducers in the direction of X-axis are arranged in the midpoint of the edge parallel to Y-axis, which are close to the same plan position to check the readings of each other. One displacement transducer is installed at the roof. Accelerometers are arranged as figure 4 and figure 5. At each corner of the bottom of the concrete core-tube is there 1 vertical strain sensor to measure the concrete stress Z-axis direction; on four columns of the bottom floor, 4 strain sensors are glued in Z-axis direction.

The prototype of the structure model is designed to resist 7-grade seismic loads. X-axis is the primary shaking direction. Three seismic waves are selected as the input waves in the test:

1) Artificial seismic wave of Shanghai (P)
2) 1940 EL-Centro wave (E)
3) San Fernando wave (S).

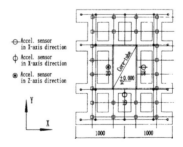

Figure 4. The accelerometers on the model base

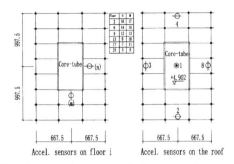

Figure 5. The accelerometers on the floors and roof

Table 2. Seismic load case of the model test

| Load | Acceleration peak-value[*] (g) | | | Earthquake |
case	X	Y	Z	intensity
1W	0.10	0.10	0.10	
2S	0.188	0.037	0.035	6 grade
3E	0.168	0.035	0.027	6 grade
4S	0.028	0.064	0.150	6 grade
5S	0.174	0.100	0.024	6 grade
6S	0.032	0.192	0.022	6 grade
7E	0.030	0.190	0.024	6 grade
8E	0.203	0.100	0.031	6 grade
9W	0.10	0.10	0.10	
10E	0.557	0.041	0.057	7 grade
11S	0.586	0.054	0.044	7 grade
12E	0.495	0.320	0.038	7 grade
13S	0.529	0.307	0.036	7 grade
14W	0.10	0.10	0.10	
15E	1.175	0.118	0.238	8 grade
16S	1.369	0.088	0.098	8 grade
17E	1.030	0.609	0.110	8 grade
18S	1.074	0.686	0.099	8 grade
19W	0.10	0.10	0.10	
20E	2.009	0.919	0.277	9 grade
21S	1.638	1.023	0.263	9 grade
22P	2.304	0.348	0.309	9 grade
23W	0.10	0.10	0.10	
24P	2.336	0.261	0.319	9 grade

When the model is loaded with an one-dimensional seismic wave, the larger peak-value component is chosen from the two horizontal directions. When the model is loaded with 2-D or 3-D seismic waves, the larger peak-value component is chosen in X-axis direction, and the real vertical wave component is still selected as the vertical wave component of the shaking table. Moreover, white noise table motion tests (W) are conducted to measure the dynamic property of the model in different phases of the test.

3.2 The damage process and feature of the model

Under 6-grade seismic loads, the model structure stays in elastic state. Under 7-grade seismic loads, minute cracks of the concrete wall at the bottom floor appear appreciably. Under 8-grade seismic loads, the cracks become more obvious (the maximum one reaches about 1.5mm) and the longitudinal reinforcements of the concealed columns begin

yielding and bulging outward. As the earthquake reaches 9-grade, many of them yield and bulge outward; the concrete at the bottom of the core-tube spalls, looses and crushes in large area as shown in figure 6. At this time, however, there are few apparent cracks in connecting beams and the walls except the bottom. Nevertheless, there are many concrete cracks around the embedded joint plates connecting the steel frame beams and the core-tube, especially at the corners of the core-tube (see figure 7). The reason is that there are considerably great internal forces in these joints under alternating loads because of the difference of stiffness, strength and ductility between SF and CT. The anchorage of the embedded joint plates is damaged by the alternative tension and compression applied on these joints, which carry the task of behaving coordination of SF and CT.

Under 9-grade seismic loads, the maximum dynamic axial stress of steel columns reaches yield stress; the maximum dynamic normal strain of concrete core-tube is not more than 0.0015, but most concrete at the bottom of the walls crushed due to the alternative normal stress. After 9-grade artificial seismic loads being applied, white noise shaking table test is done to diagnose the serving status of the model. Later as another 9-grade artificial seismic load is acted, the damage of the model becomes more serious, concentrating on the previously damaged parts while there are few newly damaged parts appearing. The displacement of the roof increases up to 122mm, about 1/40 of the total height of the model. However, the steel frames sustain in good

Figure 6. The concrete damages at the bottom of the core-tube

Figure 7. The concrete damages at the joints of SF & CT

379

condition and play a role of supporting the core-tube to serve, keeping the integrity of the whole structure, and preventing it from collapsing. The phenomenon tells that hybrid structure has good deformation ability and ductility. The test shows us that the failure of the model is focused mainly on the concrete core-tube at the bottom, where the shear walls including the concealed columns are ruptured by normal stress due to the global bending of CT.

4 ANALYSIS OF THE TEST DATA

4.1 The natural vibration and damping ratio

Table 3 gives the first 10 orders of damping ratios, the natural frequencies measured in the test, and those obtained through theoretical computation. The first and second self-power mode shapes in X-axis and Y-axis direction are similar to those of a cantilever beam. The damping ratios of the low orders are close to those of high rise building structures. Table 4 gives the first two frequencies of the model respectively after different grade seismic load. These data show that the frequencies and damping ratios increase due to the more serious damages of the model as the seismic intensity increases.

4.2 The acceleration response and dynamic amplification of the model

The acceleration amplification coefficients (AAC) of each story of the model in a certain seismic load case can be obtained according to the ratios of acceleration peak values measured through the accelerometers on the floors and those through the accelerometers of the shaking table in the corresponding direction.

Table 3. The first 10 orders of damping ratios, the natural frequencies of the model

Order	Damping ratios (%)	Frequencies measured (Hz)	Frequencies computed (Hz)	Vibrating direction
1	3.84	4.30	4.31	X
2	4.61	5.66	5.59	Y
3	4.89	7.62	7.36	θ_z
4	3.89	20.12	19.82	X
5	2.86	20.70	21.16	θ_z
6	1.60	25.39	26.94	Y
7	3.52	30.27	29.98	Z
8	1.75	37.70	35.88	Z
9	0.82	41.60	45.48	X
10	0.91	45.12	51.46	Y

Table 4. The first 2 orders of damping ratios, the natural frequencies of the model measured as seismic loads increase

Seismic grade	X-axis direction		Y-axis direction	
	Damping ratios (%)	Frequencies (Hz)	Damping ratios (%)	Frequencies (Hz)
No load	3.84	4.31	4.61	5.66
6	4.58	3.91	5.19	5.47
7	6.50	3.52	5.18	4.69
8	6.56	3.32	5.78	4.30
9	8.85	1.95	7.66	2.93

To compare the change of the dynamic amplification of the model along the height, AAC in X-axis direction of each story are demonstrated in figure 8. As the seismic intensity increases, the damages become worse and the plastic zones of the model develop. So the whole stiffness weakens and the damping augments. As a result, the AAC decrease.

The dynamic responses of the floors change as the input seismic loads vary in intensity or wave shape. Taking the roof of the model as the research object, figure 9 and figure 10 demonstrate the changing relation curves between the AAC in horizontal directions there and 2-D seismic loads different in intensity or wave shape.

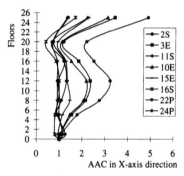

Figure 8. AAC in X-axis direction under X-axis one-dimensional earthquakes

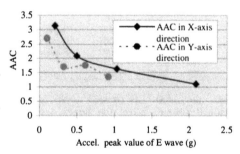

Figure 9. AAC at the top of the model under 2-D E wave

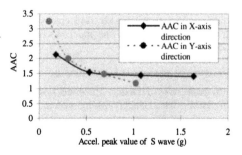

Figure 10. AAC at the top of the model under 2-D S wave

4.3 The displacement response and deformation performance of the model

Subject to the condition of the lab, in the test, besides those of the test table in three directions, the lateral displacements in X-axis direction are measured only on the sixth floor and the roof. As a result we applied integral method to deduce the unmeasured displacements from the time history records of acceleration responses. Compared with the measured displacements, the displacements by integral in frequency domain have an acceptable accuracy. Figure 11 shows the X-axis displacement coverage curves of the model under X-axis, where the displacements of each story refer to the shaking table.

The analysis indicates that the displacement responses in X-axis direction are larger that in Y-axis direction when the same intensity one-dimensional load is applied in the corresponding direction. From the distribution of regularity of the maximum displacement responses of all the stories, we can see that the whole deformation of the model under the horizontal earthquake is principally flexural deformation. The theoretical analysis and test measurement show that the maximum lateral displacement at the top of the model is about 1/600~1/1000 of the total height of the model under the commonly-happened earthquakes, and reaches 1/45 of the total height under the rarely-happened earthquakes. Nevertheless, the model structure sustains a good integrity under the rarely-happened earthquakes. This tells that hybrid structure has good ductility. The displacement responses of the structure are sensitive to the wave shape of the different input seismic waves: Under the seismic loads with the same intensity, the displacement responses of the structure under the artificial wave (P wave) are much larger than those under E wave and S wave.

5 CONCLUSION

The main conclusions may be drawn as follows:

1. The maximum lateral displacement at the top of the model is about 1/600~1/1000 of the total height of the model under the commonly-happened earthquakes, and reaches 1/45 of the total height under the rarely-happened earthquakes.

2. As the seismic effects on the hybrid structure increase, its damp and the natural periods increase due to the development of the structural damages; the damages concentrate on the bottom of the concrete core-tube.

3. The internal forces in the joints between SF and CT are complicated. The concrete around the embedded plates and the corners of CT damaged seriously, partly due to large axial forces induced by earthquakes in girders of SF connecting to CT.

4. Experimental and analytical results show that the hybrid structures designed according to the Chinese Code for Aseismic Design of Building Structures can meet the aseismic requirements of "no damages under commonly-happened earthquakes; being repairable under unusually-happened earthquakes; not collapses under rarely-happened earthquakes".

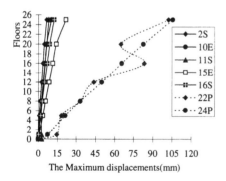

Figure 11. The maximum displacement in X-axis direction under X-axis one-dimensional earthquakes

REFERENCES

Li, G.Q. Developing Strategies of High-rise Building Steel Structures in China, *Journal of Building structures (China)*, 1997(4).

Cai, Y.Y. & Zhong, S.T. Discussion on Developing Direction of High-rise Building Steel Structure, *'98 Academic Conference of Building Steel Structural Engineering in China.*

Council on Tall Buildings, *Structural Design of Tall Steel Buildings*, ASCE, New York, 1979.

Gong, B.N. Experimental Study on Dynamic Behavior of Steel-Concrete Hybrid Structure Model, *Journal of Building Structures (China)*, 1995(6).

Sabnis, G.M. Structural Modeling and Experimental Techniques, *Prentice-hall, Inc., Englewood Cliffs*, N. J., 1983.

Li, G.Q., Zhou, X.M. & Ding, X. Review of the Modeling of Nonlinear Seismic Response Analysis of RC Shear Wall, *World Information on Earthquake Engineering (China)*, 2000(2).

Earthquake Engineering Frontiers in the New Millennium, Spencer & Hu (eds),
© *2001 Swets & Zeitlinger, ISBN 90 2651 852 8*

Advances in earthquake engineering with information technology

G.L. Fenves
University of California, Berkeley, CA, USA

ABSTRACT: Information technology is improving in capability and economy much faster than earthquake engineering applications. An important application is simulation of structural and geotechnical systems during earthquakes. A new software framework for simulation provides facilities for modeling, computation, including parallel computation, and processing of information for evaluating performance. The framework is designed using object-oriented principles and is accessible through an application program interface.

1 INTRODUCTION

An important step in the design and evaluation of structures and geotechnical systems during earthquakes is determination of the stresses, deformations, and measures of damage induced by earthquake ground motion. This process has four major ingredients: (1) development of a mathematical model representing the geometry, topology, materials, loads, and boundary conditions of the system, (2) spatial discretization to give the governing equations of motion, (3) numerical computation to solve the discretized equations, and (4) processing of equation solutions to evaluate performance of components and the system. This entire process can broadly be described as *simulation* in the sense of simulating the behavior of a system in an earthquake. This paper focuses on information technology techniques for improving simulation of structural and geotechnical systems in earthquakes.

In each of the steps in a simulation, there have been important and wide-ranging research advances over the past twenty-five years. Structural and geotechnical engineers use a variety of specialized or general software ("codes") depending on the appropriateness of the mathematical models implemented in the codes, the available computational resources, or individual and organizational experience and policies. In the research area there are tremendous needs for improvements in simulation methods, such as for example new models, computational procedures, and visualization of performance. Individual researchers often have customized versions of specialized codes or work within the limits imposed by commercial, general purpose codes.

An important question is how well does this approach to simulation work, particularly in regards to utilizing the dramatic improvements in information technology? The current software approaches make it difficult for researchers and developers to improve simulation methods that take advantage of the rapid changes in parallel and distributed processing, networking, databases, visualization, and entirely new approaches to computing such as application service providers, peer-to-peer computing and computational grids. The inability to exchange and communicate software implementations of models, computational methods, and performance evaluation methods is a significant drag on research and on transfer of new methods to industry and engineering professionals.

From the perspective of an engineer conducting a simulation, there are a number of desirable requirements for simulation. There should be the capability for using, selecting, and sharing models for materials, elements, components, and entire substructures. The models should be independent of the simulation methods used to compute the state of the model so as to provide flexibility in how simulations are performed. There should be interfaces between models, databases, and visualization tools to provide capabilities for interrogating and investigating the model and results of the simulation. Furthermore, the simulation should support capabilities for system identification studies, optimization, and reliability, all of which will become increasingly important techniques for earthquake engineering design. A scenario for such a framework is that engineers have access libraries of material models, component models, model building tools, computational resources, visualization tools, performance evaluation over the network.

The objective of this paper is to describe the design and an implementation of a software framework that provides this capability. Also described are implementations of models and computational procedures for earthquake engineering simulation. Earlier and related work can be found in Archer et al. (1999) and Rucki & Miller (1996).

2 SOFTWARE FRAMEWORKS

The methodologies for designing and implementing large software systems are part of the computer science field of software engineering. The most successful approach is based on defining *abstractions* of software components. An abstraction provides specified functionality but hides the details of how an implementation of the data and procedures provide the functionality. The key to using a software component is an unambiguous specification of the functionality. The specification is implemented as an application program interface (API), so the software component exists only through its API. The important point is that the details of the implementation are hidden behind the API.

A large software system may have hundreds to thousands of software components, but many of them are related by *associations*. Associations between software components must be carefully designed so that they provide the needed representation, yet have as few assumptions as possible so as not to limit application of the component. The balance between flexibility and specificity is central to a successful software design. A *software framework* is a collection of interrelated components, usually through associations, that can be used to build applications.

The concept of hiding data, procedures, and associations is the basis of *object-oriented* software design. Object-oriented programming languages provide language-level support for encapsulating data and binding of procedures with the data. Object-oriented approaches have well-documented advantages in improving the quality of software, aiding in re-use of code, and providing greater flexibility for modifications and extensions (Fenves 1990).

Most modern programming languages, including C++ and Java, are based on the object-oriented paradigm. Matlab supports object-oriented programming and large parts of Matlab itself (such as the graphics) are implemented with the object-oriented features. Fortran 90 and 95 do not directly support object-oriented programming but have some of the useful features. Object-oriented languages are increasingly being used for scientific computing.

3 DESIGN OF THE OPENSEES FRAMEWORK

Because of the need from improved simulation methods for performance evaluation of structural and geotechnical systems and the advantages of a software framework for sharing and re-use of soft-

ware, the Pacific Earthquake Engineering Research Center is developing the "Open System for Earthquake Engineering Simulation," or OpenSees for short. At its most basic level, OpenSees can be viewed as a set of API's for creating models, performing analyses of models, and evaluating responses of models. The framework was designed using object-oriented principles and OpenSees is implemented in C++. Details about OpenSees, including the API, are available from its website at http://opensees.berkeley.edu.

Before presenting the OpenSees framework, it is worthwhile to illustrate a well-known component. This example presents the software design for material models. At the abstract level a material model represents the mechanical relationship between a deformation measure and a force measure. More specific definition of a material model is one that represents a uniaxial force-deformation relationship and one that represents multi-dimensional relationships. Although the two specific forms of the materials are the same in the abstract, the uniaxial material is used extensively for structural applications and it is not worth the overhead of dealing with scalar measures of force and deformation as vector or tensor quantities required for multi-dimensional constitutive models. As an illustration of extensibility, since much of structural simulation uses beam-column elements, an abstraction of a material relationship between section forces (stress resultants) and section deformations (such as curvature) is useful. Dealing with these types of tradeoffs in engineering software design requires a balance between specificity (and usually computational efficiency) and generality.

The design for the material software component is illustrated in Figure 1. The graphical notation is similar to the unified modeling language (Rumbaugh et al. 1998). A rectangle represents a class, and class inheritance is represented by a hierarchy that shows the subclass ("is a") relationships. Classes may be *abstract* or *concrete* (in the software sense, not the material sense). Abstract classes are not used to create objects; objects must be created from concrete classes. The four classes illustrated in Figure 1 are abstract, and each has concrete classes that provide a specific material model. For example a concrete subclass of UniaxialMaterial is an elastic-plastic material; a subclass of SectionForceDeformation is a FiberSectionForceDeformation class that represents a fiber discretization of the cross section; a subclass of NDMaterial is J2Plasticity.

Figure 1. Class hierarchy for material models

Table 1. Operations in API for material models

Operation	Function
setTrialStrain(strain)	Set current estimate of strain in material
getStress()	Return estimate of stress
getTangent()	Return tangent
commitState()	Accept trial state

A specific material uses (or instantiates) an object from a concrete class, but from an application program viewpoint each material implements the operations in the API given in Table 1. The return type (scalar, vector, or tensor) for the operations depends on the abstract classes. Since each material is responsible for maintaining its state, the API allows an application to set a trial strain and obtain information about the material (stress and tangent) for the strain. At some point the solution is accepted and each material is told to commit its current state. For path dependent materials, the implementation of the material class must store this state upon a commit.

This excerpt of the OpenSees API for materials does not describe all the functionality. For uniaxial materials, it is useful to compose individual models in series or in parallel for more sophisticated models. Composition is one of the most useful associations because it allows factoring of specific functionality into independent objects with limited intercommunication between each other. As another example, plasticity models for materials can be considered as composition of a yield function, a flow rule, an integration rule, a hardening rule, and so on (Jeremic and Sture 1998).

The important point of the framework is that any material, no matter how it is composed, has the same abstract properties accessible through the API. An application, such as an element can use an abstract material model without concern about the details of the implementation including storage of state, which greatly simplifies application programming.

4 MODELING FACILITIES

Moving to a higher-level, the OpenSees framework includes the standard modeling components of elements, nodes, loads, boundary conditions, and constraints. Each component has abstract properties, which are accessible through the API. Figure 2 shows the major classes of objects that provide modeling facilities. One of the key objects shown in that figure is the Domain, which is an aggregation of the modeling components. As described later, domains may be defined recursively in terms of subdomains as a useful modeling feature and for domain decomposition procedures for parallel computation.

To provide support for applications that create models, there is a separate class ModelBuilder. The functionality of a ModelBuilder is to create model objects and populate domain objects. The ModelBuilder

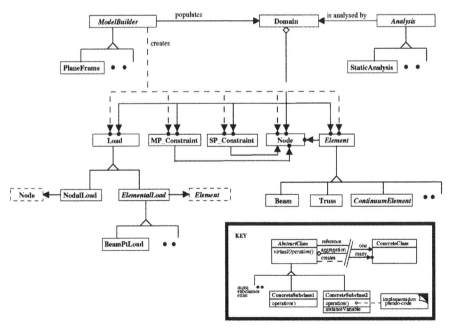

Figure 2. Domain classes in OpenSees framework.

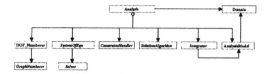

Figure 3. Analysis class as composition of components.

abstraction provides a convenient interface for mesh generation and CAD tools to access the OpenSees API and create models for simulation.

5 ANALYSIS FACILITIES

After a domain has been created with a specific model (by a ModelBuilder), the most important operation is to compute the new state of a domain caused by application of a load, change of a boundary condition, or other change in the domain (such as addition or removal of elements). The simplest approach is to provide an operation (say analyze) for the Domain that implements a specific analysis procedure. The difficulty with giving a domain the responsibility to analyze itself is that there are many types of analysis procedures depending on whether the problem is time-dependent or time-independent, whether the equations are assembled or processed locally at nodes, whether equations are solved iteratively or directly, and how constraints are enforced.

Domains should not be concerned with how the governing equations are formed and solved, which leads to the need for an analysis class distinct from the representation of the domain. This requires that the process of analysis be represented by a class, independent of domains but associated with them to provide the communication needed to advance the state of the domain. The principal tasks for an analysis are abstracted into six classes:

− AnalysisModel: a container class for elements and nodes in the domain. The primary operation is to iterate over the objects in the container.
− SolutionAlgorithm: The algorithm for advancing the solution one load step.
− Integrator: a class that defines the governing equations for one load step and updates nodal quantities after the solution for a load step is accepted.
− ConstraintHandler: procedures for enforcing constraints.
− SystemOfEqn: representation of linearized system of equations and associated solvers.
− DOFNumberer: methods for ordering degrees-of-freedom in domain.

Composition of an analysis procedure allows uncoupling solution algorithms, such as a Newton-Raphson procedure from a time integration procedure, such as Newmark's method. An advantage of

this approach is that element developers only need to provide basic functionality of state determination without concern about how the analysis procedure will compute the solution. For example elements do not form an effective or dynamic stiffness matrix because this couples the element formulation with the time integration operator.

6 PARALLEL PROCESSING

One of the motivations for a new simulation framework is to provide support for parallel and distributed processing. With the rapid change in computing, it is important that the simulation software not hard-code assumptions about parallel computer architectures. The focus of this brief section is on the software design issues, and more details are available in McKenna & Fenves (2000).

In a parallel program, the computation is broken down into tasks and these tasks are assigned to processors. There are a number of ways of doing this depending on the connectivity of the processors and whether memory is shared or distributed. The object-oriented paradigm is ideally suited to parallel programs because the tasks can be identified as the invocation of the object operations. A popular parallel object-oriented programming model is the actor model. Actors are autonomous and concurrently executing objects that execute asynchronously. An aggregate is a collection of actors.

To both minimize changes to the sequential design presented in the previous section to allow for efficient parallel computation, Shadow objects are introduced. Each shadow object is associated with an actor, or with multiple actors, an aggregate, which exist in a remote address space. The shadow object represents the remote object in the local actor's space. A message intended for a remote actor is sent to the local shadow object. The shadow object is responsible for sending an appropriate message to the remote actor(s). The remote actor(s), if required, return the result to the local shadow object, which in turn replies to the local object. The advantage of this approach over the traditional actor approach is that it allows the local shadow object to cache often used data and data that has not changed since the last call.

A key concept included in the OpenSees framework is moveable objects. All the modeling objects are moveable from one processor to another. This allows distribution of data into different memory spaces over channels, which are abstraction of communication methods, such a message-passing-interface (MPI). Once the model is distributed to processors, as determined by a separate broker class, actor and shadow objects are set up and computation can proceed at a high-level without regard to location of objects.

7 USER PERSPECTIVE

The presentation so far has dealt with modeling, forming the governing equations, and the computation to solving the equations. In addition to these aspects of the framework, the Recorder class interacts with the model objects to obtain response data and process it to determine performance measures. The Recorder objects interact with a Database object to store the results of the analysis. A Renderer class provides basic visualization of the model.

The simplest use of the framework for a specific simulation is to write a program using the API that creates the models, creates an analysis, conducts the analysis, and sends the results to a database. Although the concept of *programming* a model is very powerful, users can more easily program with an interpreted scripting language. To facilitate programming models, OpenSees is accessible through the Tcl scripting language (Ousterhout 1994), which is very convenient "glue" for creating a specific model and performing an analysis. There is a one-to-one correspondence between the classes needed for simulation and Tcl commands, and all the usual programming features such as variables, expressions, lists, and control structures are available to the user.

8 FUTURE WORK

With the OpenSees API established and modeling and analysis features available for structural and geotechnical system simulation, researchers at the Pacific Earthquake Engineering Research Center are developing new models. In addition PEER is conducting research and development on:

- general mechanism for parameterization of models and computation of gradients of response with respect to material parameters;
- extensions for reliability computation using random fields;
- distributed computation using a variety of communication mechanisms to support grid based computing;
- high-level object-oriented scripting language for model building.

Planned extension of the framework includes support for multi-physics (such as soil-fluid interaction), integration with distributed computational environments, including modeling and visualization tools, and integration with experimental frameworks for designing tests, calibrating models, and hybrid testing.

9 CONCLUSION

A new object-oriented software design for simulation of structural and geotechnical systems provides a great deal of flexibility in creating models, solving the governing equations, and processing the results of the analysis. The key aspect of the framework is the factorization into independent components with well defined interfaces accessible through an API.

10 ACKNOWLEDGEMENTS

A large number of researchers, students, and users have contributed to OpenSees, including Francis McKenna, Michael Scott, Filip C. Filippou, Jun Peng, Kincho Law, Boris Jeremic, George Turkiyyah, and Gregory Deierlein.

This work is supported by the Pacific Earthquake Engineering Research Center under grant EEC-9701568 from the National Science Foundation, Earthquake Engineering Research Center Program.

REFERENCES

Archer, G.C., Fenves, G.L. & Thewalt, C.R. 1999. A new object-oriented finite element analysis program architecture. *Computers & structures* 70(1): 63-75.

Fenves, G.L. 1990. Object-oriented programming for engineering software development. *Engineering with computers* 6(1):1-15.

Jeremic, B. & Sture, S. 1998. Tensor data objects in finite element programming. *International journal for numerical methods in engineering* 41(1): 113-126.

McKenna, F. 1998. Object-oriented finite element programming: Frameworks for analysis, algorithms and parallel computing. Ph.D. thesis. University of California, Berkeley.

McKenna, F. & Fenves, G.L. 2000. An object-oriented software design for parallel structural analysis. *Advanced technology in structural engineering.* Structural Engineering Institute, ASCE.

Ousterhout, J.K. 1994. *Tcl and the Tk Toolkit.* New York: Addison-Wesley.

Rucki, M. D. & Miller, G. R. 1996. An algorithmic framework for extensible finite element-based structural modeling. *Computer methods in applied mechanics and engineering* 136(3-4):363-384.

Rumbaugh, J., Jacobson, I. & Booch, G 1998. *The unified modeling language reference manual.* New York: Addison-Wesley.

Earthquake Engineering Frontiers in the New Millennium, Spencer & Hu (eds),
© 2001 Swets & Zeitlinger, ISBN 90 2651 852 8

Introduction of US/PRC workshop on experimental methods in earthquake engineering and its influence on seismic testing in China

Xinlin Lu & Wensheng Lu
Tongji University, Shanghai, China

ABSTRACT: The US/PRC Workshop on Experimental Methods in Earthquake Engineering was held at Tongji University in 1992, which was successful and has had an important influence on the seismic structural testing in China. The authors think even though the computer and numerical technique are advanced and effective, the structural testing methods are still the most powerful, basic and determined methods in studying structural seismic behavior, and they provide the foundation for the development of earthquake engineering. Nowadays, the researchers in China reach a relative high level on shaking table testing methods of high-rise buildings. However, there are still much attention needed to the detailed numerical analysis of the test results and optimizing the testing methods. On the other hand, the similitude theory for nonlinear behavior of structures is still incomplete. A lot of problems need to be solved in the future.

1 INTRODUCTION OF THE WORKSHOP

As part of US-PRC protocol for cooperative earthquake studies, annex 3, the US/PRC Workshop on Experimental Methods in Earthquake Engineering was held at Tongji University in Shanghai from November 10 to November 12, 1992. The goal of the workshop was to bring together the leading experts in experimental research from the US and PRC to give presentations and participate in working group discussions on issues related to improvements in methods of experimental research in earthquake engineering. Topics for discussion included advances in laboratory and field testing technologies, experimental procedure and protocols, innovative testing methods, sensors and data acquisition technology, experimental methods for specific applications.

The workshop consisted of plenary presentation sessions and discussion sessions. In the plenary sessions the participants gave presentations on the state-of-the–knowledge and new developments in experimental methods. In the working group sessions the participants focused on an assessment of the state-of–the-knowledge and future development needs in specific sub-areas of experimental methods. The conclusions of the working group sessions were discussed in a plenary session and summarized in the workshop resolution.

The workshop was held with mutual respect and friendship among all participants, and was successful. In the group discussions the following resolutions were concluded:

1) Advancements and refinements in experimental methods are needed.

2) It is important to take full advantage of the intellectual capabilities and experimental facilities existing worldwide and to promote collaboration between countries and institutions.

3) The workshop dialogue should be continued, and the cooperative experimental research on issues of interest to both countries should be encouraged.

4) Improvements in experimental methods, testing facilities, and instrumentation are needed.

5) Full advantage should be taken of recent developments made in instrumentation technology in other fields.

6) Field experimentation needs to play a more important role.

7) Research is needed to improve the reliability, accuracy, control, and realism of shaking table experimentation, pseudo-dynamic experimentation, quasi-static experimentation and field experimentation.

8) Funding is needed to improve the effectiveness and capabilities of facilities.

9) Efforts are needed to establish minimum standards for documentation of experimentation of experiments and results, archiving of test information, and dissemination of test information.

Researchers users of experimental research should convene to discuss and establish these protocols and identify technological issues to be resolved to implement these protocols.

2 INFLUENCE OF THE WORKSHOP ON SEISMIC RESEARCH IN CHINA

As above-mentioned, the workshop is one of the most successful conferences in study and discussion of seismic testing methodology. Especially the technical topics and the conclusions of the workshop are helpful to the development of seismic testing methods and research works in China. The effects can be briefly summarized as follows,

1) The workshop promoted the editing of *Specification of Testing Methods for Earthquake Resistant Building (JGJ101-96)* of China. Some technical suggestions and requirements on seismic testing methods of the workshop were adopted as items of this specification, which was issued in December 1996, and become effective from April 1, 1997. Nowadays it serves as the guideline for seismic tests and research works in China.

2) The workshop promotes the study on the test methodology of shaking table tests of high-rise building models. Before 1990, many multi-story building models had been tested by means of shaking table test method. The first batch mark test of high-rise structural models (e.g. Shanghai Oriental TV Tower, etc.) was tested at Tongji University in 1990. From then on, shaking table tests on high-rise buildings and the corresponding studies were carried out all around in China, however the testing scales and technologies were still limit. After this workshop, with rapid economic development in China, there was rather high intensity construction of infrastructure beginning from 1993. Many attentions had been paid to the seismic behavior of high-rise buildings, and gave us chances to improve the seismic testing methods in this field.

3 GENERAL SITUATION OF THE SEISMIC TESTING METHODS IN CHINA

China is one of the most earthquake active countries in the world. It is known that more than 60% area and 70% moderate and larger cities of China are located at earthquake zone. As a result of high-speed development of economy in China, many high-rise buildings have been constructed or under construction in the recent years. However, the risk of earthquake is much higher than before. Hence the safety against earthquakes is strongly required for high-rise buildings. To improve our understanding of the response of structures under earthquakes, three approaches could be adopted, i.e., site investigation of earthquake damage, theoretical analysis and structural test. Great progress on the structural test has been achieved in recent years in China.

The main purposes of seismic simulation tests are

1) To check the natural frequencies, damping ratio and vibration modes of structures;

2) To study the seismic responses of accelerations, displacements and strains of structures;

3) To determine the structural crack positions and the weakness points, to verify or find the collapse styles and failure mechanism;

4) To assess the safety reliability of main structures under different earthquake intensity, to verify the rationality and effectiveness of various earthquake-resistant countermeasures.

The possible seismic test methods for structures are pseudo-static test, pseudo-dynamic test, field test and shaking-table test. Shaking table tests are most realistic method of earthquake testing than pseudo-static and pseudo dynamic methods. Shaking tables are usually rectangular planar platforms moved by servo-hydraulic actuators to simulate earthquakes. They are used with attached specimens of interest to study earthquake resistant design of civil engineering structures, such as bridges, dams, and buildings, and to full scale qualify critical equipment like computer control systems, switching relay banks, and nuclear plant cooling pumps and turbines.

The shaking-table test is unique facility that can strictly simulate the earthquake ground motion. In 1960s the first shaking-table with dimensions of 6.1×6.1m was constructed in Berkeley, and now the largest shaking table with dimensions of 15×16m is under construction in Japan. Two large shaking table with dimensions of 6×6m is under construction in Chengdu and Beijing respectively. Along with the developing of the technology, the shaking table can vibrate in one way to three ways with six degree of freedom; the control system achieved a remarkable development, from the PID controller and three parameters controller to advanced adaptive controller. Up to now about 100 shaking tables for seismic test of structures has been installed in laboratories all over the world. More than 10 shaking table facilities have been imported or manufactured in China, the test abilities can reach to six degree of freedom, 4~6g of acceleration and 15~25 tons of static mass. Table 1 shows main performance of Large-scale high-performance shaking table facilities of Tongji University.

Shaking table can be widely used for observation the earthquake-resistant design and mathematical model of structure, especially to verify the high-rise building by small-scale model. In fact, all kinds of structures can be tested by shaking table testing method so long as the system has enough capacity to carry the specimen. However, it is difficult to model some full-scale effects, for example, shear and bond resistance in reinforcement concrete elements. Thus, the trend is toward large tables with six degree of freedom, but the problem is the cost. However, equipment costs of shaking tables are the power functions of specimen weight and dimension, so there is a desire for reduced scale model testing.

Table 1. Main performance of shaking table facilities of Tongji University

Item		Performance	Remarks
Maximum payload		25tons	
Table dimension		4m × 4m	
Exciting direction		X, Y, Z Simultaneously	X & Y: Horizontal Z: Vertical
Maximum stroke	X	±100 mm	
	Y&Z	±50 mm	
Maximum velocity	X	1000mm/s	
	Y&Z	600mm/s	
Maximum acceleration	X	4.0 g	Bear table
		1.2 g	With 15ton payload
	Y	2.0 g	Bear table
		0.8 g	With 15ton payload
	Z	4.0 g	Bear table
		0.7 g	With 15ton payload
Frequency range		0.1~50Hz	
Maximum specimen channels		96	
Degree of freedom		6	

4 THE SEISMIC STRUCTURAL MODELTESTING METHODS FOR HIGH-RISE BUILDINGS IN CHINA

The shaking table test is one economic, tangible, useful and reliable seismic proving test to assess the seismic safety and reliability of high-rise buildings. The dynamic behavior of the structure and its damage style under earthquake with great magnitude can be reproduced. As a result of this test, the structure is proved to keep safety, or too weak to resist a destructive earthquake. So the weak points of structure are determined, and suggestions and adjustment can be put forward, before the construction of prototype structures.

4.1 Purpose and similitude relationship of shaking table test

The similarity of the scaled model is an unsolved problem under severe earthquakes. The ratio between the physical quantifies of prototype structure and model is defined as similitude coefficient S. To design, construct and analyze the structural model

for shaking table tests, the most important thing is try to satisfy the similitude relationship. However it is very difficulty to meet all the requirements exactly. In order to keep the similitude of the model behavior from linear phase to nonlinear phase, the yield and ultimate strengths of the materials as well as the strain-stress relationship between the model and the prototype must be similar. Usually, some predominate similitude coefficients shall be determined first, and other coefficients can be revised to an appropriate range. Unfortunately, we cannot achieve the full similitude relationship, because each similitude coefficient cannot be determined arbitrarily in its own way. For example, the gravity acceleration similitude coefficient always equals to 1, while the earthquake acceleration similitude coefficient does not; the material modulus and stress (strength) similitude coefficients are not always in the same value, which leads to the strain similitude coefficient not always equals to 1, etc. On the other hand, in practice the design of scale model is always restricted by many conditions, such as test facilities, model materials and manufacturing techniques etc. Table 2 shows the similitude relationship of some models which have been tested recently.

4.2 Design and construction of high-rise building models

Micro-concrete and fine reinforcement is always used to construct a model. The skill of using this kind of material and its construction details is mature in China, which has been used to construct more than 50 building models during last decades. The main structural members are always entirely reproduced, and the dimension and reinforcement is reproduced carefully. Usually it takes 3 to 5 months to construct one whole building model. Figure 1 shows several structural models. Table 3 shows the data of model test of some high-rise buildings in China.

Elastic model can also be made of plastic and other kinds of materials. Those model shall keep in elastic state and the linear behavior will be mensured by sensors during test.

4.3 The procedure of shaking table tests

For choosing suitable earthquake waves to excite a high-rise building model, the soil type of construction site and the dynamic behavior of the prototype structure shall be taken into account. During a test, the exciting intensities of each earthquake waves vary from the frequently occurred earthquake to the seldom-occurred earthquake step by step, and the exciting inputs can be one dimension or three dimensions simultaneously. Sensors such as strain gauges, displacement transducers and acceleration meters are placed at critical and interesting points of models.

Table 2. Similitude coefficients of typical models tested in Tongji University

Physics Behavior	Main Coefficients	Guangzhou Tianwang Center	Guangzhou International Plaza	Shanghai BO-COM Financial Tower	Shanghai Changshou Commercial Plaza	Shanghai Pudong Reception Center
Geometry Property	Length	1/25	1/40	1/33	1/25	1/20
	Elastic modulus	1/6	1/2.6	1/2.96	1/6	1/3.8
Material Property	Stress	1/8	1/2.6	1/6.92	1/6	1/8.7
	Mass density	1	2.5	1	1.39	
	Mass	1/11194	1/25600	6.96×10^{-5}	1/11250	1/4800
Dynamic Property	Period	1/8.64	1/15.69	0.11	0.12	6.25
	Acceleration	2.99	6.25	2.64	3	1.9

4.4 Theoretic analysis methods

Many high-rise building models have been tested on the shaking table in China. The numerical analysis and field tests are also ordinary manner to support and verify the test results. Several effective methods have been applied in the analysis of high-rise buildings by researchers of Tongji University. Here after are some theoretic models for analyzing these kinds of complex structures.

1) Multi-tower with rigid podium, which is suitable for the analysis of multi-tower building with strong transfer floor and the torsion response is negligible.

2) Multi-towers with spring element podium, which is suitable for the analysis of multi-tower building with flexible podium and also the torsion response is negligible.

3) Rigid-plate mass spring model, which is suitable for the analysis of complex building. This model uses the master-slave constrain mode for the slab with relative large stiffness in slab plane, i.e. there is usually one master node for each podium floor or conjunction floor and one master node for each floor of separate tower. The torsion of the floor is considered in this model.

4) A new multi-rigid block model, in which rigid floors of each tower, flexible transform floor, rigid

Figure 1. High-rise buildings models

Table 4. The data of model test of some high-rise buildings in China

Project	Height of prototype (m)	Structural style	Main material used in model	S_L	S_a	Tested and studied by
Shanghai oriental TV tower	468	Pre-stressed RC tube	Micro-concrete	1/50	3.58	Tongji University
Futong international mansion, Hainan	175	RC tube in tube	Micro-concrete	1/25	3.125	Tongji University
Finacials mansion, Shanghai	138	RC frame-tube	Micro-concrete	1/25	8.29	Tongji University
Triumph Gate Building, Shanghai	100	Multi-tower	Micro-concrete	1/25	5.0	Tongji University
Xinhai mansion	84	Multi-tower	Micro-concrete	1/25	7.11	Tongji University
Jinguang Center, Shenzhen	128	RC frame-tube	Micro-concrete	1/25	4.0	Tongji University
International Commercial and trade plaza, Guangzhou	249	Multi-tower with large podium	plastic	1/100	6.02	Tongji University
Shanghai Grant Theatre	38	Mage structure	Micro-concrete	1/25	1.0~3.0	Tongji University
Central TV Tower	386.5	Tube	Micro-concrete	1/50	1.0~3.0	Tsinghua University
Futian rainbow city mansion, Shenzhen	119.6			1/30	3.0~9.58	CABS
Wenjing plaza, Shenzhen	109			1/75	12.5~13.5	CABS
Office building, Haikou	161.4			1/80	2.02~2.80	Tsinghua University
Shenzhen Futian buildings group	139	Frame-tube	Micro-concrete	1/38	5.55	South China Construction University
World Celebrities plaza	224	SRC-column	Micro-concrete	1/40	6.25	South China Construction University

or flexible connection and the elastic floor of large opening are considered.

The theoretical dynamic behavior is compared with the test results. Furthermore, vibration tests on site are also under going. The error between the results of testing and the results of calculation and site testing are within 10%~30%. Hence the results of shaking table testing are satisfactory.

5 CONCLUSION AND FUTURE TREND

The US/PRC Workshop on Experimental Methods in Earthquake Engineering provides a forum for all the experiences and achievements on both static and dynamic structural test methods. A lot of works have been done during the past 8 years, and many progress have been achieved in China. Many large testing facilities have been constructed or under constructed in China. Such as the new structural testing laboratory at Tsinghua University, the new reaction wall and upgrading of shaking table system at Tongji University, two 6x6m shaking tables under construcion, etc. Even more, an academic committee on the seismic testing methodology of China have founded at 1999.

1) As above-mentioned, the workshop is one of the most successful conferences in studying and discussion of seismic test methodology. It promotes the development of seismic test methodologies and research works, especially promotes the implementation of new test facilities and technologies on structural engineering in China.

2) In conclusion, the authors think even though the computer and numerical technique are advanced and effective, the structural test method is still one of the most powerful, basic and determined methods in the field of structural seismic behavior, and it provides the foundation for the developing of seismic engineering.

3) Nowadays, the researchers in China reach a relative high level in study on shaking table test methodology of high-rise buildings. The testing results are usually adapted by design and construction. However, there are still much attention needed to the detailed numerical analysis of the test results and optimizing the testing methodology.

4) The similitude theory for nonlinear behavior of structures is still imperfect. A lot of problems need to be solved in the future.

5) It is essential to enhance the cooperation between countries and research institutes on experimental methods in earthquake engineering.

REFERENCES

Helmult Krawinkler & Bolong Zhu, 1992, US/PRC workshop on experimental methods in earthquake engineering.

CODE, 1992, Seismic Design Code for Buildings (DBJ8-9-92), Shanghai.

CODE, 1997, Specification of Testing Methods for Earthquake Resistant Building (JGJ101-96), Beijing.

Xilin Lu, et al., 1997, Seismic Safety Analysis and Model Test of High-rise Building Structures, Proceedings of International Symposium on Engineering for Safety. Reliability and Availability (ESRA), 187~194.

Xilin Lu & Haibo Cheng, 1996, Research and Prospect of Dynamic Similitude Theory in Structural Model Test, Contemporary development, theory and application of Structural Engineering, China Architectural Industry Press.

S. Su, 1996, Earthquake resistance behavior and vibration control of door-shape building, paper for doctor degree.

Xilin Lu et al., 1998, Shaking Table Testing of a U-Shaped Plan Building Model with Engineering Application, Asia-Pacific Workshop on Seismic Design & Retrofit of Structures, Chinese Taipei, 114~191.

Wensheng Lu et al., 1998, Shaking Table Test of a High-rise Building Model with Multi-tower and Large Podium, The Fifth International Conference on Tall Buildings, Hong Kong, 814~819.

Andress Vogel, et al., 1998, *Earthquake prognostics strategy*, Anhui Xinhua Printing House.

Earthquake Engineering Frontiers in the New Millennium, Spencer & Hu (eds),
© 2001 Swets & Zeitlinger, ISBN 90 2651 852 8

Finite element response sensitivity analysis in earthquake engineering

J. P. Conte

University of California, Los Angeles, CA 90095-1593, USA

ABSTRACT: This paper presents a method to compute the exact sensitivities with respect to material constitutive parameters and discrete loading parameters of the computationally simulated response of a structure. Focus is placed on plasticity-based materially-nonlinear-only finite element models of structural systems and on two types of state-of-the-art analysis used in earthquake engineering, namely nonlinear static push-over analysis and nonlinear response history analysis. Implementation of the method in a general purpose nonlinear finite element analysis program (FEAP) is discussed. Application examples related to a concrete gravity dam and a moment-resisting building frame are provided, serving also to validate the exact response sensitivity results through their finite difference calculations. These application examples illustrate the fact that sensitivities with respect to both material and loading parameters of plasticity-based structural models are discontinuous in time and that the method presented is able to capture these discontinuities.

1 INTRODUCTION

Earthquake engineering relies on both physical and computational simulation of structural and geotechnical systems. The state-of-the-art in computational simulation of the seismic response of civil structures is in the nonlinear domain to account for material and geometric nonlinearities governing the complex behavior of geotechnical and structural systems, especially near their failure range. Maybe even more important than the simulated nonlinear seismic response of a civil structure is its sensitivity to various geometric, mechanical and material properties defining the structure and to earthquake loading parameters.

This paper presents an algorithm for "exact" nonlinear finite element response sensitivity calculation in the case of displacement-based finite elements within a nonlinear analysis program based on the direct stiffness method. The exact sensitivity, with respect to material constitutive parameters and loading parameters, of the computationally simulated system seismic response is obtained through exact differentiation of the finite element numerical scheme (including the material constitutive law integration scheme) with respect to the sensitivity parameters in question.

2 FORMULATION

2.1 Nonlinear structural response calculation

After spatial discretization using the finite element method, the equation of motion of a materially-nonlinear-only structural system takes the form

$$M\ddot{u}(t) + C\dot{u}(t) + R(u(t)) = F(t) \qquad (1)$$

where t = time, $u(t)$ = vector of nodal displacements, C = damping matrix, M = mass matrix, $R(u(t))$ = history dependent internal (inelastic) resisting force vector, $F(t)$ = vector of external dynamic loads, and a dot over a symbol denotes one differentiation with respect to time. In the case of earthquake ground excitation and assuming rigid-soil excitation, the dynamic load vector takes the expression $F(t) = -ML\ddot{u}_g(t)$ in which L = influence coefficient vector and $\ddot{u}_g(t)$ = ground acceleration time history (assuming here a single component of earthquake ground motion).

Expressed in residual form at discrete time $t_{n+1} = (n+1)\Delta t$, the above equation of motion becomes

$$\begin{aligned} \Psi_{n+1} &= F_{n+1} - M\ddot{u}_{n+1} - C\dot{u}_{n+1} - R(u_{n+1}) \\ &= 0 \end{aligned} \qquad (2)$$

The Newmark-beta time-stepping method interpolates the velocity and acceleration response vectors at t_{n+1} as

$$\dot{u}_{n+1} = (\Delta t)\left(1 - \frac{\alpha}{2\beta}\right)\ddot{u}_n + \left(1 - \frac{\alpha}{\beta}\right)\dot{u}_n +$$

$$\frac{\alpha}{\beta(\Delta t)}(u_{n+1} - u_n) \tag{3}$$

$$\ddot{u}_{n+1} = \left(1 - \frac{1}{2\beta}\right)\ddot{u}_n - \frac{1}{\beta(\Delta t)}\dot{u}_n +$$

$$\frac{1}{\beta(\Delta t)^2}(u_{n+1} - u_n)$$

where α, β = parameters controlling the accuracy and stability of the numerical integration scheme. The linear acceleration method and constant average acceleration method correspond to $\{\alpha = 1/2, \beta = 1/6\}$ and $\{\alpha = 1/2, \beta = 1/4\}$, respectively. Substituting Equation 3 into Equation 2 yields the following nonlinear algebraic equation in u_{n+1}:

$$\boldsymbol{\Psi}_{n+1} = \tilde{F}_{n+1} - \left[\frac{1}{\beta(\Delta t)^2}Mu_{n+1} + \right.$$

$$\left. \frac{\alpha}{\beta(\Delta t)}Cu_{n+1} + R(u_{n+1})\right] = 0 \tag{4}$$

where

$$\tilde{F}_{n+1} = F_{n+1} + M\left[\frac{1}{\beta(\Delta t)^2}u_n + \frac{1}{\beta(\Delta t)}\dot{u}_n\right.$$

$$\left. - \left(1 - \frac{1}{2\beta}\right)\ddot{u}_n\right] + \tag{5}$$

$$C\left[\frac{\alpha}{\beta(\Delta t)}u_n - \left(1 - \frac{\alpha}{\beta}\right)\dot{u}_n - (\Delta t)\left(1 - \frac{\alpha}{2\beta}\right)\ddot{u}_n\right]$$

The newton-Raphson iterative procedure is used to solve Equation 4 over time step $[t_n, t_{n+1}]$ through solving a sequence of linearized problems

$$(K_T^{dyn})_{n+1}^i \delta u_n^{i+1} = \boldsymbol{\Psi}_{n+1}^i \tag{6}$$

where

$$(K_T^{dyn})_{n+1}^i = \frac{1}{\beta(\Delta t)^2}M + \frac{\alpha}{\beta(\Delta t)}C +$$

$$(K_T^{stat})_{n+1}^i \tag{7}$$

and

$$\boldsymbol{\Psi}_{n+1}^i = \tilde{F}_{n+1} - \left[\frac{1}{\beta(\Delta t)^2}Mu_{n+1}^i + \right.$$

$$\left. \frac{\alpha}{\beta(\Delta t)}Cu_{n+1}^i + R(u_{n+1}^i)\right] \tag{8}$$

The updated nodal displacement vector u_{n+1}^{i+1} (or displacement at the end of iteration # i+1 of time step $[t_n, t_{n+1}]$ is obtained as

$$u_{n+1}^{i+1} = u_n + \Delta u_n^{i+1} = u_{n+1}^i + \delta u_n^{i+1} \tag{9}$$

where Δu_n^{i+1} and δu_n^{i+1} denote the total incremental displacement vector from the last converged step and the last incremental displacement vector, respectively. In Equations 6 and 7, K_T^{dyn} and K_T^{stat} denote the dynamic tangent stiffness matrix and the static consistent tangent stiffness matrix (obtained through consistent linearization of the numerical scheme used to integrate the constitutive equations), respectively. The vector of internal resisting forces is obtained by assembling, at the structure level, the vectors of elemental internal resisting forces as

$$R(u_{n+1}^i) = \mathop{A}_{e=1}^{Nel}\left\{\int_{\Omega_e} B^T \cdot \sigma(\epsilon_{n+1}^i) \cdot d\Omega_e\right\} \tag{10}$$

where $A\{...\}$ denotes the direct stiffness assembly operator from the element level (possibly in local element coordinates) to the structure level in global reference coordinates; B = strain (or strain resultant) - displacement transformation matrix, σ = stress (or stress resultant) vector; ϵ = strain (or strain resultant) vector. Similarly,

$$K_T^{stat}(u_{n+1}^i) = \partial R(u_{n+1}^i) / \partial u_{n+1}^i$$

$$= \mathop{A}_{e=1}^{Nel}\left\{\int_{\Omega_e} B^T \cdot D_T \cdot B \, d\Omega_e\right\} \tag{11}$$

where D_T denotes the matrix of material consistent tangent moduli.

2.2 Sensitivity of nonlinear finite element response

Assuming that u_{n+1} is the converged solution for the current time step $[t_n, t_{n+1}]$, and differentiating Equation 4 with respect to the sensitivity parameter θ (= material constitutive parameter or loading parameter) using the chain rule and recognizing that $\sigma_{n+1} = \sigma_{n+1}(\epsilon_{n+1}(\theta), \theta)$ yields

$$[K_T^{dyn}(u_{n+1})] \cdot \frac{\partial u_{n+1}}{\partial \theta} =$$

$$-\left(\frac{1}{\beta(\Delta t)^2}\frac{\partial M}{\partial \theta} + \frac{\alpha}{\beta(\Delta t)}\frac{\partial C}{\partial \theta}\right)u_{n+1} - \tag{12}$$

$$\left.\frac{\partial R(u_{n+1}(\theta), \theta)}{\partial \theta}\right|_{u_{n-1}} + \frac{\partial \tilde{F}_{n+1}}{\partial \theta}$$

where

$$\frac{\partial \tilde{F}_{n+1}}{\partial \theta} = \frac{\partial F_{n+1}}{\partial \theta} +$$

$$\frac{\partial M}{\partial \theta}\left(\frac{1}{\beta(\Delta t)^2}u_n + \frac{1}{\beta(\Delta t)}\dot{u}_n - \left(1-\frac{1}{2\beta}\right)\ddot{u}_n\right) +$$

$$M\left[\frac{1}{\beta(\Delta t)^2}\frac{\partial u_n}{\partial \theta} + \frac{1}{\beta(\Delta t)}\frac{\partial \dot{u}_n}{\partial \theta} - \left(1-\frac{1}{2\beta}\right)\frac{\partial \ddot{u}_n}{\partial \theta}\right] + \quad (13)$$

$$\frac{\partial C}{\partial \theta}\left[\frac{\alpha}{\beta(\Delta t)}u_n - \left(1-\frac{\alpha}{\beta}\right)\dot{u}_n - (\Delta t)\left(1-\frac{\alpha}{2\beta}\right)\ddot{u}_n\right] +$$

$$C\left[\frac{\alpha}{\beta(\Delta t)}\frac{\partial u_n}{\partial \theta} - \left(1-\frac{\alpha}{\beta}\right)\frac{\partial \dot{u}_n}{\partial \theta} - (\Delta t)\left(1-\frac{\alpha}{2\beta}\right)\frac{\partial \ddot{u}_n}{\partial \theta}\right]$$

In Equation 12, $\left[\partial R(u_{n+1}(\theta), \theta)/\ \partial \theta\right]\big|_{u_{n+1}}$ denotes the derivative of the internal resisting force vector $R(u_{n+1}(\theta),\ \theta)$ with respect to the sensitivity parameter θ for fixed displacement u_{n+1} and can be obtained as

$$\frac{\partial R(u_{n+1}(\theta),\ \theta)}{\partial \theta}\bigg|_{u_{n+1}} =$$

$$\underset{e=1}{\overset{Nel}{A}}\left\{\int_{\Omega_e} B^T \cdot \frac{\partial \sigma(\epsilon_{n+1}(\theta),\ \theta)}{\partial \theta}\bigg|_{\epsilon_{n+1}} \cdot d\Omega_e\right\} \quad (14)$$

where $\left[\partial \sigma(\epsilon_{n+1}(\theta),\ \theta)/\ \partial \theta\right]\big|_{\epsilon_n}$ denotes the derivative of the stress vector $\sigma(\epsilon_{n+1}(\theta),\ \theta)$ with respect to θ for fixed strain vector ϵ_{n+1}. Analytical expressions for this conditional derivative of the stress vector have been derived by Zhang and Der Kiureghian and Conte and co-workers for the constitutive J_2 (or von Mises) plasticity model and by Conte and co-workers for the constitutive cap plasticity model (Zhang & Der Kiureghian 1993, Conte & Jagannath 1995a, b, Conte & Vijalapura 1998) in the case of a return map constitutive integration algorithm (Simo and Hughes 1998). Note that once the numerical (finite element) response of the system is known at time t_{n+1}, the algebraic response sensitivity equation (Equation 12) is linear in the displacement response gradient $\partial u_{n+1}/\ \partial \theta$ and has the same left-hand-side matrix as the consistently linearized equation for the response (Equation 6) at the last iteration of the current time step $[t_n, t_{n+1}]$, namely $K_T^{dyn}(u_{n+1})$. Therefore at the end of each time step (or load increment), after convergence is achieved for the response and in order to compute the response sensitivity, a new right-hand-side vector of the response equation needs to be formed for each sensitivity parameter θ. Since the factorization of the dynamic tangent stiffness

matrix $K_T^{dyn}(u_{n+1})$ on the left-hand-side of the response sensitivity equation is already available from the response calculation, after the right-hand-side vector is computed, only the forward and backward substitution phases of solution of a linear system of equations need to be performed to obtain the response sensitivity with respect to θ, thus resulting in a very efficient scheme.

The response sensitivity computational scheme presented above was implemented in FEAP, a general-purpose nonlinear Finite Element Analysis Program developed by Prof. R.L. Taylor at the University of California at Berkeley (Taylor 1999), for both the J_2 and cap plasticity models and for several materially-nonlinear-only finite elements (truss, plane-strain quadrilateral, and 2-D beam-column elements). This computational scheme offers a very efficient utilization of the computer memory space, since at a given time step, the response sensitivities with respect to different sensitivity parameters are computed sequentially, thus avoiding the use of multiple large matrices. Moreover, sensitivity matrices such $\partial M/\ \partial \theta$ and $\partial C/\ \partial \theta$ are never evaluated at the structural level, but are computed at the element level and directly assembled into the global right-hand-side vector of the sensitivity equation in Equation 12.

3 APPLICATION EXAMPLES

3.1 Concrete Gravity Dam

The Pine Flat Dam on King's River near Fresno, California, shown in Figure 1 is analyzed when subjected to the first 15 seconds of the N-S component of the Imperial Valley earthquake of May 18, 1940, recorded at the El Centro site and displayed in Figure 2. The analysis assumes plane strain condition, empty reservoir condition, and rigid base excitation. The displacement-based, four-noded, bilinear, isoparametric finite element with four integration points is used in conjunction with the cap constitutive model. The cap

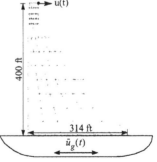

Figure 1. Finite element model of Pine Flat Dam (California) on rigid base, with empty reservoir, and subjected to earthquake ground motion

397

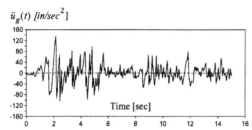

$\ddot{u}_g(t)$ [in/sec^2]

Time [sec]

Figure 2. Imperial Valley earthquake, May 18, 1940, El Centro site, component S00E (N-S)

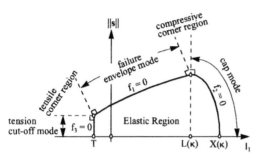

Figure 3. The yield surface of the cap model

model used in this study to model the inelastic behavior of plain concrete is a non-smooth, multi-surface, rate independent, associative plasticity model. It is defined by a convex yield surface in the subspace of the two invariants: $I_1 = \text{trace}(\sigma) = \sigma_{ii}$, the first invariant of the stress tensor σ, and $\|s\| = [2J_2]^{1/2}$ where J_2 denotes the second invariant of the deviatoric stress tensor. As shown in Figure 3, the yield surface is composed of the failure envelope, $f_1(\sigma) = 0$, a strain hardening elliptical cap, $f_2(\sigma, \kappa) = 0$, in which κ denotes the hardening parameter, and a tension cut-off region, $f_3(\sigma) = 0$. The failure envelope and tension cut-off region behave as ideal plasticity surfaces. The functional forms of f_1, f_2, and f_3 are defined as

$$f_1(\sigma) = \|s\| - F_e(I_1), \quad \text{for } T \le I_1 \le \kappa \qquad (15)$$

where $\quad F_e(I_1) = \alpha - \lambda e^{-\beta I_1} + \theta I_1 \qquad (16)$

$$f_2(\|s\|, I_1, \kappa) = F_c(\|s\|, I_1, \kappa) - F_e(\kappa), \qquad (17)$$

$\quad$ for $\kappa \le I_1 \le X(\kappa)$, where

$$F_c(\|s\|, I_1, \kappa) = \left[\|s\|^2 + \left(\frac{I_1 - L(\kappa)}{R} \right)^2 \right]^{1/2}, \quad (18)$$

$$f_3(\sigma) = T - I_1, \quad \text{for } I_1 = T \qquad (19)$$

In the above definition of the cap model, compressive stresses and compressive strains (i.e., compaction) are

taken to be positive. The coefficients α, β, λ, θ, and R are material parameters for the cap and failure envelope, and T is the tension cut-off or maximum allowable hydrostatic tension, also a material constant. The function $L(\kappa)$ is defined as $L(\kappa) = \kappa$ if $\kappa > 0$, $L(\kappa) = 0$, otherwise. The point of intersection of the cap with the I_1 axis is defined as $X(\kappa) = \kappa + RF_e(\kappa)$. The hardening parameter κ is implicitly defined in terms of the effective plastic volumetric strain, $\bar{\varepsilon}_v^p$, through

$$\bar{\varepsilon}_v^p = W(1 - e^{-DX(\kappa)}) \qquad (20)$$

in which W and D are material parameters. The effective plastic volumetric strain, $\bar{\varepsilon}_v^p$, is a history dependent functional of the volumetric plastic strain and is defined by the following hardening law expressed in rate form:

$$\dot{\bar{\varepsilon}}_v^p = \begin{cases} \dot{\bar{I}}_1^p = \dot{\varepsilon}_v^p \text{ if } \dot{\bar{I}}_1^p > 0, \text{ or } \kappa > 0 \text{ and } \kappa > I_1 \\ 0, \text{ otherwise} \end{cases} \quad (21)$$

where $\bar{I}_1 = \varepsilon_{ii}$ is the first invariant of the strain tensor ε and the superscript "p" denotes the plastic part assuming the additive decomposition of the small strain tensor into an elastic and plastic part, i.e., $\varepsilon = \varepsilon^e + \varepsilon^p$. The elastic part of the cap model (inside the yield surface) is taken to be linear elastic and fully characterized by the shear modulus G and the bulk modulus K. For a set of cap model constitutive parameter values identified from a set of real concrete stress-strain data (Conte et al. 1995a), Figure 4 shows the computed horizontal displacement response of the dam crest relative to the ground due to the El Centro 1940 record scaled up by a factor of three (gravity is applied first). The sensitivities of the computed crest displacement to the cap model constitutive parameters T and α are plotted in Figures 5 and 6, respectively. The sensitivities obtained by exact differentiation of the numerical finite element algorithm are validated by finite difference calculations of the response sensitivities using $\Delta T/T = 0.0023$ and $\Delta \alpha / \alpha = 0.0018$. The sensitivity of the computed dam crest displacement to the 50th digital value of the discretized El Centro record (= ground acceleration at 0.98 sec) is given in Figure 7 where it is also validated through finite difference calculation. The sensitivities given in Figures 5, 6, and 7 are scaled by the sensitivity parameter itself and can therefore be interpreted as 100 times the change in the crest displacement response history per percentage change in the sensitivity parameter.

398

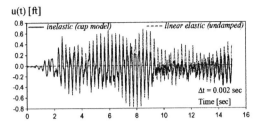

u(t) [ft]

Figure 4. Relative displacement response history of dam crest to the El Centro 1940 record scaled up by a factor of three

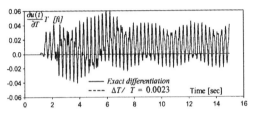

Figure 5. Sensitivity of crest displacement to T

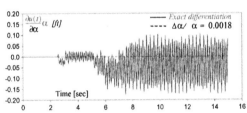

Figure 6. Sensitivity of crest displacement to α

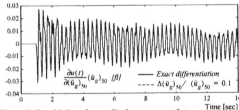

Figure 7. Sensitivity of crest displacement to $(\ddot{u}_g)_{50}$

3.2 Building Frame Structure

A second application example is provided by a five-story single bay steel moment-resisting frame, a finite element model of which is shown in Figure 8. This frame is the object of (a) a nonlinear static push-over analysis under an inverted triangular pattern of horizontal loads applied at floor levels as shown in Figure 8, and (b) a nonlinear response history analysis for earthquake base excitation. The frame is modeled using a displacement-based simplified distributed plasticity 2-D beam-column element implemented in FEAP. The source of material nonlinearity is the moment-curvature relation, which is modeled using

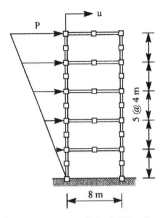

Figure 8. Five story moment-resisting building frame model

the 1-D J_2 plasticity model with linear kinematic hardening and zero isotropic hardening (i.e., the moment curvature is modeled as bilinear). The axial force - longitudinal strain relation is taken as linear elastic and uncoupled from the flexural behavior, and the effects of shear deformations are neglected. All columns and beams of the frame are W21x50 steel I-beams with a yield moment M_y = 384,200 N-m. A 20 percent post-yield to initial flexural stiffness ratio is assumed. A material mass density of 4 times the mass density of steel is used to account for typical additional masses (e.g., slabs, cross-beams, floors, ceilings, ...). The frame has an initial fundamental period of 0.52 sec.

Figure 9 shows the computed nonlinear static push-over curves for each of the floors and roof (i.e., floor or roof displacement versus roof force P).The sensitivities of these nonlinear static push-over curves to the yield moment M_y of all beams and columns and to the linear kinematic hardening modulus H_{kin} (which governs the post-yield to pre-yield flexural stiffness ratio) are plotted in Figures 10 and 11, respectively. The same figures contain finite difference calculations

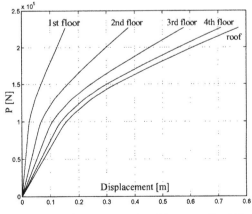

Figure 9. Floor displacements vs. lateral force at roof level (P)

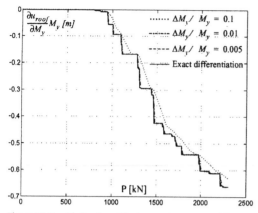

Figure 10. Sensitivity of roof displacement response to M_y

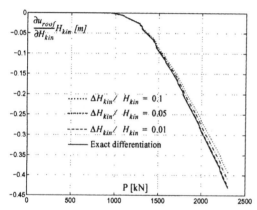

Figure 11. Sensitivity of roof displacement response to H_{kin}

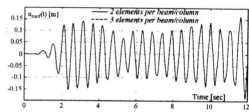

Figure 12. Roof earthquake displacement response history

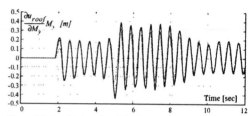

Figure 13. Sensitivity of roof displacement with respect to M_y

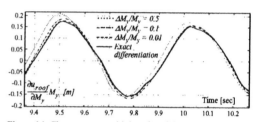

Figure 14. Close-up on sensitivity of roof displacement with respect to M_y

of the response sensitivities validating the sensitivities obtained through exact differentiation of the numerical finite element scheme.

The computed roof displacement relative to the ground when the frame is subjected to the 1940 El Centro record scaled up by a factor of three is given in Figure 12. It is observed that the two levels of discretization (2 vs. 3 frame elements per beam/column) yield identical results, indicating convergence of the global frame response with respect to the spatial discretization. The normalized sensitivity of the computed roof displacement with respect to the yield moment M_y of all columns and beams is displayed in Figure 13. A close-up is given in Figure 14, which shows convergence of the finite difference sensitivity results to the exact result as the size of the parameter increment ΔM_y decreases.

Response sensitivities with respect to material parameters as well as loading parameters of plasticity-based finite element models of structures are discontinuous in time (fictitious time for quasi-static loading). Such discontinuities are clearly shown in Figures 10 and 11 for quasi-static loading and are also

present, although more "smeared out" and therefore less visible, for dynamic loads such as earthquake excitation. These discontinuities along the time axis correspond to switchings between material states, i.e., switching between elastic and plastic state and vice versa, at Gauss points (Kleiber et al. 1997, Conte and Vijalapura 1998, 1999). The method presented is able to capture these discontinuities which result from the upward propagation of the discontinuities in derivatives of history variables at the element (or Gauss point) level. Finally, It is worth mentioning that the exact sensitivity of any computed response quantity (global or local, kinematic or static) can be obtained using the method presented above.

4 CONCLUSIONS

A method to compute the exact sensitivities with respect to material parameters and discrete loading parameters of any response quantity computed from a displacement-based finite element model of a structural system is presented. This method consists of differentiating exactly the numerical finite element algorithm with respect to the sensitivity parameters and requires extension of standard finite element analysis programs used for response calculation only. The method is efficient, since at each time/load step, it uses the factorization of the dynamic consistent tangent stiffness matrix available from the response calculation at convergence.

Several application examples are presented including a concrete gravity dam with the multi-axial state of stress/strain cap plasticity model in plane strain condition, and a two-dimensional moment-resisting structural frame with distributed plasticity formulated in stress/strain resultants using the uniaxial version of the J_2 plasticity model. The "exact" response sensitivity computation scheme and its implementation in a general-purpose nonlinear finite element analysis program (FEAP) are validated through finite difference calculations of various response sensitivities of the structural models considered.

Beside their intrinsic value in providing insight into system response, sensitivities of the computationally simulated response of a system represent an essential ingredient for gradient-based optimization methods needed in structural reliability analysis, structural optimization, structural identification, finite element model updating, and structural health monitoring. The response sensitivity computation method presented here is in the process of being integrated in the new software framework *OpenSees* developed under the auspice of the Pacific Earthquake Engineering Research (PEER) Center to simulate the seismic response of structural and geotechnical systems. See http://opensees.berkeley.edu for more information on this new computational simulation platform.

5 ACKNOWLEDGEMENTS

The work was supported by the main Italian Electricity Company (ENEL-CRIS) and by the Earthquake Engineering Research Centers Program of the National Science Foundation under Award Number EEC-9701568 to the Pacific Earthquake Engineering Research (PEER) Center at the University of California, Berkeley. Opinions and finding presented are those of the authors and do not necessarily reflect the views of the sponsor or PEER.

REFERENCES

Conte, J. P. & Jagannath, M.K. 1995a. Seismic reliability analysis of concrete gravity dams. *A Report on Research Sponsored by the Main Italian Electricity Company (ENEL)*, Dept. of Civil Engineering, Rice University, Houston, Texas, USA.

Conte, J.P. et al. 1995b. Earthquake response sensitivity analysis of concrete gravity dams. *Proc. 7th Int. Conf. on Applications of Statistics and Probability*, Paris, France, 10-13 July 1995, 395-402.

Conte, J.P. & Vijalapura, P. K. 1998. Seismic safety analysis of concrete gravity dams accounting for both system uncertainty and excitation stochasticity. *A Report on Research Sponsored by the Italian National Power Board (ENEL-CRIS)*, Dept. of Civil Engineering, Rice University, Houston, Texas, USA.

Conte, J.P. et al. 1999. Consistent finite element sensitivities in seismic reliability analysis. *CDROM Proc. 13th ASCE Eng. Mech. Conf.*, The Johns Hopkins Univ., Baltimore MD, USA, 13-16 June 1999.

Kleiber, M. et al. 1997. Parameter sensitivity in nonlinear mechanics: theory and finite element computations. Springer-Verlag.

Simo, J.C., and Hughes, T.J.R. 1998. *Computational inelasticity*, Springer-Verlag.

Taylor, R.L., 1999. FEAP User Manual (can be downloaded from http:\\www.ce.berkeley.edu/~rlt/feap/).

Zhang, Y., and Der Kiureghian, A. 1993. Dynamic response sensitivity of inelastic structures, *Comp. Methods Appl. Mech. Eng.* 108:23-36.

Earthquake Engineering Frontiers in the New Millennium, Spencer & Hu (eds),
© 2001 Swets & Zeitlinger, ISBN 90 2651 852 8

Full-range pushover analysis of a RC frame[1]

J. Qian & J. Zhou
Department of Civil Engineering, Tsinghua University, Beijing, P. R. China

ABSTRACT: To solve the ill-conditioned linear equations (ICLEs) encountered in full-range push over analysis of RC frames, composite structure method (CSM) is proposed. Compared with the methods on the basis of constructing constant differential equations, the mechanical concept of the composite structure method is clear. The absolute convergence of the CSM is proved in this paper. The numerical efficiency of the CSM is shown by 4 examples. Furthermore, push over analysis of a RC frame is carried out to obtain its full-range capacity curve with descending branch. The 2-bay by 2-story reinforced concrete frame was tested under constant vertical loads and reversed lateral loads (Feng Shiping 1985). The analytical full-range capacity curve has good agreement with the test skeleton curve.

1 INTRODUCTION

Pushover analysis is a basic analytical tool for the performance/displacement based seismic design of building structures. By pushover analysis, the base shear versus top displacement curve of the structure, usually called capacity curve, is obtained. The capacity curve obtained by most available pushover analysis programs is always ascending owing to the positive stiffness assumption of the structural elements. It is impossible to obtain the actual elastoplastic deformation capacity of the structure from the above-mentioned capacity curve. Actually while enough members of structural elements reach their load carrying capacity, the capacity curve of the structure reaches its peak and starts to descend and the structure stiffness matrix become ill-conditioned. To solve the ill-conditioned linear equations (ICLEs), composite structure method (CSM) is derived in this paper. The convergence of the CSM is proved. Three numerical examples show that the CSM is efficient. Finally, a 2-bay and 2-story reinforced concrete frame is analyzed by pushover analysis with the CSM. The obtained capacity curve with descending branch has good agreement with the experimental curve.

2 COMPOSITE STRUCTURE METHOD FOR SOLVING ICLES

2.1 *Summary of methods for solving ICLEs*

Solving ill-conditioned linear equations is an impor-

tant topic in numerical algebra. Some researchers have made a great effort in this field. These methods can be classified into two categories based on different theories: iterative methods based on constructing constant differential equations, numerical iterative methods based on directly processing the coefficient matrix of the equations. A predictive correction method for solving rigid constant differential equations was proposed in (Han Tianmin, 1966, Han Tianmin, 1976). Methods based on the constant differential equations are also given in (Wu Xinyuan, 1994, 1995, 1999). Some methods directly modify the coefficient matrix (Mao Xianjin & Yang Lingying, 1999), numerically iterate to the exact solution of the equations. These methods are applicable only to positive-determined coefficient matrix, moreover, the modification of the coefficient matrix is difficult to be determined, hence, not suit to programming. In this paper, a direct numerical method, called the composite structure method, based on structural mechanics is proposed to solve the ICLEs. The CSM constructs an auxiliary stiffness matrix from the modal structure, and superpose the modal structure stiffness matrix and the auxiliary structure stiffness matrix to numerically iterate to the exact solution vector. The absolute convergence of the CSM is derived in the paper. The method is not only conceptually clear, programmable, but also converges rapidly. It's numeric efficiency also illustrated by solving 3 numerical examples by the CSM and the method proposed in (Wu Xinyuan, 1999). The CSM could also be generalized to solve the negative stiffness problems by only slighter modification.

---

[1] Supported by Essential Project of the National Natural Science Foundation (59895410), P. R. C.

2.2 Composite structure method

To take the status variation into consideration, the inelastic structural analysis procedure is carried out incrementally. At the end of each loading step, the status of the elements is checked and modified, if it is necessary. The structural stiffness matrix of the next step is based on the accumulated displacement. Under each i th loading step, the equilibrium requirement, $K_i(\sum_{i=0}^{i-1} \bar{D}_i) \cdot \bar{D}_i = \bar{P}_i$, be satisfied, where, $\bar{P}_i$ is the load vector of the i th step; $\bar{D}_i$ is the corresponding displacement vector (or displacement field); $K_i(\sum_{i=0}^{i-1} \bar{D}_i)$ is the stiffness matrix of the structure of the i th step. The current stiffness matrix $K_i(\sum_{i=0}^{i-1} \bar{D}_i)$ changes with the variation of the inelastic status of the members of the structure. It is the function of the structure's accumulated displacement field $\sum_{i=0}^{i-1} \bar{D}_i$ of the i th step (inelastic status of the modal structure is assumed to be independent of the loading path, and is determined only by the accumulated displacement field $\sum_{i=0}^{i-1} \bar{D}_i$ of the structure. This assumption has no effect on the derivation of the CSM for solving ICLEs). From Figure 1, it can be seen that, when the incremental load approaching the peak of the capacity curve of the structure, the displacement vector $\delta \bar{D}_i$ increase significantly with only slight change of the load vector $\delta \bar{P}_i$. It means that the stiffness matrix of the structure $K_i(\sum_{i=0}^{i-1} \bar{D}_i)$ is ill-conditioned, and the structure becomes unsteady. In order to make the structure sustain load steadily, a auxiliary constraint structure is superposed to the real structure to construct a steady imaginary composite structure. The auxiliary constraint structure has 3 properties: 1) Any component of the displacement vector be identical with the counterpart in the real structure. 2) Any variation of one component should not affect the otherwise component of the displacement field; 3) The stiffness of any component is sufficient enough to maintain the composite structure as a steady one, in other words, the composite structure stiffness matrix is a non- ill-conditioned matrix. Condition 2 assures the iterative result converges to the exact displacement field. Condition 3 assures that the process for solving the ICLEs is numerically steady.

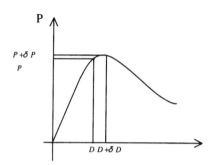

Figure 1. Structure model of the ICLEs

Let's assume that the structure matrix K_i of the i th step is positive determined (but not singular, a singular matrix theoretically has no numeric solution. In process of numerical computation, due to the presence of the rounding errors, the absolute singular matrix is impossible to exist). The eigenvalue, by ascending sequence, is:

$$\lambda_{K_i,1} \geq \lambda_{K_i,2} \geq \lambda_{K_i,3} \geq \ldots \geq \lambda_{K_i,n} > 0, \ \lambda_{K_i,n} \approx 0.$$

Under this circumstance, the loading status corresponding to the linear equations is unsteady. To make it steady, an auxiliary structure with all 3 properties above mentioned maybe constructed. The stiffness matrix of the auxiliary structure is K_i', $K_i' = \alpha \cdot I$. I is a unit matrix. α is the spectral radius of the structural stiffness matrix K_i. That is to say, $\alpha = \lambda_{K_i,1}$. The auxiliary structure constructed in this ways meets the above-mentioned requirements. The stiffness matrix of the composite structure is $K = K_i + K_i'$, the minimum eigenvalue is $\alpha + \lambda_{K_i,n} > \alpha$, the maximum eigenvalue is $\alpha + \lambda_{K_i,1} = 2 \cdot \alpha$. If the composite structure stiffness matrix K is positively symmetrical (in the practice of the FEM analysis of the engineering structure. This is often the case). The condition value of the matrix mesured by 2-norm is:

$$1 \leq cond(K) = \frac{\alpha + \lambda_{K_i,1}}{\alpha + \lambda_{K_i,n}} < 2$$

which means the condition of the stiffness matrix is healthy. The maximum eigenvalue of the structure stiffness matrix could be obtained by means of exponential iterative method.

As the equilibrium requirement, the following be satisfied in the composite structure $K \cdot \bar{D} = \bar{P}$. In order to assure the displacement field of the composite structure is identical with those of the exact one, the following iteration be conducted. The iteration difference, $\varepsilon_n = \left\| \vec{P} - K_i \cdot \vec{D}_n \right\|$, could be expressed by any of the vector norm (the latter 3 numerical examples in the paper employ the vector's 1-norm), the accuracy requirement is ε_{req}. The initial state: $\vec{D}_0 = \vec{0}$, where, $\vec{0}$ is zero vector.

$$\delta\vec{B}_0 = \vec{P} - K_i \cdot \vec{D}_0 = \vec{P} \tag{1}$$

The first iteration:

$$\left(K_i + K_i'\right) \cdot \delta\vec{D}_1 = \delta\vec{P}_0 \tag{2a}$$

$$\delta\vec{D}_1 = \left(K_i + K_i'\right)^{-1} \cdot \delta\vec{P}_0 \tag{2b}$$

$$\vec{D}_1 = \vec{D}_0 + \delta\vec{D}_1 \tag{2c}$$

The error is $\varepsilon_1 = \left\| \vec{P} - K_i \cdot \vec{D}_1 \right\|$. If $\varepsilon_1 \leq \varepsilon_{req}$, then stop the iteration procedure, otherwise, continue to the following steps:

$$\delta\vec{P}_1 = 2 \cdot \left(\vec{P} - K_i \cdot \vec{D}_1\right) \tag{3}$$

The m th iteration:

$$\left(K_i + K_i'\right) \cdot \delta\vec{D}_m = \delta\vec{P}_{m-1} \tag{4a}$$

$$\delta\vec{D}_m = \left(K_i + K_i'\right)^{-1} \cdot \delta\vec{P}_{m-1} \tag{4b}$$

$$\vec{D}_m = \vec{D}_{m-1} + \delta\vec{D}_m \tag{4c}$$

The difference is $\varepsilon_m = \left\| \vec{P} - K_i \cdot \vec{D}_m \right\|$. If $\varepsilon_m \leq \varepsilon_{req}$, then stop the iteration procedure, otherwise, continue to the following:

$$\delta\vec{P}_m = 2 \cdot \left(\vec{P} - K_i \cdot \vec{D}_m\right) \tag{5}$$

The iteration procedure continues until the accuracy requirement is satisfied. The iteration procedure absolutely converges to the exact displacement field $\vec{D}^*$. The convergence is proved as follows.

From the above mentioned iteration procedure, we have the following iteration relationship:

$$\delta\vec{P}_n = \left[I - 2 \cdot K_i \cdot \left(K_i + K_i'\right)^{-1} \right] \cdot \delta\vec{P}_{n-1}$$

Let's make $T = I - 2 \cdot K_i \cdot \left(K_i + K_i'\right)^{-1}$, then the iteration relationship could be expressed by:

$$\delta\vec{P}_n = T \cdot \delta\vec{P}_{n-1} = T^n \cdot \delta\vec{P}_0 = T^n \cdot \vec{P} \tag{6}$$

If $\lambda_{K_i,m}$ is one of the eigenvalue of the matrix K_i, and the corresponding eigenvector is $\vec{\Phi}_{K_i,m}$, the corresponding eigenvector $\vec{\Phi}_{T,m}$ of the matrix T is

identical with $\vec{\Phi}_{K_i,m}$, and the corresponding eigenvalue $\lambda_{T,m}$ is $\left(a - \lambda_{K_i,m}\right)/\left(a + \lambda_{K_i,m}\right)$. It is easy to see that $0 \leq \lambda_{T,m} < 1$ is always true, and the spectral radius $\rho(T)$ is less than 1.0. We know from the compressive mapping theory that as long as the spectral radius of the matrix T is less than 1, $\delta\vec{P}_n = 0 \Leftrightarrow \rho(T) < 1$, then the iterative procedure of the CSM above mentioned for solving ICLEs converges absolutely.

It is possible that the coefficient matrix has a negative eigenvalue $\lambda_{K_i,m}$, and the corresponding eigenvector is $\vec{\Phi}_{K_i,m}$. If the right-hand vector of the ICLEs is $\gamma \cdot \vec{\Phi}_{K_i,m}$, $\gamma > 0$, it's easy to see that the solution of the equations is $\vec{D} = \gamma/\lambda_{K_i,m} \, \vec{\Phi}_{K_i,m}$, where, $\left(\gamma/\lambda_{K_i,m}\right) < 0$. It means that although the structure is loading, the displacement decreases (the displacement vector and the force vector are assumed to have the same positive direction). This stage just corresponds to the descending phenomena of the structure. It is the negative stiffness problem encountered in the practice of the engineering analysis. When the structure is in this stage, the coefficient matrix of the linear equations has at least one negative eigenvalue, then the iteration procedure above mentioned is divergent. Under this circumstance, we could perform the following transformation, $\vec{P}' = K_i \cdot \vec{P}$, $C = K_i \cdot K_i$. If $\vec{\Phi}_{K_i,m}$ is an eigenvector of the matrix K_i, the corresponding eigenvalue is $\lambda_{K_i,m}$, $\lambda_{K_i,m}$ is non-zero and $C \cdot \vec{\Phi}_{K,m} = K_i \cdot \left(K_i \cdot \vec{\Phi}_{K,m}\right) = \lambda_{K_i,m}^2 \cdot \vec{\Phi}_{K,m}$, so that the corresponding eigenvector of the matrix C, $\vec{\Phi}_{C,m}$, is identical with $\vec{\Phi}_{K_i,m}$, and the associated eigenvalue is $\lambda_{C,m} = \lambda_{K_i,m}^2 > 0$, and it is positive. Applying the iterative procedure above mentioned to the equations $C \cdot \vec{D} = \vec{P}'$, it is easy to see that the exact solution of the equations $C \cdot \vec{D} = \vec{P}'$ is identical with that of the equations $K_i \cdot \vec{D} = \vec{P}$.

From the numerical comparison, we can see that the CSM for solving ICLEs presented in the paper is not only accurate, but also converges rapidly.

2.3 Numerical examples

The CSM is used to solve 3 numerical examples. For sake of comparison with the method proposed in (Wu Xinyuan, 1999), the difference is expressed by the 1-norm of the residual vector as well).

Example 1 Solving the ICLEs: $A \cdot \vec{X} = \vec{B}$, where,

$$A = \begin{bmatrix} 3.4336 & -0.5238 & 0.67105 & -0.15272 \\ -0.5238 & 3.28326 & -0.73051 & -0.2689 \\ 0.67105 & -0.73051 & 4.02612 & 0.01835 \\ -0.15272 & 0.01835 & 0.01835 & 2.75702 \end{bmatrix}$$

$$\vec{B} = (-1.0 \ , \ 1.5 \ , \ 2.5 \ , \ -2.0)^T$$

With CSM, after iterating for 22 steps, the results are:

$x_{22}^1 = -0.39771799265$, $x_{22}^2 = 0.510053607822$,
$x_{22}^3 = 0.782983724032$, $x_{22}^4 = -0.702916129739$,
$\varepsilon_{22} < 10^{-10}$.

In (Wu Xinyuan 1999) the step interval $h = 1.0$ is prescribed, after iterating for 29 steps, the results is:

$x_{29}^1 = -0.39771799265$, $x_{29}^2 = 0.510053607798$,
$x_{29}^3 = 0.782983724052$, $x_{29}^4 = -0.702916129751$,
$\varepsilon_{29} < 10^{-10}$.

Example 2 Solving the badly ill-conditioned Hilbert linear equations: $H \cdot \vec{X} = \vec{B}$, where,

$$h_{i,j} = \frac{1}{i + j - 1}, (i, j = 1, 2, 3, 4, 5), b_i = \sum_{j=1}^{5} h_{i,j} \cdot j.$$

With CSM, after 15852 steps, the results are:

$x_{15852}^1 = 0.99945889$, $x_{15852}^2 = 2.0108751$
$x_{15852}^3 = 2.9511235$, $x_{15852}^4 = 4.0758358$
$x_{15852}^5 = 4.9621841$, $\varepsilon_{15852} < 10^{-6}$.

In (Wu Xinyuan 1999) the step interval $h = 1.0$ is prescribed, after 200000 steps, the results are:

$x_{200000}^1 = 0.99948919$, $\qquad x_{200000}^2 = 2.0103609$
$x_{200000}^3 = 2.9531972$, $\qquad x_{200000}^4 = 4.0728545$
$x_{200000}^5 = 4.9635884$, $\qquad \varepsilon_{200000} < 10^{-6}$

The exact solution of the example 2 is:
$\vec{X}^* = (1.0 \ , \ 2.0 \ , \ 3.0 \ , \ 4.0 \ , \ 5.0)^T$

Example 3 Solving the ICLEs: $A \cdot \vec{X} = \vec{B}$, where,

$$A = \begin{bmatrix} 5.0 & 7.0 & 6.0 & 5.0 \\ 7.0 & 10.0 & 8.0 & 7.0 \\ 6.0 & 8.0 & 10.0 & 9.0 \\ 5.0 & 7.0 & 9.0 & 10.0 \end{bmatrix}$$

$$\vec{B} = (23.0 \ , \ 32.0 \ , \ 33.0 \ , \ 31.0)^T$$

With CSM, after 26469 steps, the results is :

$x_{26469}^1 = 0.9999999951$, $x_{26469}^2 = 1.000000003$
$x_{26469}^3 = 1.000000001$, $x_{26469}^4 = 0.9999999993$
$\varepsilon_{26469} < 10^{-10}$.

In (Wu Xinyuan, 1999), the step interval $h = 0.01$ is prescribed, after 165800 steps, the results are:

$x_{165800}^1 = 0.9999999901$, $x_{165800}^2 = 1.000000006$
$x_{165800}^3 = 1.000000002$, $x_{165800}^4 = 0.9999999985$
$\varepsilon_{165800} < 10^{-10}$.

The exact solution of the example 3 is:

$$\vec{X}^* = (1.0 \ , \ 1.0 \ , \ 1.0 \ , \ 1.0 \ , \ 1.0)^T$$

The computation was carried out on a P III 450 personal computer, the real is double precision. From above 4 numerical examples, we can see that the CSM converges rapidly for highly ill-conditioned linear equations. Further more, the CSM can be generalized to solve the negative stiffness problem. While analyzing a structure, it is impossible to know the exact eigenvalue of the stiffness matrix of the structure, for sake of generalization, the CSM has a bright prospect for structure analysis.

3 PUSHOVER ANALYSIS OF A RC FRAM

3.1 *The configuration of the sample frame*

A 2-bay by 2-story reinforced concrete frame (Feng Shiping 1985) is studied by the push over analysis employing the CSM. The frame's elevation and the elements' dimension are shown in Figure 2. The span of the frame is 2.025 m, the story height is 1.2 m. The measured $150 \times 150 \times 150 \, mm^3$ concrete cubic strength is 20.77 N/mm². As the elastic modulus of the concrete was not given in (Feng Shiping 1985), it is taken as 2.55×10^4 N/mm² according to (Concrete Structure Design Code (GBJ10-89)). The elements are reinforced by grade, steel bars. The elastic modulus of steel bars is 2.0×10^5 N/mm² according to the GBJ10-89. The measured yield strength of the bar is 362.9 N/mm². To assure a strong column weak beam frame, the designed resistant moment ratio of beam to column is 0.42~0.78. Moreover, to design a ductile frame, 6mm diameter ties at interval of 75 mm are adopted for beams and columns.

To simulate the dead and live loads, the frame was subjected to three downward concentrated loads which were 106 kN, 211 kN and 106 kN respectively. The concentrated loads were kept constant during the loading process. The lateral force was applied by a beam with the arm ratio of 1/2, hence the lateral force was reverse-triangle distributed.

3.2 *M - Φ relationship of element cross sections*

In most current push over analysis programs, the representative of $M - \Phi$ relationship of beam and

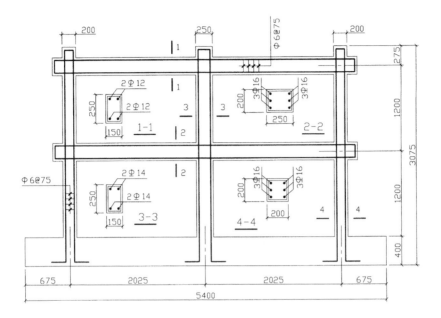

Figure 2. The configuration of the sample plane frame

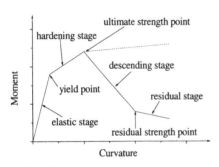

Figure 3. $M - \Phi$ relationship of the cross-section

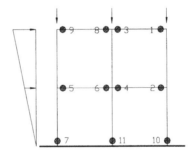

Figure 4. The failure mode of the structure

column cross sections is linearized as the dash line in Figure 3. To simulate the behavior of beams and columns more accurately, the linearized $M - \Phi$ relationship with descending branch as the solid line in Figure 3 is adopted in this paper. The slopes of the first two lines are positive, and the other two are negative. The third line represents the phenomena that the element releases its internal force that has been sustained to the other elements. It is also called softening stage. The fourth line is the residual strength stage. It is introduced for the reason that the cross section will never have negative internal forces during the monotonically increasing lateral loading. The slopes of four lines depend on many factors such as the strength grade of concrete, the longitudinal reinforcement area, the amount of the tie or stir-

rup, the axial force level, etc. For the beams and columns of the sample plane frame, the first line of linearized $M - \Phi$ relationship is from the original point to the yield point. The slope of the first line is the elastic or pre-yield bending stiffness of the cross section. To take into consideration of cracking of concrete, it is taken as 0.8 times of the product of the elastic modulus of concrete E and the moment of inertia of cross section I, i.e., $0.8EI$. The yield moment of cross section is calculated by the method proposed in (Feng Shiping 1985).

3.3 Pushover analysis result

The sequence of forming the rotating hinge (a rotate hinge is developed when the cross section of an element reaches its ultimate strength. A plastic hinge is

407

Table 1. Comparison of the pushover analysis results and the test results of a RC frame

	First plastic hinge		Ultimate strength		Ultimate state		Δ/H =1/50	
	Test	Analysis	Test	Analysis	Test	Analysis	Test	Analysis
V (kN)	-	63.6	166.6	162.7	141.6	138.3	163	161.8
Δ/H (m/m)	-	1/620	1/59	1/77	1/25	1/27	1/50	1/50
δ_1/h (m/m)	-	1/629	1/67	1/81	-	1/31	-	1/57
δ_2/h (m/m)	-	1/613	1/53	1/74	-	1/24	-	1/44

Note: V -Base shear, Δ-Top displacement, H -Total height, δ_1, δ_2-Story drift of the first story and the second story respectively, h -Story height.

developed when the cross section of the element reaches its yielding point.) of the sample frame by the push over analysis is shown in Figure. 4. The location of the rotate hinges is the same as the test result. The path through the rotating hinges is generally called weak line. The structure deforms along this path without increasing of its resistance. In current push over analysis programs, the concentrated plastic hinge model is assumed, it means that the hinge length is zero. This assumption stiffens the structure, and the analytical displacement is less than the actual value, and makes the evaluation unsafe. In this paper, the distributed plastic hinge model is adopted. The length of the hinge is assumed to be equal to the efficient height of the cross section. Figure 5 shows two full-range capacity curves.

The solid line was obtained by the push over analysis. The dash line was obtained by the test. They are in good agreement. Figure. 6 shows the story shear versus story drift ratio curves obtained by the push over analysis. The test curves are not available in reference (Feng Shiping 1985). But some characteristic points were given in table 1. Table 1 shows the comparison between the test results and the push over analysis results. In the table, the ultimate state means that the base shear descends to 85% of the maximum base shear. Due to the ductile design of the sample frame, the drift ration δ/H of the ultimate state is much larger than 1/50. To assess the seismic resistant capacity of a structure, the ductility ratio is one of the most referable measurement. The experimental and analytical top displacement ductility ratio, defined as $\mu_\Delta = \Delta_u/\Delta_y$, is 6.1 and 5.7 respectively. The experimental and analytical story drift ductility ratio of first story, defined as $\mu_\delta = \delta_u/\delta_y$, is 5.5 and 4.6 respectively. Δ_y and δ_y are yield top displacement and yield story drift which determined by so called general yield moment method. Δ_u and δ_u are top displacement and story drift at ultimate states.

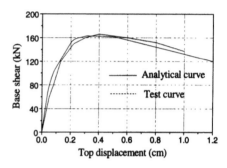

Figure 5. The full-range capacity curve

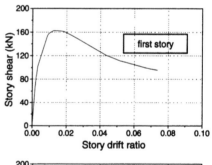

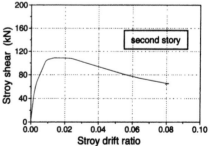

Figure 6. Analytical curves of story shear versus story drift ratio

Without the CSM for solving the ICLEs, the descending branch of the base shear versus top displacement curve, the descending branch of the story shear versus story drift ratio curve, the ultimate drift ratio of the sample frame could not be obtained by the push over analysis. In this case, it is difficult to determine whether the frame has enough elasto-plastic deformation capacity to resist the severe earthquakes.

4 CONCLUSION

By adopting CSM in the push over analysis, the full-range capacity curve of building structure can be obtained. The push over analysis is performed to a 2-bay by 2- story reinforced concrete plane frame. The analytical full-range capacity curve of the frame has good agreement with the test curve. From the descending branch of the full-range capacity curve, the ultimate deformation capacity, as well as the ductility ratio of the structure can be obtained.

REFERENCES

Concrete Structure Design Code (GBJ10-89), China Building Industry Press, P.R.C.

Feng Shiping 1985, *Study on the Post Yield Behavior of the Reinforced Concrete Structure*, Thesis for PhD degree. Tsinghua University.

Han Tianmin 1966, A method of numerical solution for ordinary differential equations, *Applied Mathematics & Mathematics Computation*, 3(3): 187-191.

Han Tianmin 1976, A method of numerical solution for initial value problem in stiff ordinary differential equations, *Science in China*, 1: 21-34.

Giuseppe Faella 1996, Evaluation of The R/C Structures Seismic Response by Means of Nonlinear Static Push Over Analyses. *Eleventh Word Conference on Earthquake Engineering*. Paper No. 1146.

Mao Xianjin & Yang Lingying 1999, Simplified iterative method for ill-conditioned linear equations. *Physical and Chemical probation Computation technology*, l21(1): p14-18.

Sang-Dae Kim et al. 1999, A Modified Dynamic Inelastic Analysis of Tall Buildings Considering Changes of Dynamic Characteristics. *The Structure Design of Tall Buildings*. 8: 57-73.

Vojko Kilar & Peter Fajfar 1996, Simplified Push Over Analysis of Building Structure. *Eleventh Word Conference on Earthquake Engineering*. Paper No. 1011.

Wu Xinyuan 1994, The Method of Ordinary Differential Equations for Solving Linear Algebraic Equations, *Numerical Algebra (Proceedings of 92' Shanghai International Numerical Algebria and Its Applications conference*, Editor Jiang Er-xiong),*China Science and Technology Press*.

Wu Xinyuan, Ouyang Zixiang 1995, Applications of ODE Method, *The Second Asian Mathematical Conference*, Thailand :Nakhon Ratchsima.

WK Tso & AS Moghadam 1998, Push Over Procedure for Seismic Analysis of Buildings. *Progress in Structural Engineering and Materials*. 1(3): 337-344.

Earthquake Engineering Frontiers in the New Millennium, Spencer & Hu (eds),
© 2001 Swets & Zeitlinger, ISBN 90 2651 852 8

Study on shaking table tests of building models

M. Zhang, X. Gao, & Q. Meng
Institute of Engineering Mechanics, China Seismological Bureau, Harbin, Heilongjiang, China

ABSTRACT: Based on the Buckingham's π theorem a general similitude law was developed. The law is useful for the design of building models without enough artificial mass and for scaling live loads and mass of non-structural components in models. Numerical simulations analyzed the test errors, which were caused by the models without enough artificial mass and by use of the non-prototype material in models. Some examples of shaking table tests of high-rise buildings were described in the paper.

1 INTRODUCTION

In the areas of seismic engineering shaking table tests are powerful method for assessing the seismic capacity of buildings. The tests can avoid a series of technique difficulties, such as the establishment of motion equations of the non-linear system and their solutions. Not only the seismic responses of buildings were obtained, but the damages of buildings can be observed by eyes also in the tests.

However the size and bearing capacity of existing shaking tables are limited, we have to design scale models in tests. In the design of scaling models the adoption of model materials, similitude law and the estimation of relative test errors should be researched. In recent ten years many building models with small scale were designed and tested in IEM, some of the research results were described in the paper.

2 THE GENERAL SIMILITUDE LAW

By Buckingham's π theorem we can transfer a dynamic equation of a building under earthquakes into a dimensionless motion equation, and then the similar ratio of physical quantities in the equation can be determined, the dimensional theorem provided a traditional method to establish the similitude law. Based above method the artificial mass law and the gravity force-neglecting law have been set up and used in shaking table tests. However because the size and bearing capacity of tables is limited, the artificial mass model (AMM) with a large number of artificial mass sometimes can not be achieved. If we use the gravity force-neglecting model (GFNM)

without any artificial mass, the model not only will cause test errors after the model structure damaged, but also require more outputs and still broader frequency band of tables. Due to above mentioned difficulties we have to design the lack artificial mass model (LAMM) without enough artificial mass, the LAMM model is situated between the AMM model and the GFNM model, so that a new similitude law have to be developed.

Considering the difference of AMM, GFNM and LAMM model is that the more or less of the attaching mass was installed on models, we can standardize three laws with the aid of the equivalent mass density of models. The equivalent mass density ratio of model to prototype d_r can be expressed as

$$d_r = (m_m + m_a + m_{om}) / [l_r (m_p + m_{op})] \qquad (1)$$

where m_m = the mass of structural members in the model; m_a = the achieved artificial mass in the model; m_{om} = the achieved mass to simulate the live loads and non-structural members in the model; l_r = length ratio of model to prototype; m_p = the mass of structural members in the prototype; m_{op} = the mass of live loads and non-structural members in the prototype.

Based on the Buckingham's theorem, selecting the length, modulus and mass density as basic physical quantities, a general similitude law can be developed as following:

It is easy to prove that the artificial mass law and the gravity force-neglecting law are only two special cases of the general similitude law. Moreover we should pay attention to simulation of live loads and non-structural members, which were contained in the general similitude law. Live loads and non-

411

Table 1. The general similitude law

Scaling ratio	By prototype material	By any material
Length l_r*	l_r	l_r
Modulus E_r	1	E_r
Mass density d_r	d_r	d_r
Time t_r	$l_r d_r^{0.5}$	$l_r d_r^{0.5} E_r^{-0.5}$
Frequency f_r	$l_r^{-1} d_r^{-0.5}$	$l_r^{-1} d_r^{-0.5} E_r^{0.5}$
Velocity v_r	$d_r^{-0.5}$	$d_r^{-0.5} E_r^{0.5}$
Acceleration a_r	$l_r^{-1} d_r^{-1}$	$l_r^{-1} d_r^{-0.5} E_r$
Strain ε_r	1	1
Stress σ_r	1	E_r

*Footnote r denotes the ratio of a quantity of model to prototype.

structural members generally have tremendous mass, neglecting their mass effects will not obtain proper estimations of the seismic capacity of buildings.

The enough artificial mass m_a^e to simulated structural members can be expressed as

$$m_a^e = m_p E_r l_r^2 - m_m \qquad (2)$$

The enough artificial mass m_{om}^e to simulated live loads and non-structural members similarly can be expressed as

$$m_{om}^e = m_{op} E_r l_r^2 \qquad (3)$$

3 TEST ERRORS OF LAMM MODEL

The similitude laws were developed only for the elastic system, if shaking table tests were carried out for linear-elastic deformation stage, the all of test results of AMM, LAMM and GFNM model are consistent and correct. However we are especially interested in non-elastic performance of buildings. In the non-elastic tests only the results of AMM models made by the prototype material are proper, the tests of LAMM and GFNM model will lead some errors because the artificial mass is not enough. Lack of artificial mass will reduce the vertical loads, so that the stress state of model members is different from those of the prototype, the seismic performance of models will change. For example, decrement of vertical stresses in brick walls will result in reduction of their capacity of resistance seismic lateral forces. Numerical simulations of non-elastic seismic responses of different models were carried out to research the errors. We assumed that simulations of seismic responses of AMM model are correct, and then compared the calculation results of AMM with LAMM models for multistories brick houses, in which the mass and stiffness of the structure is homogeneous. The analysis model of shear force - deformation for building stories and calculation results of seismic responses can be seen in figure1 and figure 2.

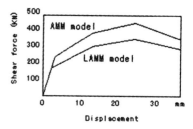

Figure 1. The constitutive curve for building stories

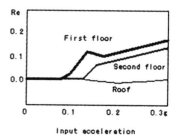

Figure 2. The errors of seismic responses of models

In figure 2 the test error ratio R_e were defined below,

$$R_e = (R_L - R_A) / R_A \qquad (4)$$

where R_L = seismic responses of LAMM model; R_A = seismic responses of AMM model. We can see in the meantime that the responses of lower stories of LAMM model were overestimated, the responses of upper stories deceased due to damage of first story. The simulation results are consistent with our experiences. It is quite evident that test errors of LAMM model can not avoid, but numerical simulations indicated that when the total mass of a LAMM model achieved the 75% of total mass of the AMM model, the test errors is a minor quantities can be neglected.

4 TEST ERRORS DUE TO MODEL MATERIALS

We always cannot use the prototype material to make the models of RC buildings because the size of structural members is smaller than the aggregate size of building concrete. In practices the Plexiglas sometimes was used as model material. Because the mechanics characteristics of Plexiglas is completely different from concrete, the use of the Plexiglas models is limited, they only can be used to measure the mode parameters and elastic seismic responses, this kid of results can be obtained simply by numerical calculations and not have to do tests.

In order to carry out the experimental study of structural damages and the non-linear seismic responses, the proper material to make small models is necessary. The micro-concrete, which was composed of cement, sand and infinitely small cobble, has been used to make RC building models. Table 2 showed the equivalent modulus of micro-concrete, the results were obtained by axial pressure tests of samples.

It can be seen in table 2 that modulus of micro-concrete is less than normal concrete. The lower modulus is very useful for design of models, because it has contributed to reduce the amount of artificial mass in models. The data in the table 3 indicated that equivalent modulus ratio of micro-concrete to normal concrete is about a constant in a widely strain range, so that the mechanics property of micro-concrete is similar to the normal concrete. We can use the micro-concrete to make models of RC buildings and then carry out non-linear seismic tests. However the test data of concrete materials make known that when the axial pressure stress is close to the extreme strength, the equivalent modulus ratio will decease slightly. The modulus ratio is not a strict constant, this phenomenon will cause difficult of model design. If we are extremely interested in

the structural damages, the modulus ratio $E_{r1.0}$ can be used to establish the similitude law, but it is regrettable that all of the structural members cannot damage simultaneously in tests. Numerical simulations indicated while the structural damage occurred in lower stories, the responses of no-damaged members in upper stories will be underestimated.

Therefore the model materials, which are not completely similar to the prototype materials, certainly caused some of test errors, we should explain the test results with errors.

5 EXAMPLES OF MODEL TESTS

5.1 Post and telecommunication building

The building is located in Shenzhen, where the basic seismic intensity is VII. The building consists of a main and a sub building, podium is between them. The main building is RC frame-tube structure of 51-story (202m). The sub building is RC frame structure of 23-story (104m). Podium is frame structure of 7-story (39m). The building model is LAMM model designed by the general similitude law. The model was made by micro-concrete, total weight of structural members is 84 KN., and achieved artificial mass is 120 KN. The similar ratios of different physical quantity were showed in table 4.

Table 2. Equivalent modulus of micro-concrete

Modulus (10^4N/mm^2)	The axial pressure strength (Mpa)		
	2.5	5.0	10
$E_{0.1}$*	0.68	1.12	2.05
$E_{0.2}$	0.66	1.09	1.99
$E_{0.3}$	0.64	1.06	1.93
$E_{0.4}$	0.62	1.02	1.87
$E_{0.5}$	0.60	0.99	1.80
$E_{0.6}$	0.57	0.94	1.72
$E_{0.7}$	0.54	0.90	1.63
$E_{0.8}$	0.51	0.84	1.53
$E_{0.9}$	0.46	0.76	1.39
$E_{1.0}$	0.38	0.61	1.13

* $E_{0.1}$ is the second modulus in 10% of axial pressure strength.

The equivalent modulus ratio of micro-concrete to normal concrete showed in table 3.

Table 3. The equivalent modulus ratio

Modulus ratio	Strength (micro-concrete / normal concrete)			
	10/30	5.0/40	2.5/40	2.5/50
$E_{r0.1}$*	0.587	0.293	0.177	0.146
$E_{r0.2}$	0.587	0.293	0.177	0.145
$E_{r0.3}$	0.587	0.292	0.177	0.144
$E_{r0.4}$	0.587	0.292	0.177	0.144
$E_{r0.5}$	0.587	0.292	0.177	0.142
$E_{r0.6}$	0.587	0.292	0.177	0.141
$E_{r0.7}$	0.587	0.291	0.177	0.139
$E_{r0.8}$	0.586	0.290	0.176	0.136
$E_{r0.9}$	0.586	0.288	0.175	0.131
$E_{r1.0}$	0.579	0.270	0.169	0.115

*$E_{r0.1}$ is the second modulus ratio of micro-concrete to normal-concrete in 10% of the axial pressure strength.

Table 4. Similar ratio of the post and telecommunication building

Physical quantities	Similar ratio
Length	0.030
Modulus	0.134
Mass density	3.310
Time	0.149
Frequency	6.706
Acceleration	1.348
Velocity	0.201
Strain	1.000
Stress	0.134

The results of mode tests indicated that the fundamental period of main building is 2.7seconds, and after damaged the period has increased by 50%. The model tests carried out under the input seismic intensity of VI, VII and VIII. Under the input acceleration of 0.25g (intensity VIII) the acceleration response in the roof of main building is 0.45g, the displacement response is 13cm, and some of members damaged. The shear tube of main building at 6-story damaged and lift wells of sub building at 9,13 and 15-story cracked. In the design of prototype building the SRC columns were adopted at 1—6 story in the main building, but other columns are RC columns with larger cross section. There are not shear walls except of lift wells in sub building. Above-mentioned design of structures produced unfavorable influences of seismic capacity.

5.2 Shenzhen world trade building

The high-rise building is RC frame-tube structure of 54-story (232m). The structures of the building is RC frame around lift well of "+" type. Micro-concrete was used to make the model. In the lower story there are big space with transfer Truss. The model is LAMM model designed by the general similitude law, and the non-constant of the modulus ratio was considered. In other words, the different similar ratios were adopted for different test stages (see table 5). The fundamental period of the building is 2.7seconds, and the period was changed to 4.5seconds after damaged. Under the inputs of intensity VII and VIII, the maxim acceleration responses are 0.305g and 0.417g, and the maxim displacement responses are 23cm and 62cm respectively.

Table 5. Similar ratio of the world trade building

Physical quantities	Similar ratio	
	Elastic	Non -elastic
Length	0.029	
Modulus	0.172	0.138
Mass density	3.886	
Time	0.138	0.154
Frequency	7.255	6.498
Acceleration	1.526	1.225
Velocity	0.210	0.188
Strain	1.000	
Stress	0.172	0.138

It is evident that the building is quite a flexible structure. There are not evident weak stories in the building, the transfer truss with support members are in good condition. Under the inputs of intensity VIII, some of outside members (shear wall, beam and column) in middle and upper stories cracked, but the damages are not serious. The stranger responses of upper stories are related to effects of high modes and torsion vibrations.

6 CONCLUSIONS AND DISCUSSIONS

The research works indicated that:

a. It is feasible and beneficial to quantitative the seismic responses of prototype buildings and to assess their seismic capacity by use the shaking table tests of small-scale models.
b. Considering the limited size and bearing capacity of the shaking tables, the general similitude law is useful in design of building models.
c. There are some difficult to simulate the exact non-elastic behavior of prototype buildings by the small-scale model. The test errors cannot be avoided, so that we should pay the attention to explain the test results.
d. The good materials for making models remain to be developed, and the simulation theorem and technique of the composite structures is a pending further research.

REFERENCES

Sedov, L.I., 1980. Similarity dimensional methods in michanics. Moscow; Russia.
Zhang, M.Z., 1997. Study on similitude laws for shaking table tests. Earthquake engineering and engineering vibration. 17(2).
Zhang, M.Z., Guo, X. & Dai,J. W., 1998. Shaking table tests of a high-rise building model. *Proc. of the Fifth Intern. Conf. on Tall Buildings*. Hong Kong.

Earthquake Engineering Frontiers in the New Millennium, Spencer & Hu (eds),
© 2001 Swets & Zeitlinger, ISBN 90 2651 852 8

Experimental research and seismic design of RC frame with segmental short columns

Q. Hu & Y. Xu
Beijing Institute of Architectural Design & Research

Y. Hao, G. Kang, & Z. Li
Tianjin University

ABSTRACT: Inferior behavior of reinforced concrete short column had been proved by past earthquake lessons. Many measures had been suggested to improve its behavior, e.g., special detailing of reinforcements and the adoption of SRC, but in case of very short column (shear-span ratio<1.5), the above measures show ineffectiveness or complexity in construction. Segmental column is a "short column", whose section is splitted into four segments within the clear height of column, gypsum boards are used for separation. Experimental researches had been carried out, relating to column element, beam-column joint and frame model. The obvious effectiveness proves its applicability in seismic design of frame structures. Design recommendations have been suggested.

1 EXPERIMENTAL STUDY OF SEISMIC BEHAVIOR OF SEGMENTAL COLUMN

1.1 Test specimens and testing instrumentation

12 specimens of short columns including 10 segmental short columns and 2 ordinary short columns had been tested under cyclic loading.

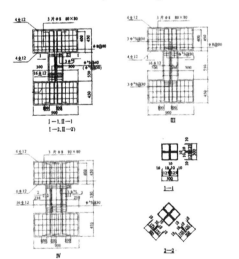

Figure 1. Test specimens of segmental columns

Test results show the displacement ductility of ordinary short column μ_Δ=1–1.9, corresponding to shear span ratio λ=1.0–1.4 and volumetric hoop ratio ρ_v=1.33. As to segmental column μ_Δ=4.1—9.0 corresponding to λ=2.2–3.0 and ρ_v=1.19–2.37.

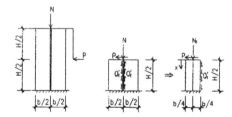

Figure 2. Testing instrumentation

1.2 Elastic displacement of segmental column

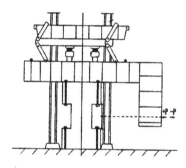

Figure 3. Decomposition of segmental column

$$Q_f' = \frac{3}{4}\beta\frac{P}{b} \tag{1}$$

Q_f' — Shear force along the height of column, due to the restraining of diaphragm board;

β — Coefficient of monolithic effect due to dia -

415

phragm. $\beta=1$, for integrate column. $\beta=0$, for segmental column without diaphragm board.

Elastic displacement of segmental column Δe due to lateral force P:

$$\Delta e = \frac{PH(4-3\beta)(H^2+3b^2)}{Ecb^4} \qquad (2)$$

Ec — Modulus of elasticity of concrete column.
The above expression disregards N-Δ effect.

1.3 Stiffness of segmental column

The stiffness degradation coefficient of segmental column with respect to integrate column can be expressed by following expression:

$$\lambda = \frac{1}{4-3\beta} \qquad (3)$$

$$\beta = \frac{4}{3} - \frac{Ecb^4\Delta}{3PH(H^2+3b^2)} \qquad (4)$$

From the above expression and test results of different specimens, the relation between β and story drift Δ/H in elastic and elasto-plastic stage can be expressed by Figure 4 and formula (5).

$$\beta = \frac{\dfrac{\Delta}{H}}{1.2(\dfrac{\Delta}{H}-0.3)^2 + \dfrac{\Delta}{H}} - 0.07 \qquad (5)$$

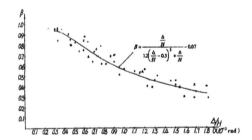

Figure 4. β–Δ/H relationship

Test values of β at ultimate strength are 0.29-0.45, in average β can be taken as 0.30. Tests show variation of hoop content and compressive stress has no obvious effect on β. Knowing β, stiffness of segmental column can be obtained. The above study explains segmental column provides elastic behavior under minor earthquake, and adequate deformability during major earthquake.

1.4 Strength of segmental column

Test specimens of segmental columns have the following characteristics:
  Shear-span ratio of unit column=2.2–3.0,
  Compressive stress ratio=N/(4f_cb$_1$h$_1$)=0.40–0.79,
  Volumetric hoop ratio ρ_v=(1.19–2.37)%,
  Length of transition section=(0–1/7.3)H.

1.4.1 Flexural strength of segmental column under compression

Experimental study explains flexural strength of segmental column ($M_c^{exp.}$) is lower than that of integrate column, but it is higher than the summation of the flexural strength of unit column without diaphragm board ($M_c^{th.}$), $M_c^{th.}/M_c^{exp.}$=0.68–0.96. For simplicity segmental column can be designed according to equal distribution of total moment and axial load among the unit columns.

1.4.2 Shear strength of segmental column

Tests show shearing strength of segmental column can be taken as the summation of strength of all unit columns.

1.4.3 Transition section of segmental columns

The variation of hoop strain within transition section shows obvious effect to stop cracking along diaphragm board.

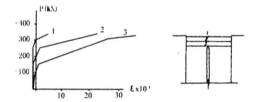

Figure 5. Hoop strain within transition section

2 EXPERIMENTAL STUDY OF BEAM COLUMN JOINT WITH SEGMENTAL COLUMN

2.1 Test specimens and results

4 interior joints and 4 exterior joints had been tested (Figure 6). Interior and exterior joints with segmental columns on top and bottom of joint cores provide better behavior against shear strength degradation of joint core. Comparing specimens with integrate column and segmental column, joint core of integrate column exhibits more serious pinching effect of hysteretic loops. Segmental column exerts greater compressive stress on longitudinal reinforcement of beam, resulting the strengthening of bond stress between concrete and reinforcement, less pinching effect of hysteretic loops and higher shear strength of joint core. Segmental column without transition section causes stress concentration at joint core and twin-core phenomena, which may affect the integrity of joint core under high axial load (Figure 8).

416

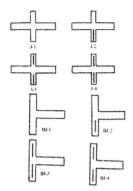

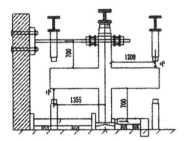

Figure 6.

Figure 7. Instrumentation of joint test

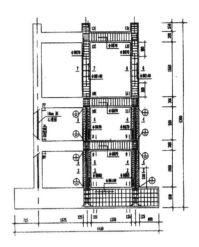

Figure 9a. Frame model

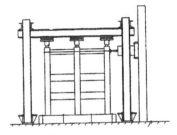

Figure 9b. Instrumentation of testing

The sequence of plastification and failure mode prove that frame with segmental columns can satisfy the requirement of "strong column weak beam" and "flexural yielding before shear failure" (Figure 10).

$$\Delta_{max.}^{exp.}/H = \frac{1}{51}, \quad \Delta_u^{th} = \frac{1}{34}, \quad u_\Delta = 6$$

$\Delta_{max.}^{exp}$ —Experimental value of top displacement;

$\Delta_u^{th.}$ — Top displacement corresponding to 0.85 P_{max};

u_Δ —Displacement ductility factor.

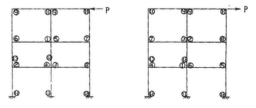

Figure 10. Sequence of plastification

Figure 8. Twin core phenomena

3 EXPERIMENTAL RESEARCH OF FRAME MODEL WITH SEGMENTAL COLUMNS

Frame model scaled to 1/3 of actual frame is designed according to the ductility requirements, seismic category 1 of Chinese code GBJ 11-89 (Figure 9).

Segmental columns are adopted in first and second story. Shear span ratio of 2nd story columns according to over all size is 1.9, length of transition section is 50mm. Stiffness of segmental column is taken as 80% of integrate column, max. compressive stress ratio of segmental column is 0.6.

417

4 DESIGN RECOMMENDATION OF FRAME STRUCTURE WITH SEGMENTAL SHORT COLUMNS

4.1 *Limitation for the adoption of segmental short column*

4.1.1 Segmental column is feasible for short column whose shear span ration $\lambda<1.5$. Segmental column should not be adopted when inflection point of column does not exist within the story.

4.1.2 Segmental column should have the same over all dimensions with columns of adjacent stories (Figure 11).

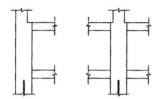

Figure 11.

4.1.3 Eccentricity between center line of column and center line of beam should not be more than ¼ width of column. When beam width is less then 1/2 width of column, horizontal haunch may be adopted. (Figure 12)

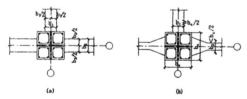

Figure 12.

4.1.4 Segmental column should not be used for boundary columns of seismic walls.

4.2 *Detailing and design of column section*

4.2.1 Unit columns of segmental column can be square or rectangular section. Side dimensions of unit column should not be less than 400mm. Side length ratio of unit column should not be greater than 1.5. Splitting of column should not be more than once in each direction (Figure 13).

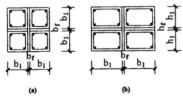

Figure 13.

4.2.2 Transition sections should be left at both ends of segmental column and depth of transition section should not be less than 100mm (Figure 14)

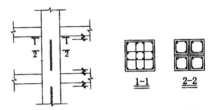

Figure 14.

4.2.3 In the elastic analysis of frame structure consisting of ordinary column and segmental columns, the stiffness of segmental column can be taken as 70% of the stiffness of integrate column having the same overall section.

4.2.4 Flexural strength of segmental column under compression can be evaluated by the summation of flexural strength of unit columns. Moment, axial load, and shear can be equally distributed between unit columns. A surplus eccentricity should be considered, taking the larger value between 15mm and 1/30 of the side dimension of unit column.

4.2.5 Compressive stress ratio, minimum amount of longitudinal reinforcement, volumetric hoop ratio of segmental column can be taken according to requirement of ordinary column.

4.2.6 Closely spaced hoops should be arranged along whole height of segmental column.

4.2.7 Axial loads exerted on unit columns of exterior column should be multiplied by magnifying factor 1.2.

4.2.8 Hoops in beam-column joint core and transition section of segmental column should consist of circumferential hoop and inner hoops (Figure 15), diameter of hoops should not be smaller than 10mm, diameter of circumferential hoop should be 2mm more than inner hoop.

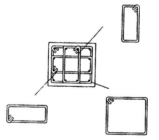

Figure 15. Hoops in joint core and transition section

418

Earthquake Engineering Frontiers in the New Millennium, Spencer & Hu (eds),
© 2001 Swets & Zeitlinger, ISBN 90 2651 852 8

Seismic design and retrofit using shape memory alloys

R. DesRoches, R. Leon, & G. Hess
Georgia Institute of Technology, Atlanta, USA

J. Ocel
University of Minnesota

ABSTRACT: This study evaluates the use of Shape Memory Alloys as passive dampers in applications in buildings and bridges. Using the *shape memory* characteristic of Nitinol, an innovative steel partially restrained connection with SMA tendons is tested. The connection is subjected to several loading cycles at large rotations, exhibiting no strength degradation. After completing a series of testing cycles, the tendons were heated through their transformation temperature and recovered their previous undeformed shape. The connection was retested and displayed nearly identical performance. Using the *superelastic* property of Nitinol, SMA restrainers are tested for application in bridges. Analytical studies show that these devices, used at intermediate hinges and abutments are effective in limiting the relative hinge displacement and overall demands on piers.

1 INTRODUCTION

Recent earthquakes in the United States and abroad have highlighted the vulnerability of structures to moderate and strong ground motion. In the United States, the October 17, 1989 Loma Prieta earthquake and the January 17, 1994 Northridge earthquake produced direct economic losses in excess of $50 billion (Eguchi et al., 1998). Similar losses resulted from the January 17, 1995 Kobe earthquake in Japan (Comartin et al., 1995). One effective method for limiting the damage to structures is through the use of base isolation and passive energy dissipation. A considerable number of applications using energy dissipation and seismic isolation exist in the US, Japan, New Zealand, and other highly seismic countries (Skinner et al., 1993; Soong and Dargush, 1997). The current devices used can be classified as friction-based dampers, viscous and viscoelastic dampers, and hysteretic dampers. While research and applications have proven effective, they have inherent shortcomings, hampering more widespread application of these technologies.

In this paper, the efficacy of using Shape Memory Alloys (SMAs) as passive dampers is investigated through a series of experimental and analytical studies.

2 BACKGROUND ON SHAPE MEMORY ALLOYS

2.1 *General properties*

Shape memory alloys are a class of alloys that display several unique characteristics, including Young's modulus-temperature relations, shape memory effects, and high damping characteristics. Unlike plastically deforming metals, the nonlinear deformation is metallurgically reversible. This unique "shape memory" characteristic is a result of a martensitic phase-change that can be temperature induced or stress-induced. Although several alloys exhibit the shape memory property, the most widely used shape memory alloy is Nitinol, which is a combination of Nickel and Titanium. Nitinol can have three different forms: martensite, stress-induced martensite, and austenite. When the material is in its martensite form, it is soft and ductile and can be easily deformed. In its austenitic form, it is quite strong and hard (similar to Titanium). The NiTi material has all these properties, their specific expression depending on the temperature in which it is used, as shown in Figure 1. Below the martensite finish temperature, M_f, the material has the *shape memory effect* – the strain can be can be recovered by heating the specimen. At a temperature above the austenite finish, A_f, and below M_d – the maximum temperature to strain induce martensite, stress-induced martensite is formed (known as the *superelstic effect*). At an operating temperature above M_d, the material is austenitic and undergoes ordinary plastic deformation, with very high strength.

Shape memory alloys possess several characteristics that make them desirable for use as a passive energy dissipation in structures. These characteristics include: (1) hysteretic damping; (2) highly reliable energy dissipation based on a repeatable solid state phase transformation; (3) excellent low - and

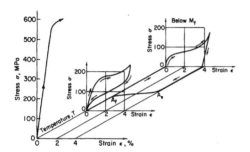

Figure 1. Three-dimensional stress-strain-temperature diagram showing the deformation and shape memory behavior of a Ni-Ti alloy (Duerig et al., 1990)

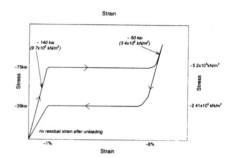

Figure 3. Typical stress-strain selationship showing superelastic characteristic ($M_d > T > A_f$)

high-cycle fatigue properties; and (4) excellent corrosion resistance (Duerig et. al., 1990). The sections below describe the characteristics of the *shape memory* and *superelastic* properties of SMAs.

2.2 *Shape memory characteristics (SME)*

The stress-strain diagram for Nitinol exhibiting the shape memory effect (SME) is shown in Figure 2. After initial elastic deformation, the material yields at a constant stress due to detwinning of crystal layers. If the material is unloaded in this region, residual strain remains. The residual strain is recovered by heating the material above its austenite finish temperature (A_f). As the material cools, it returns to its original martensitic phase. If the material is strained beyond approximately 8%, plastic flow initiates and the SME can be incomplete.

2.3 *Super-elastic characteristics (SEE)*

Super-elasticity (or pseudo-elasticity) refers to the ability of NiTi to return to its original shape upon unloading after a substantial deformation. This is based on a stress-induced martensite formation. The austenitic phase is stable before the application of a stress, but at some critical stress the martensite becomes stable, causing yielding and a stress plateau, as shown in Figure 3. Since the martensite is only stable because of the applied stress, the austenite structure again becomes stable during unloading,

and the original undeformed shape is recovered. The unloading occurs at a lower stress level due to transformational hysteresis. When the stress is released, the martensite transforms back into austenite and the specimen returns to its original shape.

3 INNOVATIVE STEEL CONNECTIONS USING THE SHAPE MEMORY CHARACTERISTIC OF NITINOL

3.1 *Background*

The extensive damage to steel moment resisting frame (SMRF) structures during the 1994 Northridge and 1995 Kobe earthquakes has led to significant research efforts to try to develop more effective SMRF structures. SAC, a joint venture of the Structural Engineers Association of California (SEAOC), the Applied Technology Council (ATC), and the California Universities for Research in Earthquake Engineering (CUREe), was formed to study the steel moment-resisting frame problems and has published several reports and interim repair and retrofit guidelines (FEMA 273, 1997).

One of the SAC research initiatives was to reinvestigate the effectiveness of various types of partially-restrained (PR) connections as an alternative to fully restrained (FR) welded connections. As an extension to this research, the use of innovative materials in connections is explored. This study looks at the behavior of a beam-column joint with SMA tendons, exhibiting the shape memory behavior. The tendons, 34.9 mm (1.38 inch) diameter rods are connected to the column from the top and bottom flanges of the beam. The specimen is tested according to SAC protocol, reheated above the austenite finish temperature, then re-tested again.

3.2 *Experimental test setup*

The connection tested consisted of a W24x94 beam connected to a W14x159 column, as shown in Figure 4. The shear tab was specially designed with

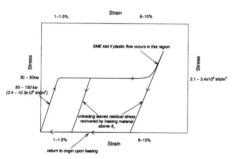

Figure 2. Typical stress-strain relationship showing shape memory characteristic ($T < M_f$)

Figure 4. Steel connection with SMA tendons

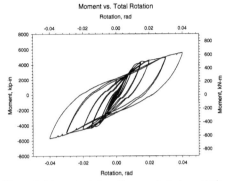

Figure 5. Moment rotation diagram of steel connection with SMA tendons

a circular hole in the center and two long-slotted holes on either side to allow the beam to freely rotate about the center bolt without bearing elongation of the shear tab.

Four 381 mm (15") long SMA tendons were threaded into anchorages specially designed to allow the tendons to resist loads in both tension and compression. The beam anchorage includes two rectangular tubes welded on three sides with a fillet to the beam flange.

The SMA material used in this test was a fully annealed, binary NiTi alloy with an austenite finish temperature of 200° F.

3.3 Loading and instrumentation

The loading protocol used for the test was taken from the SAC Steel Project protocol load history, with the intermittent fatigue cycles omitted. The deformation parameter used to determine the loading history was the interstory drift, defined as the beam deflection divided by the beam span.

The beam, column, and SMA tendons were fully instrumented with a combination of axial strain gauges, rosette gauges, LVDT's, load cells, and string potentiometers. In addition, the temperature in the SMA tendons was measured using 4 temperature gauges and 1 thermal pyrometer.

3.4 Test results

The connection was tested at increasing cycles up to 4% drift. At peak drifts during the 1.5% load step, both of the top anchorage blocks were noticed to have slipped out of plane away from the beam flange. This movement, due to the unrestrained translational degree of freedom at the top of the tendons appeared to be gradual. The movement of the top anchorage blocks away from the beam flanges was allowed for in the design to limit the bending

stresses in the tendons. At the 3% drift load case, the lack of the translational restraint of the top anchorage block allowed the compression side tendons to completely buckle in side-sway. After the first cycle at 4% drift, severe warping of the beam caused by the side-sway buckling of the tendons produced out-of-plane displacements of approximately 25.4 mm (1"). At this point, a residual displacement of 23.4 mm (0.92") of tip displacement was measured, indicating residual deformation in the SMA tendons. Figure 5 shows the moment versus total rotation curve for the first series of tests. The 3% and 4% drift cycles are much flatter than the early cycles, indicating that significant yielding in the tendons has occurred. Second, there are no pronounced stiffness increases at large rotations in the higher cycles, indicating that the hardening region of the SMA material was not encountered. Overall, the moment rotation curves show a very stable and repeatable hysteresis, with excellent energy dissipation capacity. Since the beam/column remained elastic throughout testing, the connection's deformation components at the maximum tip displacement can be proportioned as follows: approximately 95% in tendons, and 5% in both the column and beam.

3.5 SMA rebound (Memory recovery)

The SMA tendons are heated beyond their austenite finish temperature in an attempt to trigger the "memory effect", thus returning the SMA tendons to their original shape. Following the initial test, a residual tip deformation of 23.4 mm (0.92") was measured in the SMA tendons. A heat tape was used to heat the top tendons. The heat tape was used because of its ability to wrap around the specimen and uniformly heat the tendon. However, the rate of heating was slow and the maximum temperature reached was 284 °F. The heat tape was ineffective in initiating the "shape memory" effect, producing a tip displacement recovery of less than 2.54 mm (0.10"). To provide a high level of heat at a rapid rate, two

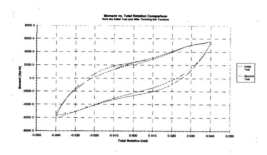

Figure 6. Comparison of first test series to second test series (after re-heating SMA bars)

Figure 7. Unseating of bridge deck during 1999 Taiwan Earthquake (Mahin et al. 1999)

butane torches were used to heat the top tendons. The tendons were heated at a temperature of 550 F. During the first 5 minutes, the actuator force decreased steadily from 5 kips to 2 kips, equivalent to a 19 kip release in the two tendons. When the bottom heating was completed, 15.2 mm (0.60") of tip displacement was recovered, giving a total of 17.8 mm (0.70") tip displacement.

After recovering most of the displacement, the connection was re-tested. The modified loading history was composed of one cycle at 1%, 2%, 3%, and twenty cycles at 4% drift. Starting in the 3% cycle, pronounced buckling of the tendons was observed. In the 4% cycles, loud pounding noises, corresponding to the tendons buckling against the beam side of the bottom rectangular tubing were heard. The testing continued through seven cycles at 4% drift when the first tendon fractured immediately below the top anchorage block. The fracture occurred directly at the fillet termination.

Figure 6 shows a comparison of the moment vs. total rotation plot before and after heating. As shown in the Figure, the hysteresis for the 1st testing series was nearly identical to that of the 2nd testing series.

4 INNOVATIVE RESTRAINERS USING SUPERELASTIC SHAPE MEMORY ALLOYS

4.1 Background

Bridge damage in recent earthquakes has highlighted the vulnerability of bridges to moderate and strong ground motion. During the 1999 Taiwan earthquake, over 40 bridges sustained major damage or collapse (Uang et al., 1999). During the Kobe earthquake, 60% of all bridges in the city were damaged, with a total cost estimated at over 10 billion US dollars for their repair (Comartin et al., 1995). Significant damage also occurred in the 1989 Loma Prieta earthquake (Benuska et al., 1990). A major problem in all these earthquakes was excessive movement at the hinges due to bearing and restrainer failure, as shown in Figure 7. These earthquakes highlight

the need for more effective unseating prevention devices (restrainers) to limit the relative displacement between bridge decks and reduce the likelihood of collapse from unseating. In this study, *superelastic* shape memory alloy restrainers are evaluated to determine their effectiveness as restrainers/dampers in bridges.

4.2 Application of SMA restrainers at hinges

Superelastic SMAs possess ideal properties for use as restrainers in bridges. They have an elastic strain capacity of nearly 6%, providing them with highly reliable energy dissipation based on a repeatable solid state phase transformation. In addition, at large strains, the material strain hardens, providing added protection against unseating.

The shape memory alloy restrainers can serve as both restraint devices and dampers and thus will serve multiple roles. First, they can limit the relativedisplacement between frames, thereby reducing the risk of collapse from unseating of frames at the hinge. Second, they can be used to reduce the negative effects of pounding between bridge frames. Finally, by concentrating damage and energy dissipation in controlled locations, these devices can be used to reduce the demand on individual frame in a multi-frame bridge.

4.3 Experimental testing of SMA restrainers

To evaluate the efficacy of SMA restrainers, a prototype restrainer was designed and tested. The restrainer consists of a 305 mm long (12") – 25.4 mm diameter (1") bar made of Nitinol. The bar was machined and subsequently heat treated to produce superelastic properties. The bars were tested under cyclical loading (tension) cycles up to 8 percent strain. Previous studies of large diameter shape memory alloys indicate that the superelastic properties of large diameter bars are not as pronounced as that of small diameter bars (Wayman, 1977). However, as shown in Figure 7, superelastic properties can be achieved in large diameter bars with a proper combination of cold working and annealing. The SMA bars in this study were subjected to several cycles at 6%, and 8% strain, with residual strains less than 1%.

4.4 Case study – multi-span simply supported bridge

The relative displacement of multiple spans simply supported bridges at the hinges and at the abutments provides an opportunity to dissipate energy through devices such as the SMA restrainers. To evaluate the effectiveness of SMA restrainers, an analytical model of a typical MSSS bridge with SMA dampers, shown in Figure 8, is developed using the DRAIN-2DX nonlinear analysis program (Prakash, ets al., 1992). Full composite action between the deck-slab and steel girders is assumed. The superstructure is modeled using linear elastic elements, since it is expected to remain linear for the cases considered. Nonlinear elements are used to model the columns, abutments, restrainers (SMA and conventional restrainers), bearings, and impact between decks.

A case study is conducted to illustrate the effectiveness of SMA restrainers compared with conventional restrainers. The conventional cable restrainers are modeled as a bilinear spring that only resists tensile forces. The SMA restrainers are modeled as trilinear elements that only act in tension. The properties are based on the results from the experimental tests of the 25.4 mm (1") diameter rods. The response of the bridge with the SMA restrainers and the conventional restrainers is compared for the Oakland Harbor Record, from the 1989 Loma Prieta earthquake, scaled to a peak ground acceleration of 0.70 g.

The results of the time history analysis of the multi-span simply supported bridge are summarized in Table 1. The first 2 columns show the effect of the SMA damper on the pier ductility demands. Typically, restrainer devices are effective in limiting the relative hinge opening at the expense of increasing forces in the piers. This is evident from Table 1 with the cable restrainers. The ductility of pier 1 is slightly increased compared to the case without restrainers. However, the SMA damper produces a 20% and 33% decrease in the ductility demand on pier 1, and pier 2, respectively. The relative displacement for the Oakland Harbor record between abutment 1 and span 1 (Δ_1), span 1 and pier 1 (Δ_2), and span 2 and pier 1 (Δ_3) are shown in the second column of Table 1. Since the bridge is symmetric about the centerline, the displacements at pier 2 and abutment 2 are similar to that at abutment 1 and pier 1. The maximum relative displacement at the abutment without any devices, Δ_1, is 111 mm (4.37"). The maximum relative displacement with conventional restrainers and SMA dampers is 55 mm

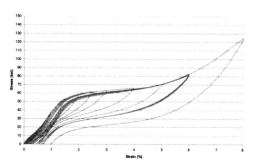

Figure 8. Stress vs. strain diagram for SMA restrainer (25.4 mm diameter bar)

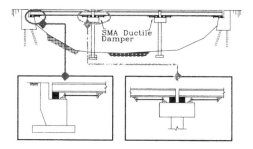

Figure 9. Multi-span simply supported bridge with SMA restrainer/damper

Table 1. Results of nonlinear time history analysis of bridge with SMA restrainers

Oakland Harbor (1989 Loma Prieta), Scaled to 0.70g					
	Pier 1 μ	Pier 2 μ	Δ_1 (mm)	Δ_2 (mm)	Δ_3 (mm)
As built	5.12	5.12	111	48	68
Typical restrainers	5.41	4.56	55	47	72
SMA restrainers	4.00	3.43	32	30	36

(2.17") and 32 mm (1.26"), respectively. Significant reductions are also shown for Δ_2 and Δ_3. The effectiveness of the SMA dampers compared with conventional restrainers is due to the superelastic properties of the SMA dampers. The rods can yield repeatedly without permanent deformation, leading to significant energy dissipation capabilities. Conventional restrainers suffer residual deformation following yielding cycles. This reduces the available energy dissipation capacity of the restrainer.

5 CONCLUSIONS

Shape Memory Alloys exhibiting either the shape memory effect or the superelastic effect possess characteristics that make them ideal for applications in passive control of building and bridge structures. Using the Shape Memory effect, innovative steel connections with SMA tendons are evaluated. The connections are tested to drifts of approximately 4%, showing stable and repeatable behavior with significant energy dissipation. Following the test series, the SMA tendons are reheated beyond their transformation temperature, and retested. The tendons recovered 80% of their original shape and the connection had nearly identical behavior to the first testing series.

Using the superelastic property of shape memory alloys, SMA restrainers in bridges are evaluated through experimental and analytical studies. Case studies of a multi-span bridge show that the SMA restrainers are much more effective than conventional restrainers. The damping provided by the SMA restrainers reduces relative hinge displacements while also reducing the demands in the columns.

6 ACKNOWLEDGEMENTS

This work has been funded in part by the Transportation Research Board IDEA program (Award N-635), E*Sorbs Corporation, and Earthquake Engineering Research Centers Program of the National Science Foundation under Award Number EEC-9701785.

REFERENCES

Benuska, L. 1990. Loma Prieta Earthquake Reconnaissance Report. *Earthquake Spectra*, (6): 149–188.

Comartin, C., Green, M. , and Tubbesing, S. 1995. The Hyogo-Ken Nanbu Earthquake, Preliminary Reconnaissance Report, EERI, Oakland, CA.

Duerig, T., W., Melton, K. N., Stockel, D., Wayman, C. M. 1990. Engineering Aspects of Shape Memory Alloys, Butterworth Heinemann, London, 1990.

FEMA 1997. Interim Guidelines Advisory No. – Supplement to FEMA 267, *Federal Emergency Management Agency*, 1997.

Soong, T. T., and Dargush, G. F. 1997. Passive Energy Dissipation Systems in Structural Engineering, Wiley, New York, New York, 1997.

Skinner, R. I., Robinson, W. H., and McVerry, G. H. 1993. *An Introduction to Seismic Isolation*, Wiley, New York, New York, 1993.

Uang, C. M., Elgamal, A., Li, W. S., Chou, C. C. 1996. Ji-Ji Taiwan Earthquake of September 21, 1999 – A Brief Reconnaissance Report. (http://www.structures.ucsd.edu/Taiwaneq /index.html)

Wayman, C. M. 1977. "*Deformation, Mechanisms, and Other Characteristics of Shape Memory Alloys*," Shape Memory Effects in Alloys, J. Perkins, Plenum Press, New York, 1977, pp. 1-27.

Earthquake Engineering Frontiers in the New Millennium, Spencer & Hu (eds),
© 2001 Swets & Zeitlinger, ISBN 90 2651 852 8

Strength reduction factor spectra based on structural damage performance

X. Zhu & Y. Ni

College of Civil Engineering and Architecture, Northern Jiaotong University, Beijing, China

ABSTRACT: It is reviewed the study on reduction factor which allow the estimation of lateral strength required to control the level of inelastic deformation during strong earthquake ground motion. Based on modified Takeda hysteretic model, including the hysteretic effects of post-yield stiffness, stiffness degradation and strength deterioration, the reasonable hysteretic parameters have been developed, and the reasonable relations between the hysteretic parameters and ductility coefficient have been derived. Based on the equivalent ductility coefficient from Park-Ang dual-damage criterion, strength reduction factor spectra respect to different damage indices have been calculated using the regression formulas suggested by L. H. Lee. The spectra can satisfy the different seismic design level. The results of comparison study have shown that the strength deterioration and the cumulative damage of the structure have a remarkable effect on the strength reduction factor and the neglect of the two factors above mentioned might lead to an unconservative design result.

1 INTRODUCTION

In present, the response spectrum method is widely used to design aseismic structures. A reduction factor was generally used to reduce the elastic demand of earthquake action in most seismic design codes. In both Chinese seismic design codes for highway and railway engineering (JTJ004-89[1] and GBJ111-87[2]), a comprehensive influence factor was used to reduce the elastic demand of earthquake action, but it is a ambiguous factor for lack of direct relation between the structure ductility and the factor.

Within the recent 30 years from 1970s, strength reduction factors had been studied by numerous researchers. In general, the strength reduction factor R_μ was defined as the ratio of elastic strength demand imposed on the SDOF system to inelastic strength demand for a given ductility level.

Based on elastic and inelastic response spectra of the NS component of the El Centro, California earthquake of May 18, 1940, as well as on previous studies of the response on simple systems to pulse-type excitations and two other recorded ground motions, Newmark and Hall (1973)[3] had presented a piece-wise curve of strength reduction factors.

J. B. Berrill et al. (1980)[4] pointed out explicitly that "equal displacement rule" can be accepted when the fundamental period T of the structure is greater than 0.7s. They suggested following formulas of strength reduction:

$$R_\mu = 1 + (\mu - 1)\frac{T}{0.7}, \text{ for } 0 \leq T \leq 0.7\text{s};\qquad(1a)$$

$$R_\mu = \mu, \qquad\qquad \text{for T>0.7s.}\qquad(1b)$$

A similar expression of behavior factor q has been suggested in Eurocode 8[5] as

$$q = 1 + (\mu - 1)\frac{T}{T_0}, \quad \text{for } T \leq T_0;\qquad(2a)$$

$$q = \mu, \qquad\qquad \text{or } T>T_0.\qquad(2b)$$

in which constant value of $T_0=0.7$ was suggested.

E.Miranda (1993)[6] have researched the main factors having an effect on strength reduction factors. A series of site- and period-dependent strength reduction factors have been evaluated. The statistical formulas of R_μ have been obtained through the regression analysis. Based on mean strength reduction factors, the following simplified expressions were proposed to estimate the reduction factors:

$$R_\mu = \frac{\mu - 1}{\Phi} + 1 \geq 1\qquad(3)$$

where Φ is a function of μ, T and the soil conditions at the site.

A perfect review about strength reduction factor was made by E. Miranda et al. (1994)[6]. He concluded that strength reduction factor was a function of structural ductility, natural period of the system and site conditions. In their works the bilinear hysteretic model was widely used, and the effects of stiffness degradation, strength deterioration and pinching were neglected.

Y.H. Chai et al. (1998)[7] have studied the strength reduction factors for different a/v(ratio of ground motion peak acceleration and peak velocity) ratios considering the stiffness degradation factor. Y.H.Chai et al. developed an expression of strength reduction factor similar to Eurocode 8. While the difference was that the characteristic period T_0 could be calculated by following equation:

$$T_0 = 0.65T_c\mu_c^{0.3} \tag{4}$$

where T_c = characteristic period of the ground motion as defined by

$$T_c = 2\pi \frac{c_v}{c_a} \frac{v_{g,max}}{a_{g,max}} \tag{5}$$

where the coefficient c_a corresponds to the ratio of spectral elastic response acceleration to peak-ground-acceleration ($a_{g,max}$) in the short period range(acceleration-controlled), and the coefficient c_v corresponds to the ratio of spectral elastic response velocity to peak-ground-velocity ($v_{g,max}$) in the medium period range(velocity-controlled). A constant value of $c_a = 2.5$ was used in their study. For the set of ground motion used to determine the coefficient c_a, however, the coefficient c_v was found to vary from 1.6 to 2.6, and the mean value of $c_v = 2.0$ was used in their study.

Based on Clough's stiffness degradation model, W.D. Zhuo (2000)[8] constructed mean strength reduction factor spectra of three different soil sites, namely that (i) firm site, (ii) medium-firm site and (iii) medium-soft site.

A composite hysteretic model considering the effects of post-yielding stiffness, stiffness degradation, strength deterioration and pinching has been used to develop the influence of all hysteretic parameters to strength reduction factor by L.H. Lee et al. (1999)[9]. A series of regression formula have been suggested to determine the strength reduction factor based on earthquake records from hard and rock site. But the appropriate values of hysteretic parameters have not been discussed.

In this paper a modified Takeda model, which was consistent with the experiment results well, has been studied. A series of semi-analytical relations between hysteretic parameters (including the effects of second slope, stiffness degradation and strength deterioration) and structural ductility ratio have been constructed. Using the equivalent ductility coefficient resulted from the damage index under the well-known Park-Ang dual-damage criterion, the strength reduction factors have been calculated for different damage index according to the regression formula suggested by L.H. Lee et al.

2 EQUIVALENT DUCTILITY COEFFICIENT

2.1 Under Park-Ang criterion

Park and Ang (1985)[10] pointed out that the structural damage under repeated loading was relied on two aspects, namely the maximum deformation (or stress) and the dissipation of the cumulative energy (or repeated cycles). This was so-called the dual-damage criterion. They suggested that the damage index (DI) could be expressed as:

$$DI = \frac{\delta}{\delta_u} + \beta \frac{E_H}{P_y \delta_u} = \frac{\mu}{\mu_u} + \beta \frac{E_H}{P_y \delta_y \mu_u} \tag{6}$$

a proper β value suggested by Williams (1995)[11] was 0.15. In this paper β value is fixed to 0.15.

A formula of equivalent ductility coefficient was presented by Farjfar et. el. (1992)[12] under Park-Ang dual-damage criterion as:

$$\mu = \frac{\sqrt{1 + 4DI\beta\gamma^2\mu_u} - 1}{2\beta\gamma^2} \tag{7}$$

where $\gamma = \frac{\sqrt{E_H/m}}{\omega\delta}$

Farjfar et al. have studied the γ value using eight hysteretic models. Their results show that γ value is a relatively stable parameter within the period of engineering interest. In this study γ value is designated to 0.8.

Y.J. Park et al. (1987)[13] suggested the following detailed classification to describe the structural damage performance:

DI<0.1   No damage or localized minor cracking,

0.1≤DI<0.25   Minor damage – light cracking throughout,

0.25≤DI<0.40   Moderate damage – severe cracking, localized spalling,

0.40≤DI<1.0   Severe damage – crushing of concrete, reinforcement exposed,

DI>1.0   Collapsed.

Subsequently, Ang et al. (1993) [14] suggested using a value of DI=0.8 to represent collapse.

The effects of maximum plastic displacement and the cumulative damage resulted from low cycle fatigue were properly took into account. And the criterion matched the experiment result well. So Park-Ang criterion is used to determine the equivalent ductility coefficient of the structure in this paper.

3 PARAMETRIC STUDY OF COMPOSITE HYSTERETIC MODEL OF PIERS

The modelization of the restoring force curve has been studied extensively since 1960s. In fact, any appearance of nonlinear restoring force of actual

426

structure always resulted from the interaction of multiple physical factors along the duration of ground motion. Therefore, an idealized way must be taking all kinds of factors into account to construct some basic models. The actual process of nonlinear restoring force and deformation could be expressed through the combination and transformation of basic models. This idea resulted in the composite hysteretic model of the restoring force. A modified Takeda model (including the hysteretic effects of second slope, stiffness degradation and strength deterioration), which is consistent with the experiment results of railway piers with low longitudinal reinforcement ratio (ratio of 1.14%) in China, will be adopted. The appropriate values of hysteretic parameters will be evaluated from the adopted model.

3.1 Second slope factor α_1

A sketch map of the modified Takeda model has been presented in Figure.1. Where, P_u and P_y are the ultimate and yielding strength of member, respectively. δ_u and δ_y are the ultimate and yielding displacement of member, respectively. K_e is the equivalent stiffness (secant stiffness). The second slope factor α_1 according to Q. H. Liu[15] could be expressed as:

$$\alpha_1 = \frac{P_u - P_y}{(\delta_u - \delta_y)K_y} = \frac{p-1}{\mu_u - 1} \quad (8)$$

where p is the strength ratio (P_u/P_y), μ_u is the ultimate ductility coefficient resulted from the monotonic loading experiment. Based on the experiment results of Q.H. Liu, we observed that the damage mode of structure would change gradually from shear damage to bend-shear damage up to flexure damage as the increase of shear span ratio and stirrup ratio, and the ductile capability of structure would increase correspondingly. In this paper the experiment results with damage format of flexure will be adopted to determine the hysteretic

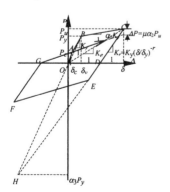

Figure 1. Sketch Map of Modified Takeda Model

parameters above mentioned. A strength ratio p of 1.33 and ultimate ductility coefficient μ_u of 10.0 will be applied in subsequent section.

3.2 Strength deterioration factor α_2

It can be seen from Figure 1 that

$$\Delta P = P_u - P = \mu \cdot \alpha_2 \cdot P_u \quad (9)$$

An alternative expression of ΔP can be obtained as:

$$\Delta P = \left(\frac{\delta - \delta_y}{\delta_u - \delta_y} \right)^\eta \Delta P_f = \left(\frac{\mu - 1}{\mu_u - 1} \right)^\eta \Delta P_f \quad (10)$$

where η is a constant, depending upon the repeated cycles of loading endured by the structure. A value of $\eta=3.0$ was suggested by Q.H. Liu. However, $\eta=3.0$ will lead to the negligible effect of strength deterioration. Through a series of trivial calculation, considering the few repeated cycles of structure endured under major earthquake ground motion, a more reasonable value of $\eta=1.0$ would be adopted. ΔP_f was the maximum decrease value of strength when the structure failed to service any more. Generally, a value of $20\% P_u$ is more acceptable.

The strength deterioration factor can be determined from the equivalent relation of equation (9) and (10) as:

$$\alpha_2 = \frac{\left(\frac{\mu - 1}{\mu_u - 1} \right)^\eta}{\mu} \cdot \frac{\Delta P_f}{P_u} \quad (11)$$

3.3 Stiffness degradation factor α_3

The stiffness degradation factor α_3 can be solved through the coordinate of the intersection of the lines of loading part OB and reloading part of BC based on the sketch map of hysteretic model (Figure.1). The equation of loading part OB of the line can be easily expressed as:

$$P = K_y \Delta \quad (12)$$

Assumed that unloading stiffness K_r as:

$$K_r = K_y \left(\frac{\delta}{\delta_y} \right)^{-r} = K_y \mu^{-r} \quad (13)$$

where r is an exponent of stiffness degradation. The model test results for a great deal of reinforcement columns and beams provided by University of California, San Diego, U.S.A.[16] is shown that when the axial compression ratio was greater than 0.15, the stiffness degradation exponent ranged from 0.5 to 0.7, while the exponent equaled to 0.25 approxi-

427

mately without axial loading. In the subsequent calculation, an exponent r of 0.5 will be adopted.

Therefore the equation of unloading part BC can be obtained through simple geometrical calculation as:

$$P = K_y \mu^{-r} \Delta + \left(1 + \alpha_1 \mu - \alpha_1 - \mu^{1-r}\right) K_y \delta_y \qquad (14)$$

Then the stiffness degradation factor α_3 can be solved from equation (12) and (14):

$$\alpha_3 = \frac{1 + \alpha_1 \mu - \alpha_1 - \mu^{1-r}}{1 - \mu^{-r}} \qquad (15)$$

It is worth noting that the factor α_3 derived from equation (15) must be rectified positively for subsequent application.

Based on the design limit state of the structure or member, for piers (or columns) with flexure damage mode, the above three hysteretic parameters can be determined according to the equation (8), (11) and (15), respectively, having acquired the equivalent ductility coefficient from corresponding damage index.

4 EVALUATION OF STRENGTH REDUCTION FACTOR, R_μ

Based on the parametric study of the composite hysteretic model, the strength reduction factor R_μ can be calculated through the regression formulas suggested by L.H. Lee et al. (1998)[12]. The factor R_μ was assumed to be a function of the characteristic parameters of each hysteretic model, target ductility coefficient and structural period. The regression formula of R_μ was

$$R_\mu = R(T, \mu) \times C_{\alpha1} \times C_{\alpha2} \times C_{\alpha3} \times C_{\alpha4} \qquad (16)$$

where $R(T,\mu)$ was the functional format of the R_μ factor of the elasto-perfectly plastic model which was treated as a basis model in their study. The factors, $C_{\alpha1}$, $C_{\alpha2}$, $C_{\alpha3}$ and $C_{\alpha4}$ were considered as correction factors accounting for the effect of post-yielding (α_1), strength deterioration (α_2), stiffness degradation (α_3), and pinching (α_4) to the factor obtained from the elasto-perfectly plastic model. The effects of each hysteretic parameter to the R_μ factor were detailed elaborated in reference (12).

Because of the mechanical flexure properties in this study, the pinching effect is neglected. Moreover, minor pinching only has negligible effect on R_μ factor. It means that the pinching factor $C_{\alpha4}$ is equal to 1.0.

Based on the studied hysteretic model, the different ductility level μ ranged from 1.717 to 5.302 have been developed corresponding to four different damage indices, namely that DI ranged from 0.25 to

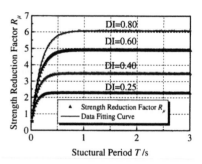

Figure 2. Curves of R_μ and their fitting results

0.80. The spectra curves of strength reduction versus period T have been constructed using equation (16) in Figure 2.

From Figure.2, it can be seen that the inflexion period values in R_μ spectra curves increased gradually as the increase of damage indices DI (or the equivalent ductility coefficient of the structure), and the constant value R_μ of on the right of the inflexion period increased correspondingly, too. It was found by the way of data fitness that the uprising partition of the R_μ spectra curve matched very well with the Boltzman function curve. The expression of Boltzman function was:

$$R_\mu = R_2 + \frac{R_1 - R_2}{1 + e^{(T - T_0)/dT}} \qquad (17)$$

Table 1. Fitting coefficients of R_μ

DI	Boltzman Function: $R_\mu = R_2 + \dfrac{R_1 - R_2}{1 + e^{(T-T_0)/dT}}$			
	R_1	R_2	T_0	DT
0.25	-36.421	2.269	-0.213	0.076
0.40	-135.51	3.464	-0.390	0.106
0.60	-82.484	4.886	-0.389	0.137
0.80	-122.16	6.076	-0.497	0.165

The fitting results of uprising partition of R_μ spectra simulated by Boltzman function for the different damage indices are presented in Table.1.

It can be seen from the spectra curves (Figure.2) with different damage indices that in the medium-period and long-period spectra regions, R_μ values for the different system period had approximately the same values. This phenomenon was consistent with the "equal displacement rule" concluded by Newmark and Hall[3]. The constant values and the inflexion periods of R_μ spectra curves vary linearly with the damage index DI (or the equivalent ductility coefficient μ). The approximate relation(see Figure.3 and Figure.4) for the inflexion period T_g and the constant value of R_μ respect to the equivalent ductility coefficient μ are simulated as following, respectively:

$$T_g = 0.217866\mu - 0.02071 \tag{18}$$

$$R_\mu = 1.01714\mu + 0.38857 \tag{19}$$

5 COMPARISON STUDY OF STRENGTH REDUCTION FACTOR

For the sake of comparison, the strength reduction factors on rock or firm site defined by different seismic code and researchers are calculated. And the R_μ spectra curves are constructed in Figure 5.

It was shown from Figure.5 that R_μ defined by Eurocode 8 equaled to the one suggested by Y.H.Chai after which the structural period T was greater than the inflexion period T_g. The major differences between them were the inflexion period.

R_μ values suggested by W. D.Zhuo were generally greater than the values of this paper within the medium and long period range. It was noted that his ductility coefficient was slightly different with others for a relatively higher value of the failure strength (85% of the initial strength, but 80% of the initial strength is adopted more widely). The R_μ spectra curve of Miranda had a typical peak point. The strength reduction factors evaluated from equation (16) were commonly larger than the R_μ values

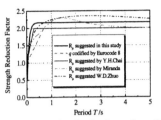

Figure 5a. Strength reduction factor for μ=2

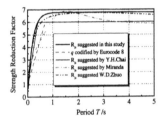

Figure 5b. Strength reduction factor for μ=4

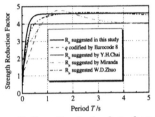

Figure 5c. Strength reduction factor for μ=6

of Y.H. Chai and Eurocode8. It can be explained that the effects of strength deterioration and cumulative damage of the structure on R_μ could not be neglected because of the remarkable reduction of the carrying capacity under the strong earthquake ground motion. It will be more conservative than the formers using the reduction factor suggested in this paper to reduce the elastic after which a descending curve was obtained. This result was not consistent with the well-known "equal displacement rule".

The strength reduction factors evaluated from equation (16) were commonly larger than the R_μ values of Y.H. Chai and Eurocode8. It can be explained that the effects of strength deterioration and be neglected because of the remarkable reduction of the carrying capacity under the strong earthquake ground motion. It will be more conservative than the formers using the reduction factor suggested in this paper to reduce the elastic earthquake demand.

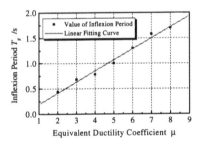

Figure 3. Inflexion period and their data fitting

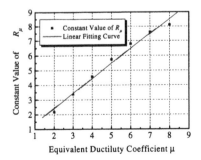

Figure 4. Constant values of R_μ and their data fitting

6 CLOSURE

The first objective of this study is to review investigation on the reduction factor which allow the estimation of lateral strength required to control the level of inelastic deformation during strong earthquake ground motion. Based on modified Takeda hysteretic model (including the hysteretic effects of post-yield stiffness, stiffness degradation and strength deterioration), the reasonable hysteretic parameters have been developed, and the semi-analytical relations between the hysteretic parameters and ductility coefficient have been derived. Using the equivalent ductility coefficient from Park-Ang dual-damage criterion, the strength reduction factors spectra respect to different damage indices have been calculated using the regression formulas suggested by L.H.Lee et al. The spectra can satisfy the different design levels of seismic resistance by selecting different damage levels. The results of comparison study have verified that a significant degradation of the carrying capacity considering the strength deterioration and the cumulative damage would occur. The strength deterioration and the cumulative damage of the structure have a remarkable effect on the strength reduction factor. The neglect of the two factors above mentioned might lead to an unconservative design result.

7 ACKNOWLEDGEMENTS

This study is sponsored by National Nature Science Foundation of China Grant No. 59978001. This support is gratefully acknowledged.

REFERENCES

Chinese Seismic Code of Highway Engineering [S], JTJ004-89
Chinese Seismic Code of Railway Engineering [S],GBJ111-87
N.M. Newmark, W.J. Hall, Seismic Design Criteria for Nuclear Reactor Facilities, Proc. 4th WCEE, Santiago, Chile: IAEE, 1969, Vol.2, 37-50

J.B. Merrill, M.J.N. Priestley and H.E. Chapman, Design Earthquake Loading and Ductility Demand, Bulletin of the New Zealand National Society for Earthquake Engineering, 1980, Vol.13(3): 232-241
Eurocode 8 (1996) [S]: Design Provisions for Earthquake Resistance of Structures, Part 1.1 General Rules— Seismic Actions and General Requirements for Structures; Part 2 Bridges
E. Miranda and V.V. Bertero, Evaluation of Strength Reduction Factors for Earthquake-Resistant Design[J], Earthquake Spectra, Vol.10, No.2, 357-379(1994)
Y.H. Chai, P. Fajfar and K.M. Romstad, Formulation of Duration-Dependent Inelastic Seismic Design Spectrum[J], Journal of Structural Engineering, Vol.124, No.8, 913-921(1998)
W.D. Zhuo, Study on Ductile Seismic Design for Highway Bridges [D], Tongji University, Shanghai, China, 2000
L. H. Lee, S.W. Han and Y.H. Oh, Determination of Ductility Factor Considering Different Hysteretic Models[J], Earthquake Engineering and Structural Dynamics, Vol.28, 957-977 (1999)
Y.J. Park, A.H-S. Ang, Mechanistic Seismic Damage Model for Reinforced Concrete[J], Journal of Structural Engineering, Vol.111, No.4, 722-739, (1985)
M.S. Williams, R.G. Sexsmith, Seismic Damage Indices for Concrete Structures: A State-of-the-Art Review, Earthquake Spectra[J], Vol.11, No.2, 319-349(1995)
P. Fajfar, Equivalent Ductility Factors, Taking Into Account Low-Cycle Fatigue, Earthquake Engineering and Structural Dynamics [J], Vol.21, 837-848 (1992)
Y.J. Park, A.H-S. Nag, Y.K. Went, Damage-limiting Seismic Design of Buildings [J], Earthquake Spectra, Vol.3, No.1, 1-26(1987)
A.H-S. Ang, W.J. Kim, S.B. Kim, Damage Estimation of Existing Bridge Structures, Structural Engineering in Natural Hazards Mitigation: Proc. ASCE Structures Congress 1993, Irvine CA, Vol.2, 1137-1142.
Liu Qinhua, Experimental and Theoretical Research about Hysteretic Model and Damage Model in Seismic Design of Reinforced Concrete Piers [D], Northern Jiaotong University, Beijing, China. 1994.
M.J.N. Priestley, F. Seibler and G.M. Calvi, Seismic Design and Retrofit of Bridges[M]. New York: John Wiley & Sons, 1996

Earthquake Engineering Frontiers in the New Millennium, Spencer & Hu (eds),
© 2001 Swets & Zeitlinger, ISBN 90 2651 852 8

Shaking table tests to simulate earthquake responses of passive control structures

Michio Yamaguchi
Graduate Student, Tokyo Institute of Technology, Yokohama, Japan

Satoshi Yamada
Associate Professor, Tokyo Institute of Technology, Yokohama, Japan

Masayoshi Nakashima
Kyoto University, Kyoto, Japan

Akira Wada
Professor, Tokyo Institute of Technology, Yokohama, Japan

ABSTRACT: In this paper, an experimental method of shaking table test on partial moment resisting steel frames is discussed. This experiment system, shown in Fig. 1, consists of a Mass, a Spring, a Loading Beam, a Specimen and a Shaking Table. By this composition, the natural period of this experiment system is about 0.7-0.8 second which value is nearly the same as middle rise steel buildings. Features of this experiment system are as follows.

1. This experiment method can estimate the response of partial frame against Real Time Speed Earthquake.
2. This method approximately reproduces an earthquake resisting behavior of the partial frame.
3. It is possible to effectively utilize the performance of small and medium size shaking table.

1 INTRODUCTION

After two big earthquakes, Northridge and Hyogo-ken-Nanbu occurred in 1994 and 1995, it became popular to require much seismic performance for structures. Moreover, in recent years steel materials for structures varied form low yield strength steel to high yield strength steel and these materials have been adapted to strengthen structures as well as viscosity and viscoelastic materials.

With this background, A. Wada is suggested the concept of Damage tolerant structures (A. Wada 1991). This structure consists of main frames and seismic members. Main frame only support the vertical load and remains elastic during earthquakes. While input earthquake energy is concentrated and absorbed by seismic members. This kind of structure has a lot of advantages such as: better seismic performance, more economic, large life span, environmentally friendly and so on. Recently, many studies for various damping devices have been taken (such as Y.Maeda 1998 and E.Saeki 1996 A.Wada 1997). However, member tests by means of fixing within the frame is not so many (E.Saeki 1996 and K.Kasai 1997). A behavior of an actual building during earthquake that is most important to grasp is normally done by means of analytic tools. Though real behavior of frames and damping devices with elastoplastic ranges are very complicated. Therefore, it needs to understand the behavior of frames under shaking table test to simulate real earth-quake.

2 METHOD OF EXPERIMENT

In common cases, shaking table tests are carried with specimens of partial frame model, input wave is modified a lot on time history domain. It is compressed on time history in order to fit in frequency spectrum for that natural period shortened. In this case, with such method, strain rate is different from that of real behavior in structures. Especially in the case with damping device that depends on velocity, the response behavior is a completely different matter.

In order to solve these problems, the experiment system using shaking table tests is proposed in this paper. The concept of this system is show in Fig 2-1. And outline of this system is shown in Fig.2-2 and Photo.2-3. This experiment system corresponds to actual buildings. This system consists of a weight, a spring, a loading beam, a specimen and a shaking table. The weight simulates the weight of the upper

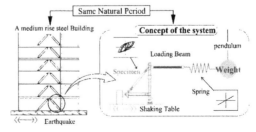

Fig. 2-1 Concept of the experiment system

part of the model building. A spring is connected to the weight and the loading beam by series methods. And a spring simulates the stiffness of upper part of the building. By these compositions, the natural period of this experiment system is about 0.7-0.8 seconds, which is nearly the same as the natural period of a medium rise steel building. Such a system decreases a scale down effect of a partial frame specimen by setting a natural period. Visualized earthquake resistant behavior makes easy to understand the superiority of damage tolerant structures because of real time speed.

The major part of weight is a row slab whose weight is about 10 tons. In addition, balance weights and safety devices are included. Total weight of the system is about 16.0 tons. The weight is hanging like a pendulum with wire ropes. This hanging technique makes shaking table release from a restriction under a gravity load, especially for small and medium size shaking tables. The weight produces inertia force while the shaking table test.

As for the spring, this spring consists of two isolators that arranged in parallel. Shear deformations are indicated in Fig.2-4. The stiffness of elastic spring is 10.0 kN/ mm. Deformation limit is 25.0cm.

The specimens are 1/2 span and half size scale model, which is picked up from a building. The beam (literal member) end is supported on a pin condition and near beam-to-column connection is supported on a pin-roller condition. A connection between a specimen and loading beam, which is wide flange beam, is used split-T that connected

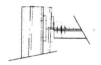

Fig. 2-4 Spring   Fig. 2-5 Pin joint

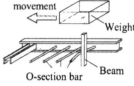

Fig. 2-6 Outline of a safety device

with bolts. (Fig. 2-5) The split-T is connected to web member of loading beam. The rotation stiffness is much smaller than that of other part of experiment system. Therefore, this connection is dealt with as a pin support. Another end of the loading beam is in connected too.

In this test, shaking table is used as single degree of freedom system. There are some devices to prevent outplate displacements in the experiment system. These are shown in follows.

1. Two pins are arranged in parallel at the end of a specimen. This technique is a role to enhance stiffness against another degree of freedom.
2. Middle part of the column, steel plate with a teflon sheet is welded. If it occurs out-plate displacement, it plays a role to restrict extra deformations.
3. The loading beam is restricted by an equal technique.
4. The weight is supported with four rollers, which are very tough and restricts extra deformations.

Attention to a safety of this shaking table tests, it is constructed a safety devices. Its outline is indicated in Fig.2-6. The mechanism is that O-section bars absorb energy of the system to be deformed by a wide flange beam connected to be the weight, if it occurred extra deformations.

3 SPECIMENS (SHAPES AND PERFORMANCE)

Tests were taken with two types of specimens, Moment Resistant Frame 1 (M.R.F.1), show in Fig.3-1, using ordinary so far and Moment Resistant Frame 2 with Hysteretic Damper (M.R.F.1 with damper), show in Fig.3-2.

Specimens are partial frame models that are divided from a vertical steel building. Considering symmetry of buildings, 1/2 span half-size scale

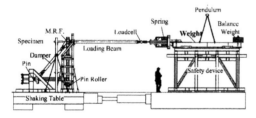

Fig. 2-2 Outline of the experiment system

Photo 2-3. Experiment system

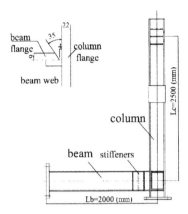

Fig. 3-1 Specimen of M.R.F.1

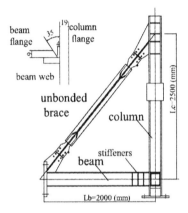

Fig. 3-2 Specimen of M.R.F.2 with damper

model is used. The length of beam member (literal member, Lb) is 2000mm; the length of column (vertical member, Lc) is 2500mm. Both members have H-section (shown in Table.3-3) The connection panel is reinforced with double plates. In order to prevent local buckling, three pieces of stiffeners are arranged with equal distance in beam-to-column connection.

Non-scallap detail is applied for beam-to-column connections. Because it is paid no attention to great earthquakes that it occurs fractures at beam-to-column connections. A damper is a Hysteretic and brace type damper. It is called "unbonded brace". The mechanism of unbonded brace is shown in Fig.3-3. The unbonded brace used in this experiment is shown in Fig.3-4. This is to prevent Euler buckling of central steel core by encasing it over its length in a steel filled concrete or mortar. Therefore unbonded brace provides a stable behavior in both sides of tension and compression.

Table.3-5 shows the mechanical properties of material of specimens obtained from tension tests.

Table 3-3. Sections of the elements

	M.R.F.1	M.R.F.2(with damper)
Beam	BH-340x200x 6x 9	BH-240x170x 6x 9
Column	BH-280x280x12x22	BH-220x220x12x19
Yielding Steel Core	-	25x16

In this test, upper limit of resisting force is as 15 ton for performance of shaking table. For a yielding steel core of the unbonded brace, it is selected low yield steel (BTLYP 100) which is able to expect an energy absorption in ranges of small amplitudes, so its yield strain is very small.

Fig. 3-6 shows resisting force-displacement relation of both specimens. The curves are obtained from analysis.

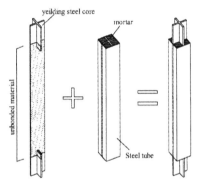

Fig. 3-3 Mechanism of unbonded braces

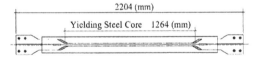

Fig. 3-4 Unbonded braces

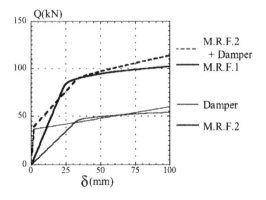

Fig. 3-6 Force-displacement relation of specimens

433

Table 3-5. Mechanical properties

Steel	Yield stress Mpa	Tensile strength Mpa	Yield strain %	Elongation %	Yield ratio
SM400A(web)	311	435	0.148	31.7	0.71
SM400A(flange)	292	451	0.142	30.8	0.65
LYP100(damper)	96	258	0.240	59.6	0.37

Design points of specimens to which we pay attention in the case designed cross section are shown as follows.

1. Shear forces at yield point of both specimens are designed equally (about 90 kN).
2. Shear ratio with horizontal resisting force between M.R.F.2 and damper is about 1:1.
3. M.R.F.2 is designed to have large elastic ranges, compared with M.R.F.1, therefore its section depth beam and column are slender.

4 ABOUT IMPUT

4.1 Shaking Table

The Shaking table, used in this test, is medium size and belongs to the Disaster Prevention Research Institute, Kyoto University. This table has six degree of freedoms. In this test, however, it is used as single degree of freedom system by means of fixing other 5 degrees of freedom. Fig.4-1 and Fig.4-2. show the performance curve of this shaking table concerned with Acceleration Period relation and with Velocity-Period relation.

At first, we assume the weight of a Mass to be 15 tons. In attention to ranges 0.7-0.8sec (in Fig.4-1. gray zone) which is natural period of this system considering an expansion of period by plasticity. We recognize that this shaking table has sufficient capacity to shake table so far as 1.0 G. In order to inflict much damage to specimens, it is required to have capacity to shake with sufficient velocity. Fig.4-2. shows that the shaking table has enough capacity to shake table as fast as 100 kine.

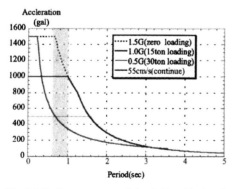

Fig. 4-1 Performance curve of the shakig table (acceleration-period ralation)

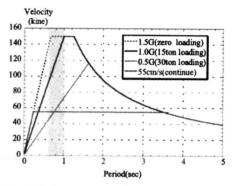

Fig. 4-2 Performance curve of the shakig table (velocity-period relation)

Table 4-3. Input waves

Year	Earthquake	Record	component	description
1940	Imperial Valley	El Centro	NS	El Centro
1968	Tokachi	Hachinohe	EW	Hachinohe
1995	Hyogoken-nanbu	Kobe JMA	NS	Kobe

4.2 Types of input waves

Considering the variety of real ground motions, we choose three earthquake records shown in Table 4-3.

Shaking time is determined at 30.0 sec, since amount of Energy input with elastoplastic analysis during first 15.0 sec is more than 90 percents of total input energy, and very little over 30.0 sec with these input waves. It is an acceleration record to control shaking table. However, those input waves are little reversed. Because integration of original acceleration records of earthquake gives phenomena such as remaining velocity of shaking table or divergence of displacement. Compensation approaches of input waves are shown in the follows.

1. Removal of long period components more than 10.0 sec.
2. Base line compensation.

Through these compensations, numerical integration was taken again. Input waves were generalized with Maximum-velocity of ground motion (shaking table), after being confirmed maximum displacement.

4.3 Input levels

Generally, in the case of designs of real buildings, the input level was established at first as a design criteria. On this experiment, the process order is reversed, decided with considering relative evaluation between performance of specimens, experiment system and intensity of input wave. Design criteria and maximum-velocity of ground motion are shown in table 4-4.

Table 4-4. Input Level

Input Level	Design Criteria	Maximum-velocity
Level_1	Protection of Faculty	25.0kine
Level_2	Protection of Property	50.0kine

In ordinary input level is depend on the structural designer's requirment. According to "the theory of Damage Tolerant Structure", the purpose of structure demanded, if it occur equivalent Level 1 earthquakes, is "Protection of Faculty". And in Level 2 is "Protection of Property". These are an outline of required seismic performance by the structures.

Deciding an approach of concrete input level is as follows: It was determined by means of elastoplastic analysis of SDOF model that is this experiment system. Fig.4-5 shows correspondence relation between input level and force-displacement relation, which is derived from preanalysis of specimen. Input levels are shown with maximum response displacement which is the results of elastoplastic analysis using El Centro earthquake.

Assumptions of behavior in both input levels are as follows: On Level 1 shaking, frames of both specimens are remained within elastic ranges, although the steel core of the damper is already yields and absorbs energy. On Level 2 shaking, M.R.F.1 is yielding a lot, although the frame of M.R.F.2 with damper remained within elastic ranges.

4.4 Test procedures

The procedure is planed and shown as follows: concerned with section 4-4-1~4-4-3, these are defined as tests to compare the seismic performance of both specimens, M.R.F.1 and M.R.F.2 with damper.

4.4.1 Pulse-wave test

Shaking table test using pulse wave with small amplitude is taken in order measure the damping ratio and natural period of this system.

4.4.2 Level 1 tests

Maximum-velocity of input wave is 25.0 kine. Using this Input level, seismic performance of M.R.F.2 with damper is compared with M.R.F.1. Behaviors

of both specimens are maintained within elastic ranges. Procedures of input waves are El Centro, Hachinohe and Kobe.

4.4.3 Level 2 tests

Maximum-velocity of Input wave is 50.0 kine. Using this Input level, seismic performance of M.R.F.2 with damper is compared with M.R.F.1. on elastoplastic ranges. Only M.R.F.1 received a lot of damage.

5 RESULTS

Fig.6-1 shows results of impulse wave tests. And equation of natural period of this experiment system is shown in Eq.6-2.

Natural period of this experiment system with both specimens are nearly the same as that of a middle rise steel building (about 0.7-0.8 seconds). About equations, it is recognized that natural period of this system depends on Ks (stiffness of spring). A difference in natural period is about 10%. Therefore, it is anticipated that amount of input energy to the system is nearly equal order for both specimens. Damping ratios in both tests were 1.7%(M.R.F1) and 2.3% respectively. These values are nearly same as the damping ratio of actual structures. Therefore it was judged that we constructed the experiment system, which is appropriate on natural period and damping ratio.

Table 6-1. Results of pulse wave tests

	Period	damping ratio
	(sec)	(%)
M.R.F.1	0.82	1.70
M.R.F.2 with damper	0.70	2.30

$$T = 2\pi\sqrt{\frac{M_{mass}}{K_{system}}} \quad (6\text{-}1) \qquad K_{system} = \frac{K_s \cdot K_{spe}}{K_s + K_{spe}} \quad (6\text{-}2)$$

T: natural period of the system, M_{mass}: mass of the weight
K_{system}: stiffness of the system, K_s: stiffness of the spring
K_{spe}: stiffness of the specimen

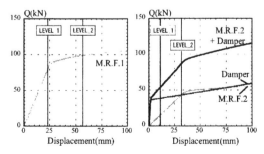

Fig. 4-5 Relationship between specimens and input levels

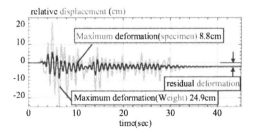

Fig. 6-3 Relative deformations time history

435

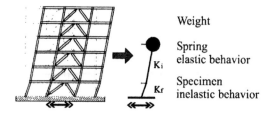

Weight

Spring
elastic behavior

K_i

Specimen
inelastic behavior

K_f

Fig. 6-4 The behavior of experiment system (concept)

As typical test results, relative deformations (weight and specimen) are shown in fig.6-3. In this case, type of specimen is M.R.F.1. And input wave and level are El Centro level 2.The correspondence concept) of experimental system and the actual structure is shown in figures of 6-4. I t is proven that both specimens and weight keep the vibration along a natural period. And, the deformation of the weight shows 24.9cm in largest deformation (8.8 cm) of the specimen, and the weight has approximately expressed the movement of center of gravity of the structure, as it is shown in fig. 6-4.

6 CONCLUSIONS

Features of experimental system proposed in this paper are shown in the following.

1. This experiment method can estimate the response of partial frame against Real Time Speed Earthquake.
2. This method approximately reproduces an earthquake resisting behavior of the partial frame.

3. It is possible to effectively utilize the performance of small and medium size shaking table.

Actual shaking table tests were carried out, and a following knowledge was obtained.

1. The experimental system, which simulates an actual structure was constructed.
2. It was confirmed that in which the movement of the weight is similar to the movement of center of gravity position of an actual structure.

REFERENCE

Akira WADA; Damage Tolerrant Structure, Proseeding of Fifth U.S. Japan Workshop on The Inprovement of Structural Design and Construction Practices, Applied Technology Council, 27-39, 1991.
Yasushi MAEDA et al; Fatigue properties of axial-yield type hysteresis dampers, J. Struct. Constr. Eng., AIJ, No.503, 109-, Jan., 1998.
Eiichiro SAEKI et al; Analytical study on the unbonded brace fixed in a frame, J. Struct. Constr. Eng., AIJ, No.489, 95, Nov., 1996.
Eiichiro SAEKI et al; A study on htsteresis and hysteresis energy characteristics of low yield strength steel, J. Struct. Constr. Eng., AIJ, No.473, 159-, Jul., 1995.
Eiichiro SAEKI et al; Experimental study of practical-scale unbonded braces, J. Struct. Constr. Eng., AIJ, No.476, 149, Oct., 1995.
Hiroshi AKIYAMA et al; Experimental method of the full scale shaking table using inertial loading equipment, J. Struct. Constr. Eng., AIJ, No.505, 139-, Mar., 1998.

Earthquake Engineering Frontiers in the New Millennium, Spencer & Hu (eds),
© 2001 Swets & Zeitlinger, ISBN 90 2651 852 8

Study on performance-based seismic design of foundations and geotechnical structures

X. Yuan
Institute of Engineering Mechanics, CSB, 9 Xuefu Road, Harbin 150080, China

ABSTRACT: The cooperative research achievements for analysis and evaluation of sand soil liquefaction in soil dynamics between USA and PRC in 1980's are reviewed. The IEM's recent progresses in soil dynamics and geotechnical earthquake engineering are introduced, including the soil nonlinear analysis under dynamic loads, determination of seismic settlements of weak foundations, evaluation of effect of saturated sand soil layers on the seismic surface ground motion and building responses, etc. Suggestions for the future cooperative research program in geotechnical earthquake engineering and soil dynamics between USA and PRC are finally presented. The suggestive topic "Study on performance-based seismic design of foundations and geotechnical structures" is divided into 4 aspects of research: 1. Dynamic properties of soils in large deformation range; 2. Classification of deformation and stability of foundations and geotechnical structures; 3. Analytical method for deformation and stability of foundations and geotechnical structures; 4. Effect of SSI on structure earthquake responses.

1 REVIEW OF PRC-USA COOPERATIVE RESEARCH IN SOIL DYNAMICS

In 1980's, sponsored by the State Seismological Bureau of China and the National Science Foundation of the Unite States, the PRC-USA cooperative research in soil dynamics was conducted. The cooperative program is mainly on the sand liquefaction criteria and risk assessment of saturated sands as well as determination of soil properties. The bilateral scientists complete the research project. One side is from the Institute of Engineering Mechanics (IEM), State Seismological Bureau of China, including Profs. J.F. Xie, Y. Liu, Z.J. Shi, Q.Z. He, etc., and another from the University of California, Davis of USA, including Profs. K. Arulanandan, C.K. Shen, Y.Z. Gu, X.S. Li, etc. The purpose of the cooperative research is to confirm and improve the sand liquefaction criteria and method at that time for evaluation of liquefaction risk by in situ and laboratory testing study on Tanshang Earthquake in 1976 and Haicheng Earthquake in 1975 in China.

One of the notable features of the two large earthquakes, Tanshang Earthquake in 1976 and Haicheng Earthquake in 1975 in China, is that sand liquefaction appears widely in the area and produces severe damages to ground surface and various buildings, and the stability of different subsoils and foundations suffer from checking under different densities of the seismic ground motion in the earthquake events. Making full use of the two earthquakes, field tests including standard penetration, cone penetra-

tion, electric probe, piezocone and wave velocity measurement are performed in Tianjing in 1981 and in 1982. The soil specimen from the borehole is sent to the laboratories in IEM and USA for testing the dynamic properties including liquefaction strength, dynamic shear modulus and damping ratio. The field tests are also conducted in Yingkou in 1984 for investigating the liquefaction potential of non-pure sand soils.

The remarkable achievements have been gained through the cooperative program in soil dynamics. The common knowledge has been formed and significant progress has been obtained in liquefaction criteria of silt soils. The scientists in China have also learned the advanced technique in geotechnical experiment at that time. The cooperative research progress is helpful in revising the aseismic design code in China and is also important for the engineering construction in USA and other countries.

2 RECENT PROGRESS IN SOIL DYNAMICS IN IEM

The notable progress in soil dynamics in IEM has been made in recent years and some of them can be summarized in the following introduction.

2.1 *Determination of nonlinear properties of soils*

Dynamic shear modulus and damping ratio are the most primary two parameters of soil dynamic properties and have significant effect on seismic ground motion. There is better understanding in the two parameters in other countries. In China, however, shortage of systematical research on the two parameters has been taken for longer time, and some mistakes related with determination of the two parameters have been made in assessment of safety of engineering sites.

Therefore, researchers in IEM have make efforts to solve the problem. The average curves, recommendable values and envelopes of dynamic shear modulus ratio G/G_{max} and damping ratio λ versus dynamic shear strain γ for 6 types of soils in China are presented using the free vibration method by resonant column apparatus. The refitted resonant column apparatus in IEM is employed for the experimental investigation and its accuracy and reliability have been verified by the tests. The undisturbed soil samples, including clay, silty clay, silt, sand, muck and mucky soil, come from about ten different regions in China. The relation of G/G_{max}-γ is fitted by two hyperbolic equations and the relation of λ-γ is also fitted by two hyperbolic equations. The parameters a and b in the equations are obtained from regression analysis of each testing data by the least squares method. For the same type of soil, the mean values, $\bar{a}$ and $\bar{b}$, are gotten by averaging all a and b for the same type of soil. The relations of G/G_{max}-γ and λ-γ for each type of soil are then derived using the mean parameters $\bar{a}$ and $\bar{b}$, and finally, the average curves and recommendable values of G/G_{max}-γ and λ-γ for the 6 types of soils are obtained. Because the laboratory test is performed under two kinds of consolidation pressures, the presented results are suitable respectively for two depths of soils, less than 10 meters as well as between 10 meters and 20 meters. The above results can be used in revising the aseismic design code in China.

2.2 Settlement of soft subsoils

According to the earthquake field investigation, differential settlement of buildings on soft subsoil is major form of foundation failure during earthquake, and has close relation to existence of buildings.

In the recent years, the major aim of research in IEM is to evaluate the uniform subsidence of subsoils. Using FEM the analytical model and method for assessing the subsidence is presented in which the parameters come from the field and laboratory tests. The presented method has been checked using the examples of earthquake field investigation. The method is based on soften model which means that seismic shock makes the soil soft and makes the soil modulus decrease, and as a result, the settlement occurs in subsoil. The researchers in IEM also present

a simplified method to assessment the subsidence. Both of the two methods have been used in practice.

The differential settlement is complicated problem and IEM's researchers have just started working on it.

The mechanism of ground lateral deformation caused by liquefaction is analyzed in IEM and a method is presented to evaluate the permanent lateral displacements of ground due to earthquake-induced liquefaction. The pore water pressure in every stress cycle is considered and the permanent lateral ground displacement is calculated using the Newmark's rigid block model. The numerical results have shown agreeable with the earthquake events.

2.3 Effect of soft soil layer on surface ground motion and building response

As supporting layer of buildings, failure of liquefied subsoils can cause damages of buildings. Also, as the medium of wave propagation, liquefied soil layer can change the form of coming waves from the bedrock in great extent. However, the current methods for analyzing liquefaction problem can generally be used in evaluating the earthquake stability. If used in analyzing the seismic ground motion the methods obtained at present should be improved.

In the recent work, researchers in IEM improve 2-D effective stress method for estimating the effect of liquefied lenticular sand zone on the seismic ground motion and earthquake response of buildings. The effect of the input form of the seismic waves and the thickness, width and depth of the sand lenticle on surface ground motion and building response is discussed. The numerical results indicate: (1) The results by the improved method is agreeable with those by the model test, which means the method here can be used to analyze the problem of sand liquefaction in case of building existence. (2) The liquefied lenticle mainly takes isolation effect on the buildings of short period and the effect of the vibration isolation increases with the increasing of the thickness, depth and width of the sand lenticle. (3) When the subjected seismic waves are the shock type, the shallow, thin and narrow liquefied lenticle may amplify the response of the buildings of short period.

2.4 Research on wave propagation in two- phase medium

Wave propagation theory in two-phase medium has been studying for nearly thirty years in IEM. Recently, the main effort is made in estimating the mechanism of liquefaction of saturated soil layer and the influence on wave propagation. Remarkable progress has been gained in this aspect.

3 SOIL DYNAMICS AND GEOTECHNICAL EARTHQUAKE ENGINEERING

Performance of foundation and earth structure in earthquake environments has close relation to seismic load definition, dynamic soil properties, field and laboratory testing technique, seismic stability of subsoils and foundations and others. Foundation analysis has been regarded as problems better understood and easy to solve. However, recent earthquake events, such as Mexico City in 1985, Loma Prieta in 1989, Northridge in 1994, Kobe in 1995 and Chichi in 1999, have obviously shown that the seismic behavior of soil-foundation system has been far away from being fully understood. In terms of performance-based design idea, the following aspects may be important in future research on soil dynamics and geotechnical earthquake engineering.

3.1 Dynamic properties of soils in large deformation range

Soil dynamic properties have great influence and in many cases are dominant factors on seismic surface ground motion and earthquake damages of structures. Although there are many problems to be solved, two aspects may be obvious shortage of research if new trends of research on soil dynamics and needs from engineering are considered. One is that it is not very clear in the dynamic properties for soft soil, such as saturated sand and soft clay soil, and other special soils, such as garbage soil, reclaimed soil, frozen soil, reinforced soil and non-saturated soil. Another is that feature of deformation of above soft and special soils is not clear, particularly in large deformation range, and the current theory of pore water pressure and deformation under seismic loads is based on equiamplitude cyclic loads. However, uniform cyclic loads are far away from the real earthquake loads and there are big gap between them for nonlinear soils, especially when large deformation appears. The known results in the laboratory tests have proved that the difference of permanent deformation resulting from uniform cyclic and random seismic loads is quite obvious. Therefore, three research aspects should be considered: (1) Dynamic parameters of soft soils and special soils; (2) Residual strain of soft clay soils under seismic loads; (3) Pore water pressure and residual strain of saturated sands due to seismic loads.

3.2 Classification of deformation and stability of foundations and geotechnical structures

It has been shown again in recent earthquake events that large deformation and stability loss are the main reason for structure damages relating to soils. In terms of idea of performance-based design the rule of deformation control should be taken into account in geotechnical aseismic design. So, calculation for permanent deformation of soft soil becomes very important. According to different functions of buildings the different requirements will be presented to the subsoils and foundations. So, classification of deformation and stability of foundations is essential in research work. For geotechnical structures, such as earth dams, retaining walls, quays, slops and embankments, it is impossible not to allow any deformation. So, according to the different function of earth structures, studying on how much deformation is allowable is important work. In terms of the classification, aseismic check contents for different foundations and earth structures will be different, and more research should be made on basis of the current codes.

3.3 Analytical method for deformation and stability of foundations and geotechnical structures

Compared with the building problem, analytical methods for assessment of seismic behavior of foundations and geotechnical structures are not well developed. Particularly when the large deformation of subsoils and earth structures appears, the current analytical methods should be greatly improved. In the procedure of evaluation of seismic performance of foundations and geotechnical structures, the rules for determination of failure and stability loss should be carefully considered when various buildings and structures exist. Using the large-scale earthquake geotechnical model experiments such as centrifuge and shaking table tests, the seismic responses and damages of soil-foundation-structure system can be reproduced in some degree. So, the large-scale testing studies on earthquake geotechnical problems have been gradually used in investigating the mechanism, checking the numerical method and determining the parameter. The prototype or large-scale model experiments should be emphasized in the future research in soil dynamics and geotechnical earthquake engineering. Besides, as one of important factors on damage of ground and structures, lateral spreading and permanent displacements of soil ground should further be studied.

3.4 Effect of SSI on structure earthquake responses

One of the influences of soil structure interaction will make settlements and permanent deformation of structures increase as the soil compressibility is high. Another influence is to modify the free field ground motion. The foundation-structure and its supporting medium are the natural source of consuming energy and in many cases will greatly reduce the seismic responses of structures. The nonlinear properties of soils are the very important factor in the two influences, but have not been taken into account enough. If the effect of SSI can be well considered and adequately applied in aseismic design, reasonable engineering methods for analysis of SSI should be developed, and the effect of soil nonlinear properties and soft soil layer on surface ground motion and structure responses also should be carefully studied.

Although there are many developments in analytical tools and laboratory testing techniques, there is still a notable gap between the analytical model of SSI and reality. In order to know the information on real seismic responses of soil-foundation-structure system, seismic monitoring of foundations should be carried out to record the real phenomena on foundation soil, on the different components of foundation and at interface of soil-foundation during earthquake. The measured information will be helpful in knowing the mechanism, in modifying the analytical models and methods. Until now, however, only small number of foundations has been instrumented in the world to record the real responses of soil-foundation-structure system during earthquake.

Besides, knowledge-based technique and ANN (artificial neural network) technique gradually are used to solve input-output problems in geotechnical engineering. Some inverse methods such as GAS (genetic algorithm search) and NION (normalized input-output minimization) also are employed in finding soil properties using acceleration records. However, these advanced techniques and methods just begin to apply in soil dynamics and there are many problems to be solved in the future because soil-foundation-structure is a quite complicated system.

4 CONCLUSIONS

The cooperative research in soil dynamics between USA and PRC in 1980's is fruitful and some key problems such as silt liquefaction criteria have been solved. According to needs from engineering and new trends of development in soil dynamics and geotechnical earthquake engineering, emphasis of future studies should be focused on soil dynamic properties in large deformation range, on the effect of soil nonlinear behavior and local soft soil layers on surface ground motion and building responses, on hazard analysis of liquefaction and settlement of subsoils and foundations under considering SSI, as well as on utilization of the advanced techniques from other fields.

REFERENCES

Shen, C.K., Gu, Y.Z. and Li, X.S. 1982, Dynamic property of Tianjin soils (Chinese), Proc. US-PRC Bilateral Workshop on Earthquake Engineering, Harbin, China.

Arulanandan, K., Douglas, B.J., et al. 1982, Evaluation of earthquake induced liquefaction in Tiensin during Tangshan Earthquake P.R.C. (Chinese), Proc. US-PRC Bilateral Workshop on Earthquake Engineering, Harbin, China.

Liu, Y., Xie, J.F. and Shi, Z.J. 1984, Field investigation on liquefaction (Chinese), PRC-USA Cooperative Research Report, Harbin, China.

Liu, Y., Xie, J.F., He, Q.Z. and Shi, Z.J. 1984, 1984, Experiments of soil liquefaction and assessments of silt liquefaction in Tanshan Earthquake in 1976 (Chinese), PRC-USA Cooperative Research Report, Harbin, China.

Hardin, B.O. and Drnevich, V.P. 1972, Shear modulus and damping in soils design equation and curves, J. Soil Mech. Found. Div. ASCE, 98(SM7).

Seed, H.B. and Idriss, I.M. 1970, Soil moduli and damping factors for dynamic response analysis, Report No. EERC 70-10, Earthquake Engineering Research Center, University of California, Berkeley.

Yuan, X.M. and Sun, R. 2000, Laboratory determination of dynamic shear modulus and damping ratio (Chinese), Earthquake Engineering and Engineering Vibration, 20(4).

Shi, Z.J., et al. 1988, Analysis of settlements in Tangguo area (Chinese), J. Civil Engineering.

Yuan, X et al. 2000, A procedure for evaluation of differential settlements of buildings on weak soil foundations, Proc. 12WCEE.

Jin, L.P. and Wang, S.B. 1996, Study on ground lateral deformation induced by liquefaction (Chinese), Earthquake Engineering and Engineering Vibration, 16(3).

Sun R., et al. 1998, 2-D finite element analysis of effective stress of seismic liquefaction (Chinese), World Information on Earthquake Engineering, 14(1).

Yuan, X.M. and Sun, R. 2000, Effect of liquefaction of lenticular saturated sand zone on seismic response of buildings (Chinese), Earthquake Engineering and Engineering Vibration, 20(1).

Men, F.L. 1992, Wave propagation and liquefaction in saturated medium (Chinese), Proc. Research on Earthquake Engineering, Earthquake Publishing house, China.

Li. X.J. 1993, Study on the method for analyzing the earthquake response of nonlinear site (Chinese), Doctoral Dissertation, Institute of Engineering Mechanics, SSB.

Ishihara, K., et al. 1984, Strength of a cohesive soil in irregular loading, Proc. 8WCEE.

Chen, G.X. 1993, Research on earthquake behavior analysis of soil-structure systems (Chinese), Doctoral Dissertation, Institute of Engineering Mechanics, SSB.

Earthquake Engineering Frontiers in the New Millennium, Spencer & Hu (eds),
© 2001 Swets & Zeitlinger, ISBN 90 2651 852 8

Nonlinear soil-pile-structure interaction for structures resting on a 2×2 pile group under earthquake excitations

K.T. Chau & X. Yang
Department of Civil and Structural Engineering, Hong Kong Polytechnic University, Hong Kong

ABSTRACT: This paper summarizes a new continuum model for the nonlinear soil-pile-structure interaction under earthquake excitations. The earthquake excitation is modeled as a harmonic shear wave, and the soil is considered to compose of an inner nonlinear zone around the pile and an outer linear elastic zone. The pile-soil-pile interaction is considered by using a Winkler model with its spring constant estimated from the wave-interference. The difference of shaking for the coupled system and the free field may be up to three times when λ is at the first natural frequency of the structure or at high frequency. There exists a critical ground shaking at which the seismic response is the largest. The response of pin-piled structures can either larger or smaller than that of fixed-piled structures. In general, the interactions are highly dependent of the excitation frequency, the pile-to-pile distance, the amplitude of ground excitation, and the nonlinearity of soil.

1 INTRODUCTION

Remains of piles have been discovered in ruined ancient cities, and thus there is strong evidence that piles have been used for supporting structures for thousands of years. Static theoretical stress analysis for piles started about 100 years ago, while more intense development of the pile dynamics happened in the last thirty years or so. Our understanding of the soil-pile-structure interaction under earthquake excitations is, however, far from mature (see the reviews by Novak, 1991; Hadjian et al., 1992). For example, many pile damages have been observed in the recent Kobe earthquake in 1995 (e.g. Matsui and Oda, 1996; Tokimatsu et al., 1996; Finn et al., 1996), as well as in other recent earthquakes. The main difficulties in predicting the response of the soil-pile-structure system arise from the highly nonlinear interactions of soil-pile, pile-soil-pile, and soil-pile-structure. Vibrations of soil, pile and structure are hardly independent and they are coupled together through the pile cap. However, such full coupling between soil, pile and structure has not been investigated comprehensively.

In Hong Kong, most of the tall buildings built on reclaimed land (such as those in Wan Chai, Central, and Tsim Sha Tsui) are founded on piles. Most of these piles rest on the bedrock. Some of these piles were more than 50 meters long, and nearly all of them are end-bearing. In addition, the normal practice for bored piles in Hong Kong requires that the piles must be socketed to the bedrock, while driven piles (such as mini-piles and H-piles) can only rest

upon the bedrock. Thus, bored-piles can be modeled as fixed-tip piles while driven piles as pinned-tip piles. Both of them are commonly used in Hong Kong. In addition, most of these tall commercial or residential blocks require a large number of piles to be constructed in order to spread the loads from the superstructure to the ground. Inevitably, pile group has to be used.

Due to the historical reason that Hong Kong was a colony of Great British, no seismic provision is included in the building codes in Hong Kong. That is, nearly all of the tall buildings were built strictly according to their self-weight and wind load. No consideration has been made to consider the seismic performance of these pile group foundations.

Recently, the Building Department of Hong Kong has obtained an initial fund to carry out a preliminary study to examine whether there is a genuine need to incorporate a seismic code into the current building regulations. If the result of such preliminary study turns out to be positive, all pile foundations need to be designed for seismic excitations. Thus, a simple analytic model will be extremely useful for the purpose of preliminary design and for the purpose of identifying the key parameters of the problem of soil-pile-structure interaction under earthquake actions. Most of the owners of these tall buildings would certainly prefer a cheap and quick initial estimation of the seismic performance of their proposed piled-structures. Very expensive computations would be their last resort to the problem, due to the expensive computations of a realistic finite element model incorporating soil nonlinearity (e.g. Cai et al., 2000).

A thorough literature search discovers that such a simple analytical method is still not available. Thus, in view of this the authors have recently completed a series of analyses (Chau and Yang, 2000a,b; Yang and Chau, 2000a,b) which target to provide a efficient and reliable method that can handle soil nonlinearity and the full coupling of soil-pile-structure interactions. Such a simple model would also be very useful for the construction industry in mainland China because many tall buildings are being constructed in Shanghai, Beijing and other big cities. In the case of Shanghai, all of these tall buildings are going to be built on pile foundations through a thick layer of soft soil. The effect of soil-pile-structure is inevitably important.

The purpose of this conference paper is to summarize the main ideas and results of the recent analyses by Chau and Yang (2000a,b) and Yang and Chau (2000a,b).

In particular, a nonlinear inner soil zone will be introduced to model soil nonlinearity in the vicinity of the pile. An equivalent linearization technique discussed by Angelides and Roesset (1981) and by Chau and Yang (2000b) will be used. The reduced modulus and damping of the inner soil are modeled as functions of the average strain by assuming a hyperbolic stress-strain model (Hardin and Drnevich, 1972; Idriss et al., 1978). Within the framework of such a model, an iterative approach has to be used to estimate the equivalent modulus in terms of the strain level. This model is generalized from those previous studies of an inner zone with "constant reduced modulus" by Novak and Sheta (1980), Veletsos and Dotson (1986, 1988), and Novak and Han (1990). To account for the pile group effect or the pile-soil-pile interaction, we also generalize the "superposition approach" of Gazetas and Makris (1991), Makris and Gazetas (1992), and Gazetas et al. (1991, 1992) in the present study.

2 THEORETICAL FORMULATION

2.1 Seismic wave input

As shown in Figure 1, we consider a soil-pile-structure system consisting of a 2×2 pile group of spacing D, a rigid pile cap and a superstructure. The soil is isotropic and of thickness H. A steady-state harmonic shear wave excitation is assumed traveling from the underlying bedrock:

$$\overline{U}_g = U_g e^{i\omega t} \tag{1}$$

where U_g is the displacement amplitude along the x-axis, ω is the circular frequency of excitation, and $i = \sqrt{-1}$ is the imaginary constant.

2.2 Nonlinear soil resistance

In particular, although the shear modulus G_n and hysteretic material damping ξ_n are assumed constant within the inner zone, they are assumed as functions of the average shear strain $\overline{\gamma}$ (Angelides and Roesset, 1981):

$$G_n = \frac{G}{1 + \alpha \overline{\gamma}}, \quad \xi_n = \xi + \xi_{max}\left(1 - \frac{G_n}{G}\right) \tag{2}$$

and

$$\overline{\gamma} = \frac{1}{\pi H (R_1^2 - R_0^2)} \int_0^H \int_0^{2\pi} \int_{R_0}^{R_1} |\gamma_{r\theta}| r \, dr d\theta \, dz \tag{3}$$

where α and ξ_{max} are material constants of the soil, and G and ξ are the constitutive constants of the outer soil. Therefore, the corresponding parameters for the outer region can be recovered from (2) as a special case by setting $\alpha=0$. Expression (2) can also be interpreted by using a hyperbolic stress-stain relation (e.g. Hardin and Drnevich, 1972). An iterative procedure is needed to evaluate G_n and ξ_n as a function of the current strain level (Chau and Yang, 2000b).

To solve for soil displacement within the inner soil zone ($\tilde{u}_r^i$ and $\tilde{u}_\theta^i$), we can use a wave potential approach as demonstrated by Nogami and Novak (1977). More specifically, Yang and Chau (2000b) showed that the equations of motion of the inner soil can be satisfied if the following wave potentials, Φ^i and, Ψ^i, are introduced:

$$\Phi^i = H^2 \cos\theta$$
$$\times \sum_{k=1}^{\infty}\left[C_{1k}K_1(q_k \frac{r}{H}) + C_{2k}I_1(q_k \frac{r}{H})\right]\sin(\alpha_k \frac{z}{H}), \tag{4}$$

$$\Psi^i = H^2 \sin\theta \tag{5}$$
$$\times \sum_{k=1}^{\infty}\left[C_{3k}K_1(s_k \frac{r}{H}) + C_{4k}I_1(s_k \frac{r}{H})\right]\sin(\alpha_k \frac{z}{H});$$

where C_{ik} $(i=1,2,3,4)$ are unknown constants, K_1 and I_1 are the modified Bessel function of the first and second kinds of the first order (Abramowitz and Stegun, 1965), and q_k and s_k are given by

$$q_k^2 = \frac{1}{\eta_n^2}\left[\alpha_k^2 - \frac{1}{1+2\xi_n i}\left(\frac{\omega H}{V_s^i}\right)^2\right],$$
$$s_k^2 = \alpha_k^2 - \frac{1}{1+2\xi_n i}\left(\frac{\omega H}{V_s^i}\right)^2, \quad \eta_n = \frac{V_l^i}{V_s^i} \tag{6}$$

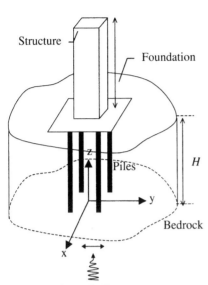

Structure

Foundation

Piles

Bedrock

H

y

x

Figure 1. A sketch for a soil-pile-structure system

In addition, we have also used the following results:

$$\alpha_k = \frac{2k-1}{2}\pi, \qquad k = 1,2,3,\ldots \tag{7}$$

$$V_s^i = \sqrt{G_n/\rho_n} \tag{8}$$

$$V_l^o = \sqrt{\frac{2(1-\nu)}{1-2\nu}\frac{G_n}{\rho_n}}, \tag{9}$$

where ν is the Poisson's ratio.

To express the displacement field using the wave potentials (Φ^i and Ψ^i), we can substitute (4-5) into the following equations (e.g. Nogami and Novak, 1977; Chau and Yang, 2000a):

$$\tilde{u}_r^i = \frac{\partial \Phi^i}{\partial r} + \frac{1}{r}\frac{\partial \Psi^i}{\partial \theta},$$

$$\tilde{u}_\theta^i = \frac{1}{r}\frac{\partial \Phi^i}{\partial \theta} - \frac{\partial \Psi^i}{\partial r}. \tag{10}$$

Similar solution technique can also be applied to the outer soil zone. The full detail is referred to Yang and Chau (2000b).

2.3 Soil resistance on the pile subjected to ground shaking

By considering the displacement compatibility between the pile and soil at $r = R_0$, the pile deflection $u_p(z)$ can be expressed as (Yang and Chau, 2000b):

$$u_p(z) = U_g\left[1 + \sum_{k=1}^{\infty} a_k^{if}\sin\left(\alpha_k \frac{z}{H}\right)\right]$$

$$+ H\sum_{k=1}^{\infty} U_k \sin\left(\alpha_k \frac{z}{H}\right), \tag{11}$$

where U_k is an unknown constant to be determined.

The continuity conditions of displacements and stresses between the inner and outer soil zones (at $r = R_1$) as well as the continuity conditions of displacements between the inner soil and pile (at $r = R_0$) require:

$$\begin{cases} u_r^i(r,\theta,z) = u_p(z), & r = R_0, \theta = 0 \\ u_\theta^i(r,\theta,z) = -u_p(z); & r = R_0, \theta = \dfrac{\pi}{2} \end{cases} \tag{12}$$

$$\begin{cases} u_r^i(r,\theta,z) = u_r^o(r,\theta,z), \\ u_\theta^i(r,\theta,z) = u_\theta^o(r,\theta,z), \\ \sigma_r^i(r,\theta,z) = \sigma_r^o(r,\theta,z), \\ \tau_{r\theta}^i(r,\theta,z) = \tau_{r\theta}^o(r,\theta,z). \end{cases} \qquad r = R_1 \tag{13}$$

These equations provide a set of six simultaneous equations for the unknown constants $C_{1k}, C_{2k}, C_{3k}, C_{4k}, C_{5k}$ and C_{6k} $(k = 1,2,3,\cdots)$. Thus, these unknown constants can be evaluated by solving this linear system of equations.

Finally, the displacement u_x^o of the outer soil can be written as

$$u_x^o(r,\theta,z) = U_g\left[1 + \sum_{k=1}^{\infty} a_k^{of}\sin\left(\alpha_k \frac{z}{H}\right)\right]$$

$$+ U_g\sum_{k=1}^{\infty} \Gamma_{xk}^g(r,\theta)\sin\left(\alpha_k \frac{z}{H}\right) \tag{14}$$

$$+ H\sum_{k=1}^{\infty} \Gamma_{xk}^u(r,\theta)U_k \sin\left(\alpha_k \frac{z}{H}\right),$$

Physically, the first term on the right hand side of (14) is from the free field response of the linear outer zone, the second term from the free field of the nonlinear inner soil zone, and the third term due to the interaction between soil and pile. Equation (14) will be employed in the next section to estimate the effect of pile-to-pile interaction through the use of superposition, as employed by Kaynia and Kausel (1982), Gazetas and Markis (1991) and Markis and Gazetas (1992) for earthquake-excited pile group. The explicit forms of these interaction factors (Γ_{xk}^g and Γ_{xk}^u) are referred to Yang and Chau (2000b).

The horizontal soil resistance $P(z)e^{i\omega t}$ due to the pile vibration field $u_p(z)e^{i\omega t}$ can be evaluated through the force equilibrium:

$$P(z) = -\int_0^{2\pi} \left[\sigma_r^i(R_0,\theta,z)\cos\theta - \tau_{r\theta}^i(R_0,\theta,z)\sin\theta\right]R_0 d\theta$$

$$= \pi G H$$

$$\times \sum_{k=1}^{\infty} T_k^u U_k \sin(\alpha_k \frac{z}{H}) + \pi G U_g \sum_{k=1}^{\infty} T_k^g \sin(\alpha_k \frac{z}{H}),$$

$$(15)$$

where T_k^u and T_k^g are the dimensionless horizontal soil resistance factor for the k-th vibration mode and the dimensionless resistance factor for the k-th wave mode due to the effect of nonlinearity of inner soil zone. They are given by Yang and Chau (2000b).

2.4 Pile response

Assuming a simple beam model for the vibration of the pfile $\bar{u}_p(z,t) = u_p(z)e^{i\omega t}$, we have

$$E_p^* I_p \frac{\partial^4 \bar{u}_p}{\partial z^4} + m_p \frac{\partial^2 \bar{u}_p}{\partial t^2} = -P_s(z)e^{i\omega t} \qquad (16)$$

where $E_p^* = E_p(1 + 2\xi_p i)$ is the complex elastic modulus of the pile and $P_s(z)$ is given by subtracting an equivalent Winkler spring stiffness (due to pile-soil-pile interaction) from the pile deflection.

As shown by Yang and Chau (2000b), the pile head deflection (U_0) can be expressed in terms of the magnitude of ground shaking (U_g) and the shear force applied from the pile cap (P_0), which remains as unknown of the problem:

$$U_0 = u_p(H) = H\left\{S_P \frac{H^2 P_0}{E_p I_p} + S_g \frac{U_g}{H}\right\}, \qquad (17)$$

where S_P and S_g are given in Yang and Chau (2000b). Note that all pile-soil-pile interactions are incorporated into these factors. This result will be used to couple the soil-pile system to the structure.

3 SOIL-PILE-STRUCTURE SYSTEM

Our next step is to couple the pile group with the structure to formulate the couple soil-pile-structure system. First, harmonic vibrations of the structure and the pile cap are assumed $\bar{u}_b(z,t) = u_b(z)e^{i\omega t}$ and $\bar{U}_0(t) = U_0 e^{i\omega t}$. Further assuming a beam model, the governing equations and boundary conditions for the structural response $\bar{u}_b(z,t)$ are

$$\begin{cases} E_b^* I_b \dfrac{\partial^4 \bar{u}_b}{\partial z^4} + m_b \dfrac{\partial^2 \bar{u}_b}{\partial t^2} = 0, \\[2mm] \bar{u}_b(z,t) = \bar{U}_0(t), \quad \dfrac{\partial}{\partial z}\bar{u}_b(z,t) = 0, \quad \text{at } z = 0 \\[2mm] \dfrac{\partial^2}{\partial z^2}\bar{u}_b(z,t) = 0, \quad \dfrac{\partial^3}{\partial z^3}\bar{u}_b(z,t) = 0. \quad \text{at } z = L \end{cases} \qquad (18)$$

The solution for u_b can be expressed in terms of sine, cosine, hyperbolic sine and hyperbolic cosine functions (Yang and Chau, 2000a).

The equation of motion of the rigid pile cap subjected to forces $4\bar{P}_0$ (from the piles) and $\bar{Q}_0$ (from the structure) is

$$4\bar{P}_0 + \bar{Q}_0 + M_f \frac{d^2 \bar{U}_0}{dt^2} = 0, \qquad (19)$$

Substitution of (17) into (19) gives the relation between U_0 and P_0

$$4P_0 = -\frac{E_b I_b}{L^3} K_f U_0, \qquad (20)$$

where K_f is given by Yang and Chau (2000). By solving (20), we have P_0 in terms of the ground vibration amplitude U_g as:

$$\frac{H^2}{E_p I_p} P_0 = -\frac{K_f S_g}{K_f S_P + 4\left(\dfrac{E_p I_p}{E_b I_b}\right)\left(\dfrac{L}{H}\right)^3} \frac{U_g}{H}. \qquad (21)$$

Subsequently, substitution of (21) into (20) gives the deflection of the rigid pile cap in terms of the ground vibration amplitude U_g

$$U_0 = \frac{4\left(\dfrac{E_p I_p}{E_b I_b}\right)\left(\dfrac{L}{H}\right)^3 S_g}{K_f S_P + 4\left(\dfrac{E_p I_p}{E_b I_b}\right)\left(\dfrac{L}{H}\right)^3} U_g \qquad (22)$$

And, finally, the structural response in terms of the ground vibration amplitude U_g is

$$U_b = u_b(L) = [f(\lambda_b)]\{A_b\}U_0$$

$$= [f(\lambda_b)]\{A_b\}\frac{4\left(\dfrac{E_p I_p}{E_b I_b}\right)\left(\dfrac{L}{H}\right)^3 S_g}{K_f S_P + 4\left(\dfrac{E_p I_p}{E_b I_b}\right)\left(\dfrac{L}{H}\right)^3} U_g. \qquad (23)$$

By now we have obtained the dynamic responses of the soil-pile-structure by incorporating the nonlinear soil-pile interaction under harmonic seismic SH-wave excitation. The next section will present numerical results obtained using these formulas.

4 NUMERICAL RESULTS

In this section, we will illustrate some of the results obtained by Yang and Chau (2000b) and Chau and Yang (2000b). The full detail will not be reported here due to space limitation. The typical example considered by Gazetas et al. (1992) is adopted here. We should note that the results for linear soil case obtained from the current method are in good agreement with those of Gazetas et al. (1992). The parameters used were basically the same as those given in Table 3 of Gazetas et al. (1992). Note that the structure is modeled as a beam in the present paper, instead of using a single-degree-of-freedom oscillator as by Gazetas et al. (1992).

The seismic amplification is examined through the normalized amplitudes for the free field, the pile cap, and the top of the structure:

$$\frac{|U_f|}{U_g} = \frac{|U_f^0(H)|}{U_g}, \frac{|U_0|}{U_g}, \frac{|U_b|}{U_g} \qquad (24)$$

Figure 2 shows a typical plot for $|U_0|/U_g$ versus the dimensional excitation frequency λ, which is defined as $\lambda = R_0\omega/V_s^o$, with $U_g/H = 0.005$. This figure is for nonlinear soil response. It is shown that the soil-pile-structure to the ground level shaking is not significant.

Figure 3 is for $|U_0|/U_g$ versus the dimensional excitation frequency with λ $U_g/H = 0.005$. The effect of soil-pile-structure interaction to the structural re

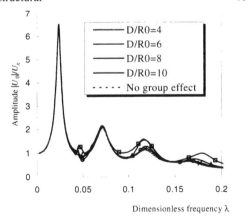

Figure 2. Seismic amplitude ratio versus excitation frequency at the pile cap level for various D/R_0 for $U_g/H = 0.005$

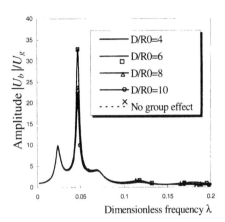

Figure 3. The structural response versus excitation frequency for various D/R_0 for $U_g/H = 0.005$

sponse is clearly more significant at the first natural frequency of the structure (at $\lambda \approx 0.05$).

5 CONCLUSIONS

We have summarized a new analytical model for investigating the interaction of a soil-pile-structure system subjected to earthquake actions, incorporating the effect of soil nonlinearity and pile group interaction. Soil nonlinearity is modeled by introducing a nonlinear inner soil region in the neighborhood of each pile. Continuum mechanics is used to estimate the nonlinear soil resistance. The pile-soil-pile interaction is modeled by using a simple superposition or "wave-interference" method through the introduction of a Winkler spring. The overall stiffness of the pile-soil-pile (or pile group) is then incorporated with the structure through the force equilibrium at the pile cap level. Both the pile and structure are modeled as simple beams. The simplest case of a 2×2 pile group is formulated.

Although the full detail is not given here, Yang and Chau (2000b) and Chau and Yang (2000b) showed that the nonlinear soil behavior alone does not lead to significant changes in structural responses, but the coupling between the pile group effect and the soil nonlinearity effect can lead to a double in the structural response. When soil nonlinearity is neglected, the solutions are the same as those obtained by Gazetas et al. (1992). In general, the effect of the pile-soil-pile interaction is stronger than that of the soil nonlinearity. When the spacing between the piles decreases, the seismic response of the structure at the natural frequency of the structure can increase drastically. When the magnitude of bedrock shaking increases, the structural responses do not necessarily increases. There exists a critical level of ground shaking, at which the structural response is most severe.

6 ACKNOWLEDGEMENT

This research was supported by "ASD:Earthquake resistance appraisal and Design" Project No. A202 of the Hong Kong Polytechnic University.

REFERENCES

Abramowitz, M., Stegun, I.A., 1965. *Handbook of Mathematical Functions*, Dover, New York.

Angelides, D.C., Roesset, J.M., 1981. Nonlinear lateral dynamic stiffness of piles. *J. Geotech. Eng. Div.* 107(GT11), 1443-1460.

Cai, Y.X., Gould, P.L., Desai, C.S., 2000. Nonlinear analysis of 3D seismic interaction of soil-pile-structure systems and application. *Engineering Struct.* 22(2), 191-199.

Chau, K.T., Yang, X., 2000a. Nonlinear interaction of soil-pile in horizontal vibration Part I: Soil resistance. Submitted to *Earthquake Eng. Struct. Dyn.*

Chau, K.T., Yang, X., 2000b. Nonlinear interaction of soil-pile-structure under earthquake excitations Part II-pile tip condition. Submitted to *Int. J. Solids Structures.*

Finn W.D.L., Byrne, P.M., Evans, S., Law T., 1996. Some geotechnical aspects of the Hyogo-ken Nanbu (Kobe) earthquake of January 17, 1995. *Can. J. Civ. Eng.* 23, 778-796.

Gazetas, G. and Makris, N., 1991. Dynamical pile-soil-pile interaction. I: Analysis of axial vibration. *Earthquake Eng. Struct. Dyn* 20, 115-132.

Gazetas, G., Fan, K., Kaynia, A. and Kausel, E., 1991. Dynamic interaction factor for floating pile groups. *J. Geotech. Eng. ASCE* 117(10), 1531-1548.

Gazetas, G., K. Fan, Tazoh, T., Shimizu, K., Kavvadas, M., Makris, N., 1992. Seismic pile-group-structure interaction. In *Piles Under Dynamic Loads*. ASCE, September 13-17, 1992, New York, N.Y., 56-93.

Hadjian A.H., Fallgren R.B., Tufenkjian M.R., 1992. Dynamic soil-pile-structure interaction the state-of-practice. *Piles under Dyanmics Loads* (ed. S. Parkash), Geotechnical Special Publication, 34. 1-25.

Hardin, B.O., Drnevich, V.P., 1972. Shear modulus and damping in soils: Design equations and curves. *J. Soil Mech. Found. Div. ASCE* 98(SM7), 667-692.

Idriss, I.M., Dorby, R., Singh, R.D., 1978. Nonlinear behavior of soft clays during cyclic loading. *J. Geotech. Eng. Div. ASCE* 104(GT12), 1427-1447.

Kaynia, A.M., Kausel, E., 1982. Dynamic stiffness and seismic response of pile groups. *Research Report R82-03*, Massachusetts Institute of Technology, Cambridge, Mass.

Makris, N., Gazetas, G., 1992. Dynamical pile-soil-pile interaction. II: Lateral and seismic response. *Earthquake Eng. Struct. Dyn.* 21, 145-162.

Matsui, T., Oda, K., 1996. Foundation damage of structures. Special issue of *Soils and Foundations,* 189-200.

Novak, M., 1991. Piles under dynamic loads. *Proc. Second Int. Conf. on Recent Advances in Geotech. Earthquake Eng. and Soil Dyn.*, St. Louis, Missouri, pp. 2433-2457.

Novak, M., Han, Y. C., 1990. Impedances of soil layer with boundary zone. *J. Geotech. Eng. ASCE* 116, 1008-1015.

Novak, M., Sheta, M., 1980. Approximate approach to contact effects of piles. In *Dynamic Response of Pile Foundations: Analytical Aspects*, ASCE, (ed. O'Neill and Sobry), 53-79.

Tokimatsu, K., Mizuno, H., Kakurai, M., 1996. Building damage associated with geotechnical problems. Special issue of *Soils and Foundations* 219-234.

Veletsos, A.S., Dotson, K.W., 1986. Impedances of soil layer with disturbed boundary zone. *J. Geotech. Eng. ASCE* 112(3), 363-368.

Veletsos, A.S., Dotson, K.W., 1988. Vertical and torsional vibration of foundations in inhomogeneous media. *J. Geotech. Eng. ASCE* 114(9), 1002-1021.

Yang, X., Chau, K.T., 2000a. Nonlinear interaction of soil-pile in horizontal vibration: Part II- soil-pile interaction. Submitted to *Earthquake Eng. Struct. Dyn.*

Yang, X., Chau, K.T., 2000b. Nonlinear interaction of soil-pile-structure under earthquake actions: Part I- Pile group effect. Submitted to *Int. J. Solids Structures.*

Earthquake Engineering Frontiers in the New Millennium, Spencer & Hu (eds),
© 2001 Swets & Zeitlinger, ISBN 90 2651 852 8

Displacement-loading modes of low-cycle test simulating earthquake response of structure

Z.M. Huang & Y.W. Wang
Civil Engineering Department, Chongqing University, Chongqing, P.R. China

B.J. Yin
China Academy of Building Research, Beijing, China

ABSTRACT: Earthquake ground motions are classified into 5 types base on their duration, Fourier spectrum and effects on displacement responses of SDOF systems in this paper. Features of nonlinear displacement responses of SDOF systems to different types of earthquake ground motions are then investigated respectively. Based on the results, displacement-loading modes of low-cycle test simulating structural responses to earthquake ground motions of types S, L-1 and M-1 are presented.
Key words: seismic response; low-cycle test; displacement-loading modes.

1 INTRODUCTION

Structural dynamic tests are usually done with small-scale models of prototype structures due to the limitation of capability of test facilities. Therefor it is very common to use low-cycle test to investigate the seismic performance of structures or structural elements. Generally, the displacement-loading mode is used when the structure is in inelastic phase. However, the accumulated damage of structures, especially reinforced concrete structures, subjected to cycling loads, are related not only to the total dissipated energy but also to some other factors such as the magnitude and occurrence sequence of peak displacement responses, and the offset of the displacements. The same total dissipated energy but different displacement-loading traces (or patterns) may result in remarkably different damage of structures. Many efforts have been made to find appropriate displacement-loading modes in low-cycle tests for simulating structural earthquake responses. Unfortunately, it is very difficult to present a general displacement-loading mode that can consider all the factors mentioned above.

Earthquake ground motions are indeed complicated, but their rough categories exist. The structural responses caused by different types of ground motions may follow different rules corresponding to the excitations. It may be both simple and feasible to establish different displacement-loading modes for different types of earthquake ground motions, instead of pursuing a general loading mode.

2 CLASSIFICATION OF EARTHQUAKE GROUND MOTIONS

Generally, earthquake ground motions can be classified into three types, i.e., motions of pulse style with short epicenter distance, motions of medium duration without clear predominant period and motions of long duration with clear predominant period (Newmark & Rosenbluth 1971). The features of non-linear displacement response of SDOF systems are considered here for classifying earthquake ground motions, in addition to their duration and spectral characteristics. The quantitative indexes are presented for the identification of types of ground motions. More than 200 earthquake records with peak acceleration of 0.2 g or greater, are used in the analysis.

2.1 *Duration of strong earthquake ground motions*

There are many definitions of strong earthquake duration (Trifunac & Brady 1975), and the definition of energy duration is used here. From the results of inelastic responses of SDOF systems, the 70% energy duration T'_{70} is a good choice.

$$T_{(70)}' = t_{(80)} - t_{(10)} \tag{1}$$

where $t_{(80)}$ and $t_{(10)}$ are the time points at which 80% and 10% accumulated energy of the ground motion reached respectively. $T'_{(70)}$ matches the strong motion phase of the earthquakes better than the usually used definition of $T_{(70)}=t_{(85)}-t_{(15)}$.

2.2 The consideration of the spectral characteristics

Many ground motion records with long duration have clear predominant periods, and their Fourier spectra mainly distribute over relatively narrow frequency bands. On the contrary, the pulse type ground motions usually contain relatively ample frequencies, and their spectra distribute over a wider frequency band. But there are some exceptions, such as Chile earthquake (Mar. 3, 1985) Llolleo record and Imperial Valley earthquake (May 18, 1940) EL Centro record. Their Fourier spectra distribute over a relatively wider range without clear predominant period while their $T_{(70)}$ are 23.01 sec and 13.28 sec respectively, which belong to long duration. A high magnitude earthquake recorded on rock bed or on rigid ground, may have a longer duration but a wider frequency spectrum band. It is not enough using the duration only as the index to classify earthquake ground motions. The spectral characteristic of earthquake ground motions is also a very important factor in the classification.

The length between the first and last intersection point of a Fourier spectrum curve and the horizontal line with an ordinate of 1/10 peak value of the Fourier spectrum is defined as Fourier spectrum width (W_{FS}). The results of inelastic earthquake responses of SDOF systems show that displacement responses excited by most ground motions with $W_{FS} < 4$ Hz show much more regularity than those excited by ground motions with $W_{FS} > 4$ Hz. Therefore, the ground motions with $W_{FS} < 4$ Hz can be treated as the ground motions having clear predominant period, and the others ($W_{FS} > 4$ Hz) are treated as irregular motions without clear predominant period.

2.3 Classification of earthquake ground motions

Based on the principle that the envelope appearances of inelastic displacement responses of SDOF systems should be similar if subjected to the same type of ground motions, the ground motions are classified into three types, i.e., short duration pulse (type S), medium duration (type M) and long duration (type L). Generally, the ground motions of type S occurred close to the hypocenter have wider Fourier spectra width and shorter duration with few exceptions. It is not necessary to subdivide this kind of ground motions according to their spectral characteristics. While the Fourier spectra-widths of ground motions of type M and type L change over a wide range, and their frequency features remarkably influence the structure responses due to their longer duration. Therefor it is necessary to subdivide them into two sub-types, one with clear predominant period (type M-1 and type L-1) and the other without clear predominant period (type M-2 and type L-2), as shown in Table 1.

3 RESPONSE FEATURES OF STRUCTURES SUBJECTED TO DIFFERENT TYPES OF EARTHQUAKES

3.1 Research method

79 earthquake records of different types with peak accelerations of 0.2g or greater are selected as input, and the inelastic displacement time-history responses of SDOF systems are investigated in order to find the structural response rules under different types of earthquake ground motions above mentioned. The distribution of the earthquake records among these types is shown in Table 2.

Table 1. The classification of earthquake ground motions

Classification	type-S	type-M		type-L	
		M-1	M-2	L-1	L-2
W_{FS}	....	< 4Hz	≥4Hz	<4Hz	≥4Hz
$T_{(70)}$	<4s	4s≤$T_{(70)}$<10s		$T_{(70)}$≥10s	

Table 2. The distribution of the earthquake records

Type	S	L-1	M-1	L-2	M-2
Numbers	19	17	19	8	16

The natural period of SDOF systems indicated as T_0 is selected to be 0.5s, 1.0s and 2.0s respectively, and the target ductility demand μ of the systems takes 2.0, 4.0 and 6.0. Reighley damping is adopted with a constant damping ratio 0.05, and Clough's hysterestic model is used in the analysis.

For each displacement response time-history, the following response indexes are taken to describe its offset and envelope shape. They are the normalized average displacement offset X, the ratio of maximum displacement in positive direction to that in negative direction X_{max}/X_{min}, the normalized peak-occurrence number (NPON) diagram and its kurtosis coefficient μ_4. The occurrence sequence of the peak displacement is analyzed through the time-history.

3.2 The rules of structural displacement responses under ground motions of type S

The typical displacement time-history curve of SDOF systems under ground motions of type S is shown in Figure 1.

Figure 2 is the superposed normalized peak-occurrence number (NPON) diagram of displacement responses of SDOF systems with $T_0 = 0.5$ s and μ= 4 under 19 ground motion records of type S. The horizontal axis indicates the normalized peak values, and the vertical axis refers to the occurrence numbers of related peak values. It can be easily seen from Figure 2 that the displacement responses of the systems are asymmetry, and remarkable displacement offset exists. The kurtosis μ_4 of NPON diagram

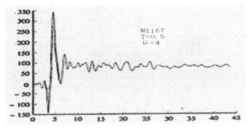

Figure 1. A typical response displacement time-history under ground motions of type S

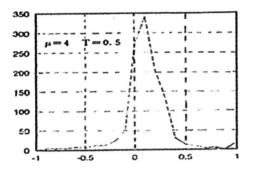

Figure 2. Superposed NPON under ground motions of type S

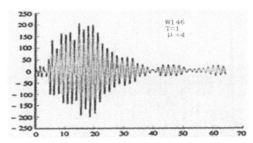

Figure 3. A typical response displacement time-history under ground motions of type L-1

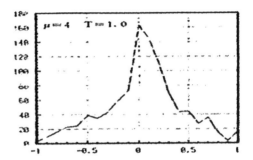

Figure 4. Superposed NPON under ground motions of type L-1

is very large (about 41.1), the diagram is concentrated over small displacements range. Each input record corresponds to one large displacement peaks only, that is, if the input ground motion is of pulse type, so is the structural displacement response.

From the statistic results, under the ground motion of type S, the relative offset of structural displacements is between 0.1 and 0.23, and the ratio of the positive to negative maximum displacement is between 2 and 4. The displacement offset is related to T_0 and μ. Larger T_0 and/or μ will cause larger offset.

In 171 computed samples, 82% structural displacement histories get their maximum peaks first, and then, a few of medium amplitude vibrations (0 to 5 times) as shown in Figure 1. After that the displacement attenuates quickly. The others undergo a few (1 to 4 times) medium amplitude oscillations first, then the peak one, thereafter the same as the first case.

3.3 The rules of structural displacement responses under ground motions of type L-1

Figure 3 and Figure 4 are typical displacement response time-history and the superposed NPON of the displacement responses of SDOF systems with period $T_0=1.0$s and ductility demand $\mu= 4$ under ground motions of type L-1.

Compared with that under ground motions of type S, μ_4 is very small under ground motions of type L-1, that is, the diagram is not so concen trated as it is.

The diagram is basically symmetry about the vertical axis through zero point. It means that the distribution of the displacement amplitudes is relatively even during the oscillation of the system, and the systems vibrate mainly around the initial balance place with very small offset. The envelope of the displacement time-history goes up gradually and then attenuates slowly along the time-axis, forming a smooth curve. The envelopes of displacement time-histories will have two peaks, if the envelopes of input earthquake records have two peaks, that is, two strong motion phases.

Within the range investigated in this paper, the displacement offset has little to do with T_0 and μ. The average normalized offset X is generally within 0.1, and X_{max}/X_{min} lies between 1.2 and 1.6.

3.4 The rules of structural displacement responses under ground motions of type M-1

Under the excitation of ground motions of type M-1, the envelopes of the displacement response time-histories and the superposed displacement NPON diagrams of the systems are similar to those under ground motions of type L-1. The same is true to the structural response indexes used here. In other words, the structural displacement responses under ground motions of type M-1 and L-1 have similar rules.

What should be mentioned here is that the equations of envelopes of the displacement time-histories that will be used later, under the two types of ground

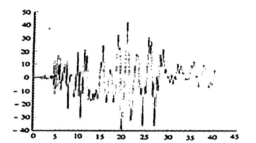

Figure 5. A typical response displacement time-history under ground motions of type L-2 & M-2

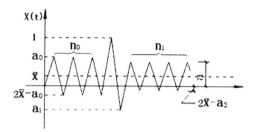

Figure 6. Displacement-loading modes for ground motions of type S

motions, are different, because of their differences in duration. In addition, the envelope of displacement time-histories is usually of single peak under ground motions of type M-1 due to their shorter strong duration.

3.5 *The rules of structural displacement responses under ground motions of type L-2 and M-2*

The typical displacement time-history under ground motions of type L-2 and M-2 is shown in Figure 5. The displacement responses are irregular due to the irregularity of the input ground motions.

In 216 computed samples, 22% displacement responses undergo considerable offset, and the offset is related not only to the structural systems but also to the ground motions. The irregularity is also reflected in the occurrence sequence of different displacement amplitudes (Figure 5).

More over, the response frequency of a SDOF system fluctuates considerably under these two types of ground motions.

Therefore, it is hard to conclude some regularity of the structural responses under the ground motions of type L-2 and M-2.

4 THE RECOMMENDATION OF DISPLACEMENT-LOADING MODES OF LOW-CYCLE TEST

Base on the results of displacement response rules of SDOF systems under different types of ground motions, the displacement-loading modes of low-cycle test are presented.

4.1 *Displacement-loading modes for ground motions of type S*

Two suggested displacement-loading modes of low-cycle test are given for ground motions of type S. Here the displacements are normalized by their maximum value.

Mode 1: a single cycle of positive amplitude of 1 and negative amplitude of a_1 is loaded first as the pulse action, and then followed by n_1 cycles of amplitude of a_2. The average displacement offset is X.

Mode 2: n_0 cycles of positive amplitude of a_0 with offset of X is loaded before Mode 1.

The two modes are shown in Figure 6 with the parameter values in Table 3.

Table 3. The parameters of Figure 6.

a_0	a_1	a_2	n_0	n_1	x
0.6-0.9	0.2-0.5	0.4-0.65	1-4	0-5	0.1-0.25

4.2 *Displacement-loading modes for ground motions of type L-1*

The duration of ground motions of type L-1 is longer; therefore it is necessary to provide the equations of envelopes of the displacement time-histories of systems to establish displacement-loading modes. For doing this, the points before cracking are taken out in every displacement response time-history, and after being normalized, the peak values of each displacement response time-history under ground motions of type L-1 are put in one single coordinate system. By polynomial-regression, the parameters of equations of envelopes of the displacement responses of structures with different ductility demands and natural periods can be determined.

Here the used polynomial is

$$x = a_0 + a_1t + a_2t^2 + a_3t^3 + a_4t^4 + a_5t^5 \qquad (2)$$

The case of single-peak envelopes of displacement responses is discussed first. The nine displacement response envelopes, in direction of positive and negative respectively, for three different ductility demands and three different natural periods of systems under ground motions of type L-1 are shown in Figures 7 (a)&(b).

The positive envelopes vary little as the natural periods and ductility demands of systems change, hence one single envelope is regressed for general use, with its parameters shown in Table 4.

The negative envelopes vary considerably with the natural periods and ductility demands of systems due to the displacement offsets. For convenience, only two negative envelopes are chosen in practice: the upper boundary and the lower boundary of these envelopes.

The upper boundary corresponds to the vibrations with larger displacement offsets, while the lower one corresponds to the symmetry vibrations that adapt to systems with shorter natural periods and/or smaller ductility demands. The actual negative envelopes of displacement of systems lie generally between the two boundaries. The parameters of the two negative boundaries are presented in Table 4.1.

Table 4.1. The parameters for single-peak envelope of displacement responses of systems under GMs of type L-1

Envelope	a_0	a_1	a_2	a_3	a_4	a_5
P	0.531	0.081	-0.006	1.3E-4	-1.4E-6	5.68E-9
N-U	-0.40	-0.065	-0.0053	-1.2E-4	-1.3E-7	2.34E-8
N-L	-0.53	-0.08	0.0056	-1.3E-4	1.42E-6	-5.7E-9

P: for positive envelope; N-U: for negative-upper boundary; N-L: for negative-lower boundary.

Table 4.2. The parameters for double-peak envelope of displacement responses of systems under GMs of type L-1

Envelope	a_0	a_1	a_2	a_3	a_4	a_5
P-L	0.059	0.099	0.0344	-9.2E-3	8.25E-4	-3.1E-5
P-R	-1.593	0.343	-0.02	5.24E-4	-6.3E-6	2.91E-8
N-L	-0.104	0.006	-0.0374	6.74E-3	-4.1E-4	8.42E-6
N-R	1.354	-0.171	7.97E-4	3.9E-4	-1.3E-5	1.2E-7

P-L: positive-left part; P-R: positive-right part; N-L: negative-left part; N-R: negative-right part.

Now the displacement-loading mode of type L-1 with single peak envelops can be described as: the positive displacement amplitudes of loading are regulated by the positive loading envelope, and the negative displacement amplitudes go between the two negative envelopes. The graphic expression of the loading mode is shown in Figure 8, where the maximum positive-displacement is k_0a_0, with k_0 varying between 1.2 and 1.6.

In the case of double-peaks envelopes of structural displacement responses, T_0 and μ have little effect on the envelope of the displacement time-

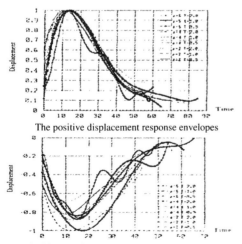

The positive displacement response envelopes

(b) The negative displacement response envelopes

Figure 7. Displacement response envelopes of systems under ground motions of type L-1

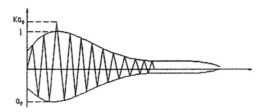

Figure 8. The displacement-loading mode of type L-1 with single-peak envelopes

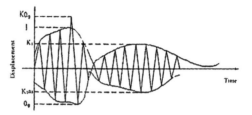

Figure 9. The displacement-loading mode of type L-1 with double-peaks envelopes

histories. So only one envelope of displacement responses is regressed for different T_0 and μ.

The graphic expression of the loading mode for double-peaks ground motions of type L-1 is shown in Figure 9. The envelope consists of two parts, the main-peak part (left part) and the sub-peak part (right part). The parameters for the envelope are given in Table 4.2 by polynomial-regression. The ratio of positive to negative maximum displacements is between 1.2 and 1.4. The sub-peak is about 0.5 to 1.0 times of the main-peak. The average displacement offset is ignorable.

4.3 Displacement-loading modes for ground motions of type M-1

The displacement response envelops regressed corresponding to different system ductility demands and natural periods under the ground motions of type M-1 are shown in Figure 10.

From Figure 10, all the displacement envelopes are similar in shape except for a translation. The translation depends on the natural periods of systems. The envelope shifts right for a longer period of system.

For the sake of application, two envelope boundary curves are given in positive and negative direction respectively, that is, the left boundary curves suitable for shorter period of systems, and the right boundary curves suitable for longer period of systems. The regressed parameters of the four envelope boundary curves are presented in Table 5.

The graphic expression of displacement-loading mode for ground motions of type M-1 is the same as that of type L-1 (Figure 8) except that the parameters of envelopes are given in Table 5.

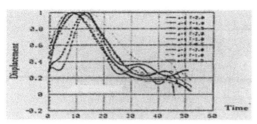

Figure 10(a). The positive displacement response envelopes under ground motions of type M-1

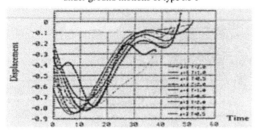

Figure 10(b). The negative displacement response envelopes under ground motions of type M-1

Table 5. The parameters of envelopes of displacement responses of systems under GMs of type M-1

Direction	a_0	a_1	a_2	a_3	a_4	a_5
P-L	0.433	0.166	-0.016	5.0E-4	-6.83E-6	3.23E-8
P-R		0.085	-0.001	-2.4E-4	8,68E-6	-8.19E-8
N-L	-0.45	-0.133	0.0135	-4.5E-4	6.34E-6	-3.08E-8
N-R	-0.37	-0.069	0.0002	3.5E-4	-1.31E-5	1.35E-7

P-L: positive left-boundary; P-R: positive right-boundary;
N-L: negative left-boundary; N-R: negative right-boundary

5 CONCLUSIONS

Earthquake ground motions are classified into 5 types, i.e., short duration pulse (type S), medium duration with and without clear predominant period (type M-1 and type M-2), and long duration with and without predominant period (type L-1 and type L-2), according to their strong duration, Fourier spectrum feature and effects on SDOF systems. The rules of displacement responses of SDOF systems with different natural periods and ductility demands under different types of ground motions are investigated. The results show that the displacement responses observe relatively obvious rules under ground motions of type S, L-1 and M-1. But no general rules can be drawn under ground motions of type L-2 and M-2. Based on the results, recommendations of displacement-loading modes of low-cycle test simulating structural responses under ground motions of type S, L-1 and M-1 are presented.

REFERENCES

Newmark, N.M. & Rosenbluth, E. 1971. *Fundamentals of Earthquake Engineering*, Prentice-hall.

Trifunac, M.D & Brady, A.G. 1975. A study of the duration of strong earthquake ground motion. *Bulletin of Seismologica Society of America*, 65: 581-626.

Earthquake Engineering Frontiers in the New Millennium, Spencer & Hu (eds),
© 2001 Swets & Zeitlinger, ISBN 90 2651 852 8

Loading method for quasi-static and pseudodynamic test

Fawei Qiu, Peng Pan, Wenfeng Li, & Jiaru Qian
Department of Civil Engineering, Tsinghua University, 100084, Beijing, China

ABSTRACT: In this paper, three structural test methods along with software developments and applications are introduced. (1) Quasi-static test and control software TUST, three kinds of loading methods were employed in the quasi-static test: force control loading, displacement control loading and hybrid (force-displacement) control loading. The force-displacement control mode switchover can be performed smoothly during testing. (2) Multi-dimensional quasi-static test and control software TUMT, Six kinds of special loading paths were considered in the software TUMT on the bases of displacement control mode. (3) Pseudodynamic test and control software TUT (include one-dimensional, two-dimensional and substructuring techniques. The α-method and PC-Newmark method were employed in TUT.). All of the three tests control software that is based on the platform of Windows NT and Visual C++ has been put into practice in structural laboratory of Tsinghua University.

1 INTRODUCTION

The primary purpose of test study is to improve the understanding of material properties and structural loading behaviors. Several test methods are available for studying structural inelastic seismic performance. Quasi-static test is the most popular test used in structure or component performance research, which adopts load control or displacement control to proceed with low cycle load on the specimens, making them destroyed from elastic stage. There are two kinds of quasi-static test: humdrum loading and cyclic loading. Quasi-static test exploits the utmost use of specimens to provide all kind of fundamental information, such as carrying capacity, stiffness, deformation capacity, energy dissipation and damage characteristic, and it is the most wildly used one among all the structure engineering tests. Although quasi-static test is versatile, the loading history on specimens may not be the representative of the loading history of earthquake. The most realistic simulation of earthquake ground motions is shaking table test. However, significant limitations on size and weight make the test results not so realistic. Pseudodynamic test combines the economy and flexibility of static test with realism of shaking table test (Takanashi 1974), and this method has been employed by many researchers for their individual studies on the seismic behaviors of various structural systems.

2 QUASI-STATIC TEST

Quasi-static test is a test method that the loading procedure is cycle with variable amplitude. The monotonic loading is only a special case. Many researchers use it to performance the test to investigate the structures or components under loading (Hwang 1984, Kobayashi 1984, Leon 1996, Low 1987). Quasi-static test method can be performed for almost every kind of structure or component; its advantage is convenience comparing with the shaking table and pseudodynamic test (Standard of People's Republic of China 1997). The actuator is not necessary to be carried out such test. In addition, the loading rate is negligible, so it is possible to compare the test results that come from individual laboratory. Many kinds of loading facilities can be chosen for the quasi-static test. The jacks and hydraulic jacks were popular in laboratories once upon a time. The shortcoming of the facilities is their instability and inaccuracy. Nowadays, the servo hydraulic loading systems are being employed in many laboratories.

2.1 *Loading rule and control*

Three loading mode is in common use: loading by force control, loading by displacement control and loading by mixed force-displacement control. The loading amplitude is shown in Fig. 1. It is prescribed in Chinese aseismic construction code JGJ 101-96 that force control load on different levels should be put into use before the specimen yield and the loading value of each step should be reduced when reaching the crack value or yield value. After yield point of specimen, displacement control loading should be adopted. The initial value of this stage is the displacement of specimen when it become yield and the following control value use multiple of the yield value. The cycle of each loading value refers to the objection of the test. Commonly, the cycle number each loading amplitude is one time before yield point and three times after that. But there are some discrepancies in the prescription above. First, how to determine the crack load? Now, artificial inspect method is widely used and even loading level by level can't find the value accurately. Another problem is lack of standards to determine the yield point. People find the yield point by their experience. So the yield point used now is inaccurate. Some specimen doesn't have obvious yield point.

Based on the Windows NT and Visual C++, quasi-static test software TUST was developed (Qiu 2001a). The force-displacement control mode switchover can be performed smoothly during testing. Considering current situation, an artificial inspection is still used to find yield point in TUST software. The process is described in Fig. 2. Initial force control is used and then the displacement move to the point D_y in hysteretic curve, which is found by experience, the control mode changed from force to displacement. To high-rise building or MDOF structure, switch mode use similar rule as single point tests. When the displacement of first DOF reaches to yield point D_y, control mode of all the loading point will switch from force to displacement and displacement distribution has the same shape as that of force at beginning of test.

2.2 *Multi-dimensional loading rule and control*

According to the field investigation and test results, the structural damage is more severe under multi-dimensional earthquake excitation. The reason is that the damage in one direction has an effect on the structural damage in the other direction. The multi-dimensional loading test is fundamental to find out the behavior of structures or components under complex stresses. Fig. 3 shows six loading rules used in multi-dimensional test. Based on the six loading paths, quasi-static test software TUMT was developed for carried out multiple dimensional loading on structure (Qiu 2001b). Lai (1984) proposed a multi-spring model for reinforce concrete column. Bousias *et al.* (1995) carried out column biaxial bending with 12 load-path. But how to appraise the test results obtained the different loading rule and which loading rule is more reasonable and should be employed in the test were considered in the future.

3 PSEUDODYNAMIC TEST

3.1 *Substructuring technique and software*

It is well known that the laboratory test of large structures is limited by its size, loading equipment and instrumentation, and cost. On the other hand, structural damage caused by seismic force is local. The behavior of the damaged part is more attractive to us. In pseudodynamic test it is possible to take the potential damage portion of the structure as a specimen while a computer simulates the rest of the structure simultaneously. The tested part is named testing substructure and the rest is called calculating substructure. Both tested and calculated substructures are connected together in the whole structure dynamic equation. The concept of substructure technique shows in Fig. 4.

Nakashima (1985) and Dermitzalds *et al.* (1985) studied the substructure integration algorithm, error accumulation, and loading control method. Iemure (1988) carried out substructure pseudodynamic test for a three story RC frame and a five story RC frame, which the tested part, is only the first story. With the PC-Newmark integration, Qiu *et al.* (1994, 1995, 1996) performed substructure pseudodynamic test for a multi-story RC industrial building, a 3-story RC frame and a 6-story RC frame. The implicit a-method which first proposed by Thewalt and Mahin (1987) is employed in pseudodynamic test. Thereafter, Shing and Vannan (1990, 1991a 1991b) gave the solution and error estimation of α-method in more details. Kanada (1995, 1996) also discussed the implicit Newmark method in substructure pseudodynamic test. With RC frame structures, Qiu (1995, 1997a, 1997b) conducted substructure pseudodynamic test with implicit a-method. According to the test results, the use of substructure pseudodynamic test method can perfectly simulate the seismic nonlinear response. For the response of the whole structure-equipment or non-structures, the specimen should include both main structure and equipment or non-structures when the shaking table test is employed. However, with substructure pseudodynamic test, only the equipment (named secondary system) needs to be tested and the main structure is simulated by computer. The interaction between secondary system and main structure is incorporated in whole structural dynamic equations. Up till now, substructure pseudodynamic test is more suitable for studying earthquake response of complex structures (Ohi 1996 and Takanashi 1996).

A pseudodynamic test software TUT based on the substructure technique developed in Department of Civil Engineering of Tsinghua University (Qiu 2000). TUT can perform four kinds of structure tests: type 1, one-dimensional pseudodynamic test; type 2, one-dimensional pseudodynamic test with substructure technique; type 3, two-dimensional

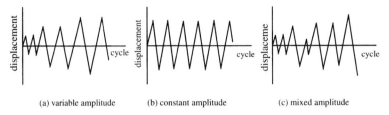

(a) variable amplitude   (b) constant amplitude   (c) mixed amplitude

Figure 1. Loading amplitude

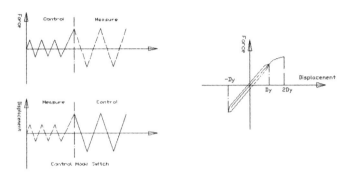

Figure 2. Loading process of single DOF structure

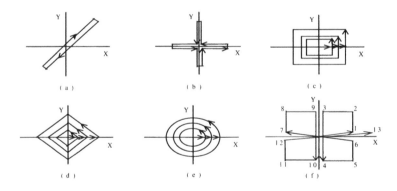

Figure 3. Loading rules for multi-dimension test

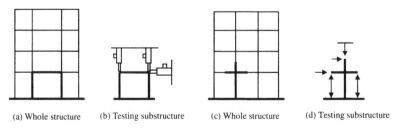

(a) Whole structure   (b) Testing substructure   (c) Whole structure   (d) Testing substructure

Figure 4. Testing substructure and calculating substructure

pseudodynamic test with substructure technique; type 4, two-dimensional pseudodynamic test, which includes torsion, with substructure technique. TUT has a friendly graphic user interface for the structural arguments input, equipment channel configuration, and real-time monitoring. The vibration modes and frequency of the structure tested can be calculated and displayed before the test. The test procedure can also be simulated with a linear model to predict the test results; the α-method of integration algorithm is employed in type 1. The PC-Newmark method and substructure technique were employed in type 2, type 3 and type 4.

3.2 Others

Large scale buildings, long span bridges, dams and pipe systems are supported at more than one point and are simultaneously subjected to different ground motions which have different amplitude, phase and response spectrum (Hu 1988 and Wang 1984). Dynamic response of structures subjected to a multiple excitation is studied by the expensive shaking table array (Adachi 1988). A much lower cost pseudodynamic test method can simulate the response of multi-dimensional and multiple excitation input (Qiu 1999).

Pseudodynamic test is also able to study the soil structure interaction (SSI). The primary difficulty in SSI pseudodynamic test is that the interaction depends not only on the structure, but also on the soil dynamic stiffness that is related to the loading frequency and velocity. Toki (1990, 1991) proposed a numerical integration algorithm that can take account of the frequency dependence and developed a testing program HENESSI (hybrid experiments on nonlinear earthquake induced soil structure interaction). The pseudodynamic test was also employed in Seismic Response of Layer (Katada 1988, Peek 1991, Kusakabe 1995 and Adachi 1996). The layer can be simplified as a MDOF system. The nonlinear layer part is tested by pseudodynamic method and the rest is simulated in computer. Kusakabe and Morio (1995) conducted a test with six DOF layer.

In pseudodynamic test, the actuator applied the response displacement to the model structure. Hence, a major concern with the result of such test is the rate of loading. Yamazaki (1986) and Kitagawa (1984) gave a comparison between shaking table test and pseudodynamic test. Shing and Mahin (1988) suggested taking the rate of loading into integration algorithm. Nakashima etc (1992, 1999) developed a real time pseudodynamic test method. Horiuchi et al. (1996) proposed a real time hybrid experimental method, in which output from an actuator-excited vibration experiment and response calculations are combined on-line and conducted simultaneously.

A SDOF system was tested with effective force testing (EFT) by Murceck et al. (1996). The loading

equipment is an actuator. But the actuator should have enough dynamic ability in order to keep the output following the target perfectly.

4 CONCLUSION

New developments and applications of quasi-static test and pseudodynamic test are discussed in this paper. The test control software is the fundamental to carry out these tests. Details, formulae and some examples are neglected here to highlight the concepts of test method and techniques. It is shown that the quasi-static tests give the constitutive relation of components. The pseudodynamic test is a very efficient approach to study structural seismic response. Although the pseudodynamic test method itself is necessary to improve in some way, it is still the strongest method to simulate the seismic response for large-scale and complex structures today.

REFERENCE

Adachi, H. et al. 1988. Dynamic response of structures subjected to multiple excitation. *Proc. of 9th world conference on earthquake engineering*, Tokyo-Kyoto, Japan (vol. V).

Adachi, T. et al. 1996. Pseudo-dynamic test on intensity of earthquake ground motions at occurrence of liquefaction, Paper No.366, 11th WCEE.

Bousias, S.N., Verzeletti, G., Fardis, M.N. & Guitierrez, E. 1995. Load-path effects in column biaxial bending with axial force. *Journal of engineering mechanics*, 121(5):96–605.

Dermitzalds, S.N. & Mahin, S.A. 1985. Development of substructuring techniques for on-line computer controlled seismic performance testing, *Report No. UCB/EERC 85-04*, University of California, Berkeley, CA.

Horiuchi, T. et al. (1996). Development of a real-time hybrid experimental system with actuator compensation. Paper No.660, 11th WCEE.

Hu, Yuxian 1988. *Earthquake Engineering*, Earthquake Press.

Hwang, T.H. & Scribner, C.F. 1984. Effect of load history variation on cyclic response of R/C flexural members, *Proc. of 8th WCEE*, San Francisco, 6:13–420.

Iemura, H. et al. 1988. Testing R/C specimens by a substructure based hybrid earthquake loading system. *Proc. of 9th WCEE*, Tokyo-Kyoto Japan, 4:5–40.

Kanda, M. et al. 1995. Implicit integration scheme based on initial stress method for substructure on-line test, (in Japanese). *J. structural construction engineering*, AIJ, No.473, 75–84.

Kanda, M. et al. 1996. Simulation of dynamic behaviors of structures by means of substructure on-line test with implicit integration scheme, (in Japanese). *J. structural construction engineering*, AIJ, No.487, 63–72.

Katada, T. & Hakuno, M. 1988. Nonlinear analysis of surface ground motion by digital controlled on-line experimental method. *Proc. of 8th world conference on earthquake engineering*, San Francisco, Calif., 3:1033–1040.

Kitagawa, Y. et al. 1984. Correlation study on shaking table tests and pseudodynamic tests by R.C. models. *Proc. of 8th WCEE*, San Francisco, California, U.S.A. 6:667–674.

Kobayashi, K. et al. 1984. Study on the restoring force characteristics of RC column to bi-directional deflection history. *Proc. of 8th WCEE*, San Francisco, Vol.6, 537–544.

Kusakabe, S. & Morio, S. 1995. The development of a substructure on-line testing system for seismic response analysis of a geotechnical system, Soils and foundations. Vol.35, No.2, 117–125. (Japanese society of soil mechanics and foundation engineering).

Lai, S.S., Will, G.T. & Otani, S. 1984. Model for inelastic biaxial bending of concrete members. *Journal of structural engineering*, 110(11):2563–2584.

Leon, R.T. & Deierlein, G.G. 1996. Considerations for the use quasi-static testing. *Earthquake spectra*, 12(1):87–109.

Low, S.S. & Moehle J.P. 1987. Experimental study of reinforced concrete columns subjected to multi-axial loading. *Report No. UCB/EERC-87/14*, University of California, Berkeley, California.

Murcek, J. et al. 1996. Effective force seismic simulation for the earthquake engineering laboratory, Paper No.460, 11th WCEE.

Nakashima, M. & Takai, H. 1985. Use of substructures in pseudodynamic testing, Bld. Res. Inst. of Japan, Vol. 111, Ministry of Construction, Tsukba, Japan.

Nakashima, M. et al. 1992. Development of real-time pseudodynamic testing. *Earthquake engineering and structural dynamics*, 21:79–92.

Nakashima, M. (1999). Real-time on-line testing for MDOF. *Earthquake engineering and structural dynamics*, 21:79–92.

Ohi, K., Lin, X. & Nishida, A. 1996. Sub-structuring pseudodynamic test on semi-rigidly jointed steel frames. Paper No.410, 11th WCEE.

Peek, R. (1991), Some possibilities in on-line analysis-test procedurce for earth structures. *Soil dynamics and earthquake engineering*, 10(6):303–312.

Qiu, Fawei, Guo, Mingchao, & Li Xuan 1994. A pseudodynamic testing with PC computer control. *Earthquake engineering and engineering vibration*, 14(3):91–96.

Qiu, Fawei (1995a). On-line testing with substructuring techniques. *Journal of experimental mechanics*, 10(4):335–342.

Qiu, Fawei 1995b. Pseudodynamic testing with implicit integration scheme. *World information on earthquake engineering*, 2:44–48.

Qiu, Fawei, Lu, Xilin & Lu, Wensheng 1996. Pseudodynamic testing method and application. Report, State Key Laboratory for Civil Engineering Prevent Disaster.

Qiu, Fawei 1997a. Pseudodynamic testing with implicit integration scheme and substructuring techniques. *China civil engineering journal*, 30(2):27–33.

Qiu, Fawei 1997b. Application of an unconditional stable numeric integration algorithm in pseudodynamic testing. *Journal of experimental mechanics*, 12(4):579–586.

Qiu, Fawei & Qian, Jiaru 1999. Pseudodynamic test under multidimensional and multiple excitation input. *China civil engineering journal*, 32(5).

Qiu, Fawei, Pan, Peng, Qian, Jiaru & Song, Yiyan 2000. Development and application of pseudodynamic testing software. *Journal of building structures*, 21(5):22–32.

Qiu, Fawei, Pan, Peng, Song, Yiyan & Qian, Jiaru 2001a. Quasi-static test loading method and control on structure. Accepted by China Civil Engineering Journal.

Qiu, Fawei, Pan, Peng, Song, Yiyan & Qian, Jiaru 2001b. Multidimensional loading method and control on structure. Accepted by China Civil Engineering Journal.

Shing, P.B. & Mahin, S.A. 1988. Rate of loading effects on pseudodynamic test. *J. of structural engineering*, ASCE, 114(11):2403–2420.

Shing, P.B & Vannan M.T. 1990. On the accuracy of an implicit algorithm for pseudodynamic tests. *Earthquake engineering and structural dynamics*, 19:631–651.

Shing, P.B., Vannan, M.T. & Cater E. 1991a. Implicit time integration for pseudodynamic tests. *Earthquake engineering and structural dynamics*, 20:551–576.

Shing, P.B. & Vannan, M.T. 1991b. Implicit time integration for pseudodynamic tests: convergence and energy dissipation. *Earthquake engineering and structural dynamics*, 20:809–819.

Standard of People's Republic of China 1997. Method and code for the building seismic experimental (JGJ 101-96), China building industrial press, Beijing, China.

Takanashi, K. et al. 1974. Seismic failure analysis of structures by computer-pulsator on-line system. *Journal of the institute science*, University of Tokyo, 26(11).

Takanashi, K., Lin, X.G. & Lee, S.J. 1996. Substructuring on-line test on industrial buildings. Paper No.411, 11th WCEE.

Thewalt, C.R. & Mahin, S.A. 1987. Hybrid solution techniques for generalized pseudodynamic testing. *Report No. UCB/EERC 87-09*, University of California, Berkeley, CA.

Toki, K. et al. 1990. Hybrid experiments on non-linear earthquake-induced soil-structure interaction. *Earthquake engineering and structural dynamics*, 19:709–723.

Toki, K. et al. 1991. Seismic behavior of pile groups by hybrid experiments, *Earthquake engineering and structural dynamics*, 20:895–909.

Wang, Qianxin et al. 1984. Arch structure response under multi-point earthquake input. *Earthquake engineering and engineering vibration*, 4(2): 49–81.

Yamazaki, Y. et al. 1986. Correlation between shaking table and pseudodynamic test on steel structural models, BRI Research paper No.119.

Author index